Mathematics in Science and Technology

Mathematical Methods, Models and Algorithms in Science and Technology

Proceedings of the Satellite Conference of ICM 2010

Mathematics *in* Science and Technology

Mathematical Methods, Models and Algorithms in Science and Technology

India Habitat Centre & India Islamic Cultural Centre, New Delhi, India

14 – 17 August 2010

Editors

A H Siddiqi
Sharda University, India

R C Singh
Sharda University, India

P Manchanda
Gurunanak Dev University, India

World Scientific

NEW JERSEY • LONDON • SINGAPORE • BEIJING • SHANGHAI • HONG KONG • TAIPEI • CHENNAI

Published by

World Scientific Publishing Co. Pte. Ltd.
5 Toh Tuck Link, Singapore 596224
USA office: 27 Warren Street, Suite 401-402, Hackensack, NJ 07601
UK office: 57 Shelton Street, Covent Garden, London WC2H 9HE

British Library Cataloguing-in-Publication Data
A catalogue record for this book is available from the British Library.

Proceedings of the Satellite Conference of ICM 2010
MATHEMATICS IN SCIENCE AND TECHNOLOGY
Mathematical Methods, Models and Algorithms in Science and Technology

ISBN-13 978-981-4338-81-3
ISBN-10 981-4338-81-8

Printed in Singapore by B & Jo Enterprise Pte Ltd

PREFACE

This volume is based on selected talks — invited, thematic and contributory — delivered during the Satellite Conference of the International Congress of Mathematicians 2010 on Mathematics in Science and Technology, held in New Delhi, India, 14–17 August 2010, sponsored by ICM 2010; Department of Science and Technology (DST); National Board of Higher Mathematics (NBHM); Council of Scientific and Industrial Research (CSIR); Defense Research and Development organization (DRDO), Government of India; Sharda University, Greater Noida, India; Abdus Salam International Centre for Theoretical Physics, Trieste, Italy (a UNESCO centre); International Council of Industrial and Applied Mathematics (ICIAM); Commission of Development and Exchanges (CDE-IMU), and the Indian Society of Industrial and Applied Mathematics (ISIAM). The volume is divided into three parts; Part A contains Chaps. 1–8 based on invited talks by international experts, who have made valuable contributions in their fields of research. Part B, comprising of Chaps. 9–16, is based on thematic review papers by scholars actively engaged in the study of related areas. Four peer reviewed contributory talks are included as Chaps. 17–20 forming Part C.

Chapter 1 is the Dr. Zakir Husain award lecture by the recipient of the Dr. Zakir Husain award 2010, Prof. M. Zuhair Nashed, University of Central Florida, USA. In the technical part of his chapter, he presents the common thread among inverse problems, signal analysis and moment problems. This effort is related to the recovery of an object (function, signal or picture) from partial or indirect information about the object. He has provided a broad prospective on some aspects of this interaction with emphasis on ill-posed problems in signal processing.

In Chap. 2, Prof. Phoolan Prasad, along with one of his research collaborator, K. R. Arun, has presented a beautiful account of Kinetic Conservation Laws (KCL), or equations of evolutions for curves and surfaces with their application to a nonlinear wave front giving numerical simulation. In Chap. 3, Prof. V. Mehrmann, jointly with his co-workers J. Heiland and M. Schmidt has presented an interesting framework for the direct discretization of the input/output map of dynamical systems governed by linear partial differential equations with distributed inputs and outputs. Global error estimates and applications to the optimal control of partial differential equations, particularly for the 2D heat equation, are discussed. In Chap. 4, Prof. Pavel Exner has given a delightful description of the physical meaning of quantum graph models through analysis of their vertex coupling approximations.

Prof. R. Lozi of France, known for the strange attractor as the Lozi map, has given a lucid presentation on "Complexity Leads to Randomness in Chaotic Systems" in Chap. 5. In Chap. 6, Prof. N. Rudraiah has drawn the attention to certain important real world problems, where mathematical concepts can play important role. He highlights, among other topics, the role of mathematical modeling in nanotechnology. Prof. O. P. Bhutani, jointly with Lipika Chowdhary, presents a study on equivalence transformations of a Helmholtz-type equation in Chap. 7.

In Chap. 8, Prof. U. B. Desai and his research associates have focused on emerging areas of communication technology and the challenging mathematical problems in the area. Cognitive Radio (CR) is a challenging field of wireless communication. In this chapter, the authors have investigated the optimal power allocation problem for an orthogonal frequency division multiplexing based CR.

Part B comprises of eight thematic reviews, Chaps. 9–16. These are: 1. Inverse Problems of Parameter Identification in Partial Differential Equations (B. Jadamba, A. A. Khan and M. Sama); 2. Finite Element Methods for HJB Equations (M. Boulbrachene); 3. Dynamics and Control of Underactuated Space Systems (K. D. Kumar and Godard); 4. Some New Classes of Inverse Coefficient Problems in Engineering Mechanics and Computational Material Science Based on Boundary Measured Data (A. Hasanov); 5. Some Recent Developments on Mathematical Aspects of Wavelets (P. Manchanda and Meenakshi); 6. Relevance of Wavelets and Inverse Problems to Brain (A. H. Siddiqi, H. K. Sevindir, Z. Aslan and C. Yazici); 7. Wavelets and Inverse Problems (K. Goyal and M. Mehra); 8. Optimization Models for a Class of Structured Stochastic Games (S. K. Neogy, S. Sinha, A. K. Das and A. Gupta).

Chapter 9 deals with inverse problems which is a vibrant and fast progressing theme that has found numerous significant applications. Twelve methods to solve a class of inverse problems along with applications are discussed. New avenues of research in this field are indicated. The theme is quite relevant to emerging problems of science and technology. In Chap. 10, recent progress in the finite element approximation of HJB (Hamilton–Jacobi–Bellman) equations are reviewed and open problems are mentioned. Chapter 11 is devoted to the feasibility of achieving reliable formation control without the need for thrust in the radical or along-track direction. The advantages of the control method considered in this paper are validated via numerical simulations. Updated results are reviewed on the theme of relative motion control of multiple space craft formations using thrusters in fully actuated configuration.

In Chap. 12, three classes of inverse coefficient problems arising in engineering mechanics and computational material science are considered. The first problem is related to the determination of unknown elasto-plastic properties of a beam from limited number of torsional experiments. The second problem is related to the identification of elasto-plastic properties of a 3D body from spherical identification test while the third one relates to identification of unknown coefficient in the nonlinear bending equation. Besides the solutions of these problems, their applications are also discussed.

In Chap. 13, three theoretical aspects of wavelets are considered. These are (i) The effect of replacing the set of integers of translation parameters by its subset that is not a group in the definition of scaling function, an important ingredient of multi-resolution analysis known as the heart of wavelet theory. (ii) Consequences of replacing the set of real numbers by the set of positive real numbers in the definition of multi-resolution analysis. (iii) Properties of wavelets obtained by vector-valued multi-resolution analysis. Results obtained by these modifications are reviewed. Applications of wavelets and inverse problems, particularly to EEG (signal representing functioning of brain) are reviewed in Chap. 14. In Chap. 15, certain aspects of inverse problems and wavelets are discussed. Chapter 16 presents an updated description of optimization models for a class of Structured Stochastic Games.

Part C consists of four peer reviewed contributory chapters, 17–20. Chapter 17 by Q. J. Khan and M. Al-Lawatia develops a mathematical model of an interesting real world problem, namely Predator-Prey relations for mammals in a special situation. Chapter 18 by G. Röst deals with SEI model with varying transmission and mortality rate. Chapter 19 by B. S. Kushvah presents the study of Trajectories and Stability Regions. In Chap. 20, Wasu and Rajvanshi present their study of MHD flow past an infinite plate under the effect of gravity modulation.

A complete list of dignitaries and persons who participated actively in the conference is given at the end of this volume.

We take this opportunity to thank all the funding agencies mentioned above, especially the authorities of Sharda University, honorable Mr. P. K. Gupta (Chancellor), Mr. Y. K. Gupta (Pro-Chancellor) and Dr. R. P. Singh (Vice-Chancellor). It will not be out of place to mention that the organization of the conference on such a big scale and compilation of this volume would not have been possible without the whole-hearted support of Vice-Chancellor Singh and Dr. M. Al-Lawatia, Head of the Department of Mathematics and Statistics, Sultan Qaboos University, Oman. We also express our gratitude to all those who reviewed the papers and provided us their valuable assistance at all stages, especially to Ms. Meenakshi, Ms. Noor-e-Zahra, Dr. Mani Mehra, Prof. Mushahid Husain, Dr. Khwaja Shahid and Prof. Q. H. Ansari. We also take this opportunity to thank Prof. M. S. Raghunathan (Chairman, Executive Organizing Committee of ICM 2010) and Prof. S. Kesavan (Convener Satellite Conferences of ICM 2010). Valuable cooperation of Ms. E. H. Chionh, World Scientific Publishing, is highly appreciated.

A. H. Siddiqi, R. C. Singh and P. Manchanda
30 December 2010

FOREWORD

As scientific disciplines go, Physics, Chemistry, Biology and Mathematics are reasonably well-organized in India. Applied Mathematics, on the other hand, has not yet received its share of recognition. The reason is partly that it continues to be seen as watered down version of Mathematics, or outdated Theoretical Physics, or a somewhat distant — and sometimes irrelevant — cousin of Engineering and Technology. This should change: as the need for modeling and understanding of the increasingly complex phenomena of our times becomes more pressing, so should the role of Applied Mathematics expand. Applied Mathematics as a discipline includes the application of known mathematics to practical and scientific problems as well as the invention of new mathematics with applications in mind.

Applied Mathematics often feeds Mathematics proper, and, indeed, many first-rate mathematics departments and research institutes in the world have begun to appreciate this fact. Indeed, some never wavered in this belief. Needless to say, Applied Mathematics has a vital role to play in diverse areas of engineering, energy, material science, geological and geophysical sciences, biological sciences including medicine, social sciences including economics, and so forth. This is why the work of the Indian Society of Industrial and Applied Mathematics (ISIAM), is very important — especially at a time when the country is engaged in technological rejuvenation. This is indeed why the Society deserves, and needs, your support.

Thanks to Professor Siddiqi, I got involved in this society a few years ago, and wish to take this opportunity to say a few words on the progress made and the work that still remains to be done.

We greatly appreciate the support of the International Mathematical Union, whose main Congress was held in Hyderabad and whose satellite conference was held in Delhi. This volume is the result of the latter. The collection of articles presented in this volume is a measure of the vitality of the community, and of the standard of this conference in particular. We also thank Sharda University, various government agencies which have supported this meeting, the constant background support of ICIAM, the support of the international and national communities of mathematicians, and the support of the community as a whole.

Owing to the efforts of the office bearers of this Society, which includes Professors Siddiqi, Gupta, Dikshit, Manchanda and others from the past, the scientific meetings of ISIAM are constantly improving: the invited talks are usually first-rate, Zakir Husain awardees are of high caliber, the fund-raising has become less stochastic, ISIAM's reach and scope have been expanding, and so forth. I congratulate the Society on its emergence as an increasingly professional organization which has been forging ever strengthening links within India and internationally. Let us not forget, however, that there are many aspects which need improvement. It is these aspects to which I will now draw attention, while being fully aware that the outlook towards reaching these goals appears good.

The following aspects have not changed much. Applied Mathematics is still not yet a well-organized and well-knit community within the country, in comparison with its importance. The participation of the members of this community in the technological development within the country is yet to reach a critical level; and applied mathematics graduate students, whose numbers seem to have increased recently, do not find satisfying jobs quickly enough. Applied Mathematics in India has not yet made central contributions to the grand challenges of our times: climate change, new materials, energy, problems of megacities, ground water depletion, spread of infectious diseases, ever-changing economic environment, security, etc. The overall quality of research in Applied Mathematics, whether on the applications front or in feeding mathematics proper, has not yet reached the high level that it should. The connection between the methods of Applied Mathematics and the advances on the computational front is still not strong. There is a real opportunity here for the Applied Mathematics community because of the confluence of the importance of the subject and the timeliness and availability of resources. We could make this an exciting time for us and for the subject.

Please don't get me wrong or think of me as being overly critical. I believe that there are accomplished applied mathematicians in this country. Indeed, ISIAM has itself honored a number of them — including some at this meeting. There are also some first-rate students in applied mathematics. My point is simply that the quality is not sufficiently uniform, that the average level is not sufficiently high, the numbers too small and the visibility too low. We should strive to improve the situation on all fronts.

In this task, the role of senior mathematicians is to open up new avenues, inspire younger people and mentor them well, in both their research and careers, and eventually build groups in which rigor, accomplishment, creativity and quality are valued above all, and the potential of the younger people can be explored fully and without hurdles. There are some wonderful people of this sort in the country but there are not enough of them. The role of young mathematicians is to take serious interest in their subject, develop technical skills of high order, and do competitive mathematics. They should understand that mathematics is not a local activity in which to dabble, but is an international arena in which one has to play and excel. Recognition will come in due course: as in other arenas of human activity, there are many quirks that may seem discouraging, but one has to look past them.

There is no ready-made recipe for this situation to work itself out; it requires hard and dedicated work. The first important need, however, is the spirit of learning, of placing one's own work as part of a bigger landscape, and the willingness to work single-mindedly towards creating one's own landscape. Hardly are great things possible without this spirit and hard work — especially in mathematics. While it is not possible to do good Applied Mathematics without knowing Mathematics, it is not enough to know good Mathematics; one has to move in circles where applications present themselves as opportunities.

The second important element is to ensure that there is adequate support for younger researchers. This is where ISIAM can and should do more. It ought to build not only the spirit of doing Applied Mathematics and create a culture in which ideas and their interplay with applications are valued, but also enable better support structure for the discipline. It will have done very well then. I have no doubt that its work will be amply rewarded.

K. R. Sreenivasan
New York University
December 31, 2010

CONTENTS

Part B Thematic Reviews

Part C Contributory Talks

Affiliation of Contributors

S.No.	Name	Address
1.	Prof. M. Zuhair Nashed	Department of Mathematics, University of Central Florida, Orlando, Florida, USA Email:znashed@mail.ucf.edu
2.	Prof. Phoolan Prasad	Department of Mathematics, Indian Institute of Science, Bangalore-560012, India. Email: prasad@math.iisc.ernet.in
	Dr. K. R. Arun	Department of Mathematics, Indian Institute of Science, Bangalore-560012, India
3.	Prof. V. Mehrmann	Institute of Mathematics , Technical University Berlin 10623 Berlin, Germany Email: mehrmann@math.tu-berlin.de
	Dr. J. Heiland	Institute of Mathematics , Technical University Berlin 10623 Berlin, Germany Email: heiland@math.tu-berlin.de
	Dr. M. Schmidt	GE Global Research Centre,85748 Garching bei Munchen, Germany. Email: mail@schmidt-michael.de
4.	Prof. Pavel Exner	Doppler Institute for Mathematical Physics and Applied Mathematics, Czech Technical University, Brehova7,11519 Prague, Czech Republic & Department of Theoretical Physics, Nuclear Physics Institute, Czech Academy of Sciences, 25068 Rez near Prague, Czech Republic Email: exner@ujf.cas.cz
5.	Prof. R. Lozi	Laboratory J.A.Dieudonne, UMR CNRS6621, and Institute Universitaire de Formation des Maitres Celestin Freinet, University of Nice Sophia-antipolis, France. Email: R.Lozi@unice.fr
6.	Prof. N. Rudraiah	National Research Institute of Applied Mathematics(NRIAM) JayaNagar Banalore-560070 and UGC-CAS in Fluid Mechanics, Department of Mathematics, Bangalore University, Bangalore-560070. Email: rudraiahn@hotmail.com
7.	Prof. O. P. Bhutani	Honorary Scientist(INSA), B-1/1057, Vasant Kunj, New Delhi- 110070, India.
	Dr. L. R. Chowdhury	B1K, 136 # 101-40, Vishan Ring Road, Singapore
8.	Dr. Taskeen Nadkar	Department of Electrical Engineering, IIT Bombay, Maharashtra, India. Email: taskenn@ee.iitb.ac.in
	Dr. Vinay Thumar	Department of Electrical Engineering, IIT Bombay, Maharashtra, India. Email: vinay_thumar@ee.iitb.ac.in
	Dr. Aaqib Patel	Department of Electrical Engineering, IIT

		Bombay, Maharashtra, India. Email: aaqib@ee.iitb.ac.in
	Dr. Md. Zafar Ali Khan	IIT Hyderabad, Ordnance Factory Estate, Andhra Pradesh, India. Email: zafar@iith.ac.in
	Prof. U. B. Desai	IIT Hyderabad, Ordnance Factory Estate, Andhra Pradesh, India. Email: ubdesai@iith.ac.in
	Dr. S. N. Merchant	Department of Electrical Engineering, IIT Bombay, Maharashtra, India. Email: merchant@ee.iitb.ac.in
9.	Dr. A. A. Khan	School of Mathematical Sciences, Rochester Institute of Technology, Rohester, New York,14623,USA Email: aaksma@rit.edu
	Dr. B. Jadamba	School of Mathematical Sciences, Rochester Institute of Technology, Rohester, New York,14623,USA. Email: bxjsma@rit.edu
	Dr. M. Sama	Department de Mathematica Applicado Universidad National de Educacion a Distancia Madrid, Spain. Email: msama@ind.uned.es
10.	Dr. M. Boulbrachene	Department of Mathematics and Statistics, Sultan Qaboos University, Muscat, Oman. Email: boulbrac@squ.edu.om
11.	Prof. K. D. Kumar	Department of Aerospace Engineering, Ryerson University, 350 Victoria Street, Toronto, Ontorio M5b2K3,Canada. Email:kdkumar@ryerson.ca
	Dr. Godard	Department of Aerospace Engineering, Ryerson University, 350 Victoria Street, Toronto, Ontorio M5b2K3,Canada
12.	Prof. Alemdar Hasanov	Department of Mathematics and Computer Science, Izmir University,35350,Izmir,Turkey Email: alemdar.hasanoglu@izmir.edu.tr
13	Prof. P. Manchanda	Department of Mathematics, Guru Nanak Dev University, Amritsar, India Email: pmanch2k1@yahoo.co.in
	Ms. Meenakshi	Dev Samaj College for Women, Ferozepur, India Email: meenakshi_wavelets@yahoo.com
14.	Prof. A. H. Siddiqi	Deparment of Mathematics, Sharda University, Greater Noida, India Email:siddiqi.abulhasan@gmail.com,
	Dr. H. Kodal Sevindir	Department of Mathematics, University of Kocaeli, Kocaeli, Turkey Email: hkodal@kocaeli.edu.tr
	Prof. Z. Aslan	Department of Computer Engineering, Istanbul Aydin University, Istanbul, Turkey. Email: zaferaslan@aydin.edu.tr
	Dr. C. Yazici	Department of Mathematics, University of Kocaeli, Kocaeli, Turkey Email: cuneyt.yazici@kocaeli.edu.tr

15.	Ms Kavita Goyal	Department of Mathematics, Indian Institute of Technology Delhi, Hauz Khas,New-Delhi-110116 Email; kavita-ma@student.iitd.ac.in
	Dr. Mani Mehra	Department of Mathematics, Indian Institute of Technology Delhi, Hauz Khas,New-Delhi-110116 Email: mmehra@maths.iitd.ac.in
16.	Prof. S. K. Neogy	Indian Statistical Institute Institute, 7 SJS Sansanwal Marg, New-Delhi -110016, India Email: skn@isid.ac.in
	Prof. Sagnik Sinha	Jadavpur University,Kolkata-700032,India Email:sagnik62@yahoo.co.in
	Dr. A. K. Das	Indian Statistical Institute 203,B.T.Road,Kolkata-700108,India Email:akdas@isical.ac.in
	Dr. A. Gupta	ISI,Kolkata. Email: agupta@isical.ac.in
17.	Dr. Q. J. A. Khan	Department of Mathematics and Statistics, Sultan Qaboos University, P.O.Box, Al-Khodh123, Sultanate of Oman
	Dr. M. Al-Lawatia	Department of Mathematics and Statistics, Sultan Qaboos University, P.O.Box, Al-Khodh123, Sultanate of Oman
18.	Dr. Gergely Rost	Analysis and Stochastics Research Group Hungarian Academy of Sciences, Bolyai Institute, University of Szeged,Hungary, Aradi vertanuk tere1,H-6725,Szeged,Hungary, Email: rost@math.u-szeged.hu
19.	Dr. Badam Singh Kushvah	Department of Applied Mathematics, Indian School of Mines,Dhanbad-826004,Jharkhand,India Email: bskush@gmail.com, kushvah.bs.am@ismdhanbad.ac.in
20.	Ms Sargam Wasu	Department of Applied Sciences, Gurukal Vidyapeeth Institute of Engineering and Technology, Sector 7, Banur, Dist.Patiala, Punjab, India. Email: sargam15@rediffmail.com
	Prof. S. C. Rajvanshi	Department of Applied Sciences, Gurukal Vidyapeeth Institute of Engineering and Technology, Sector 7, Banur, Dist.Patiala, Punjab, India.

IN APPRECIATION OF DR ZAKIR HUSAIN AWARD

M. ZUHAIR NASHED

Department of Mathematics, University of Central Florida, Orlando, FL 32816

During 21–25 January 2001, an International Conference combined with the 6th biennial meeting of the Indian Society of Industrial and Applied Mathemaitcs (ISIAM) was organized at Guru Nanak Dev University, Amritsar, India. The conference was very well organized, thanks to the superb organizational skills and generous hospitality of Professors A. H. Siddiqi and P. Manchanda. I had the pleasure of being one of the invited speakers from outside India. At that meeting, the first Dr. Zakir Husain award was given to Professor Jagat Narain Kapur. I was sitting in the audience when tribute was paid to the late Dr. Zakir Husain, the third President of the Republic of India, a distinguished scholar, and great humanitarian. It never occurred to me while listening to the citation for Dr. Zakir Husain Award to Professor Kapur that one day I might be a recipient of this award. But here I am 10 years later: honored and delighted! On such occasion, one is tempted to look up the roster of previous recipients of the award:

- Professor Jagat Narain Kapur, Indian Institute of Technology Kanpur, India
- Professor Helmut Neunzert, University of Kaiserslautern, Germany
- Professor Katepalli R. Sreenivasan, Director of Abdus Salam International Centre of Theoretical Physics, Italy
- Professor Roddam Narasimha, FRS, Indian Institute of Sciences Banglore, India
- Professor Hanuman Prasad Dikshit, Good Governance & Planning, Government of MP, India

I am grateful to the distinguished members of the Selection Committee comprising of Professor K. R. Sreenivasan (Courant Institute, New York University), Professor H. P. Dikshit (Ex-President of ISIAM), Professor N. K. Gupta (Vice President of the Indian National Science Academy and current President of ISIAM), Professor P. Manchanda (Joint Secretary, ISIAM) for their recommendation.

Over the past three decades, it has been my pleasure to interact and collaborate with many Indian mathematicians at major institutions in India. I served as mentor or external examiner of the Ph. D. dissertations of several young mathematicians at various IITs and other universities in India. I encouraged them to explore modern areas of applied and computational mathematics, including inverse and ill-posed problems, integral equations, optimization theory, and various applications of functional analysis. I have visited India many times. I gave invited/plenary lectures at seven international conferences, and delivered lectures at IIT Bombay, IIT Kanpur, Guru Nanak Dev University Amritsar, JMI New Delhi, University of Delhi, and Goa University.

I have co-edited a volume of invited papers dedicated to the memory of Arum Kumar Varma (1934–1994), and a volume dedicated to the memory of Ambikeshwar Sharma (1920–2003). I have collaborated with K. M. Furati and A. H. Siddiqi in editing a volume on mathematical models for real world systems.

It has been a privilege for me to make contributions to promote the aims and objectives of the Indian Society of Industrial and Applied Mathematics since its inception.

I am very grateful to ISIAM for the Dr. Zakir Husain award. It is a great honor for me. Thank you.

Technical Lecture

Inverse Problems, Moment Problems, Signal Processing: Un Menage a Trois

Keywords: Inverse Problems; Moment Problems; Signal Processing.

1. Introduction

Inverse problems deal with determining, for a given input-output system, an input that produces an observed output, or determining an input that produces a desired output (or comes as close to it as possible), often in the presence of noise. Most inverse problems are ill-posed. Signal analysis/processing deals with digital representations of signals and their analog reconstructions from digital representations. Sampling expansions, filters, reproducing kernel spaces, various function spaces, and techniques of functional analysis, computational and harmonic analysis play pivotal roles in this area.

Moment problems deal with recovery of a function or signal from its moments, and the construction of efficient stable algorithms for determining or approximating the function. Again this is an ill-posed problem. Interrelated applications of inverse problems, signal analysis and moment problems arise, in particular, in image analysis and recovery, and in many areas of science and technology. Several decades ago the connections among these areas (inverse problems, signal processing, and moment problems) was rather tenuous. Researchers in one of these areas were often unfamiliar with the techniques and relevance of the other two areas.

The situation has changed drastically in the last 25 years. The common thread among inverse problems, signal analysis, and moment problems is a canonical problem: recovering an object (function, signal, picture) from partial or indirect information about the object. In this talk, we will provide perspectives on some aspects of this interaction with emphasis on ill-posed problems in signal processing. We will show that function spaces, in particular reproducing kernel spaces, shift-invariant spaces and translation-invariant spaces, play a pivotal role in sampling expansions.

2. Inverse Problems

Given two normed spaces X, Y and a mapping $A : X \longmapsto Y$, we consider three problems:

(1) **Direct problem**: Given $x \in X$, find $Ax \in Y$.
(2) **Inverse problem**: Given an observed output y, find an input x that produces it (i.e., $y = Ax$), or that produces an output as "close" to the output y as possible (i.e., $x = \text{argmin}_{u \in X} \|Au - y\|_2$).
(3) **Identification or modelling problem**: Determine or estimate the mapping A from a collection of input-output information.

We consider next the notion of a *well-posed problem*

$$A: \ X \in x \longmapsto Ax = y \in Y. \tag{1}$$

The problem (1) is said to be *well-posed* if for each "data" y in the data space Y, the above equation (1) has one and only one solution, and the solution depends continuously on y. A well-posed problem is also known as a properly-posed problem. Pillars of well-posedness of an inverse problem include existence, uniqueness and continuous dependence.

The *ill-posed problems*, or *improperly-posed problems*, violate one of the three requirements of existence, uniqueness or continuous dependence. Many ill-posed problems arise in partial differential equations and integral equations of the first kind.

The definition of ill-posed problems is due to Jacques Hadamard (1865–1963). He dismissed improperly-posed problems as irrelevant to physics or real world applications, but he was proven to be wrong four decades after his declaration. Once called "the living legend of mathematics", Jacques Hadamard had a tremendous influence on the development of mathematics, see *Jacques Hadamard, A Universal Mathematician* by Vladimir Mazya and Tatyana Shaposhnikova for a fascinating biography on his life and legendary contributions.

Fig. 1. The cover page of the book *Jacques Hadamard, A Universal Mathematician* by Vladimir Mazya and Tatyana Shaposhnikova, American Mathematical Society and London Mathematical Society, 1998. The cover page is downloaded from http://www.maa.org/reviews/hadamard.html

The following example dramatizes the difference between direct and inverse problems.

(1) **Direct problem**: *You ask a question and hear an answer.*
(2) **Inverse problem**: *You heard an answer. What is the question?*

J. G. Keller in the interesting paper* published in the American Mathematical Monthly gave the following example of this dramatization: You heard the answer "*Washington, George*".

What is the question?

If you know the history of the United States, you will definitely say "*Who was the first president of the United States?*"

In this case as Keller writes, the question is different. President George Bush (Senior), during his term as Vice President, asked Nancy Reagan, "*What is the capitol of the USA? Nancy.*"

This example also illustrates the importance of using a priori information in the possible resolution of ill-posed inverse problems.

Here is a serious example of an interesting inverse problem. In 1966, Marc Kac posed the question "*Can you hear the shape of a drum?*"† More precisely, can you deduce the shape of a plane region by knowing the frequencies at which it resonates (where, as in a physical drum, the boundary is assumed to be held fixed)? Long before Kac posed this question, mathematicians had been investigating the analogous question in higher dimensions: Is a Riemannian manifold (possibly with boundary) determined by its spectrum?

The problem was first settled, in the negative, in higher dimensions. In 1964, John Milnor found two distinct 16-dimensional manifolds with the same spectrum (Milnor received the Field medal for his contributions). But the problem for plane regions remained open until 1991, when Carolyn Gordon, David Webb, and Scott Wolpert found examples of distinct plane "drums" which "sound" the same, see Figure 2. The story of the problem and its solution can be found in the article *You Can't Always Hear the Shape of a Drum* by Barry Cipra, which appeared in Volume 1 of What's Happening in the Mathematical Sciences, http://www.ams.org/samplings/math-history/happening-series.

Marc Kac received the Lester Ford Award and the Chauvenet Prize from the Mathematical Association of America in 1967 and 1968 respectively. He gave lectures at many universities on the theme "Can you hear the shape of a drum?" At one lecture he gave in Amsterdam, there were many psychologists in the audience, who never attended mathematical colloquia. The colloquium chair mathematician approached the psychologists

*J. G. Keller, Inverse problems, *The American Mathematical Monthly*, Vol. 83, 1976, 107–118.

†M. Kac, Can one hear the shape of a drum? *The American Mathematical Monthly*, Vol. 73, 1966, pp. 1–23.

Fig. 2. David Webb (left) and Carolyn Gordon (right) hold paper models of a pair of "sound-alike" drums. The picture is downloaded from http://www.ams.org/ samplings /feature-column/fcarc-199706

and asked "How are you interested in this talk?" They replied that the title of the talk in the announcement they received is "Can you hear the shape of a `dream`?"

3. Recovery problems from partial or indirect information

Let f be a signal belonging to an n-dimensional Hilbert space H of functions on a domain Ω,

$$f(x) = \sum_{i=1}^{n} c_i u_i(x), \ x \in \Omega \tag{1}$$

where $\{u_1, \ldots, u_n\}$ is a basis for H.

Consider two simple recovery problems. The first problem is *the recovery of a signal f from its moments.* Suppose we know the moments

$$\alpha_j = \langle f, w_j \rangle, \quad j = 1, 2, \ldots, n. \tag{2}$$

Then it follows from (1) and (2) that

$$\sum_{i=1}^{n} c_i \langle u_i, w_j \rangle = \alpha_j, \quad j = 1, \ldots, n,$$

where $\{w_1, \ldots, w_n\}$ is a given linear independent set. In particular, if the two systems $\{u_i\}_{i=1}^n$ and $\{w_j\}_{j=1}^n$ are bi-orthogonal, or if we assume that $w_i = u_i, 1 \le i \le n$, and $\{u_1, \ldots, u_n\}$ is orthonormal, then $c_i = \langle f, u_i \rangle$. This leads to the Fourier expansion

$$f(x) = \sum_{i=1}^{n} \langle f, u_i \rangle u_i(x). \tag{3}$$

The second problem is *the recovery of a signal f from its evaluations*. Suppose that $\{f(x_j)\}_{j=1}^m$ are given. Note that

$$f(x_j) = \sum_{i=1}^{n} c_i u_i(x_j), \quad j = 1, \dots, n$$

by (1). If we assume that $\{u_i\}_{i=1}^n$ is a discrete orthonormal sequence; i.e., $u_i(x_j) = \delta_{ij}$ where δ_{ij} stands for the Kronecker symbol, then we have the sampling expansion

$$f(x) = \sum_{i=1}^{n} f(x_i) u_i(x). \tag{4}$$

We notice that both the Fourier expansion (3) and the sampling expansion (4) are of the form

$$f(x) = \sum_{i=1}^{n} l_i(f) u_i(x), \tag{5}$$

where $l_i, 1 \leq i \leq n$, are continuous linear functions.

These two problems take markedly different generalizations for infinite dimensional spaces. In the next section, we consider such a generalization for the sampling expansion of band-limited functions on the real line.

4. Whittaker-Shannon-Kotelnikov sampling theorem

Consider the problem of ideal sampling

$$f \longmapsto (f(\gamma))_{\gamma \in \Gamma},$$

where Γ is a set at which the signal f is sampled. A special case is the uniform sampling on the real line:

$$f \longmapsto (\cdots, f(-2), f(-1), f(0), f(1), f(2), \cdots).$$

Denote the space of square-integrable functions band-limited to $[-\pi, \pi]$ by B_π; i.e.,

$$\begin{aligned} B_\pi &:= \{f \in L^2(\mathbf{R}) : \ \mathrm{supp}\hat{f} \subset [-\pi, \pi]\} \\ &= \left\{ \int_{-\pi}^{\pi} e^{-ixt} g(x) dx : \ g \in L^2(-\pi, \pi) \right\}, \end{aligned}$$

where $\hat{f}$ denotes the Fourier transform of f. The linear space B_π is also known as the Paley-Wiener space. The following classical Whittaker-Shannon-Kotelnikov sampling theorem says that any signal band-limited to $[-\pi, \pi]$ can be stably recovered from its samples on integers.

Theorem 4.1. *If a square-integrable function f is bandlimited to $[-\pi, \pi]$, then f can be reconstructed from its samples, $f(k)$, that are taken at the equally spaced nodes k on the time axis* $\mathbf{R}$. *Moreover,*

$$\|f\|_2^2 = \sum_{k \in \mathbf{Z}} |f(k)|^2, \tag{1}$$

and

$$f(t) = \sum_{k=-\infty}^{\infty} f(k) \frac{\sin(t-k)\pi}{(t-k)\pi}, \quad t \in \mathbf{R}, \tag{2}$$

where the series is absolutely and uniformly convergent on any compact set of the real line.

Proof. The following proof is very elementary but does not reveal the core of sampling theory. Let $f \in B_\pi$. Then

$$f(t) = \frac{1}{2\pi} \int_{\mathbf{R}} F(w) e^{-iwt} d\omega = \frac{1}{2\pi} \int_{-\pi}^{\pi} F(w) e^{-iwt} d\omega$$

for some square-integrable function F supported on $[-\pi, \pi]$. Extending F periodically to $(-\infty, \infty)$ and then using the complex Fourier expansion of the extension lead to

$$f(t) = \frac{1}{2\pi} \int_{-\pi}^{\pi} \Big(\sum_{n=-\infty}^{\infty} c(n) e^{i\omega n} \Big) e^{-i\omega t} d\omega,$$

where

$$c_n = \frac{1}{2\pi} \int_{-\pi}^{\pi} F(\omega) e^{-i\omega n} d\omega = f(n), \; n \in \mathbf{Z}.$$

Interchanging the summation and the integral then gives

$$\begin{aligned} f(t) &= \frac{1}{2\pi} \sum_{n=-\infty}^{\infty} f(n) \int_{-\pi}^{\pi} e^{i\omega n} e^{-i\omega t} d\omega \\ &= \sum_{n=-\infty}^{\infty} f(n) \frac{\sin(t-k)\pi}{(t-k)\pi}. \end{aligned}$$

This proves the reconstruction formula (2). □

Harry Nyquist, Arne Beurling and Paul Butzer have also made fundamental contributions to sampling theory and applications.

Fig. 3. Edmund Taylor Whittaker (left, 1873–1956) Claude Elwood Shannon (middle, 1916-2001), and and Vladimir Aleksandrovich Kotelnikov (right, 1908–2005). The pictures are downloaded from http://www-history.mcs.st-and.ac.uk/Ledermann/Ch5.html; http://en.wikipedia.org/wiki/Claude_Shannon; and http://www.mentallandscape.com/V_Biographies.htm respectively.

Fig. 4. Harry Nyquist (left, 1889–1976), Arne Carl-August Beurling (middle, 1905-1986) and Paul Butzer (right). The pictures are downloaded from http://www.s9.com/Biography/Nyquist-Harry, http://www-history.mcs.st-and.ac.uk/Mathematicians/Beurling.html and http://versita.com/butzer/ respectively

5. Engineering approach to Whittaker-Shannon-Kotelnikov sampling theorem

Engineers look at the Whittaker-Shannon-Kotelnikov sampling theorem by using two operations: the *sampler* $S : f(t) \longmapsto f^*(t)$ and the *low-pass filter* $P : f^*(t) \longmapsto f(t)$, where

$$f^*(t) = \sum_{n=-\infty}^{\infty} f(n)\delta(t-n) \tag{1}$$

and δ is the delta distribution (or impulse function at the origin). A low-pass filter passes all frequencies of absolute value less than π and stops all others, which converts f^* back into f.

This approach has mathematical difficulties which are not resolved in formal engineering approach.

(1) The sampler S takes f into f^* which is out of the space of band-limited functions, indeed, f^* is not a signal with finite energy.
(2) In what sense does the above series (1) for f^* converge? The series (1) has the convergence in $\mathcal{S}'$, the space of all tempered distributions.
(3) The map P, which corresponds to passing the "signal" f^* through a low-pass filter, recovers f at least formally since

$$P(\delta(t-n)) = \frac{1}{2\pi}\int_{-\pi}^{\pi} e^{-i\omega t}e^{i\omega n}\,d\omega = \frac{\sin\pi(t-n)}{\pi(t-n)},$$

and if P is continuous, then

$$(Pf^*)(t) = \sum_{n=-\infty}^{\infty} f(n)P(\delta(t-n)) = \sum_{n=-\infty}^{\infty} f(n)\frac{\sin\pi(t-n)}{\pi(t-n)}.$$

However, since $\mathcal{S}'$ is not a Hilbert space, we do not know if P is a continuous projector.
(4) Still another difficulty! P is not well-defined on $\mathcal{S}'$. Indeed,

$$Pg = \mathcal{F}^{-1}(\chi_{[-\pi,\pi]}\hat{g}),\ g \in \mathcal{S}',$$

where $\mathcal{F}^{-1}$ is the inverse Fourier transform. So P corresponds under the Fourier transform to the multiplication of the Fourier transform $\hat{g}$ in $\mathcal{S}'$ by the characteristic function of $[-\pi,\pi]$. Unfortunately, such functions are not multipliers in $\mathcal{S}'$. Hence we need to restrict ourselves to a subspace of $\mathcal{S}'$ in which the characteristic function is a multiplier.

These issues have been resolved in my joint work with Gilbert Walter.[†] We obtain a rigorous proof of the engineering approach in appropriate function spaces. This actually leads to a more general unifying approach for sampling theorems in reproducing kernel Hilbert spaces, that will be discussed in the next section.

[‡]M. Z. Nashed, and G. G. Walter, General sampling theorems for functions in reproducing kernel Hilbert spaces, *Math. Control Signals Systems*, Vol. 4, 1991, pp. 363–390.

6. Sampling in reproducing kernel Hilbert spaces

Before we state one of those theorems, we recall the definitions of two function spaces.

The *Sobolev space* $H^r, r \in \mathbf{R}$, is the set of all tempered distributions $f(t)$ whose Fourier transform $\hat{f}(\omega)$ are square-integrable with respect to the weight function $(1 + |\omega|)^r$; i.e.,

$$\int_{\mathbf{R}} |\hat{f}(\omega)|^2 (1 + |\omega|^2)^r d\omega < \infty.$$

One may show that the delta "function" δ belongs to H^r for $r < -1/2$, and that if $\{f(n)\}_{n=-\infty}^{\infty}$ converges to zero as n tends to infinity sufficiently rapidly, then the series $\sum_{n=-\infty}^{\infty} f(n)\delta(t-n)$ converges in H^{-1}

Fig. 5. Sergei L. Sobolev (1908-1989). The picture is downloaded from http://en.wikipedia.org/wiki/Sergei_Sobolev

Let $\mathcal{F}$ be a family of functions with domain S and t be a given point in S. The linear map $E_t : f \longmapsto f(t)$ is called the *evaluation functional* at t. For computing and numerical analysis it is important that the evaluation functional is also continuous, but this is not always the case, for instance, $\mathcal{F} = L^2(S)$.

A *reproducing kernel Hilbert space*, or *RKHS* for short, is a Hilbert space H of functions on a set S in which all the evaluation functionals E_t, for each fixed t in S, are continuous; i.e.,

$$|f(t)| \leq C_t \|f\|$$

where the constant C_t is independent on $f \in H$. From the Riesz representation theorem, there exists $k_t \in H$ such that $f(t) = \langle f, k_t \rangle$. Define the reproducing kernel (RK) by

$$k(t,s) = \langle k_t, k_s \rangle.$$

Now we state a representative sampling theorem for signals in a reproducing kernel Hilbert space.

Theorem 6.1. *Let H_Q be a reproducing kernel Hilbert space that is a subspace of $L^2(\mathbf{R})$ and is closed in the Sobolev space H^{-1} and under differentiation. Let the reproducing kernel $k(s,t)$ of the space H_Q be continuous and have the zero sequence $\{t_k\}$ which is a set of uniqueness for H_Q, and assume that $\{t_n\}$ tends to infinity as n tends to infinity. If $f \in H_Q$ satisfies $f(t)/k(t,t) = O(t^{-2})$, then the sampled sequence*

$$f^*(t) = \sum_n f(t_n)\delta(t - t_n)/k(t_n, t_n)$$

converges in the sense of H^{-1} and its orthogonal projection onto H_Q equals to $f(t)$, and the series

$$f(t) = \sum_n f(t_n)k(t_n, t)/k(t_n, t_n)$$

converges uniformly on sets for which $k(t,t)$ is bounded.

Fig. 6. M. Zuhair Nashed (left) and Gilbert G. Walter (right). The pictures are downloaded from http://www.math.ucf.edu/ znashed/ and https://pantherfile.uwm.edu/ggw/www/ respectively.

The Paley-Wiener space B_π of functions has many interesting properties which are not exploited or even used in the classical proof the Whittaker-Shannon-Kotelnikov sampling theorem. Among those properties, we mention that

(1) B_π is a reproducing kernel Hilbert space with the reproducing kernel $k(t,s) = \frac{\sin \pi(t-s)}{\pi(t-s)}$.
(2) The sequence $\{S_n\}_{n\in\mathbf{Z}}$ where $S_n = k(t,n)$ constitutes an orthonormal basis for B_π.
(3) The sequence $\{S_n\}_{n\in\mathbf{Z}}$ has the discrete orthogonality property: $S_n(m) = \delta_{nm}$ for all $m, n \in \mathbf{Z}$.
(4) $f(\cdot - c) \in B_\pi$ and $\|f(\cdot - c)\|_2 = \|f\|_2$ for any $f \in B_\pi$ and $c \in \mathbf{R}$. Hence B_π is a unitarily translation-invariant subspace of $L^2(\mathbf{R})$.

Abstraction of some of these properties are used as pivotal elements in general approaches to nonuniform sampling, as we shall see in the following section.

7. Sampling in unitarily translation-invariant spaces

We describe another approach of sampling theorem, where the unitary translation-invariance property of the space plays a pivotal role.

Within this framework we are able to exploit Gershgorin's theorem and results on Toëplitz matrices. To be more specific, we study sampling problems on reproducing kernel Hilbert spaces with reproducing kernel

$$k(t,u) = \int_{-\infty}^{\infty} \phi(x-t)\phi(x-u)dx,$$

where ϕ is integrable and square-integrable and its Fourier transform $\hat{\phi}$ does not have real zeros.

Given a function f from a reproducing kernel Hilbert space H with kernel $k(t,u)$ and an infinite set S of distinct sampling points $\{t_j\}_{j\in J}$, under mild conditions, the sampling map $f \longmapsto \{f(t_j)\}_{j\in J}$ and the recovery map $\{f(t_j)\}_{j\in J} \longmapsto f$ are both continuous. Hence there are positive constants C_1 and C_2 such that

$$C_1\|f\| \le \Big(\sum_{j\in J} |f(t_j)|^2\Big)^{1/2} \le C_2\|f\| \quad \text{for all } f \in H. \tag{1}$$

The preservation of the above inequalities (1) is essential to the derivation of a sampling expansion. The requirement that f belongs to a reproducing kernel Hilbert space H allows us to state the inequalities in terms of the boundedness and strict positivity of the so-called symbol function.

A sequence $\{f_j\}_{j\in J}$ of a separable Hilbert space H is said to be a *Riesz basis* in H if it is obtained from an orthonormal basis in H by applying a boundedly invertible linear operator. Note that for a Riesz basis $\{f_j\}_{j\in J}$

of a separable Hilbert space H there are positive constants A and B such that

$$A\sum_{j\in J}|c_j|^2 \le \Big\|\sum_{j\in J} c_j f_j\Big\|^2 \le B\sum_{j\in J}|c_j|^2 \quad \text{for all } (c_j)_{j\in J}\in \ell^2(J).$$

Given a reproducing kernel Hilbert space H with reproducing kernel $k(\cdot,\cdot)$ and an infinite set S of distinct sampling points $\{t_j\}_{j\in J}$, define the Gram matrix $G := (\langle k_{t_i}, k_{t_j}\rangle)_{i,j\in J} = (k(t_i,t_j))_{i,j\in J}$. Then one may verify that the Gram matrix G is a positive semi-definite linear operator on $\ell^2(J)$. We state a representative sampling theorem for translation-invariant reproducing kernel Hilbert spaces.

Theorem 7.1. *Let $\{t_j\}_{j\in J}$ be infinite many sampling points in S, H be a reproducing kernel Hilbert space of functions on S with reproducing kernel $k(\cdot,\cdot)$, and let $k_t(\cdot) = k(t,\cdot)$. Then the following statements are equivalent:*

(i) There exist positive constants A and B such that the inequalities (1) *hold and no such relation holds for any proper subset of the sampling points.*
(ii) $\{k_{t_j}\}_{j\in J}$ is a Riesz basis for H.
(iii) The sequence of functions $\{k_{t_j}\}_{j\in J}$ is complete and the Gram matrix $(k(t_i,t_j))_{i,j\in J}$ is a bounded self-adjoint operator and strictly positive.

If any one of the above three conditions holds, we have the following sampling expansion

$$f(t) = \sum_{j\in J} f(t_j)\frac{k(t_j,t)}{k(t_j,t_j)}.$$

8. Other extensions

There are various extensions of the Whittaker-Shannon-Kotelnikov sampling theorem. In the following, we consider the sampling expansion for signals in a shift-invariant space and for signals with finite rate of innovation.

8.1. *Sampling in shift-invariant spaces*

First we recall the definition of a shift-invariant space,

$$V_2(\phi) := \Big\{\sum_{n\in\mathbf{Z}} c(n)\phi(\cdot - n) : \sum_{n\in\mathbf{Z}}|c(n)|^2 < \infty\Big\},$$

generated by a square-integrable function ϕ. Usually we assume that $\{\phi(\cdot - n) : n \in \mathbf{Z}\}$ is a Riesz basis for $V_2(\phi)$; i.e., there exist positive constants A and B such that

$$A \sum_{n \in \mathbf{Z}} |c(n)|^2 \leq \Big\| \sum_{n \in \mathbf{Z}} c(n)\phi(\cdot - n) \Big\|_2^2 \leq B \sum_{n \in \mathbf{Z}} |c(n)|^2.$$

The Paley-Wiener space B_π is a shift-invariant space generated by the sinc function,

$$B_\pi = V_2(\mathrm{sinc}) = \Big\{ \sum_{n \in \mathbf{Z}} c(n)\mathrm{sinc}(\cdot - n) : \sum_{n \in \mathbf{Z}} |c(n)|^2 < \infty \Big\},$$

where $\mathrm{sinc}(t) = \frac{\sin \pi t}{\pi t}$. The following is a sampling theorem for signals in a shift-invariant space $V_2(\phi)$:

$$f(t) = \sum_{n \in \mathbf{Z}} f(n)\tilde{\phi}(t - n), \quad f \in V_2(\phi)$$

where $\tilde{\phi} \in V_2(\phi)$.[§] Some fundamental contributions to sampling theory in shift-invariant spaces have been made by Akram Aldroubi, Karlheinz Gröchenig, Michael Unser and others.

Fig. 7. Akram Aldroubi (left), Karlheinz Gröchenig (middle) and Michael Unser (right). The pictures are downloaded from http://www.math.vanderbilt.edu/people/aldroubi, http://www.univie.ac.at/nuhag-php/eucetifa/members.php and http://sti.epfl.ch/page-1698.html respectively.

[§]G. G. Walter, A sampling theorem for wavelet subspaces, *IEEE Trans. Inform. Theory*, Vol. 38, 1992, pp. 881–884.

8.2. *Sampling signals with finite rate of innovation*

A signal is said to be have *finite rate of innovation* if it has finite number of degrees of freedom per unit of time, the number of samples per unit of time to specify it.¶

Prototypical examples of signal with finite rate of innovation include delta pulses, non-uniform splines, narrow pulses in ultrawide band communication, mass spectrometry data in medical diagnosis etc. It also includes band-limited signals in the Paley-Wiener space and time signals in shift-invariant spaces.

Signals with finite rate of innovation have a parametric representation with a finite number of degrees of freedom per unit time. A function space modeling signals with finite rate of innovation is the following:

$$V_2(\Phi) = \Big\{ \sum_{\lambda \in \Lambda} c(\lambda)\phi_\lambda : \sum_{\lambda \in \Lambda} |c(\lambda)|^2 < \infty \Big\},$$

where ϕ_λ is the response of the impulse at the location $\lambda \in \Lambda$.‖ The vector $\Phi = (\phi_\lambda)_{\lambda \in \Lambda}$ is known as the freedom generator, and the rate of innovation on a ball B is given by the cardinality $\#(\Lambda \cap B)$ of the set Λ in the ball B.

The following is a sampling theorem for signals with finite rate of innovation:

$$f = \sum_{\gamma \in \Gamma} f(\gamma)\tilde{\psi}_\gamma \quad \text{for all } f \in V_2(\Phi)$$

where for each $\gamma \in \Gamma$, the function $\tilde{\psi}_\gamma$ reflects the characteristic of the displayer device at the sampling location γ.** The concept of signals with finite rate of innovation is introduced and studied by Martin Vetterli and his school.

9. References

Sampling in reproducing kernel Hilbert spaces

1. M. Z. Nashed, and G. G. Walter, General sampling theorems for functions in reproducing kernel Hilbert spaces, *Math. Control Signals Systems*, Vol. 4, 1991, pp. 363–390.

¶M. Vetterli, P. Marziliano, and T. Blu, Sampling signals with finite rate of innovation, *IEEE Trans. Signal Proc.*, Vol. 50, 2002, pp. 1417–1428.

‖Q. Sun, Frames in spaces with finite rate of innovation, *Adv. Comput. Math.*, Vol. 28, 2008, pp. 301–329.

**Q. Sun, Non-uniform average sampling and reconstruction of signals with finite rate of innovation, *SIAM J. Math. Anal.*, Vol. 38, 2006, pp. 1389–1422.

Fig. 8. Martin Vetterli (left) and Qiyu Sun (right). The picture is downloaded from http://lcavwww.epfl.ch/ vetterli/ and provided by Qiyu Sun

2. M. Z. Nashed, and G. G. Walter, Reproducing kernel Hilbert spaces from sampling expansions, *Contemporary Math.*, Vol. 190, 1995, pp. 221–226.
3. C. van der Mee, M. Z. Nashed, S. Seatzu, Sampling expansions and interpolation in unitarily translation invariant reproducing kernel Hilbert spaces. *Adv. Comput. Math.*, Vol. 19, 2003, pp. 355–372.
4. D. Han, M. Z. Nashed, and Q. Sun, Sampling expansions in reproducing kernel Hilbert and Banach spaces, *Numerical Functional Analysis and Optimization*, Vol. 30, 2009, pp. 971–987.
5. M. Z. Nashed, Applications of wavelets and kernel methods in inverse problems, In *Integral Methods in Science and Engineering Theoretical and Practical Aspects*, C. Constanda, Z. Nashed, and D. Rollins eds, Birkhuser, 2006, pp. 189-197.
6. M. Z. Nashed and Q. Sun, Sampling and reconstruction of signals in a reproducing kernel subspace of $L^p(\mathbf{R}^d)$, *J. Funct. Anal.*, Vol. 258, 2010, pp. 2422–2452.

Sampling in shift-invariant spaces

7. A. Aldroubi and K. Gröchenig, Nonuniform sampling and reconstruction in shift-invariant spaces, *SIAM Rev.*, Vol. 43, 2001, pp. 585–620.
8. M. Unser, Sampling - 50 years after Shannon, *Proceedings of the IEEE*, Vol. 88, 2000, pp. 569–587.

Sampling signals with finite rate of innovation

9. M. Vetterli, P. Marziliano, and T. Blu, Sampling signals with finite rate of innovation, *IEEE Trans. Signal Proc.*, Vol. 50, 2002, pp. 1417–1428.

10. P.L. Dragotti, M. Vetterli and T. Blu, Sampling moments and reconstructing signals of finite rate of innovation: Shannon meets Strang-Fix, *IEEE Trans. on Signal Processing*, Vol. 55, 2007, pp. 1741–1757.
11. Q. Sun, Non-uniform average sampling and reconstruction of signals with finite rate of innovation, *SIAM J. Math. Anal.*, Vol. 38, 2006, pp. 1389–1422.
12. N. Bi, M. Z. Nashed, and Q. Sun, Reconstructing signals with finite rate of innovation from noisy samples, *Acta Applicandae Mathematicae*, Vol. 107, 2009, pp. 339–372.

Three volumes in the series *Contemporary Mathematics* published by American Mathematical Society

13. M. Ismail, Z. Nashed, A. Zayed and A. Ghaleb, eds., Mathematical Analysis, Wavelets and Signal Processing, Vol. 190, 1995.
14. M. Z. Nashed and O. Scherzer eds., Inverse Problems, Image Analysis and Medical Imaging, Vol. 313, 2002.
15. D. Larson, P. Massopust, Z. Nashed, M. C. Nguyen, M. Papadakis, A. Zayed eds., Frames and Operator Theory in Analysis and Signal Processing, Vol. 451, 2008.

Books, edited collection of papers, and surveys on sampling theorems and related topics

16. J. A. Jerri, The Shannon sampling theorem – its various extensions and applications: A tutorial review, *Proc. IEEE*, Vol. 65, 1977, pp. 1565–1596.
17. P. L. Butzer, A survey of the Whittaker-Shannon sampling theorem and some of its extensions, *J. Math. Res. Exposition*, Vol. 3, 1983, pp. 185–212.
18. J. R. Higgins, Five short stories about the cardinal series, *Bull. Amer. Math. Soc.*, Vol. 12, 1985, 45–89.
19. P. L. Butzer, W. Splettstößer, and R. L. Stens, The sampling theorem and linear prediction in signal analysis, *Jahresber. Deutsch. Math.-Verein.*, Vol. 90, 1988, pp. 1–60.
20. R. J. Marks II, *Introduction to Shannon Sampling and Interpolation Theory*, Springer Verlag, 1991.
21. P. L. Butzer and R. L. Stens, Sampling theory for not necessarily band-limited functions: a historical overview, *SIAM Review*, Vol. 34, 1992, pp. 40–53.

22. Robert J. Marks, *Advanced Topics in Shannon Sampling and Interpolation Theory*, Springer-Verlag Berlin and Heidelberg, 1993.
23. A. I. Zayed, *Advances in Shannon's Sampling Theory*, CRC Press, 1993.
24. J. R. Higgins, *Sampling Theory in Fourier and Signal Analysis Volume 1: Foundations*, Oxford University Press, 1996.
25. J. R. Higgins and R. L. Stens, *Sampling Theory in Fourier and Signal Analysis: Volume 2: Advanced Topics*, Oxford Science Publications, 2000.
26. J. J. Benedetto and P. J. S. G. Ferreira (editors), *Modern Sampling Theory: Mathematics and Applications*,. Birkhuse, Boston, 2001.
27. F. A. Marvasti (editor), *Nonuniform Sampling: Theory and Practice (Information Technology: Transmission, Processing, and Storage)*, Plenum Pub Corp, 2001.
28. J. J. Benedetto and A. I. Zayed (editors), *Sampling, Wavelets, and Tomography*, Birkhauser Boston, 2003.

KINEMATICAL CONSERVATION LAWS (KCL): EQUATIONS OF EVOLUTION OF CURVES AND SURFACES

K. R. ARUN and PHOOLAN PRASAD*

Department of Mathematics,
Indian Institute of Science,
Bangalore, 560012, India
**E-mail: prasad@math.iisc.ernet.in*
http://math.iisc.ernet.in/~prasad

The d-dimensional (d-D) kinematical conservation laws (KCL) are the equations of evolution of a moving surface Ω_t in $\mathbb{R}^d$. The KCL are derived in a specially defined ray coordinates $(\xi_1, \xi_2, \dots, \xi_{d-1}, t)$, where $\xi_1, \xi_2, \dots, \xi_{d-1}$ are surface coordinates on Ω_t and $t > 0$ is time. We discuss various properties of 2-D and 3-D KCL systems. We first review the important properties of 2-D KCL and some of its applications. The KCL are the most general equations in conservation form, governing the evolution of Ω_t with special type of singularities, which we call kinks. The kinks are points on Ω_t when Ω_t is a curve in $\mathbb{R}^2$ and curves on Ω_t when it is a surface in $\mathbb{R}^3$. Across a kink the normal $\boldsymbol{n}$ to Ω_t and amplitude w on Ω_t are discontinuous. From 3-D KCL we derive a system of six differential equations and show that the KCL system is equivalent to the ray equations for Ω_t. The six independent equations and an energy transport equation for small amplitude waves in a polytropic gas involving an amplitude w (related to the normal velocity m of Ω_t) forms a completely determined system of seven equations. We have determined eigenvalues of the system by a novel method and find that the system has two distinct nonzero eigenvalues and five zero eigenvalues and the dimension of the eigenspace associated with the multiple eigenvalue zero is only four. For an appropriately defined m, the two nonzero eigenvalues are real when $m > 1$ and pure imaginary when $m < 1$. Finally, we have presented an an application of the theory to get evolution of a nonlinear wavefront by solving the conservation laws numerically.

Keywords: ray theory; kinematical conservation laws; nonlinear waves; conservation laws; shock propagation; hyperbolic and elliptic systems; Fermat's principle

1. Introduction

Geometrical features of a curved nonlinear wavefront or a shock front differ significantly from those of a linear wavefront. Difference is not only in the nature of singularities which appear on the fronts but also in the shapes of the fronts topologically. In two-dimensional space, when a linear wavefront moving with a constant normal velocity is concave to the direction of motion, a caustic is formed and the wavefront folds as seen in Fig. 1. A moderately weak converging nonlinear wavefront or a shock front ultimately takes a topologically different shape without any fold but with special singularities such that when we move on the front, the normal direction suffers jump discontinuity across each of these singularities as seen in Fig. 2. The wave amplitude w of a linear front tends to infinity at a cusp but that of a nonlinear front or a shock front remains finite everywhere and jumps across the special singularity, see[1] for more details.

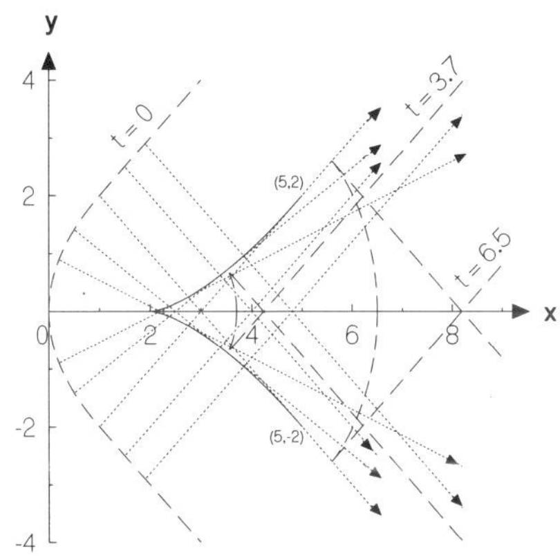

Fig. 1. Linear wavefront is shown by broken line, continuous line is the caustic and dotted lines are rays. Central part of the initial wavefront is a parabola extended on two sides by tangents. The caustic is of finite extent.

The geometrical features of a nonlinear wavefront and a shock front in three-dimensional space will be far more complex than what we observe in Fig. 2 and both fronts will possess curves of discontinuities (what we called special singularities) across which the normal direction to the fronts and the amplitude distribution on them will suffer discontinuities. These are discontinuities of the first kind, i.e. the limiting values of the discontinuous functions and their derivatives on the front as we approach a curve

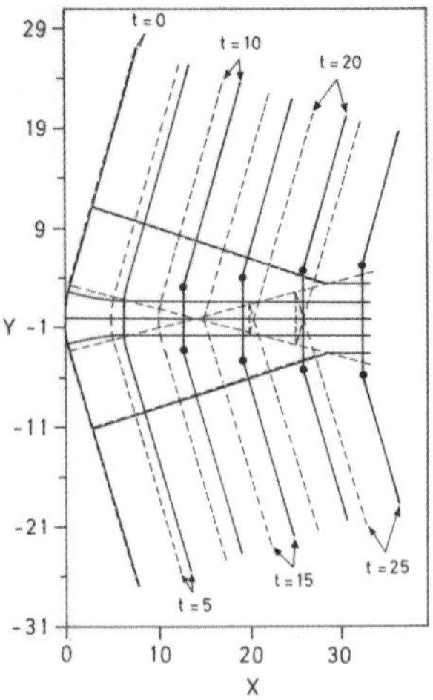

Fig. 2. Difference in geometrical features of a linear and moderately weak nonlinear wavefront in 2-D. Broken lines: linear wavefront and rays, continuous lines: nonlinear wavefronts and rays.

of discontinuity from either side are finite. Such a discontinuity was first analysed by Whitham[2] in 1957, who called it shock-shock, meaning shock on a shock front. However, as we shall see later that a discontinuity of this type is geometric in nature and can arise on any propagating surface Ω_t, and we give it a general name kink.

The linear theory of wavefront propagation governed by a linear hyperbolic system, say the wave equation is well known, where small amplitude and high frequency approximations are made. To trace the successive positions of a nonlinear wavefront governed by a quasi-linear hyperbolic system we still make the above two approximations to get a tractable system of approximate equations. In addition, in to order to get the equations which take into account nonlinear diffraction of rays due to a nonzero gradient of the wave amplitude w along the front, we need to construct a special perturbation scheme. This method, different from that of Choquet-Bruhat,[3] captures the wave amplitude in the eikonal equation itself. Following Gubkin,[4] derivation of such approximate equation for a general hyperbolic system was given by Prasad.[5,6] We quote here 2-D equations of this weakly nonlinear ray theory (WNLRT) for a polytropic gas initially in uniform state and at rest in non-dimensional[a] variables

$$
\begin{aligned}
\frac{\mathrm{d}x_1}{\mathrm{d}t} &= mn_1, \quad \frac{\mathrm{d}x_2}{\mathrm{d}t} = mn_2, \\
\frac{\mathrm{d}\theta}{\mathrm{d}t} &= -\left(-n_2\frac{\partial}{\partial x_1} + n_1\frac{\partial}{\partial x_2}\right) m, \\
\frac{\mathrm{d}m}{\mathrm{d}t} &= -\frac{1}{2}\left(\frac{\partial n_1}{\partial x_1} + \frac{\partial n_2}{\partial x_2}\right) m,
\end{aligned}
\tag{1}
$$

where $\boldsymbol{n} = (\cos\theta, \sin\theta)$ is the the unit normal to the nonlinear wavefront Ω_t and $m = 1 + \frac{\gamma+1}{2}w$ its velocity of propagation. Given initial position Ω_0 of the wavefront and a suitable amplitude distribution w_0 on it, the successive positions of the wavefront Ω_t can be obtained by numerically solving the system of equations (1). The extensive numerical results of Ramanathan[7] shows a strong diffraction of the nonlinear rays due to non-zero gradient of the wave amplitude along Ω_t in the caustic region and possibility of formation of kinks. It is to be pointed out that the numerical computation with (1) cannot be continued when a kink starts appearing on Ω_t since these equations in differential form are not valid at a point of discontinuity. However, it is possible to follow the computation for some more time and trace the path of a kink by a bit cumbersome procedure as done by Henshaw *et. al*, Kevlahan.[8,9] But in order to follow the formation and propagation of kinks automatically for a very long time, we need a new formulation of the equations (1). This was the situation when Morton, Prasad and Ravindran were looking for physically realistic conservation form of equations of (1). They derived such conservation laws in a specially defined ray coordinates (ξ, t) and since the conservation laws are derived purely on geometrical consideration, we call them 2-D kinematical conservation laws or simply 2-D KCL.[10] The mapping from (ξ, t)-plane to (x_1, x_2)-plane is continuous, a $\xi =$ const line maps onto a ray and a $t =$ const line onto the front Ω_t. When a discontinuous solution of the 2-D KCL system in the ray coordinates has a shock satisfying Rankine-Hugoniot conditions, the image of the shock in (x_1, x_2)-plane is a point singularity on Ω_t, which we call, kink. We derive 2-D KCL in the next paragraph.

Consider a one parameter family of curves Ω_t in (x_1, x_2)-plane, where the subscript t is the parameter whose different values give different positions of a moving curve. We assume that successive positions of the curve

[a]We assume that all variables, both dependent and independent, used in this paper are non-dimensional. There is one exception, the dependent variables in the first paragraph in section 4 are dimensional.

has been obtained by motion of its points in a velocity field. Let us call the associated velocity $\boldsymbol{\chi} = (\chi_1, \chi_2)$ of a point in this field as ray velocity, which depends not only on $\boldsymbol{x}, t$ and $\boldsymbol{n}$ but also on a function $w(\boldsymbol{x}, t)$. We assume that motion of this curve Ω_t is isotropic so that we take the ray velocity $\boldsymbol{\chi}$ in the direction of $\boldsymbol{n}$ and write it as

$$\boldsymbol{\chi} = m\boldsymbol{n}. \tag{2}$$

We further assume that the scalar function m depends on $\boldsymbol{x}$ and t but is independent of $\boldsymbol{n}$. The normal velocity m of Ω_t is non-dimensionalised with respect to a characteristic velocity, say the sound velocity a_0 in a uniform ambient medium, in the case Ω_t is a wavefront in such a medium.

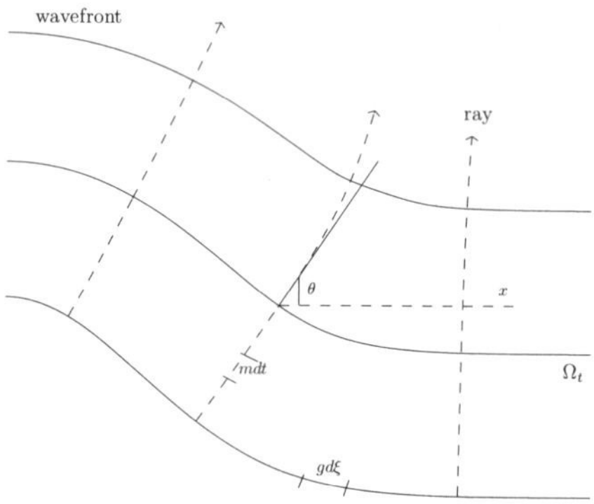

Fig. 3. 2-D ray coordinate system (ξ, t) associated with Ω_t. Continuous lines are successive positions of Ω_t and broken lines are rays. Ray direction makes an angle θ to the x-axis so that $\boldsymbol{n} = (\cos\theta, \sin\theta)$.

With the help of the velocity field, we introduce a ray coordinate system (ξ, t) such that $t =$ const represents the curve Ω_t and $\xi =$ const represents a ray,[11] see Fig. 3 for a schematic representation of a ray coordinate system in 2-D. Then mdt is an element of distance along a ray, i.e. m is the metric associated with the variable t. Let g be the metric associated with the variable ξ. We assume for the derivation of KCL that this gives a mapping: $(\xi, t) \to (x_1, x_2)$ which is smooth. Let $(\mathrm{d}\xi, \mathrm{d}t)$ be an arbitrary displacement of a point (ξ, t) in the ray coordinate plane, then the corresponding displacement $\mathrm{d}\boldsymbol{x}$ in (x_1, x_2)-plane is given by

$$\mathrm{d}\boldsymbol{x} = (g\boldsymbol{u})\mathrm{d}\xi + (m\boldsymbol{n})\mathrm{d}t, \tag{3}$$

where $\boldsymbol{u}$ is the tangent vector to Ω_t, i.e. $\boldsymbol{u} = (-n_2, n_1)$. Equating $(x_1)_{\xi t} = (x_1)_{t\xi}$ and $(x_2)_{\xi t} = (x_2)_{t\xi}$, we get the 2-D KCL[b]

$$(gn_2)_t + (mn_1)_\xi = 0, \quad (gn_1)_t - (mn_2)_\xi = 0. \tag{4}$$

From the conservation laws (4), we can derive the jump relations across a shock $\xi = \xi_s(t)$ in (ξ, t)-plane. This leads to the shock velocity $K = \mathrm{d}\xi_s/\mathrm{d}t$ as

$$K = \pm \left(\frac{m_-^2 - m_+^2}{g_+^2 - g_-^2} \right)^{1/2}, \tag{5}$$

where the symbols $+$ and $-$ refer to the states on the two sides. Fig. 4 shows the geometry of rays and Ω_t on two sides of a kink path which is the image in (x_1, x_2)-plane of a shock path in (ξ, t)-plane.

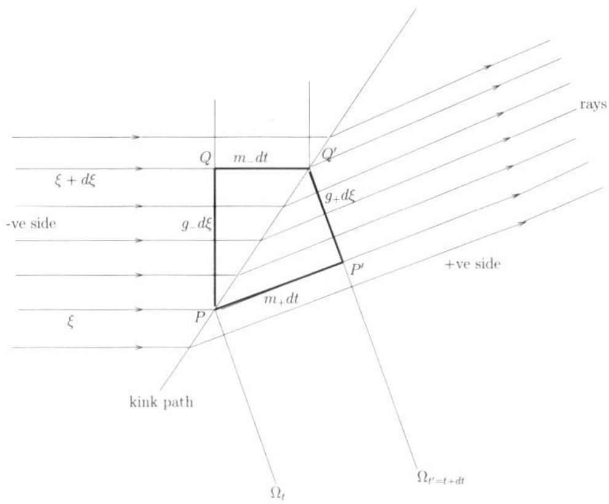

Fig. 4. Kink phenomenon in (x_1, x_2)-plane

Using Pythagoras theorem on two right angle triangles in Fig. 4 we derive

$$m_+^2(\mathrm{d}t)^2 + g_+^2(\mathrm{d}\xi)^2 = (PQ')^2 = m_-^2(\mathrm{d}t)^2 + g_-^2(\mathrm{d}\xi)^2, \tag{6}$$

which immediately gives the expression (5) for K. This result is intimately related to the theorem 2.1 in section 2 saying that KCL is physically realistic as it implies conservation of distance.

[b]We observe that the derivation of KCL remains valid even if m depends on $\boldsymbol{n}$ or the curvature or some more properties of Ω_t.

The 2-D KCL system (4) is under-determined since it contains only two equations in three variables θ, m and g. It is possible to close it in many ways. One possible way is to close it by a single conservation law

$$(gG^{-1}(m))_t = 0, \tag{7}$$

where G is a given function of m. Baskar and Prasad[12] have studied the Riemann problem for the system (4) with the closure (7) assuming some physically realistic conditions on $G(m)$. For a weakly nonlinear wavefront[1] in a polytropic gas, the third equation in (1) can be used to derive conservation of energy along a ray tube with an appropriate choice of ξ,

$$G = G_p(m) := (m-1)^{-2}e^{-2(m-1)}. \tag{8}$$

Thus, the complete set of conservation laws governing the evolution of a weakly nonlinear wavefront is given by

$$(g\sin\theta)_t + (m\cos\theta)_\xi = 0, \ (g\cos\theta)_t - (m\sin\theta)_\xi = 0, \tag{9}$$

$$\left((m-1)^2e^{2(m-1)}g\right)_t = 0. \tag{10}$$

The system of conservation laws (9)-(10) is very interesting. Its eigenvalues are

$$\lambda_1^{(2-D)} = \sqrt{\frac{m-1}{2G_p^2}}, \quad \lambda_2^{(2-D)} = -\sqrt{\frac{m-1}{2G_p^2}}, \quad \lambda_3^{(2-D)} = 0. \tag{11}$$

Hence, the system (9)-(10) is hyperbolic for $m > 1$ and has elliptic nature for $m < 1$. Prasad and his collaborators have used the 2-D KCL with suitable closure relations to solve several interesting problems and obtained many new results.[12–16] We shall discuss some results below, which contain an essence of their results.

We present two solutions arising out of the following Riemann problem for the system of conservation laws (9)-(10) with the initial data

$$\begin{aligned} m(\xi,0) &= m_0 > 1, \\ g(\xi,0) &= (m_0-1)^{-2}e^{-2(m_0-1)}, \\ \theta(\xi,0) &= \begin{cases} \theta_0, & \xi < 0, \\ -\theta_0 & \xi > 0, \end{cases} \end{aligned} \tag{12}$$

where θ_0 is constant. We take $g(\xi,t) = G_p(m)$. The coordinate ξ on Ω_0 is chosen in such a way that when the initial wavefront Ω_0 has a wedge shaped geometry convex to the x-direction $\theta_0 < 0$, and when it is concave to the x-direction $\theta_0 > 0$, as seen in Fig. 5. In linear propagation of a wavefront, the ray equations and amplitude equation decouple. The ray equations (1) with $m = 1$ simply become

$$\frac{\mathrm{d}x}{\mathrm{d}t} = n_1, \quad \frac{\mathrm{d}y}{\mathrm{d}t} = n_2, \quad \frac{\mathrm{d}\theta}{\mathrm{d}t} = 0 \tag{13}$$

and the position and geometry of Ω_t is not affected by the amplitude variation along the rays given by (13). The linear wavefronts for $\theta_0 < 0$ and $\theta_0 > 0$ are depicted in Fig. 5, where the circular arcs in the central parts of Ω_t are obtained by Huygens method.

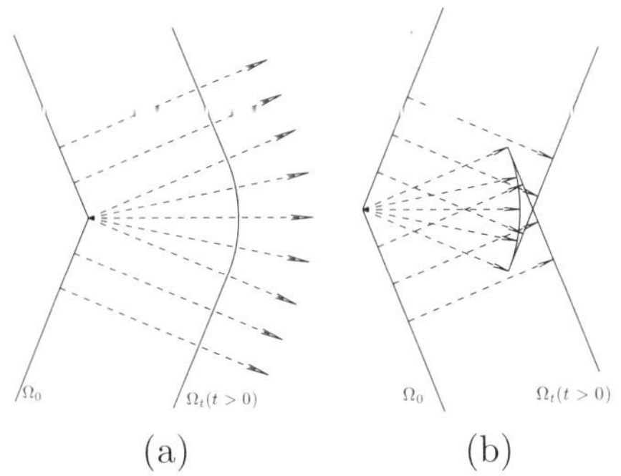

Fig. 5. (a): Linear wavefront produced by a convex wedged shaped piston. The wavefront from the corner is a circle and and the rays from there are straight radial lines. (b): Linear wavefront produced by a concave wedged shaped piston. The wavefront from the corner is a circle and and the rays from there are straight radial lines.

The solution of the Riemann problem for (9)-(10) with (12) for $\theta_0 < 0$ can be easily obtained by the procedure given in.[12] First we notice that $g = G_p$. The solution for (m, θ) is given by

$$(m,\theta)(\xi,t) = \begin{cases} (m_0, \theta_0), & \xi < t\lambda_2^{(2-D)}(m_0), \\ (m,\theta)\left(\frac{\xi}{t}\right), & t\lambda_2^{(2-D)}(m_0) < \xi < t\lambda_2^{(2-D)}(m_i), \\ (m_i, 0), & t\lambda_2^{(2-D)}(m_i) < \xi < t\lambda_1^{(2-D)}(m_i), \\ (m,\theta)\left(\frac{\xi}{t}\right), & t\lambda_1^{(2-D)}(m_i), < \xi < t\lambda_1^{(2-D)}(m_0), \\ (m_0, -\theta_0), & \xi > t\lambda_1^{(2-D)}(m_0), \end{cases} \tag{14}$$

where

$$m_i = \frac{1}{8}\{1 + (\theta_0 + 8(m_0 - 1))^2\} \tag{15}$$

and the expressions for the centred simple waves in $t\lambda_2^{(2-D)}(m_0) < \xi < t\lambda_2^{(2-D)}(m_i)$ and $t\lambda_1^{(2-D)}(m_i) < \xi < t\lambda_1^{(2-D)}(m_0)$ can be easily written down. Note that the solution (14) consists of three constant states separated by two centred rarefaction waves symmetrically situated about the t-axis. The solution, when mapped onto the (x_1, x_2)-plane with the help of first two equations in (1), gives Fig. 6.

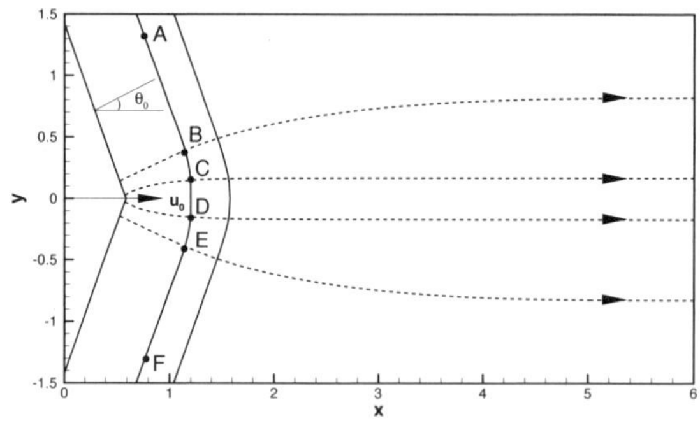

Fig. 6. Nonlinear wavefront (continuous lines) and rays (broken lines) produced by a convex wedged shaped piston. All rays ultimately tend to become parallel to the x-axis.

When $\theta_0 > 0$ and $m_0 - 1$ is not too small, the solution of the Riemann problem consists of three constant states separated by two shocks

$$(m, \theta)(\xi, t) = \begin{cases} (m_0, \theta_0), & \xi < st \\ (m_i, 0), & -st < \xi < st \\ (m_0, -\theta_0), & \xi > st, \end{cases} \tag{16}$$

where m_i and s can be easily determined, see.[1] This solution when mapped onto the (x_1, x_2)-plane is shown in the Fig. 7 and represents resolution of the caustic of the Fig. 5.

We have just presented the essence of the effect of genuinely nonlinearity in multi-dimensions, namely nonlinear diffraction of rays due to a nonzero gradient of the wave amplitude along Ω_t and its interplay with the curvature of Ω_t. In the above Riemann problem, there was a discontinuity on Ω_0 at $\xi = 0$, which implied infinite curvature at that point. This resulted in instantaneous resolution of the discontinuity in Fig. 6 and instantaneous breaking of the initial discontinuity into two kinks in Fig. 7. We take help of the results given in above examples to describe the corrugational stability of a nonlinear wavefront,[16] see also[15] for a discussion of corrugational stability of a shock front. Suppose we have a periodic wavefront Ω_t with initial shape $\Omega_0\colon x = \sin y$. The part of Ω_t concave to positive x-direction, will have a

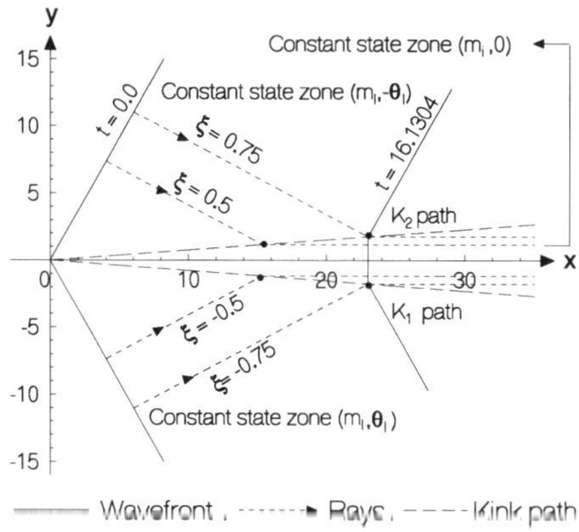

Fig. 7. Nonlinear wavefront and rays produced by a concave wedged shaped piston. All rays ultimately tend to become parallel to the x-axis. K_1 and K_2 are kinks.

tendency to bulge out with two kinks as in the Fig. 7 and the part convex to the positive x-direction will flatten as in the Fig. 6. The rays will ultimately become parallel to the x-axis. This implies that Ω_t will ultimately become plane leading to corrugational stability.

Baskar and Prasad[13] have used KCL to get the geometry and successive positions of the crest line of a curved solitary wave on the surface of a shallow water. Out of many closure equations for this case, the the physically realistic one is again (7) but now $G = (m-1)^{-\frac{3}{2}}e^{-\frac{3}{2}(m-1)}$ which is different from that given in (8).

To close the KCL for a shock front, we need an infinite system of equations.[17,18] This being too complex to handle mathematically, Prasad and Ravindran proposed in 1990 a new theory of shock dynamics[19] in which the infinite system is truncated to get a finite system of closure equations. The same closure relations for a weak shock can be obtained in a very simple manner from the WNLRT equations (1), see.[14,15] For a shock front we denote m, θ and g by M, Θ and G. Now, the complete system of approximate equations governing the evolution of a shock front moving in to a polytropic gas in uniform state at rest Ω_t is[14]

$$(G\sin\Theta)_t + (M\cos\Theta)_\xi = 0,\ (G\cos\Theta)_t - (M\sin\Theta)_\xi = 0, \tag{17}$$

$$(G(M-1)^2e^{2(M-1)})_t + 2M(M-1)^2e^{2(M-1)}GV = 0, \tag{18}$$

$$(GV^2e^{2(M-1)})_t + GV^3(M+1)e^{2(M-1)} = 0, \tag{19}$$

where one more variable, V representing the gradient of the state of the flow in the normal direction $\boldsymbol{n}$, appears.

The WNLRT equations (9)-(10) and the shock ray theory (SRT) equations (17)-(19) have been used together to give a new formulation of the sonic boom problem[20] in terms of a one parameter family of Cauchy problems. In this interesting formulation, nonlinear wavefronts in the front part of the sonic boom and the leading shock are given by solutions of Cauchy problems for hyperbolic systems and nonlinear wavefronts in the rear part and the trailing shock are given by those of elliptic systems. This helps to show an important property that the leading shock in a sonic boom may develop kinks but the trailing shock would remain smooth. Finding successive positions of nonlinear wavefronts in the rear part of the sonic boom and the trailing shock now requires solution of an ill-posed Cauchy problem for an elliptic system.

The KCL is widely applicable to research problems in many areas: from propagation of various types of wavefronts to motion of interfaces appearing in crystal growth and oil extraction industry. In the case of the motion of an interface separating a crystal from its surrounding medium, the velocity of the interface may depend only on $\boldsymbol{n}$ and in this case it may be simple to deal with KCL (4). However when m depends on the curvature, i.e. $-\frac{1}{g}\frac{\partial\theta}{\partial\xi}$, the flux functions in (4) depend also on the first derivatives of θ. This would require some new mathematical development and a new numerical method. One of many advantages of 2-D KCL theory is that the number of independent variables is reduced from three (namely, x, y and t) of the original problem to two (namely, ξ and t) in KCL. This reduction of one independent variable will be seen in 3-D KCL also.

2. Kinematics of a propagating surface and 3-D KCL

As in the last section, we assume that the successive positions of a moving surface Ω_t in $\boldsymbol{x} = (x_1, x_2, x_3)$-space is given by rays with ray velocity $\boldsymbol{\chi} = m\boldsymbol{n}$, where $\boldsymbol{n}$ is the unit normal to Ω_t. The initial position Ω_0 of the surface Ω_t can be parametrically represented in terms of two parameters ξ_1 and ξ_2. The position of Ω_t at any time t can be obtained by solving the ray equation

$$\frac{\mathrm{d}\boldsymbol{x}}{\mathrm{d}t} = m\boldsymbol{n} \tag{20}$$

with appropriate initial conditions, see also Prasad.[21] Thus, we have introduced a ray coordinate system (ξ_1, ξ_2, t) in $\boldsymbol{x}$-space such that $t =$ const represents the surface Ω_t. The surface Ω_t in $\boldsymbol{x}$-space is now generated by a one parameter family of curves such that along each of these curves ξ_1

varies and the parameter ξ_2 is constant. Similarly Ω_t is generated by another one parameter family of curves along each of these ξ_2 varies and ξ_1 is constant. Through each point (ξ_1, ξ_2) of Ω_t there passes a ray orthogonal to the successive positions of Ω_t. Thus, rays form a two parameter family. Given ξ_1, ξ_2 and t, we uniquely identify a point P in $\boldsymbol{x}$-space. For the derivation of KCL, we assume that the mapping from (ξ_1, ξ_2, t)-space to (x_1, x_2, x_3)-space is one to one and smooth. On Ω_t let $\boldsymbol{u}$ and $\boldsymbol{v}$ be unit tangent vectors of the curves $\xi_2 =$ const and $\xi_1 =$ const respectively and $\boldsymbol{n}$ be unit normal to Ω_t. We take $(\boldsymbol{u}, \boldsymbol{v}, \boldsymbol{n})$ to be a right handed set of vectors, then

$$\boldsymbol{n} = \frac{\boldsymbol{u} \times \boldsymbol{v}}{\|\boldsymbol{u} \times \boldsymbol{v}\|}. \tag{21}$$

Let an element of length along a curve $\xi_2 = \text{const}, t = \text{const}$ be $g_1 \mathrm{d}\xi_1$ and that along a curve $\xi_1 = \text{const}, t = \text{const}$ be $g_2 \mathrm{d}\xi_2$. The element of length along a ray $\xi_1 = \text{const}, \xi_2 = \text{const}$ is $m\mathrm{d}t$. A displacement $\mathrm{d}\boldsymbol{x}$ in $\boldsymbol{x}$-space due to arbitrary increments $\mathrm{d}\xi_1, \mathrm{d}\xi_2$ and $\mathrm{d}t$ is given by (this is an extension of the result (3))

$$\mathrm{d}\boldsymbol{x} = (g_1 \boldsymbol{u})\mathrm{d}\xi_1 + (g_2 \boldsymbol{v})\mathrm{d}\xi_2 + (m\boldsymbol{n})\mathrm{d}t. \tag{22}$$

This gives the Jacobian

$$J := \frac{\partial(x_1, x_2, x_3)}{\partial(\xi_1, \xi_2, t)} = g_1 g_2 m \sin\chi, \quad 0 < \chi < \pi, \tag{23}$$

where $\chi(\xi_1, \xi_2, t)$ is the angle between the $\boldsymbol{u}$ and $\boldsymbol{v}$, i.e.

$$\cos\chi = \langle \boldsymbol{u}, \boldsymbol{v} \rangle. \tag{24}$$

As explained after (44) in section 4, we shall like to choose $\sin\chi = \|\boldsymbol{u} \times \boldsymbol{v}\|$ which requires the restriction $0 < \chi < \pi$ on χ. For a smooth moving surface Ω_t, we equate $\boldsymbol{x}_{\xi_1 t} = \boldsymbol{x}_{t\xi_1}$ and $\boldsymbol{x}_{\xi_2 t} = \boldsymbol{x}_{t\xi_2}$, and get the 3-D KCL of Giles, Prasad and Ravindran[22]

$$(g_1 \boldsymbol{u})_t - (m\boldsymbol{n})_{\xi_1} = 0, \tag{25}$$

$$(g_2 \boldsymbol{v})_t - (m\boldsymbol{n})_{\xi_2} = 0. \tag{26}$$

We also equate $\boldsymbol{x}_{\xi_1 \xi_2} = \boldsymbol{x}_{\xi_2 \xi_1}$ and derive 3 more scalar equations contained in

$$(g_2 \boldsymbol{v})_{\xi_1} - (g_1 \boldsymbol{u})_{\xi_2} = 0. \tag{27}$$

Equations (25)-(27) are necessary and sufficient conditions for the integrability of the equation (22), see Courant and John.[23] When we have constructed coordinates (ξ_1, ξ_2) on Ω_t at any time t they would satisfy (27).

These coordinates dynamically evolve in time according to (25)-(26) in such a way that they continue to remain coordinates on Ω_t when t changes.

From the equations (25)-(26) we can show that $(g_2\boldsymbol{v})_{\xi_1} - (g_1\boldsymbol{u})_{\xi_2}$ does not depend on t. If any choice of coordinates ξ_1 and ξ_2 on Ω_0 implies that the condition (27) is satisfied at $t = 0$ then it follows from (25)-(26) that (27) is automatically satisfied for all $t > 0$. It is very interesting to note that the constraint (27) is analogous to the solenoidal condition in the equations of ideal magnetohydrodynamics. To see this, let us introduce three vectors $\mathfrak{B}_k, k = 1, 2, 3$, in $\mathbb{R}^2$ via

$$\mathfrak{B}_k := (g_2 v_k, -g_1 u_k). \tag{28}$$

Using this definition of $\mathfrak{B}_k$, (27) can be recast in an equivalent form

$$\mathrm{div}(\mathfrak{B}_k) = 0, \;\; k = 1, 2, 3. \tag{29}$$

Therefore, we infer that all the three vectors $\mathfrak{B}_k$ are divergence-free at any time t if they are so at time $t = 0$. Note that there are three scalar constraints in (29) analogous to the solenoidal condition in the equations of two-dimensional ideal magnetohydrodynamics. We shall refer to (27) (or (29)), as 'geometric solenoidal constraint'.

The 3-D KCL is a system of six scalar evolution equations (25) and (26). However, since $\|\boldsymbol{u}\| = \|\boldsymbol{v}\| = 1$, there are seven dependent variables in (25)-(26): two independent components of each of $\boldsymbol{u}$ and $\boldsymbol{v}$, the front velocity m of Ω_t, g_1 and g_2. Thus, KCL is an under-determined system and can be closed only with the help of additional relations or equations, which would follow from the nature of the surface Ω_t and the dynamics of the medium in which it propagates.

The system (25)-(26) consists of equations which are conservation laws, so its weak solution may contain shocks which are surfaces in (ξ_1, ξ_2, t)-space. Across these shock surfaces m, g_1, g_2 and vectors $\boldsymbol{u}, \boldsymbol{v}$ and $\boldsymbol{n}$ will be discontinuous. The image of a shock surface into $\boldsymbol{x}$-space will be another surface, let us call it a kink surface, which will intersect Ω_t in a curve, say kink curve $\mathscr{K}_t$. Across this kink curve or simply the kink, the normal direction $\boldsymbol{n}$ of Ω_t will be discontinuous as shown in Figure 8. As time t evolves, $\mathscr{K}_t$ will generate the kink surface. We assume that the mapping between (ξ_1, ξ_2, t)-space and (x_1, x_2, x_3)-space continues to be one to one even when a kink appears.

The plane $t =$ const is mapped onto Ω_t with kinks, see Figure 8. It has been shown in[24] that the 3-D KCL (25)-(26) is physically realistic in the sense that they give kinks across which distance in three independent directions are preserved. We now state a theorem of[22,24] without proof.

Theorem 2.1. *The jump relations of the 3-D KCL (25)-(26) imply conservation of distance in x_1, x_2 and x_3 directions and hence in any arbitrary direction in $\boldsymbol{x}$-space in the sense that the expressions for a vector displacement $(\mathrm{d}\boldsymbol{x})_{\mathscr{K}_t}$ of a point of the kink line $\mathscr{K}_t$ in an infinitesimal time interval* $\mathrm{d}t$, *when computed in terms of variables on the two sides of a kink surface, have the same value. This displacement of the point is assumed to take place on the kink surface and that of its image in (ξ_1, ξ_2, t)-space takes place on the shock surface such that the corresponding displacement in (ξ_1, ξ_2)-plane is with the shock front.*

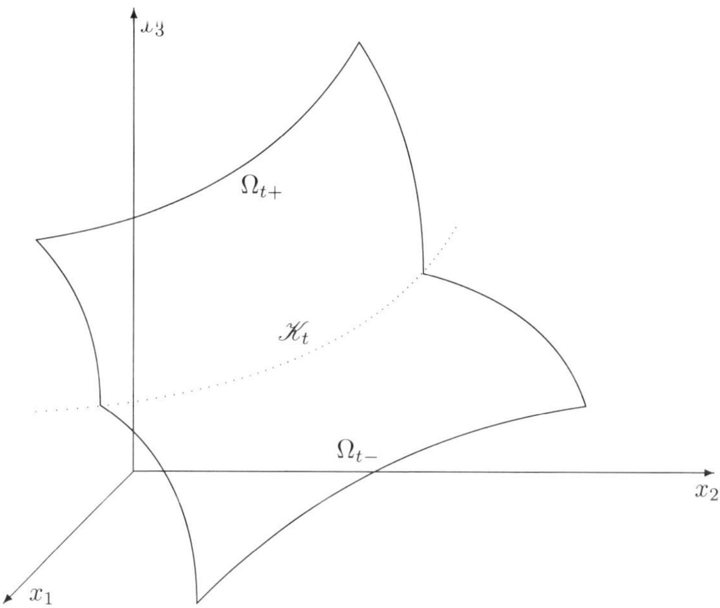

Fig. 8. Kink curve $\mathscr{K}_t$ (shown with dotted lines) on $\Omega_t = \Omega_{t+} \cup \Omega_{t-}$

3. Explicit differential form of KCL and equivalence to the ray equations

The eikonal equation governing the propagation of a front in an isotropic media is

$$\varphi_t + m\|\nabla\varphi\| = 0. \tag{30}$$

For m as a given function of $\boldsymbol{x}$ and t, bicharacteristic equations or the ray equations of (30),[1,21] are

$$\frac{\mathrm{d}\boldsymbol{x}}{\mathrm{d}t} = m\boldsymbol{n}, \ \|\boldsymbol{n}\| = 1, \tag{31}$$

$$\frac{\mathrm{d}\boldsymbol{n}}{\mathrm{d}t} = -\mathbf{L}m := -\left(\nabla - \boldsymbol{n}\langle \boldsymbol{n}, \nabla\rangle\right) m. \tag{32}$$

Here, $\mathrm{d}/\mathrm{d}t$ denotes the convective derivative

$$\frac{\mathrm{d}}{\mathrm{d}t} = \frac{\partial}{\partial t} + m\langle \boldsymbol{n}, \nabla\rangle \tag{33}$$

which is the time rate of change along a ray. It is to be noted that the bicharacteristic equations in the above form are obtained from the Charpit's equations of (30) and then using

$$\boldsymbol{n} = \frac{\nabla\varphi}{\|\nabla\varphi\|}. \tag{34}$$

Carrying out the differentiations in (25)-(26) and simplifying the resulting expressions yields the differential form of 3-D KCL, which is the following system of quasilinear equations

$$g_{1t} = -m\langle \boldsymbol{n}, \boldsymbol{u}_{\xi_1}\rangle, \tag{35}$$

$$g_{2t} = -m\langle \boldsymbol{n}, \boldsymbol{v}_{\xi_2}\rangle, \tag{36}$$

$$g_1\boldsymbol{u}_t = m_{\xi_1}\boldsymbol{n} + m\langle \boldsymbol{n}, \boldsymbol{u}_{\xi_1}\rangle\boldsymbol{u} + \frac{m}{\|\boldsymbol{u}\times\boldsymbol{v}\|}\left\{(\boldsymbol{u}\times\boldsymbol{v})_{\xi_1} + \frac{\boldsymbol{n}\langle \boldsymbol{u}\times\boldsymbol{v}\rangle}{\|\boldsymbol{u}\times\boldsymbol{v}\|}\langle \boldsymbol{u}, \boldsymbol{v}\rangle_{\xi_1}\right\}, \tag{37}$$

$$g_1\boldsymbol{v}_t = m_{\xi_2}\boldsymbol{n} + m\langle \boldsymbol{n}, \boldsymbol{v}_{\xi_2}\rangle\boldsymbol{v} + \frac{m}{\|\boldsymbol{u}\times\boldsymbol{v}\|}\left\{(\boldsymbol{u}\times\boldsymbol{v})_{\xi_2} + \frac{\boldsymbol{n}\langle \boldsymbol{u}\times\boldsymbol{v}\rangle}{\|\boldsymbol{u}\times\boldsymbol{v}\|}\langle \boldsymbol{u}, \boldsymbol{v}\rangle_{\xi_2}\right\}. \tag{38}$$

It is to be noted that since $\|\boldsymbol{u}\| = \|\boldsymbol{v}\| = 1$, there are only two independent equations in each of (37) and (38). We require another set of long calculations to show that the expressions on the right hand sides of (35)-(38) are equal to those of (31)-(32). This establishes the equivalence of the differential forms (35)-(38) of the 3-D KCL to the equations of the ray theory. We state this important result as a theorem and refer to[24] for more details.

Theorem 3.1. *For a given smooth function m of $\boldsymbol{x}$ and t, the ray equations (31)-(32) are equivalent to the 3-D KCL as long as their solutions are smooth.*

4. Energy transport equation from WNLRT and the complete set of equations

In this section we shall derive a closure relation in conservation form for the 3-D KCL so that we get a completely determined system of conservation laws. Let the mass density, fluid velocity and gas pressure in a polytropic gas be denoted by $\varrho, \boldsymbol{q}$ and p. Consider now a forward facing small amplitude curved wavefront Ω_t running into the gas in a uniform state and at rest $\varrho_0 = \text{const}, \boldsymbol{q} = \boldsymbol{0}$ and $p_0 = \text{const}$. A perturbation in the state of the gas on Ω_t can be expressed in terms of an amplitude w and is given by[1]

$$\varrho - \varrho_0 = \left(\frac{\varrho_0}{a_0}\right) w, \ \boldsymbol{q} = \boldsymbol{n} w, \ p - p_0 = \varrho_0 a_0 w. \tag{39}$$

where a_0 is the sound velocity in the undisturbed medium $= \sqrt{\gamma p_0/\varrho_0}$ and w is a quantity of small order, say $\mathcal{O}(\epsilon)$. Let us remind, what we stated in the section 1, all dependent variables are dimensional in this (and only in this) paragraph. Note that w here has the dimension of velocity.

The amplitude w is related to the non-dimensional normal velocity m of Ω_t by

$$m = 1 + \frac{\gamma+1}{2}\frac{w}{a_0}. \tag{40}$$

Note that the operator $\mathrm{d}/\mathrm{d}t$ defined in (33) in space-time becomes simply the partial derivative $\partial/\partial t$ in the ray coordinate system (ξ_1, ξ_2, t). Hence, the energy transport equation of the WNLRT[1] in non-dimensional coordinates becomes

$$m_t = (m-1)\mathit{\Omega} = -\frac{1}{2}(m-1)\langle \nabla, \boldsymbol{n} \rangle, \tag{41}$$

where the italic symbol $\mathit{\Omega}$ is the mean curvature of the wavefront Ω_t. The ray tube area $\mathcal{A}$ of any ray system[1,25] is related to the mean curvature $\mathit{\Omega}$ (we write here in non-dimensional variables) by

$$\frac{1}{\mathcal{A}}\frac{\partial \mathcal{A}}{\partial l} = -2\mathit{\Omega}, \ \frac{\partial}{\partial l} \text{ in ray coordinates}, \tag{42}$$

where l is the arc length along a ray. In non-dimensional variables $\mathrm{d}l = m\mathrm{d}t$. From (41)-(42) we get

$$\frac{2m_t}{m-1} = -\frac{1}{m\mathcal{A}}\mathcal{A}_t. \tag{43}$$

This leads to a conservation law, which we accept to be physically realistic,

$$\left\{(m-1)^2 e^{2(m-1)}\mathcal{A}\right\}_t = 0. \tag{44}$$

Integration gives $(m-1)^2 e^{2(m-1)}\mathcal{A} = F(\xi_1, \xi_2)$, where F is an arbitrary function of ξ_1 and ξ_2. The ray tube area $\mathcal{A}$ is given by $\mathcal{A} = g_1 g_2 \sin\chi$, where χ is the angle between the vectors $\boldsymbol{u}$ and $\boldsymbol{v}$. In order that $\mathcal{A}$ is positive, we need to choose $0 < \chi < \pi$. Now the energy conservation equation becomes

$$\left\{(m-1)^2 e^{2(m-1)} g_1 g_2 \sin\chi\right\}_t = 0. \tag{45}$$

To readers used to linear ray theory, appearance of the factor $e^{2(m-1)}$ in the above equation may look a little unfamiliar. This term appears here due to nonlinear stretching of the rays. Though the term $(m-1)^2 e^{2(m-1)}$ is approximately equal to $(m-1)^2$ for small $m-1$, we need the factor $e^{2(m-1)}$ for consistency of the equations.

Hence, the complete set of conservation laws for the weakly nonlinear ray theory (WNLRT) for a polytropic gas are: the six equations in (25)-(26) and the equation (45). The equations (27) need to be satisfied at any fixed t, say at $t = 0$. Moreover, the complete set of equations of WNLRT in a differential form for the unknown $V = (u_1, u_2, v_1, v_2, m, g_1, g_2)^T$ can be derived from (35)-(38) and (45) and are written in the usual matrix form as

$$AV_t + B^{(1)} V_{\xi_1} + B^{(2)} V_{\xi_2} = 0, \tag{46}$$

where the expressions for the matrices A, $B^{(1)}$ and $B^{(2)}$ are given in.[24] The hyperbolicity of (46) will depend on its eigen-structure. Arun and Prasad[24,26] have studied the eigenvalues and eigenvectors of (46) in detail. In what follows we briefly sketch their final result in the form of a theorem

Theorem 4.1 (Theorem 8.2 of[24]). *The system (46) has seven eigenvalues λ_1, $\lambda_2(= -\lambda_1)$, $\lambda_3 = \lambda_4 = \ldots = \lambda_7 = 0$, where λ_1 is given by*

$$\lambda_1 = \left[\frac{m-1}{2}\left\{(\gamma_1^2+\gamma_2^2)\frac{e_1^2}{g_1^2} + 2(\gamma_1\delta_1+\gamma_2\delta_2)\frac{e_1}{g_1}\frac{e_2}{g_2} + (\delta_1^2+\delta_2^2)\frac{e_2^2}{g_2^2}\right\}\right]^{1/2}. \tag{47}$$

The eigenvalues λ_1 and λ_2 are real for $m > 1$ and purely imaginary for $m < 1$. Further, the dimension of the eigenspace corresponding to the multiple eigenvalue zero is four.

5. Numerical approximation and further remarks

The aim of this section is to set up an initial value problem for the conservation laws of 3-D WNLRT, i.e. (25)-(26) and energy conservation equation

(45). We notice that the set of equations can be recast in the divergence form

$$W_t + F_1(W)_{\xi_1} + F_2(W)_{\xi_2} = 0, \tag{48}$$

with the conserved variables W and the fluxes $F_i(W)$ given as

$$\begin{aligned} W &= \left(g_1\boldsymbol{u}, g_2\boldsymbol{v}, (m-1)^2 e^{2(m-1)} g_1 g_2 \sin\chi\right)^T, \\ F_1(W) &= (m\boldsymbol{n}, \boldsymbol{0}, 0)^T, \\ F_2(W) &= (\boldsymbol{0}, m\boldsymbol{n}, 0)^T. \end{aligned} \tag{49}$$

In order to formulate an initial value problem for (48) we take the initial position of a weakly nonlinear wavefront Ω_t as

$$\Omega_0\colon x_3 = f(x_1, x_2). \tag{50}$$

On Ω_0, we choose $\xi_1 = x_1, \xi_2 = x_2$, then

$$\Omega_0\colon x_{10} = \xi_1,\ x_{20} = \xi_2,\ x_{30} = f(\xi_1, \xi_2). \tag{51}$$

Therefore,

$$g_{10} = \sqrt{1 + f_{\xi_1}^2},\ g_{20} = \sqrt{1 + f_{\xi_2}^2}, \tag{52}$$

$$\boldsymbol{u}_0 = \frac{(1, 0, f_{\xi_1})}{\sqrt{1 + f_{\xi_1}^2}},\ \boldsymbol{v}_0 = \frac{(0, 1, f_{\xi_2})}{\sqrt{1 + f_{\xi_2}^2}}. \tag{53}$$

We can easily check that the geometric solenoidal constraint (29) is satisfied by this choice of the initial values. The unit normal $\boldsymbol{n}_0$ of Ω_0 is given by

$$\boldsymbol{n}_0 = -\frac{(f_{\xi_1}, f_{\xi_2}, -1)}{\sqrt{1 + f_{\xi_1}^2 + f_{\xi_2}^2}} \tag{54}$$

in which the sign is so chosen that $(\boldsymbol{u}_0, \boldsymbol{v}_0, \boldsymbol{n}_0)$ form a right handed system. Let the distribution of the front velocity be given by

$$m = m_0(\xi_1, \xi_2). \tag{55}$$

We have now completed formulation of the initial data for the system of conservation laws (48). The problem is to find a solution of (48) satisfying the above initial data. Having solved these equations, we can get Ω_t by solving the first part of the ray equations, namely (31) at least numerically for a number of values of ξ_1 and ξ_2.

Since we have an incomplete set of eigenvectors for the system of conservation laws of 3-D WNLRT, an initial value problem is not well-posed in the strong hyperbolic sense and is likely to be more sensitive than regular hyperbolic systems from computational point of view. Numerical as well as theoretical analysis indicates that the solution may not belong to BV spaces and can only be measure valued. Despite theoretical difficulties, we have been able to develop numerical codes for 3-D WNLRT using simple but robust central schemes. For a weakly hyperbolic system the central schemes are much easily applicable than any characteristic-based scheme. Moreover, the simplicity of central finite volume schemes makes it convenient to employ them for the numerical solution of the complex system of conservation laws of WNLRT. In[27,28] we have presented a numerical approximation of the balance laws using a Kurganov-Tadmor type semi-discrete central scheme.[29] In order to get second order accuracy we use standard MUSCL type reconstructions and TVD Runge-Kutta time stepping procedures. However, the high resolution central scheme need not respect the geometric solenoidal constraint (29) and hence we use a constrained transport technique to enforce it. Note that the equation (29) implies the existence of three potentials $\mathbb{A}_1, \mathbb{A}_2, \mathbb{A}_3$ so that the components of the vectors $g_1\boldsymbol{u}$ and $g_2\boldsymbol{v}$ are derivable from the potentials.[27] With the aid of the 3-D KCL system (25)-(26) we obtain the evolution equations for the potentials. These equations form a coupled system of three nonlinear equations of the Hamilton-Jacobi type. The evolution equations for the potentials are discretised on a staggered grid. The updated values of the potentials are used to get the corrected values of $g_1\boldsymbol{u}$ and $g_2\boldsymbol{v}$ at the next time step, which satisfy a discrete version of the constraint (29). In each time step, after solving the conservation laws we update the ray equations using a second order Runge-Kutta scheme which gives the successive positions of the front Ω_t.

It well known from the literature that the solution to the Cauchy problem for a weakly hyperbolic system contains a Jordan which grows polynomially in time. However, the numerical solutions of the conservation laws of 3-D WNLRT do not exhibit any such component. The reason for disappearance of the Jordan mode is the constraint (29) which is preserved by system as well as the numerical scheme. Due to the complexity of the equations of 3-D KCL we have not been able to establish this result for the full nonlinear system. In order to justify our assertion we first linearise the system of conservation laws of 3-D WNLRT about a constant state, viz. a planar wavefront. We solve the linearised system analytically and show that for the linearised system the Jordan mode does not appear when the geometric solenoidal constraint is satisfied. We strongly believe the same result to hold true also for the full nonlinear system.

6. An example of propagation of a nonlinear wavefront in three dimensions

We give in this section the results of a numerical experiment from[27] showing the time evolution of a nonlinear wavefront in three dimensions. We choose initial wavefront Ω_0 in a such a way that it is not axisymmetric. The front Ω_0 has a single smooth dip. The initial shape of the wavefront is given by

$$\Omega_0 \colon x_3 = \frac{-\kappa}{1 + \frac{x_1^2}{\alpha^2} + \frac{x_2^2}{\beta^2}}, \tag{56}$$

where the parameter values are set to be $\kappa = 1/2, \alpha = 3/2, \beta = 3$. The ray coordinates (ζ_1, ζ_2) are chosen initially as $\xi_1 = x_1$ and $\xi_2 = x_2$. Therefore, using (56), the initial wavefront can be represented in a parametric form

$$x_1 = \xi_1,\ x_2 = \xi_2,\ x_3 = \frac{-\kappa}{1 + \frac{\xi_1^2}{\alpha^2} + \frac{\xi_2^2}{\beta^2}}. \tag{57}$$

With the aid of (57) the initial values $g_1, g_2, \boldsymbol{u}$ and $\boldsymbol{v}$ are calculated, cf. section 5. The normal velocity is prescribed as a constant $m_0 = 1.2$ everywhere on the initial wavefront Ω_0.

The computational domain $[-20, 20] \times [-20, 20]$ is divided into 401×401 mesh points. The simulations are done up to $t = 2.0, 6.0, 10.0$. We have set non-reflecting boundary conditions everywhere.

In Figure 9 we plot the initial wavefront Ω_0 and the successive positions of the wavefront Ω_t at times $t = 2.0, 6.0, 10.0$. It can be seen that the wavefront has moved up in the x_3-direction and the dip has spread over a larger area in x_1- and x_2-directions. The lower part of the front moves up leading to a change in shape of the initial front Ω_0. It is very interesting to note that two dips appear at the centre of the wavefront, which are clearly visible at $t = 6.0$ and $t = 10.0$. These two dips are separated by an elevation almost like a wall parallel to the x_2-axis. There is a pair of kink lines, which are also parallel to the x_2-axis and are more clearly seen in Figure 10.

To explain the results of convergence of the rays we also give in Figure 10 the slices of the wavefront in $x_2 = 0$ section and $x_1 = 0$ section from time $t = 0.0$ to $t = 10.0$. Due to the particular choice of the parameters α and β in the initial data (56), the section of the front Ω_0 in $x_2 = 0$ plane has a smaller radius of curvature than that of the section in $x_1 = 0$ plane. This results in a stronger convergence of the rays in $x_2 = 0$ plane compared to those in $x_1 = 0$ plane as evident from Figure 10. In the diagram on the top in Figure 10, we clearly note a pair of kinks at times $t = 3.0$ onwards in

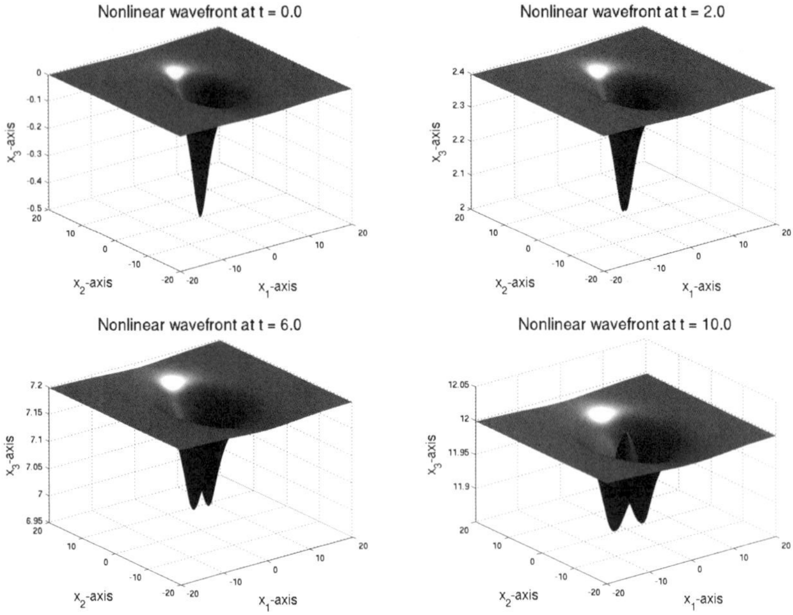

Fig. 9. The successive positions of the nonlinear wavefront Ω_t with an initial smooth dip which is not axisymmetric.

the $x_2 = 0$ section. However, there are no kinks in the bottom diagram in Figure 10 in $x_1 = 0$ section.

We give now the plots of the normal velocity m in (ξ_1, ξ_2) plane along ξ_1- and ξ_2-directions in Figure 11. It is observed that m has two shocks in the ξ_1-direction which correspond to the two kinks in the x_1-direction.

We plot the divergence of $\mathfrak{B}_1$ at time $t = 10.0$ in Figure 12. It is evident that the geometric solenoidal condition is satisfied with an error of 10^{-15}. The divergences of $\mathfrak{B}_2$ and $\mathfrak{B}_3$ also show the same trend.

7. Comparison of 2-D and 3-D KCL results

In this section we present a comparison of the 2-D and 3-D KCL results. Let us consider the 3-D axi-symmetric initial wavefront Ω_0 with a single smooth dip given by

$$\Omega_0 \colon x_3 = \kappa \left(1 - e^{-\frac{r^2}{a^2}}\right), \tag{58}$$

where $r = \sqrt{x_1^2 + x_2^2}$ is the distance from the x_3-axis. The propagation of this 3-D nonlinear wavefront Ω_t is axi-symmetric and hence the problem

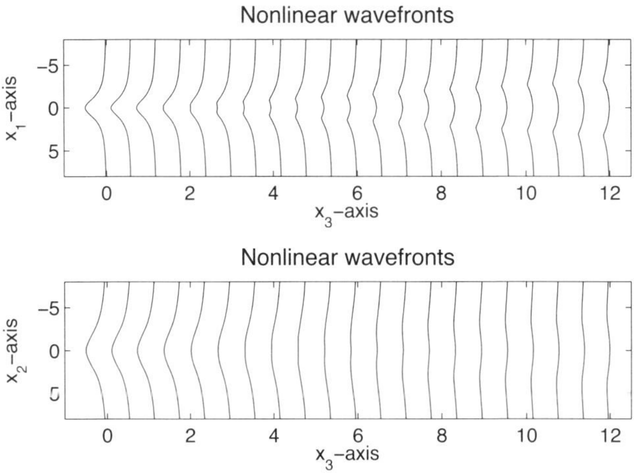

Fig. 10. The sections of the nonlinear wavefront at times $t = 0.0, \ldots, 10.0$ with a time step 0.5. On the top: along $x_2 = 0$ plane. Bottom: in $x_1 = 0$ plane.

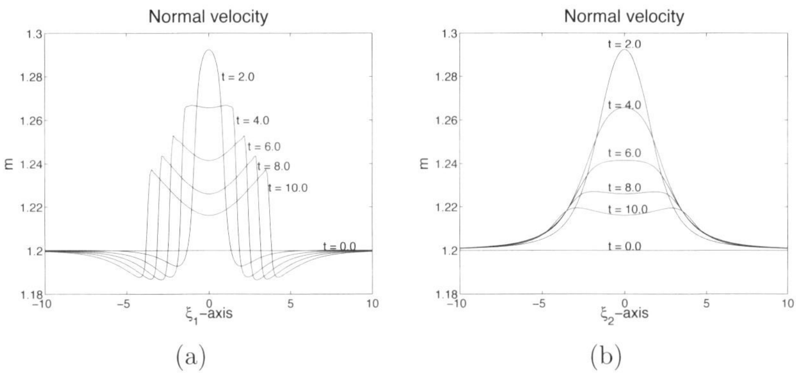

Fig. 11. The time evolution of the normal velocity m. (a): along ξ_1-direction in the section $\xi_2 = 0$. (b): along ξ_2-direction in the section $\xi_1 = 0$.

reduces to essentially 2-D. We have used the 2-D KCL to study the time evolution of this 2-D wavefront Ω_t. In order to illustrate the genuinely 3-dimensional effects of geometrical convergence, we plot the corresponding results obtained from the 2-D KCL and the 3-D KCL in Figure 13. In this figure, the solid lines represent the successive positions of the nonlinear wavefront obtained by the 3-D KCL whereas the dotted lines represents the corresponding 2-D wavefront obtained using 2-D KCL simulations. It can be observed that both the results agree qualitatively. But the 2-D and

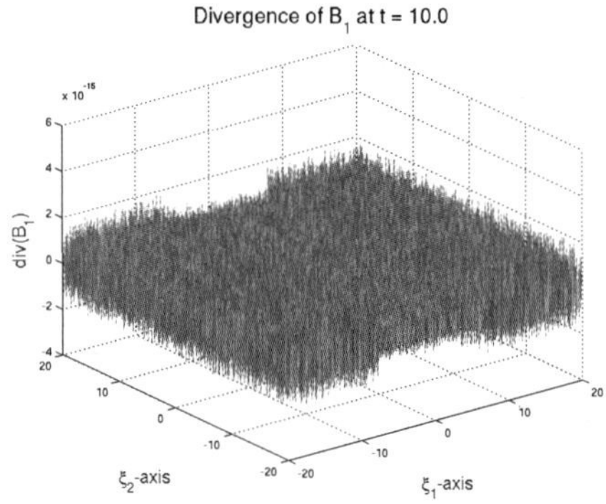

Fig. 12. The divergence of $\mathfrak{B}_1$ at $t = 10.0$. The error is of the order of 10^{-15}. The vertical axis is magnified 10^{15} times.

3-D wavefronts coincide only for small times, the 3-D wavefronts moves faster than the 2-D ones. This shows the effect of truly three dimensional geometrical convergence.

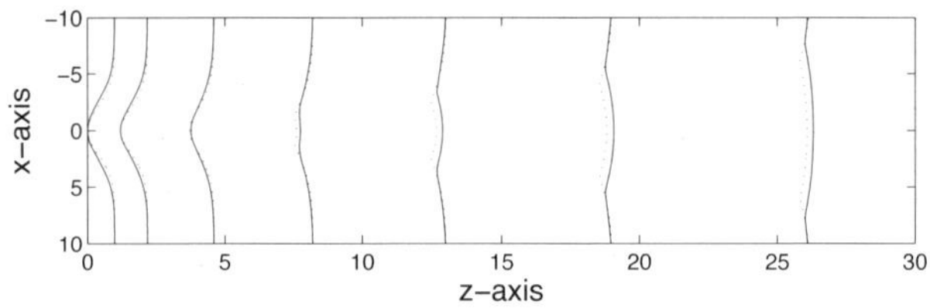

Fig. 13. Comparison of 3-D KCL and 2-D KCL: the solid lines represents the slices of 3-D wavefronts and the dotted lines are the 2-D wavefronts at times $t = 0.0$ to $t = 6.0$. The parameter $a = 4$.

8. Concluding Remarks

The 3-D KCL is a quite complex system of conservation laws. But as far as we know this is the only system which gives correct location and geometry of a moving surface - which has singularities and topologically correct shapes.[24,27,28] This article following the original results in references,[24,27] puts the theory of KCL in firm formulation. There is a lot of scope for doing theoretical investigation on KCL systems and particularly, it is very challenging to get good numerical approximations for 3-D KCL. Attempts in this direction will be fruitful as the results will be applicable also to moving interfaces in chemical and biological systems, where the theory may reveal many physically realistic results.

Acknowledgement

K. R. A is supported by the Indian Institute of Science under the junior research associateship. P. P's research is supported by the Department of Atomic Energy, Government of India under Raja Ramanna Fellowship Scheme. The authors gratefully acknowledge their respective grants. Department of Mathematics of IISc is partially funded by UGC under DSA-SAP, Phase IV.

References

1. P. Prasad, *Nonlinear Hyperbolic Waves in Multi-dimensions* (Chapman and Hall/CRC, London, 2001).
2. G. B. Whitham, *J. Fluid Mech.* **2**, 146 (1957).
3. Y. Choquet-Bruhat, *J. Math. Pures. Appl.* **48**, 117 (1969).
4. K. E. Gubkin, *PMM J. Appl. Math. Mech.* **22**, 787 (1958).
5. P. Prasad, *J. Math. Anal. Appl.* **50**, 470 (1975).
6. P. Prasad, *Proc. Indian Acad. Sci. Math. Sci.* **110**, 431 (2000).
7. T. M. Ramanathan, Huygen's method of construction of weakly nonlinear wavefronts and shockfronts with application to hyperbolic caustics, PhD thesis, Indian Institute of Science, (Bangalore, 1985).
8. W. D. Henshaw, N. F. Smyth and D. W. Schwendeman, *J. Fluid Mech.* **171**, 519 (1986).
9. N. K.-R. Kevlahan, *J. Fluid Mech.* **327**, 161 (1996).
10. K. W. Morton, P. Prasad and R. Ravindran, *Conservation forms of nonlinear ray equations*, tech. rep., Department of Mathematics, Indian Institute of Science (Bangalore, 1992).
11. M. P. Lazarev, R. Ravindran and P. Prasad, *Acta Mech.* **126**, 135 (1998).
12. S. Baskar and P. Prasad, *IMA J. Appl. Math.* **69**, 391 (2004).
13. S. Baskar and P. Prasad, Kinematical conservation laws applied to study geometrical shapes of a solitary wave, in *Wind over Waves II: Forecasting and Fundamentals*, eds. S. Sajjadi and J. Hunt (Horwood Publishing Ltd, 2003).
14. S. Baskar and P. Prasad, *J. Fluid Mech.* **523**, 171 (2005).
15. A. Monica and P. Prasad, *J. Fluid Mech.* **434**, 119 (2001).
16. P. Prasad and K. Sangeeta, *J. Fluid Mech.* **385**, 1 (1999).
17. M. A. Grinfel'd, *PMM J. Appl. Math. Mech.* **42**, 958 (1978).
18. V. P. Maslov, *J. Sov. Math.* **13**, 119 (1980).
19. R. Ravindran and P. Prasad, *Appl. Math. Lett.* **3**, 77 (1990).
20. S. Baskar and P. Prasad, *Proc. Indian Acad. Sci. Math. Sci.* **116**, 97 (2006).
21. P. Prasad, *Indian J. Pure Appl. Math.* **38**, 467 (2007).
22. M. B. Giles, P. Prasad and R. Ravindran, *Conservation forms of equations of*

three dimensional front propagation, tech. rep., Department of Mathematics, Indian Institute of Science (Bangalore, 1995).
23. R. Courant and F. John, *Introduction to Calculus and Analysis* (John Wiley and Sons, New York, 1974).
24. K. R. Arun and P. Prasad, *Wave Motion* **46**, 293 (2009).
25. G. B. Whitham, *Linear and Nonlinear Waves* (John Wiley, New York, 1974).
26. K. R. Arun and P. Prasad, *Appl. Math. Comput.* **217**, 2285 (2010).
27. K. R. Arun, *A numerical scheme for three-dimensional front propagation and control of Jordan mode*, tech. rep., Department of Mathematics, Indian Institute of Science (Bangalore, 2010).
28. K. R. Arun, M. Lukáčová-Medviďová, P. Prasad and S. V. Raghurama Rao, *SIAM. J. Appl. Math.* **70**, 2604 (2010).
29. A. Kurganov and E. Tadmor, *J. Comput. Phys.* **160**, 241 (2000).

SYSTEMATIC DISCRETIZATION OF INPUT/OUTPUT MAPS AND CONTROL OF PARTIAL DIFFERENTIAL EQUATIONS

J. HEILAND and V. MEHRMANN

Institut für Mathematik, TU Berlin, 10623 Berlin, Germany
Email: {heiland,mehrmann}@math.tu-berlin.de
www.tu-berlin.de

M. SCHMIDT

GE Global Research, 85748 Garching bei München, Germany,
Email: mail@schmidt-michael.de

We present a framework for the direct discretization of the input/output map of dynamical systems governed by linear partial differential equations with distributed inputs and outputs. The approximation consists of two steps. First, the input and output signals are discretized in space and time, resulting in finite dimensional spaces for the input and output signals. These are then used to approximate the dynamics of the system. The approximation errors in both steps are balanced and a matrix representation of an approximate input/output map is constructed which can be further reduced using singular value decompositions. We present the discretization framework, corresponding error estimates, and the SVD-based system reduction method. The theoretical results are illustrated with some applications in the optimal control of partial differential equations.

Keywords: input/output maps, discretization, control of partial differential equations

1. Introduction

The real-time control of complex physical systems is a major challenge in many engineering applications as well as in mathematical research. Typically, these control systems are modeled by infinite-dimensional state space systems on the basis of (instationary and nonlinear) partial differential equations (PDEs). The challenge arises from the fact that on the one hand, space-discretizations resolving most of the state information typically lead to very large semi-discrete systems, on the other hand, popular design techniques for real-time controllers like optimal and robust control techniques require models of very moderate size.

Numerous approaches to bridge this gap are proposed in the literature.[1,2] In many applications it is sufficient to approximate the high-order model by a low-order model that captures the essential state dynamics. To determine such low-order models one can use physical insight[3–5] and/or mathematical methods like proper orthogonal decomposition[6] or balanced truncation.[1,7] In this paper we focus on the situation, where for the design of appropriate controllers it is sufficient to approximate the *input/output (I/O) map* of the system, schematically illustrated in Figure 1.

For such configurations, empirical or simulation-based black-box system identification,[8,9] and mathematical model reduction techniques like balanced truncation,[10] moment matching[11] or recent variants of proper orthogonal decomposition[12] are common tools to extract appropriate low-order models. Typically, the bottleneck in these methods is the computational effort to compute the reduced order model from the semi-discretized model which often is of very high order.

In contrast to this, we present a new approach to construct low-order I/O maps (with error estimates) directly from the I/O map

$$\mathbb{G} : \mathcal{U} \to \mathcal{Y}, \qquad u = u(t, \theta) \mapsto y = y(t, \xi)$$

of *original* infinite-dimensional system. We suggest a new framework for the direct discretization of $\mathbb{G}$ for a general class of *infinite dimensional linear time-invariant state space systems* (introduced in Section 2). Here u and y are input and output signals from Hilbert spaces $\mathcal{U}$ and $\mathcal{Y}$, respectively, which may vary in time t and space $\theta \in \Theta$ and $\xi \in \Xi$, with appropriate spatial domains Θ and Ξ. The framework consists of two steps.

(1) *Approximation of signals (cf. Section 3).* We choose finite-dimensional

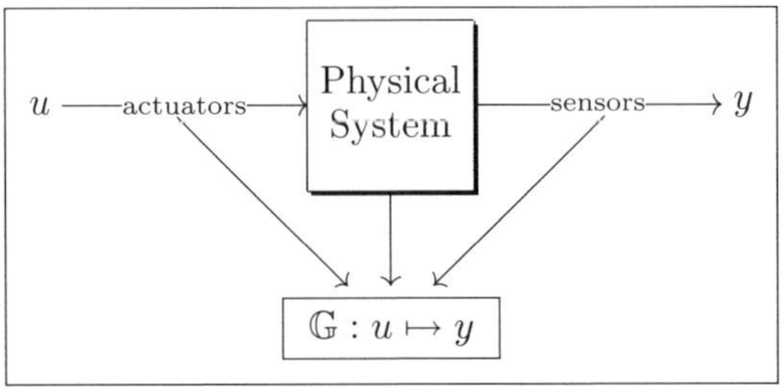

Fig. 1. Schematic illustration of the I/O map corresponding to a physical system.

subspaces $\bar{\mathcal{U}} \subset \mathcal{U}$ and $\bar{\mathcal{Y}} \subset \mathcal{Y}$ with bases $\{u_1, \ldots, u_{\bar{p}}\} \subset \bar{\mathcal{U}}$ and $\{y_1, \ldots, y_{\bar{q}}\} \subset \bar{\mathcal{Y}}$, and denote the corresponding orthogonal projections onto these subspaces by $\mathbb{P}_{\bar{\mathcal{U}}}$ and $\mathbb{P}_{\bar{\mathcal{Y}}}$, respectively. Then, the approximation

$$\mathbb{G}_S = \mathbb{P}_{\bar{\mathcal{Y}}} \mathbb{G} \mathbb{P}_{\bar{\mathcal{U}}}$$

has a matrix representation $\mathbf{G} \in \mathbb{R}^{\bar{q} \times \bar{p}}$.

(2) *Approximation of the system dynamics (cf. Section 4).* Since $\mathbb{G}$ arises from a linear state space model, the components $\mathbf{G}_{ij} = (y_i, \mathbb{G}u_j)_{\mathcal{Y}}$ can be approximated by *numerically simulating* the state space model successively for inputs u_j, $j = 1, \ldots, \bar{p}$ and by testing the resulting outputs against all $y_1, \ldots, y_{\bar{q}}$.

We discuss several features of this framework.

Error estimation (cf. Section 5). The total error ϵ_{DS} of the approximation can be estimated by combining the *signal* approximation error ϵ_S and the *dynamical* approximation error ϵ_D, i.e.,

$$\underbrace{\|\mathbb{G} - \mathbb{G}_{DS}\|}_{=:\epsilon_{DS}} \leq \underbrace{\|\mathbb{G} - \mathbb{G}_S\|}_{=:\epsilon_S} + \underbrace{\|\mathbb{G}_S - \mathbb{G}_{DS}\|}_{=:\epsilon_D},$$

where the norms still have to be specified. Here $\mathbb{G}_{DS}$ denotes the numerically estimated approximation of $\mathbb{G}_S$. Theorem 5.1 shows how to choose $\bar{\mathcal{U}}$ and $\bar{\mathcal{Y}}$ and the accuracy tolerances for the numerical solutions of the underlying PDEs such that ϵ_S and ϵ_D are balanced and that $\epsilon_S + \epsilon_D < \texttt{tol}$ for a given tolerance `tol`. Choosing hierarchical bases in $\bar{\mathcal{U}}$ and $\bar{\mathcal{Y}}$, the error ϵ_S can be progressively reduced by adding further basis functions $u_{\bar{p}+1}, u_{\bar{p}+2}, \ldots$ and $y_{\bar{q}+1}, y_{\bar{q}+2}, \ldots$ resulting in additional columns and rows of the matrix representation.

Applications and examples in control design (cf. Section 6). We explicitly construct the error estimates for the control problem associated with a 2D heat equation. Furthermore, we show how the matrix representation $\mathbf{G} = [\mathbf{G}_{ij}]$ may directly be used in control design, or a state realization of the I/O model $\mathbb{G}_{DS}$ can be used as basis for many classical control design algorithms.

Notation

For $\Omega \subset \mathbb{R}^d$, $d \in \mathbb{N}$, $L^2(\Omega)$ denotes the usual Lebesgue space of square-integrable functions, and $H^\alpha(\Omega)$, $\alpha \in \mathbb{N}_0$ denotes the corresponding Sobolev spaces of α-times weakly differentiable functions. We interpret functions v, which vary in space and time, optionally as classical functions $v : [0,T] \times \Omega \to \mathbb{R}$ with values $v(t;x) \in \mathbb{R}$, or as *abstract* functions $v : [0,T] \to X$ with values in a function space X such as $X = H^\alpha(\Omega)$. Correspondingly, $H^\alpha(0,T; H^\beta(\Omega))$, with $\alpha, \beta \in \mathbb{N}_0$, denotes the space of equivalence classes of functions $v : [0,T] \to H^\beta(\Omega)$ with $t \mapsto \|v\|_{H^\beta(\Omega)}$ being α-times weakly differentiable.[13] We introduce Hilbert spaces[14]

$$H^{\alpha,\beta}((0,T) \times \Omega) := H^\alpha(0,T; L^2(\Omega)) \cap L^2(0,T; H^\beta(\Omega)),$$
$$\|v\|_{H^{\alpha,\beta}((0,T)\times\Omega)} := \|v\|_{H^\alpha(0,T;L^2(\Omega))} + \|v\|_{L^2(0,T;H^\beta(\Omega))}.$$

By $C([0,T]; X)$ and $C^\alpha([0,T]; X)$ we denote the space of functions $v : [0,T] \to X$ which are continuous or α-times continuously differentiable. For two normed spaces X and Y, $\mathscr{L}(X,Y)$ denotes the set of bounded linear operators $X \to Y$, and we abbreviate $\mathscr{L}(X) := \mathscr{L}(X,X)$. For $\alpha \in \mathbb{N}$, $L^\alpha(0,T; \mathscr{L}(X,Y))$ denotes the space of operator-valued functions $K : [0,T] \to \mathscr{L}(X,Y)$ with $t \mapsto \|K(t)\|_{\mathscr{L}(X,Y)} = \sup_{x\neq 0}\|K(t)x\|_Y / \|x\|_X$ lying in $L^\alpha(0,T)$. Vectors, often representing a discretization of a function v, are written in corresponding small bold letters $\mathbf{v}$, whereas matrices, often representing a discrete version of an operator like $\mathbb{G}$ or G, are written in bold capital letters $\mathbf{G}$. By $\mathbb{R}^{\alpha\times\beta}$ we denote the set of real $\alpha \times \beta$ matrices, and $\mathbf{A} \otimes \mathbf{B}$ denotes the Kronecker product of matrices $\mathbf{A}$ and $\mathbf{B}$.

2. I/O maps of ∞-dimensional LTI state space systems

We consider infinite-dimensional, linear, time-invariant systems of first order

$$\partial_t z(t) = Az(t) + Bu(t), \quad t \in (0,T], \tag{1a}$$
$$z(0) = z^0, \tag{1b}$$
$$y(t) = Cz(t), \quad t \in [0,T]. \tag{1c}$$

Here for every time $t \in [0,T]$, the state $z(t)$ is supposed to belong to a Hilbert space Z like $Z = L^2(\Omega)$, where Ω is a subset of $\mathbb{R}^{d_\Omega}$ with $d_\Omega \in \mathbb{N}$. A is a densely defined unbounded operator $A : Z \supset D(A) \to Z$, generating a C^0-semigroup $(S(t))_{t\geq 0}$ on Z. The control operator B belongs to $\mathscr{L}(U,Z)$ and the observation operator C to $\mathscr{L}(Z,Y)$, where $U = L^2(\Theta)$ and $Y = L^2(\Xi)$ with subsets $\Theta \subset \mathbb{R}^{d_1}$ and $\Xi \subset \mathbb{R}^{d_2}$, $d_1, d_2 \in \mathbb{N}$.

Let us recall how a linear bounded I/O-map $\mathbb{G} \in \mathscr{L}(\mathcal{U},\mathcal{Y})$ with

$$\mathcal{U} = L^2(0,T;U) \quad \text{and} \quad \mathcal{Y} = L^2(0,T;Y)$$

can be associated[15] to (1). It is well-known that for initial values $z_0 \in D(A)$ and controls $u \in C^1([0,T]; Z)$, a unique *classical solution* $z \in C([0,T]; Z) \cap$

$C^1((0,T);Z)$ of (1) exists. For $z_0 \in Z$ and $u \in \mathcal{U}$, the well-defined function

$$z(t) = S(t)z_0 + \int_0^t S(t-s)Bu(s)\,ds, \quad t \in [0,T], \tag{2}$$

is called a *mild solution* of (1). A mild solution of (1) is unique, belongs to $C([0,T];Z)$, and is the uniform limit of classical solutions.[15] Hence, the output signal $y(t) = Cz(t)$ is well-defined and belongs to $\mathcal{Y} \cap C([0,T];Y)$. In particular, the output signals $y(u) \in \mathcal{Y}$ arising from input signals $u \in \mathcal{U}$ and zero initial conditions $z_0 \equiv 0$ allow to define the linear I/O-map $\mathbb{G} : \mathcal{U} \to \mathcal{Y}$ of the system (1) by $u \mapsto y(u)$. It is possible to represent $\mathbb{G}$ as a convolution with the kernel function $K \in L^2(-T,T;\mathscr{L}(U,Y))$,

$$K(t) = \begin{cases} C\mathcal{S}(t)D, & t \geq 0 \\ 0, & t < 0 \end{cases}.$$

Lemma 2.1. *The I/O-map $\mathbb{G}$ of* (1) *has the representation*

$$(\mathbb{G}u)(t) = \int_0^T K(t-s)u(s)\,ds, \quad t \in [0,T], \tag{3}$$

belongs to $\mathscr{L}(\mathcal{U},\mathcal{Y}) \cap \mathscr{L}(\mathcal{U}, C([0,T],\mathcal{Y}))$, and satisfies

$$\|\mathbb{G}\|_{\mathscr{L}(\mathcal{U},\mathcal{Y})} \leq \sqrt{T}\|K\|_{L^2(0,T;\mathscr{L}(U,Y))}. \tag{4}$$

Proof. Since C is bounded, the representation of $y = Cz$ based on (2) can be reformulated as in (3), calling on the theory of Bochner integrals.[13] For general $K \in L^2(-T,T;\mathscr{L}(U,Y))$, using a generalized Hölder inequality implies that for fixed $t \in [0,T]$ the function $s \to K(t-s)u(s)$ belongs to $L^1(0,T;\mathscr{L}(U,Y))$ with

$$\|(\mathbb{G}u)(t)\|_Y \leq \|u\|_{\mathcal{U}}\|K(t-\cdot)\|_{L^2(0,T;\mathscr{L}(U,Y))},$$

and by integrating over $[0,T]$ we obtain (4). □

Remark 2.1. The I/O-map $\mathbb{G}$ is *causal* in the sense that $y(t)$ only depends on $u_{|[0,t)}$ for all $t \in [0,T]$, and $\mathbb{G}$ is *time-invariant* in the sense that if $y = \mathbb{G}u$ then $\sigma_\tau y = \mathbb{G}(\sigma_\tau u)$ for all $\tau \in [0,T]$. Here σ_τ is a time-shift operator with $(\sigma_\tau u)(t) = u(t-\tau)$ for $t \in [\tau,T]$ and $(\sigma_\tau u)(t) = 0$ for $t \in [0,\tau)$.

Example 2.1. As prototypical system, we consider the heat equation with homogeneous Dirichlet boundary conditions and assume that Ω has a C^2-boundary. In this case, $Z = L^2(\Omega)$ and the operator A in (1) coincides with the Laplace operator

$$A = \triangle : D(A) = H^2(\Omega) \cap H_0^1(\Omega) \subset Z \to Z.$$

Since A is the infinitesimal generator of an analytic C^0-semigroup of contractions $(S(t))_{t\geq 0}$, the mild solution z of (1) exhibits the following stability and regularity properties.[15,16]

(i) If $z_0 = 0$ and $u \in \mathcal{U}$, then $z \in H^{1,2}((0,T) \times \Omega)$ with

$$\|z\|_{H^{1,2}((0,T)\times\Omega)} \leq c\|u\|_{\mathcal{U}}. \tag{5}$$

(ii) Assume that $u \equiv 0$. For $z_0 \in D(A)$ we have $z \in C^1([0,T]; D(A))$, but for $z_0 \in Z$ we only have $z \in C^1((0,T]; D(A))$.

We will consider concrete choices of Ω, B and C in Section 6. We note that if the observation preserves the inherent state regularity in the sense that

$$C_{|H^2(\Omega)} \in \mathscr{L}(H^2(\Omega), H^2(\Xi)),$$

then $\mathbb{G} \in \mathscr{L}(\mathcal{U}, \mathcal{Y}_s)$ and also

$$\mathbb{G}_{|\mathcal{U}_s} \in \mathscr{L}(\mathcal{U}_s, \mathcal{Y}_s), \quad \text{with } \mathcal{U}_s = H^{1,2}((0,T) \times \Theta), \quad \mathcal{Y}_s = H^{1,2}((0,T) \times \Xi). \tag{6}$$

In fact, for $u \in \mathcal{U}_s$, we have $\|u\|_{\mathcal{U}} \leq \|u\|_{\mathcal{U}_s}$, and for $u \in \mathcal{U}$ we have

$$\|\mathbb{G}u\|_{\mathcal{Y}_s} \leq c'\|z\|_{H^{1,2}((0,T)\times\Omega)} \leq c\,c'\|u\|_{\mathcal{U}},$$

where $c' = \max\{\|C\|_{\mathscr{L}(L^2(\Omega),L^2(\Xi))}, \|C\|_{\mathscr{L}(H^2(\Omega),H^2(\Xi))}\}$ and c is the constant in (5).

Remark 2.2. Many other linear time-invariant systems with distributed controls and observations admit a representation of the I/O map via (3) and exhibit properties similar to (6). This is, for instance, the case for the heat equation with homogeneous Neumann boundary conditions, and also for more general parabolic equations.[14,17] Wave equations with second order time derivatives can be represented in the form of (1) and (3) by means of an order reduction. Though hyperbolic systems do not have the smoothing properties of parabolic systems, they preserve the regularity of the data and results similar to (6) can be obtained by restricting the input signals to be of higher regularity in time.[14]

The presented framework can also be used for linearized flow systems. For the Stokes equation, results similar to (3) and (6) are obtained by working with appropriate subspaces of divergence-free functions[18] and for the spatially discretized Oseen equations, which arise as linearizations of the Navier-Stokes equations, it has been shown in[19,20] how the framework can be extended to linear time invariant descriptor systems.

Note, however, that systems with boundary control or pointwise observations do not fit directly into the setting (1).

3. Discretization of Signals

In order to discretize the input signals $u \in \mathcal{U}$ and $y \in \mathcal{Y}$ in space and time, we choose four families $\{U_{h_1}\}_{h_1>0}$, $\{Y_{h_2}\}_{h_2>0}$, $\{\mathcal{R}_{\tau_1}\}_{\tau_1>0}$ and $\{\mathcal{S}_{\tau_2}\}_{\tau_2>0}$ of subspaces $U_{h_1} \subset U$, $Y_{h_2} \subset Y$, $\mathcal{R}_{\tau_1} \subset L^2(0,T)$, and $\mathcal{S}_{\tau_2} \subset L^2(0,T)$ of finite dimensions $p(h_1) = \dim(U_{h_1})$, $q(h_2) = \dim(Y_{h_2})$, $r(\tau_1) = \dim(\mathcal{R}_{\tau_1})$ and $s(\tau_2) = \dim(\mathcal{S}_{\tau_2})$. We then define

$$\mathcal{U}_{h_1,\tau_1} = \{u \in \mathcal{U} : u(t;\cdot) \in U_{h_1},\ u(\cdot;\theta) \in \mathcal{R}_{\tau_1}, \quad t \in [0,T] \text{ a.e. }, \theta \in \Theta\},$$
$$\mathcal{Y}_{h_2,\tau_2} = \{y \in \mathcal{Y} : y(t;\cdot) \in Y_{h_2},\ y(\cdot;\xi) \in \mathcal{S}_{\tau_2}, \quad t \in [0,T] \text{ a.e. }, \xi \in \Xi\}.$$

We denote the orthogonal projections onto these subspaces by $P_{\mathcal{S},\tau_2} \in \mathscr{L}(L^2(0,T))$, $\mathbb{P}_{\mathcal{U},h_1,\tau_1} \in \mathscr{L}(\mathcal{U})$, and $\mathbb{P}_{\mathcal{Y},h_2,\tau_2} \in \mathscr{L}(\mathcal{Y})$. As first step of the approximation of $\mathbb{G}$, we define

$$\mathbb{G}_S = \mathbb{G}_S(h_1,\tau_1,h_2,\tau_2) = \mathbb{P}_{\mathcal{Y},h_2,\tau_2}\mathbb{G}\mathbb{P}_{\mathcal{U},h_1,\tau_1} \in \mathscr{L}(\mathcal{U},\mathcal{Y}).$$

In order to obtain a matrix representation of $\mathbb{G}_S$, we introduce families of bases $\{\mu_1,\ldots,\mu_p\}$ of U_{h_1}, $\{\nu_1,\ldots,\nu_q\}$ of Y_{h_2}, $\{\phi_1,\ldots,\phi_r\}$ of $\mathcal{R}_{\tau_1}$, and $\{\psi_1,\ldots,\psi_s\}$ of $\mathcal{S}_{\tau_2}$ and corresponding mass matrices $\mathbf{M}_{U,h_1} \in \mathbb{R}^{p\times p}$, $\mathbf{M}_{Y,h_2} \in \mathbb{R}^{q\times q}$, $\mathbf{M}_{\mathcal{R},\tau_1} \in \mathbb{R}^{r\times r}$ and $\mathbf{M}_{\mathcal{S},\tau_2} \in \mathbb{R}^{s\times s}$, for instance via

$$[\mathbf{M}_{U,h_1}]_{ij} = (\mu_j,\mu_i)_U, \qquad i,j = 1,\ldots,p.$$

These mass matrices induce, via

$$(\mathbf{v},\mathbf{w})_{\mathbb{R}^p;w} = \mathbf{v}^T\mathbf{M}_{U,h_1}\mathbf{w} \quad \text{for all } \mathbf{v},\mathbf{w} \in \mathbb{R}^p,$$

weighted scalar products and corresponding norms in the respective spaces, which we indicate by a subscript w, like $\mathbb{R}^p_w$ with $(\cdot,\cdot)_{\mathbb{R}^p;w}$ and $\|\cdot\|_{\mathbb{R}^p;w}$, in contrast to the canonical spaces like $\mathbb{R}^p$, with $(\cdot,\cdot)_{\mathbb{R}^p}$ and $\|\cdot\|_{\mathbb{R}^p}$. We represent signals $u \in \mathcal{U}_{h_1,\tau_1}$ and $y \in \mathcal{Y}_{h_2,\tau_2}$ as

$$u(t;\theta) = \sum_{k=1}^{p}\sum_{i=1}^{r} \mathbf{u}_i^k \phi_i(t)\mu_k(\theta), \qquad y(t;\xi) = \sum_{l=1}^{q}\sum_{j=1}^{s} \mathbf{y}_j^l \psi_j(t)\nu_k(\xi),$$

where $\mathbf{u}_i^k$ are the elements of a block-structured vector $\mathbf{u} \in \mathbb{R}^{pr}$ with p blocks $\mathbf{u}^k \in \mathbb{R}^r$, and the vector $\mathbf{y} \in \mathbb{R}^{qs}$ is defined similarly. Then

$$\|u\|_{\mathcal{U}} = \|\mathbf{u}\|_{\mathbb{R}^{pr};w}, \quad \text{and} \quad \|y\|_{\mathcal{Y}} = \|\mathbf{y}\|_{\mathbb{R}^{qs};w},$$

where $\|\cdot\|_{\mathbb{R}^{pr};w}$ and $\|\cdot\|_{\mathbb{R}^{qs};w}$ denote the weighted norms with respect to the mass matrices

$$\mathbf{M}_{\mathcal{U},h_1,\tau_1} = \mathbf{M}_{U,h_1} \otimes \mathbf{M}_{\mathcal{R},\tau_1} \in \mathbb{R}^{pr\times pr}, \quad \mathbf{M}_{\mathcal{Y},h_2,\tau_2} = \mathbf{M}_{Y,h_2} \otimes \mathbf{M}_{\mathcal{S},\tau_2} \in \mathbb{R}^{qs\times qs},$$

i.e., the corresponding coordinate isomorphisms $\kappa_{\mathcal{U},h_1,\tau_1} \in \mathscr{L}(\mathcal{U}_{h_1,\tau_1}, \mathbb{R}^{pr}_w)$ and $\kappa_{\mathcal{Y},h_2,\tau_2} \in \mathscr{L}(\mathcal{Y}_{h_2,\tau_2}, \mathbb{R}^{qs}_w)$ are unitary.

Finally, we obtain a matrix representation $\mathbf{G}$ of $\mathbb{G}_S$ by setting

$$\mathbf{G} = \mathbf{G}(h_1, \tau_1, h_2, \tau_2) = \kappa_{\mathcal{Y}} \mathbb{P}_{\mathcal{Y}} \mathbb{G} \mathbb{P}_{\mathcal{U}} \kappa_{\mathcal{U}}^{-1} \in \mathbb{R}^{qs \times pr}, \tag{7}$$

where the dependencies on h_1, τ_1, h_2, τ_2 have been partially omitted. Considering

$$\mathbf{H} = \mathbf{H}(h_1, \tau_1, h_2, \tau_2) := \mathbf{M}_{\mathcal{Y},h_2,\tau_2} \mathbf{G} \in \mathbb{R}^{qs \times pr}$$

as a block-structured matrix with $q \times p$ blocks $\mathbf{H}^{kl} \in \mathbb{R}^{s \times r}$ and block elements $\mathbf{H}^{kl}_{ij} \in \mathbb{R}$, we obtain the representation

$$\mathbf{H}^{kl}_{ij} = [\mathbf{M}_{\mathcal{Y}} \kappa_{\mathcal{Y}} \mathbb{P}_{\mathcal{Y}} \mathbb{G}(\mu_l \phi_j)]^k_i = (\nu_k \psi_i, \mathbb{G}(\mu_l \phi_j))_{\mathcal{Y}}. \tag{8}$$

To have a discrete analogon of the $\mathscr{L}(\mathcal{U}, \mathcal{Y})$-norm, for given $\mathcal{U}_{h_1,\tau_1}$ and $\mathcal{Y}_{h_2,\tau_2}$, we introduce the weighted matrix norm

$$\begin{aligned}\|\mathbf{G}(h_1, \tau_1, h_2, \tau_2)\|_{\mathbb{R}^{qs} \times \mathbb{R}^{pr};w} &:= \sup_{\mathbf{u} \in \mathbb{R}^{pr}} \frac{\|\mathbf{G}\mathbf{u}\|_{\mathbb{R}^{qs};w}}{\|\mathbf{u}\|_{\mathbb{R}^{pr};w}} \\ &= \|\mathbf{M}^{1/2}_{\mathcal{Y},h_2,\tau_2} \mathbf{G} \mathbf{M}^{-1/2}_{\mathcal{U},h_1,\tau_1}\|_{\mathbb{R}^{qs} \times \mathbb{R}^{pr}},\end{aligned}$$

and we write $(h_1', \tau_1', h_2', \tau_2') \le (h_1, \tau_1, h_2, \tau_2)$ if the inequality holds componentwise.

Lemma 3.1. *For all $(h_1, \tau_1, h_2, \tau_2) \in \mathbb{R}^4_+$, we have*

$$\|\mathbf{G}(h_1, \tau_1, h_2, \tau_2)\|_{\mathbb{R}^{qs} \times \mathbb{R}^{pr};w} = \|\mathbb{G}_S(h_1, \tau_1, h_2, \tau_2)\|_{\mathscr{L}(\mathcal{U},\mathcal{Y})} \le \|\mathbb{G}\|_{\mathscr{L}(\mathcal{U},\mathcal{Y})}. \tag{9}$$

If the subspaces $\{\mathcal{U}_{h_1,\tau_1}\}_{h_1,\tau_1>0}$ and $\{\mathcal{Y}_{h_2,\tau_2}\}_{h_2,\tau_2>0}$ are nested, in the sense that

$$\mathcal{U}_{h_1,\tau_1} \subset \mathcal{U}_{h_1',\tau_1'}, \quad \mathcal{Y}_{h_2,\tau_2} \subset \mathcal{Y}_{h_2',\tau_2'} \quad \textit{for } (h_1', \tau_1', h_2', \tau_2') \le (h_1, \tau_1, h_2, \tau_2), \tag{10}$$

then $\|\mathbf{G}(h_1, \tau_1, h_2, \tau_2)\|_{\mathbb{R}^{qs} \times \mathbb{R}^{pr};w}$ is monotonically increasing for decreasing $(h_1, \tau_1, h_2, \tau_2) \in \mathbb{R}^4_+$, and $\|\mathbf{G}(h_1, \tau_1, h_2, \tau_2)\|_{\mathbb{R}^{qs} \times \mathbb{R}^{pr};w}$ is convergent for $(h_1, \tau_1, h_2, \tau_2) \searrow 0$.

Proof. In order to show (9), we calculate

$$\|\mathbb{G}_S\|\mathscr{L}(\mathcal{U}, \mathcal{Y}) = \sup_{u \in \mathcal{U}_{h_1,\tau_1}} \frac{\|\mathbb{P}_{\mathcal{Y},h_2,\tau_2} \mathbb{G} u\|_{\mathcal{Y}}}{\|u\|_{\mathcal{U}}} \le \sup_{u \in \mathcal{U}_{h_1,\tau_1}} \frac{\|\mathbb{G} u\|_{\mathcal{Y}}}{\|u\|_{\mathcal{U}}} \le \|\mathbb{G}\|_{\mathscr{L}(\mathcal{U},\mathcal{Y})},$$

and observe that for $u \in \mathcal{U}_{h_1,\tau_1}$ and $\mathbf{u} = \kappa_{\mathcal{U},h_1,\tau_1} u \in \mathbb{R}^{pr}$, we have

$$\|\mathbb{G}_S u\|_{\mathcal{Y}} = \|\mathbf{G}\mathbf{u}\|_{\mathbb{R}^{qs};w} \leq \|\mathbf{G}\|_{\mathbb{R}^{qs}\times\mathbb{R}^{pr};w} \|u\|_{\mathcal{U}} \quad \text{and}$$
$$\|\mathbb{G}_S u\|_{\mathcal{Y}} \leq \|\mathbb{G}_S\|_{\mathscr{L}(\mathcal{U},\mathcal{Y})} \|\mathbf{u}\|_{\mathbb{R}^{pr};w}.$$

If (10) holds, then since $\|\mathbb{P}_{\mathcal{Y},h_2,\tau_2} y\|_{\mathcal{Y}} \leq \|\mathbb{P}_{\mathcal{Y},h_2',\tau_2'} y\|_{\mathcal{Y}}$ for all $y \in \mathcal{Y}$, we have

$$\begin{aligned}\|\mathbb{G}_S(h_1,\tau_1,h_2,\tau_2)\|_{\mathbb{R}^{qs}\times\mathbb{R}^{pr};w} &\leq \sup_{u\in\mathcal{U}_{h_1',\tau_1'}} \frac{\|\mathbb{P}_{\mathcal{Y},h_2',\tau_2'}\mathbb{G}u\|_{\mathcal{Y}}}{\|u\|_{\mathcal{U}}} \\ &= \|\mathbb{G}_S(h_1',\tau_1',h_2',\tau_2')\|_{\mathbb{R}^{q's'}\times\mathbb{R}^{p'r'};w}.\end{aligned}$$

Hence, (9) ensures the convergence of $\|\mathbb{G}_S(\mathbf{h})\|_{\mathbb{R}^{qs}\times\mathbb{R}^{pr};w}$. □

3.1. *Signal discretization via finite elements*

There are many possibilities to choose the finite dimensional subspaces in U, Y. As an example, consider the case $U = Y = L^2(0,1)$, choose U_{h_1} and Y_{h_2} as spaces of continuous, piecewise linear functions and $\mathcal{R}_{\tau_1}$ and $\mathcal{S}_{\tau_2}$ as spaces of piecewise constant functions, all with respect to equidistant grids.

For $p \in \mathbb{N}$, $p \geq 2$ and $h_1(p) = 1/(p-1)$, let $\mathscr{T}_{h_1} = \{I_k\}_{1\leq k\leq p-1}$ be the equidistant partition of $(0,1]$ into intervals $I_k = ((k-1)h_1, kh_1]$. The corresponding space U_{h_1} is spanned by the nodal basis

$$\{\mu_1^{(h_1)}, \ldots, \mu_{p(h_1)}^{(h_1)}\} \subset U_{h_1}, \quad \text{with } \mu_l^{(h_1)}(kh_1) = \delta_{l-1}(k), \quad k = 0, \ldots, p.$$

The subspaces $\{U_{h_1}\}$ are nested if the choice is restricted to $h_1 \in \{2^{-n}\}_{n\in\mathbb{N}_0}$ and $p \in \{2^n + 1\}_{n\in\mathbb{N}_0}$. Since the *nodal* bases of U_{h_1} and $U_{h_1'}$ do not have any common element for $h_1 \neq h_1'$, one may prefer to choose a *hierarchical* basis of finite element functions[21,22] $\hat{\mu}_l$, as in Fig. 2. Then, $U_{h_1} = \text{span}\{\hat{\mu}_1, \ldots, \hat{\mu}_{p(h_1)}\}$ for all $h_1 \in \{2^{-n}\}_{n\in\mathbb{N}_0}$ with basis functions $\hat{\mu}_k$ independent of h_1. For $r \in \mathbb{N}$ and $\tau_1 = T/r$, let $\Gamma_{\tau_1} = \{I_j\}_{1\leq j\leq r}$ be

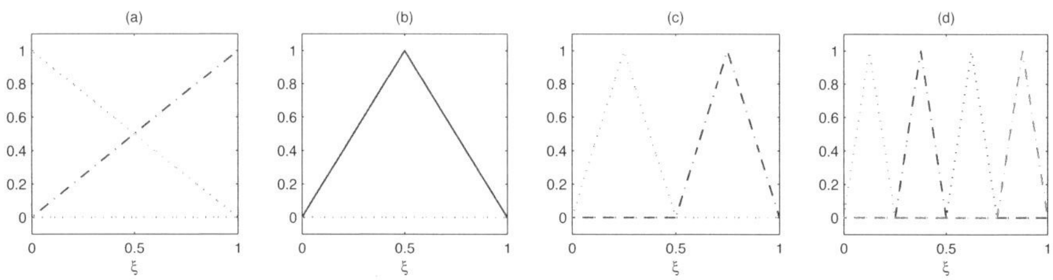

Fig. 2. Hierarchical basis for $L^2(0,1)$-subspaces of piecewise linear functions: (a) μ_1 and μ_2 (b) μ_3 (c) μ_4 and μ_5 (d) $\mu_6, \ldots, \mu_9$.

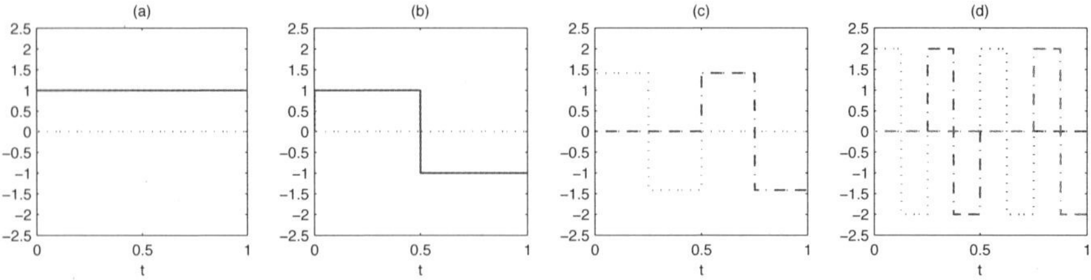

Fig. 3. Haar wavelet basis for $L^2(0,1)$-subspaces of piecewise constant functions: (a) ϕ_1 (b) ϕ_2 (c) ϕ_3 and ϕ_4 (d) $\phi_5,\dots,\phi_8$.

the equidistant partition of $(0,T]$ into intervals $I_j = ((j-1)\tau_1, j\tau_1]$. The corresponding space $\mathcal{R}_{\tau_1}$ of piecewise constant functions is, for instance, spanned by the nodal and orthogonal basis

$$\{\phi_1^{(\tau_1)}(t),\dots,\phi_r^{(\tau_1)}(t)\}, \quad \text{with } \phi_j^{(\tau_1)}(t) = \chi_{I_j}(t), \qquad j=1,\dots,r, \tag{11}$$

with χ_{I_j} denoting the characteristic function on I_j. The spaces are nested by requiring $\tau_1 \in \{2^{-n}T\}_{n\in\mathbb{N}_0}$. An orthonormal hierarchical basis for $\mathcal{R}_{\tau_1}$ is obtained by choosing ϕ_j as Haar-wavelets, cf. Fig. 3 and.[23]

Denoting the orthogonal projections onto U_{h_1} and $\mathcal{R}_{\tau_1}$ by P_{U,h_1} and $P_{\mathcal{R},\tau_1}$, respectively, the Poincaré-Friedrich inequality shows that there exist constants $c_U = 1/2$ and $c_{\mathcal{R}} = 1/\sqrt{2}$, independent of h_1, τ_1 and T, such that[24,25]

$$\begin{aligned} \|u - P_{U_{h_1}}u\|_{L^2(0,1)} &\le c_U h_1^2 \|\partial_\xi^2 u\|_{L^2(0,1)} \quad \text{for } u \in H^2(0,1),\\ \|v - P_{\mathcal{R}_{\tau_1}}v\|_{L^2(0,T)} &\le c_{\mathcal{R}} \tau_1 \|\partial_t v\|_{L^2(0,T)} \quad \text{for } v \in H^1(0,T). \end{aligned}$$

By the Fubini theorem, it follows that the corresponding projection $\mathbb{P}_{\mathcal{U},h_1,\tau_1}$ onto $\mathcal{U}_{h_1,\tau_1} = \{u \in \mathcal{U},\ u_{|I_j} \equiv u^{(j)},\ u^{(j)} \in U_{h_1},\quad j=1,\dots,r\}$ satisfies

$$\|u - \mathbb{P}_{\mathcal{U},h_1,\tau_1}u\|_{\mathcal{U}} \le (c_U h_1^2 + c_{\mathcal{R}}\tau_1)\|u\|_{\mathcal{U}_s} \quad \text{for all } u \in \mathcal{U}_s = H^{1,2}((0,T)\times(0,1)). \tag{13}$$

We define Y_{h_2},$\mathcal{R}_{\tau_2}$ and $\mathcal{Y}_{h_2,\tau_2}$ accordingly and a corresponding estimate as (13) holds for the projection $\mathbb{P}_{\mathcal{Y},h_2,\tau_2}y$ of elements $y \in \mathcal{Y}_s$.

Remark 3.1. Estimates similar to (13) also exist for domains $\Theta \subset \mathbb{R}^d$ with $d > 1$ and are classical results from the interpolation theory in Sobolev spaces.[24] Note that the interpolation constants then often have to be estimated numerically. Estimates with higher approximation order can be obtained, if ansatz functions of higher polynomial degree are used and if the input and output signals exhibit corresponding higher regularity in space and time.

4. Approximation of the system dynamics

Let us now discuss the efficient approximation of $\mathbb{G}_S$ and its matrix representation $\mathbf{G} = \mathbf{M}_{\mathcal{Y}}^{-1}\mathbf{H}$, respectively. For time-invariant systems with distributed control and observation, this task reduces to the approximation of the convolution kernel $K \in L^2(0,T;\mathscr{L}(U,Y))$.

4.1. *Kernel function approximation*

Inserting (3) in (8), by a change of variables we obtain

$$\mathbf{H}_{ij}^{kl} = \int_0^T \int_0^T \psi_i(t)\phi_j(s)(\nu_k, K(t-s)\mu_l)_Y \, ds \, dt = \int_0^T \mathbf{W}_{ij}(t)\mathbf{K}_{kl}(t)\, dt,$$

with matrix-valued functions $\mathbf{W} : [0,T] \to \mathbb{R}^{s\times r}$ and $\mathbf{K} : [0,T] \to \mathbb{R}^{q\times p}$,

$$\mathbf{W}_{ij}(t) = \int_0^{T-t} \psi_i(t+s)\phi_j(s)\, ds, \qquad \mathbf{K}_{kl}(t) = (\nu_k, K(t)\mu_l)_Y,$$

and thus

$$\mathbf{H} = \mathbf{M}_{\mathcal{Y}}\mathbf{G} = \int_0^T \mathbf{K}(t) \otimes \mathbf{W}(t)\, dt. \tag{15}$$

Remark 4.1. $\mathbf{W}(t)$ can be exactly calculated if piecewise polynomial ansatz functions $\psi_i(t)$ and $\phi_j(t)$ are chosen. For the special choice (11), we see in this way that $\mathbf{W}(t) \in \mathbb{R}^{r\times r}$ is a lower triangular Toeplitz matrix for all $t \in [0,T]$, and hence the matrices $\mathbf{H}_{ij} = \int_0^T \mathbf{W}_{ij}(t)\mathbf{K}(t)\, dt \in \mathbb{R}^{q\times p}$ satisfy $\mathbf{H}_{ij} = \mathbf{H}_{i-j}$ for $1 \le i,j \le r$ and $\mathbf{H}_{ij} = 0$ for $1 \le i < j \le r$.

For systems of the form (1), the matrix-valued function $\mathbf{K}$ is given by

$$\mathbf{K}_{kl}(t) = (\nu_k, CS(t)B\mu_l)_Y = (c_k^*, S(t)b_l)_Z,$$

where $c_k^* = C^*\nu_k \in Z$ and $b_l = B\mu_l$ for $k = 1,\dots,q$ and $l = 1,\dots,p$. Hence, the entries of $\mathbf{K}$ can be calculated by solving the homogeneous systems

$$\dot z_l(t) = Az_l(t), \qquad t \in (0,T], \tag{16a}$$

$$z_l(0) = b_l, \qquad l = 1,\dots,p, \tag{16b}$$

since (16) has the mild solution $z_l(t) = S(t)b_l \in C([0,T];L^2(\Omega))$. We obtain an approximation $\tilde{\mathbf{H}}$ of $\mathbf{H}$ by replacing $z_l(t)$ by numerical approximations $z_{l,\mathtt{tol}}(t)$, i.e.,

$$\tilde{\mathbf{H}} = \int_0^T \tilde{\mathbf{K}}(t) \otimes \mathbf{W}(t)\, dt, \tag{17}$$

with $\tilde{\mathbf{K}}_{kl}(t) = (\nu_k, Cz_{l,\mathtt{tol}}(t))_Y = (c_k^*, z_{l,\mathtt{tol}}(t))_Z$. Here the subscript `tol` indicates that the error $z_l - z_{l,\mathtt{tol}}$ is assumed to satisfy some tolerance criterion which will be specified later. The corresponding approximation $\mathbb{G}_{DS}$ of $\mathbb{G}_S$ is given by

$$\mathbb{G}_{DS} = \kappa_{\mathcal{Y}}^{-1}\tilde{\mathbf{G}}\kappa_{\mathcal{U}}\mathbb{P}_{\mathcal{U}}, \quad \text{with } \tilde{\mathbf{G}} = \mathbf{M}_{\mathcal{Y}}^{-1}\tilde{\mathbf{H}}, \tag{18}$$

and depends on h_1, h_2, τ_1, τ_2 and `tol`.

Remark 4.2. The matrix function $\mathbf{K}$ is approximated *columnwise*. The kernel may also be calculated *rowwise* by solving an adjoint autonomous system, which may be preferable if $q < p$ or if the output approximation is successively improved by adding further basis functions $\nu_{q+1}, \nu_{q+2}, \ldots$.

Remark 4.3. The calculation of $\tilde{\mathbf{H}}$ can be parallelized in an obvious way by calculating the p solutions $z_{l,\mathtt{tol}}$ in parallel and we note that no state trajectories have to be stored. In general, the matrix $\tilde{\mathbf{H}}$ is not sparse, such that the memory requirements become significant if a high resolution of the signals in space and time is required, and the question of a data-sparse representation arises. Recalling Remark 4.1, the blocks $\tilde{\mathbf{H}}^{kl}$ are lower triangular Toeplitz matrices for the special choice of time basis funtions (11) and thus only $q \cdot p \cdot r$ elements have to be stored. Another approach to obtain data-sparse representations uses approximate factorizations $\tilde{\mathbf{K}}_{kl}(t-s) = \sum_{m,n=1}^{M} \alpha_{mn} L_m(t) L_n(s)$ for $s, t \in [0, T]$ with suitable ansatz functions[26] $L_n(t)$.

4.2. *The approximation error for the dynamics*

The following proposition relates the error ϵ_D in the system dynamics to the errors made in solving the PDE (16) for $l = 1, \ldots, p$.

Prop 4.1. The error $\epsilon_D := \|\mathbb{G}_S - \mathbb{G}_{DS}\|_{\mathscr{L}(\mathcal{U},\mathcal{Y})}$ in the system dynamics satisfies

$$\begin{aligned}\epsilon_D &\leq \sqrt{T}\|\mathbf{K} - \tilde{\mathbf{K}}\|_{L^2(0,T;\mathbb{R}_w^{q\times p})} \\ &\leq p\sqrt{T}\sqrt{\frac{\lambda_{\max}(\mathbf{M}_{Y,h_2})}{\lambda_{\min}(\mathbf{M}_{U,h_1})}} \max_{1\leq l\leq p} \|\mathbf{K}_{:,l} - \tilde{\mathbf{K}}_{:,l}\|_{L^2(0,T;\mathbb{R}^q)}.\end{aligned} \tag{19}$$

Here $\mathbf{K}_{:,l}$ and $\tilde{\mathbf{K}}_{:,l}$ denote the l-th column of $\mathbf{K}(t)$ and $\tilde{\mathbf{K}}(t)$, respectively, $\lambda_{\max}(\mathbf{M}_{Y,h_2})$ is the largest eigenvalue of $\mathbf{M}_{Y,h_2}$ and $\lambda_{\min}(\mathbf{M}_{U,h_1})$ the smallest eigenvalue of $\mathbf{M}_{U,h_1}$. Similar as before, $\mathbb{R}_w^{q\times p}$ denotes the space of real $q \times p$ matrices equipped with the weighted matrix norm $\|\mathbf{M}\|_{\mathbb{R}^{q\times p};w} = \sup_{\mathbf{u}\neq 0} \|\mathbf{M}\mathbf{u}\|_{\mathbb{R}^q;w} / \|\mathbf{u}\|_{\mathbb{R}^p;w}$.

Proof. The matrix $\mathbf{K}$ is the representation of the space-projected kernel function $K_m : [-T,T] \to \mathscr{L}(U,Y)$ with $K_m(t) = P_{Y,h_2}K(t)P_{U,h_1}$, where P_{Y,h_2} and P_{U,h_1} are the orthogonal projections onto the subspaces Y_{h_2} and U_{h_1}, respectively. Introducing the corresponding I/O-map $\mathbb{G}_m = \mathbb{G}_m(h_1,h_2)$,

$$(\mathbb{G}_m u)(t) = \int_0^T K_m(t-s)u(s)\,ds, \qquad t \in [0,T]. \tag{20}$$

we note that $\mathbb{G}_S = \mathbb{P}_{\mathcal{Y},h_2,\tau_2}\mathbb{G}_m\mathbb{P}_{\mathcal{U},h_1,\tau_1}$. Similarly, we associate with $\tilde{\mathbf{K}}(t)$ the kernel function $\tilde{K} : [-T,T] \to \mathscr{L}(U,Y)$ with $\tilde{K}(t) = \kappa_{Y,h_2}^{-1}\tilde{\mathbf{K}}(t)\kappa_{U,h_1}P_{U,h_1}$, and with corresponding I/O-map

$$(\mathbb{G}_D u)(t) = \int_0^T \tilde{K}(t-s)u(s)\,ds, \qquad t \in [0,T].$$

We observe that $\mathbb{G}_{DS}$ as defined in (18) satisfies $\mathbb{G}_{DS} = \mathbb{P}_{\mathcal{Y},h_2,\tau_2}\mathbb{G}_D\mathbb{P}_{\mathcal{U},h_1,\tau_1}$ by showing via (7)-(15) that the matrix representation of $\mathbb{P}_{\mathcal{Y},h_2,\tau_2}\mathbb{G}_D\mathbb{P}_{\mathcal{U},h_1,\tau_1}$ coincides with (17). Then $\|K_m(t)\|_{\mathscr{L}(U,Y)} = \|\mathbf{K}(t)\|_{\mathbb{R}^{q\times p};w}$ and $\|\tilde{K}(t)\|_{\mathscr{L}(U,Y)} = \|\tilde{\mathbf{K}}(t)\|_{\mathbb{R}^{q\times p};w}$ for all $t \in [0,T]$. Lemma 2.1 yields

$$\|\mathbb{G}_m - \mathbb{G}_D\|_{\mathscr{L}(\mathcal{U},\mathcal{Y})} \le \sqrt{T}\|K_m - \tilde{K}\|_{L^2(0,T;\mathscr{L}(U,Y))} = \sqrt{T}\|\mathbf{K} - \tilde{\mathbf{K}}\|_{L^2(0,T;\mathbb{R}_w^{q\times p})}.$$

Defining $\mathbf{E}(t) = \mathbf{K}(t) - \tilde{\mathbf{K}}(t)$, for $\mathbf{u} \in \mathbb{R}^p$ with $\|\mathbf{u}\|_{\mathbb{R}^p} = 1$ and $t \in [0,T]$, by using the equivalence vector norms in $\mathbb{R}^p$ we have that

$$\|\mathbf{E}(t)\mathbf{u}\|_{\mathbb{R}^q} \le \sum_{l=1}^{p} |\mathbf{u}_l| \|\mathbf{E}_{:,l}(t)\|_{\mathbb{R}^q} \le \sqrt{p}\left(\sum_{l=1}^{p} \|\mathbf{E}_{:,l}(t)\|_{\mathbb{R}^q}^2\right)^{1/2},$$

and hence

$$\|\mathbf{E}\|_{L^2(0,T;\mathbb{R}^{q\times p})}^2 \le p\sum_{l=1}^{p}\int_0^T \|\mathbf{E}_{:,l}(t)\|_{\mathbb{R}^q}^2\,dt \le p^2 \max_{l=1,\dots,p}\int_0^T \|\mathbf{E}_{:,l}(t)\|_{\mathbb{R}^q}^2\,dt,$$

which concludes the proof. □

Remark 4.4. Calculating the columns of $\mathbf{K}$ directly and estimating ϵ_D via (19), the quotient of the eigenvalues of the mass matrices $\mathbf{M}_{U,h_1}$ and $\mathbf{M}_{Y,h_2}$ has to be compensated by the approximation accuracy of $\mathbf{K}_{:,l}$. This may be problematic if hierarchical basis functions are chosen, since the quotient grows unboundedly with decreasing h_1 and h_2. One may circumvent this problem by calculating $\mathbf{K}$ with respect to different bases. Approximating the columns of $\mathbf{K}^w(t) = \mathbf{M}_{\mathcal{Y}}^{1/2}\mathbf{K}(t)\mathbf{M}_{\mathcal{U}}^{-1/2}$ via an adapted problem (16), we have $\epsilon_D \le p\sqrt{T}\max_{1\le l\le p}\|\mathbf{K}_{:,l}^w - \tilde{\mathbf{K}}_{:,l}^w\|_{L^2(0,T;\mathbb{R}^q)}$. Note that the necessary back transformations have to be carried out with sufficient accuracy.

4.3. *Error estimation for the homogeneous PDE*

In order to approximate the system dynamics, the homogeneous PDE (16) has to be solved via a fully-discrete numerical scheme for p different initial values. A *first* goal in error control is to choose the time and space grids (and possibly other discretization parameters) such that

$$\|\mathbf{K}_{:,l} - \tilde{\mathbf{K}}_{:,l}\|_{L^2(0,T;\mathbb{R}^q)} < \texttt{tol} \quad \text{resp.} \quad \|\mathbf{K}^w_{:,l} - \tilde{\mathbf{K}}^w_{:,l}\|_{L^2(0,T;\mathbb{R}^q)} < \texttt{tol} \tag{21}$$

is *guaranteed* for a given $\texttt{tol} > 0$ by means of reliable error estimators. A *second* goal is to achieve this accuracy in a *cost-economic* way. A special difficulty in solving (16) numerically is the handling of initial values b_l, which, in general, only belong to Z but not necessarily to $D(A)$. Considering the example heat equation, this means that the space and time derivatives of the exact solution $z_l \in C^1((0,T], H^2(\Omega) \cap H^1_0(\Omega))$ may become very large for small t, but decay quickly for $t > 0$. In fact, in general we only have the analytic bound

$$\|\partial_t z(t)\|_{L^2(\Omega)} = \|\triangle z(t)\|_{L^2(\Omega)} \leq \frac{c}{t}\|z^0\|_{L^2(\Omega)} \quad \text{for all } t \in (0,T],$$

with some constant $c > 0$ independent of z_0 and T, cf. [27, p. 148]. Adaptive space and time discretizations on the basis of a posteriori error estimates are the method of choice to deal with these difficulties.[28] Discontinuous Galerkin time discretizations in combination with standard Galerkin space discretizations provide an appropriate framework to derive corresponding (a priori and a posteriori) error estimates, also for the case of adaptively refined grids which are in general no longer quasi-uniform.[27,29,30] We distinguish two types of error estimates.

Global state error estimates measure the error $(z_l - z_{l,\texttt{tol}})$ in some global norm. For parabolic problems, a priori and a posteriori estimates for the error in $L^\infty(0,T;L^2(\Omega))$ and $L^\infty(0,T;L^\infty(\Omega))$ can be found in.[29] Such results permit to guarantee (21) in view of

$$\|\mathbf{K}_{:,l} - \tilde{\mathbf{K}}_{:,l}\|_{L^2(0,T;\mathbb{R}^q)} \leq \|C\|_{\mathscr{L}(Z,Y)} \left(\sum_{i=1}^{q} \|\nu_i\|_Y^2\right)^{1/2} \|z - z^{(l)}_{\texttt{tol}}\|_{L^2(0,T;Z)}. \tag{22}$$

Goal-oriented error estimates can be used to measure the error $\|\mathbf{K}_{:,l} - \tilde{\mathbf{K}}_{:,l}\|_{L^2(0,T;\mathbb{R}^q)}$ directly. This may be advantageous, since (22) may be very conservative: the error in the *observations* $\mathbf{K}_{:,l}$ can be small even if some norm of the *state* error is large. The core of these error estimation techniques is an exact error representation formula, which can be evaluated if one knows the residual and the solution of an auxiliary dual PDE. This leads to the *dual-weighted residuals* (DWR) approach, see e.g.[31–39] and the references therein.

The previous discussion justifies the following assumption.

Assumption 4.1. *Given a tolerance* $\texttt{tol} > 0$, *we can ensure (by using appropriate error estimators and mesh refinements) that the solutions* z_l *of* (16) *and the solutions* $z_{l,\texttt{tol}}$ *calculated by means of an appropriate fully-discrete numerical scheme satisfy*

$$\|\mathbf{K}_{:,l} - \tilde{\mathbf{K}}_{:,l}\|_{L^2(0,T;\mathbb{R}^q)} < \texttt{tol}, \qquad l = 1, \ldots, p. \tag{23}$$

5. Total Error Estimates

We present estimates for the total error in the approximation of $\mathbb{G}$. Using general-purpose ansatz spaces $\mathcal{U}_{h_1,\tau_1}$ and $\mathcal{Y}_{h_2,\tau_2}$ for the signal approximation, we only obtain error results in a weaker $\mathscr{L}(\mathcal{U}_s, \mathcal{Y})$-norm.

Theorem 5.1. *Consider the I/O map* $\mathbb{G} \in \mathscr{L}(\mathcal{U}, \mathcal{Y})$ *of the infinite-dimensional linear time-invariant system* (3) *and assume that*

(i) $\mathbb{G}_{|\mathcal{U}_s} \in \mathscr{L}(\mathcal{U}_s, \mathcal{Y}_s)$ *with spaces of higher regularity in space and time*

$$\mathcal{U}_s = H^{\alpha_1,\beta_1}((0,T)\times\Theta), \quad \mathcal{Y}_s = H^{\alpha_2,\beta_2}((0,T)\times\Xi), \quad \alpha_1, \beta_1, \alpha_2, \beta_2 \in \mathbb{N}.$$

(ii) The families of subspaces $\{\mathcal{U}_{h_1,\tau_1}\}_{h_1,\tau_1}$ *and* $\{\mathcal{Y}_{h_2,\tau_2}\}_{h_2,\tau_2}$ *satisfy*

$$\begin{aligned} \|u - \mathbb{P}_{\mathcal{U},h_1,\tau_1} u\|_{\mathcal{U}} &\leq (c_{\mathcal{R}} \tau_1^{\alpha_1} + c_U h_1^{\beta_1}) \|u\|_{\mathcal{U}_s}, & u \in \mathcal{U}_s, \\ \|y - \mathbb{P}_{\mathcal{Y},h_2,\tau_2} y\|_{\mathcal{Y}} &\leq (c_{\mathcal{S}} \tau_2^{\alpha_2} + c_Y h_2^{\beta_2}) \|y\|_{\mathcal{Y}_s}, & y \in \mathcal{Y}_s, \end{aligned}$$

with positive constants $c_{\mathcal{R}}$, $c_{\mathcal{S}}$, c_U *and* c_Y.

(iii) The error in solving for the state dynamics can be made arbitrarily small, i.e., Assumption 4.1 holds.

Let $\delta > 0$ *be given. Then one can choose subspaces* $\mathcal{U}_{h_1^*,\tau_1^*}$ *and* $\mathcal{Y}_{h_2^*,\tau_2^*}$ *such that*

$$\tau_1^* < \left(\frac{\delta}{8c_{\mathcal{R}}\|\mathbb{G}\|_{\mathscr{L}(\mathcal{U},\mathcal{Y})}}\right)^{1/\alpha_1}, \quad h_1^* < \left(\frac{\delta}{8c_U\|\mathbb{G}\|_{\mathscr{L}(\mathcal{U},\mathcal{Y})}}\right)^{1/\beta_1}, \tag{24a}$$

$$\tau_2^* < \left(\frac{\delta}{8c_{\mathcal{S}}\|\mathbb{G}\|_{\mathscr{L}(\mathcal{U}_s,\mathcal{Y}_s)}}\right)^{1/\alpha_2}, \quad h_2^* < \left(\frac{\delta}{8c_Y\|\mathbb{G}\|_{\mathscr{L}(\mathcal{U}_s,\mathcal{Y}_s)}}\right)^{1/\beta_2}, \tag{24b}$$

and one can solve the PDEs (16) *numerically for* $l = 1, \ldots, p(h_1)$ *such that one of the following conditions holds.*

$$(i) \quad \|\mathbf{K}^w_{:,l} - \tilde{\mathbf{K}}^w_{:,l}\|_{L^2(0,T;\mathbb{R}^q)} < \frac{\delta}{2\sqrt{T}p(h_1^*)}, \tag{25a}$$

$$(ii) \quad \|\mathbf{K}_{:,l} - \tilde{\mathbf{K}}_{:,l}\|_{L^2(0,T;\mathbb{R}^q)} < \frac{\delta}{2\sqrt{T}p(h_1^*)}\sqrt{\frac{\lambda_{\min}(\mathbf{M}_{U,h_1^*})}{\lambda_{\max}(\mathbf{M}_{Y,h_2^*})}}, \tag{25b}$$

$$(iii) \quad \|z_l - z_{l,\mathtt{tol}}\|_{L^2(0,T;Z)} < \frac{\delta}{2\sqrt{T}p(h_1^*)}\sqrt{\frac{\lambda_{\min}(\mathbf{M}_{U,h_1^*})}{\lambda_{\max}(\mathbf{M}_{Y,h_2^*})}}\|C\|^{-1}_{\mathscr{L}(Z,Y)}\left(\sum_{i=1}^{q(h_2^*)}\|\nu_i\|_Y^2\right)^{-1/2}. \tag{25c}$$

In this case,

$$\|\mathbb{G} - \mathbb{G}_{DS}\|_{\mathscr{L}(\mathcal{U}_s,\mathcal{Y})} < \delta.$$

Moreover, the signal error $\epsilon_S' := \|\mathbb{G} - \mathbb{G}_S\|_{\mathscr{L}(\mathcal{U}_s,\mathcal{Y})}$ *and the system dynamics error* $\epsilon_D := \|\mathbb{G}_S - \mathbb{G}_{DS}\|_{\mathscr{L}(\mathcal{U},\mathcal{Y})}$ *are balanced in the sense that* $\epsilon_S', \epsilon_D < \delta/2$.

Proof. For $u \in \mathcal{U}_s$, we have

$$\begin{aligned}
\|\mathbb{G}u - \mathbb{G}_S u\|_{\mathcal{Y}} \le& \|\mathbb{G}u - \mathbb{P}_{\mathcal{Y},h_2,\tau_2}\mathbb{G}u\|_{\mathcal{Y}} + \|\mathbb{P}_{\mathcal{Y},h_2,\tau_2}\mathbb{G}u - \mathbb{P}_{\mathcal{Y},h_2,\tau_2}\mathbb{G}\mathbb{P}_{\mathcal{U},h_1,\tau_1}u\|_{\mathcal{Y}}, \\
\le& (c_{\mathcal{S}}\tau_2^{\alpha_2} + c_Y h_2^{\beta_2})\|\mathbb{G}u\|_{\mathcal{Y}_s} \\
&+ (c_{\mathcal{R}}\tau_1^{\alpha_1} + c_U h_1^{\beta_1})\|\mathbb{P}_{\mathcal{Y}}\|_{\mathscr{L}(\mathcal{Y})}\|\mathbb{G}\|_{\mathscr{L}(\mathcal{U},\mathcal{Y})}\|u\|_{\mathcal{U}_s}, \\
\le& \Big\{(c_{\mathcal{S}}\tau_2^{\alpha_2} + c_Y h_2^{\beta_2})\|\mathbb{G}\|_{\mathscr{L}(\mathcal{U}_{rs},\mathcal{Y}_s)} \\
&+ (c_{\mathcal{R}}\tau_1^{\alpha_1} + c_U h_1^{\beta_1})\|\mathbb{G}\|_{\mathscr{L}(\mathcal{U},\mathcal{Y})}\Big\}\|u\|_{\mathcal{U}_s},
\end{aligned}$$

and thus (24) ensures that $\epsilon_S' = \|\mathbb{G} - \mathbb{G}_S\|_{\mathscr{L}(\mathcal{U}_s,\mathcal{Y})} < \delta/2$. Proposition 4.1 in combination with (25) and in view of (22) ensures that $\epsilon_D = \|\mathbb{G}_S - \mathbb{G}_{DS}\|_{\mathscr{L}(\mathcal{U},\mathcal{Y})} < \delta/2$, which concludes the proof. □

6. Applications and numerical results

6.1. *Test problems*

As test cases, we consider two heat equations on different domains $\Omega \subset \mathbb{R}^2$ as depicted in Fig. 4. In both cases the control and observation operators are defined on rectangular subsets of Ω $\Omega_c = (a_{c,1}, a_{c,2}) \times (b_{c,1}, b_{c,2})$ and $\Omega_m =$

$(a_{m,1}, a_{m,2}) \times (b_{m,1}, b_{m,2})$, where the control is active and the observation takes place, respectively.

Test case 6.1. Setting $U = Y = L^2(0,1)$, we define $C \in \mathscr{L}(L^2(\Omega), Y)$ and $B \in \mathscr{L}(U, L^2(\Omega))$ by

$$(Bu)(x_1, x_2) = \begin{cases} u(\theta_1(x_1))\omega_c(x_2), & (x_1, x_2) \in \Omega_c \\ 0, & (x_1, x_2) \notin \Omega_c \end{cases}$$

and

$$(Cz)(\xi) = \int_{a_{m,1}}^{b_{m,1}} \frac{z(x_1, \theta_2(\xi))}{b_{m,1} - a_{m,1}}\, dx_1,$$

where $\omega_c \in L^2(a_{c,2}, b_{c,2})$ is a weight function and $\theta_1 : [a_{c,1}, b_{c,1}] \to [0,1]$ and $\theta_2 : [0,1] \to [a_{m,1}, b_{m,1}]$ are affine-linear transformations.

Note that C preserves an inherent spatial state regularity, i.e., $C_{|H^2(\Omega)} \in \mathscr{L}(H^2(\Omega), H^2(0,1))$.

For the state equation we consider a heat equation with homogeneous Dirichlet boundary conditions on $(0,T] \times \Omega$ with $T = 1$ and $\Omega = (0,1)^2$. We choose $\Omega_c = \Omega$, $\Omega_m = (0.1, 0.2) \times (0.1, 0.9)$ and $\omega_c(x_2) = \sin(\pi x_2)$. In this case, the output obtained by inputs of the special form $u(t;\theta) = \sin(\omega_T \pi t)\sin(m\pi\theta)$ with $\omega_T, m \in \mathbb{N}$ can be explicitly computed in terms of the eigenfunctions of the Laplace operator.

Test case 6.2. As second test case, we consider two infinitely long plates of width 5 and height 0.2, which are connected by two rectangular bars as shown in the cross section in Fig. 4. We assume that the plates are surrounded by an insulating material and that we can heat the bottom plate and measure the temperature distribution in the upper plate.

The input operator is chosen as in Test case 6.1 for the output operator, we just switch the variables in the definition of C.

As state equation we consider a heat equation with homogeneous Neumann boundary conditions on $(0,T] \times \Omega$ with $T = 1$ and Ω as in Fig. 4, and choose $\Omega_c = (0.05, 4.95) \times (0.05, 0.15)$, $\Omega_m = (0.05, 4.95) \times (0.85, 0.95)$ and $\omega_c(x_2) = \sin(\pi(x_2 - 0.05)/0.1)$.

The matrix approximations $\tilde{\mathbf{G}}$ of the I/O-maps $\mathbb{G}$ corresponding to the test cases have been calculated by means of a heat equation solver, which is based on the C++ FEM software library DEAL.II.[40] It realizes a discontinuous Galerkin scheme with adaptive space and time grids and applies goal-oriented DWR-based error control to ensure (21).

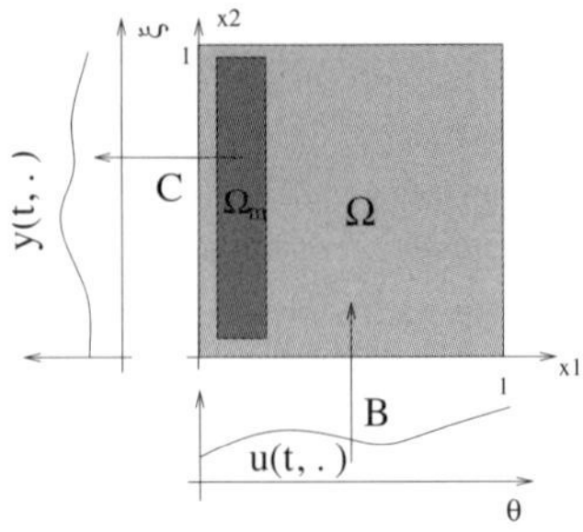

(a) Test case 6.1

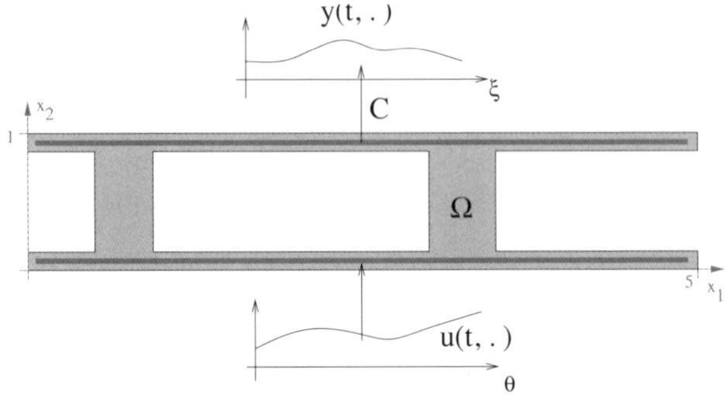

(b) Test case 6.2

Fig. 4. Test cases heat equation: (a) with homogeneous Dirichlet boundary conditions, (b) with homogeneous Neumann boundary conditions.

6.2. *Tests of convergence*

The following numerical convergence tests have all been carried out with approximations $\mathbb{G}_{DS}(h_1, \tau_1, h_2, \tau_2, \mathtt{tol})$ of the I/O-map $\mathbb{G}$ corresponding to Test case 6.1. Hierarchical linear finite elements in U_{h_1} and Y_{h_2} and Haar wavelets in $\mathcal{R}_{\tau_1}$ and $\mathcal{S}_{\tau_2}$ have been chosen. The tolerance `tol` refers to the estimate (23).

Convergence of single outputs. Considering Test case 6.1 with inputs $u(t;\theta) = \sin(\omega_T \pi t)\sin(m\pi\theta)$, and exactly known outputs $y = \mathbb{G}u$, we investigate the relative error $\|y - \tilde{y}\|_{\mathcal{Y}} / \|u\|_{\mathcal{U}_s}$, with $\tilde{y} = \mathbb{G}_{DS}(h_1, \tau_1, h_2, \tau_2, \mathtt{tol})u$, for varying discretization parameters h_1, τ_1, h_2, τ_2 and `tol`. Choosing, e.g., $m = 5$ and $\omega_T = 10$, we observe a quadratic convergence in $h_1 = h_2$ (cf. Fig. 6.2-a) and a linear convergence in $\tau_1 = \tau_2$ (cf. Fig. 6.2-b) in correspondence to Thm. 5.1. However, the error does not converge to zero but to a positive plateau value, which is due to the system dynamics error and which becomes smaller for lower tolerances `tol`. For input signals with $m > 5$ and $\omega_T > 10$ the convergence order can only be observed for smaller discretization parameters h_1, h_2, τ_1 and τ_2.

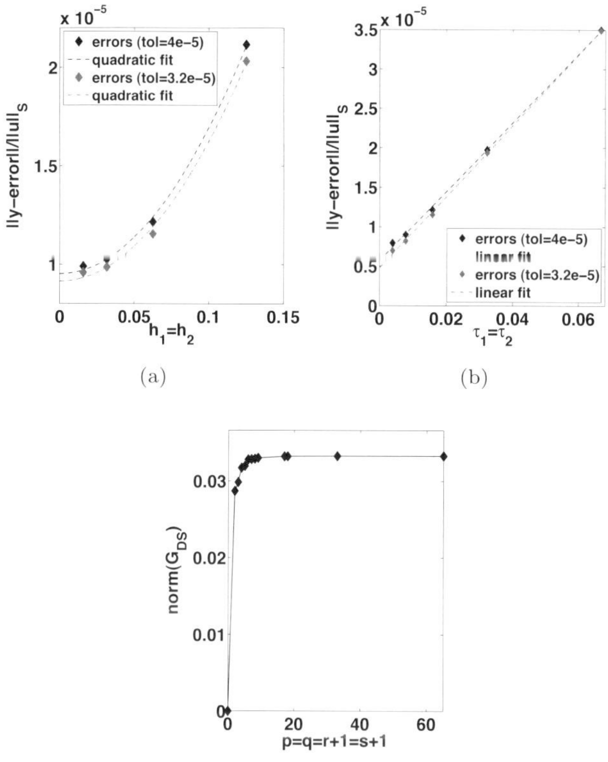

Fig. 5. (a) Relative output errors for input $u(t;\theta) = \sin(10\pi t)\sin(5\pi\theta)$, varying $h_1 = h_2$ and fixed $\tau_1 = \tau_2 = 1/64$. (b) Relative output errors for input $u(t;\theta) = \sin(10\pi t)\sin(5\pi\theta)$, varying $\tau_1 = \tau_2$ and fixed $h_1 = h_2 = 1/17$. (c) Norm $\|\mathbb{G}_{DS}(\mathbf{h})\|_{\mathscr{L}(\mathcal{U},\mathcal{Y})}$ for synchronously increasing approximation space dimensions $p = q = r + 1 = s + 1$ and fixed tolerance `tol` $= 4.0e - 5$.

Convergence of the norm $\|\mathbb{G}_S(h_1, \tau_1, h_2, \tau_2)\|_{\mathscr{L}(\mathcal{U},\mathcal{Y})}$ *for nested subspaces.* Successively improving the signal approximation by adding additional basis functions, the norm $\|\mathbb{G}_S(h_1, \tau_1, h_2, \tau_2)\|_{\mathscr{L}(\mathcal{U},\mathcal{Y})}$ converges, cf. Lemma 3.1. We approximate $\|\mathbb{G}_S\|_{\mathscr{L}(\mathcal{U},\mathcal{Y})}$ by $\|\mathbb{G}_{DS}\|_{\mathscr{L}(\mathcal{U},\mathcal{Y})}$, where $\mathbb{G}_{DS}$ has been calculated with `tol` $= 4.0e - 5$. In Fig. 6.2-c, the approximations $\|\mathbb{G}_S(h_1, \tau_1, h_2, \tau_2)\|_{\mathscr{L}(\mathcal{U},\mathcal{Y})} = \|\mathbb{G}_S(\frac{1}{p-1}, \frac{1}{r}, \frac{1}{q-1}, \frac{1}{s})\|_{\mathscr{L}(\mathcal{U},\mathcal{Y})}$ are plotted for increasing subspace dimensions $p = q = r+1 = s+1 = 2, 3, \dots, 65$.

6.3. *Matrix reduction on the basis of SVDs*

In order to resolve the input and output signal spaces accurately by means of general purpose basis functions, a large number of basis functions is needed in general. In order to reduce the large size of the resulting I/O-matrices $\tilde{\mathbf{G}}$, we apply a reduction method known as *Tucker decomposition* or *higher order singular value decomposition* (HOSVD).[41] It is based on singular value decompositions (SVDs) and preserves the space-time tensor structure of the input and output signal bases.

Considering $\tilde{\mathbf{G}} \in \mathbb{R}^{qs \times pr}$ as a fourth-order tensor $\tilde{\mathbf{G}} \in \mathbb{R}^{s \times r \times q \times p}$ with $\tilde{\mathbf{G}}_{ijkl} = \tilde{\mathbf{G}}_{ij}^{kl}$, it is shown in[41] that there exists a HOSVD

$$\tilde{\mathbf{G}} = \mathbf{S} \times_1 \mathbf{U}^{(\psi)} \times_2 \mathbf{U}^{(\phi)} \times_3 \mathbf{U}^{(\nu)} \times_4 \mathbf{U}^{(\mu)}. \tag{26}$$

Here $\mathbf{S} \in \mathbb{R}^{s \times r \times q \times p}$ is a so-called *core tensor*, satisfying some orthogonality properties, $\mathbf{U}^{(\psi)} \in \mathbb{R}^{s \times s}$, $\mathbf{U}^{(\phi)} \in \mathbb{R}^{r \times r}$, $\mathbf{U}^{(\nu)} \in \mathbb{R}^{q \times q}$, $\mathbf{U}^{(\mu)} \in \mathbb{R}^{p \times p}$ are unitary matrices and $\times_1, \dots, \times_4$ denote tensor-matrix multiplications. We define a so-called *matrix unfolding* $\tilde{\mathbf{G}}^{(\psi)} \in \mathbb{R}^{s \times rqp}$ of the tensor $\tilde{\mathbf{G}}$ by

$$\tilde{\mathbf{G}}_{im}^{(\psi)} = \mathbf{G}_{ijkl}, \quad m = (k-1)ps + (l-1)s + i,$$

i.e., we put all elements belonging to $\psi_1, \psi_2, \dots, \psi_s$ into one respective row, and we define the unfoldings $\tilde{\mathbf{G}}^{(\phi)} \in \mathbb{R}^{r \times qps}$, $\tilde{\mathbf{G}}^{(\nu)} \in \mathbb{R}^{q \times psr}$ and $\tilde{\mathbf{G}}^{(\mu)} \in \mathbb{R}^{p \times srq}$ in a similar cyclic way. Then, $\mathbf{U}^{(\psi)}$, $\mathbf{U}^{(\phi)}$, $\mathbf{U}^{(\nu)}$ and $\mathbf{U}^{(\mu)}$ in (26) can be calculated by means of four SVDs of the respective form

$$\tilde{\mathbf{G}}^{(\psi)} = \mathbf{U}^{(\psi)} \Sigma^{(\psi)} (\mathbf{V}^{(\psi)})^T,$$

where $\Sigma^{(\psi)}$ is diagonal with entries $\sigma_1^{(\psi)} \geq \sigma_2^{(\psi)} \geq \dots \sigma_s^{(\psi)} \geq 0$ and $\mathbf{V}^{(\psi)}$ is columnwise orthonormal. The $\sigma_i^{(\psi)}$ are so-called *n-mode singular values* (or in our case ψ-mode singular values) of the tensor $\tilde{\mathbf{G}}$ and correspond to the Frobenius norms of certain subtensors of the core tensor $\mathbf{S}$.

On the basis of (26) we can define an approximation $\hat{\mathbf{G}} \in \mathbb{R}^{s \times r \times q \times p}$ of $\tilde{\mathbf{G}}$ by discarding the smallest n-mode singular values $\{\sigma_{\hat{s}+1}^{(\psi)}, \dots, \sigma_s^{(\psi)}\}$, $\{\sigma_{\hat{r}+1}^{(\phi)}, \dots, \sigma_r^{(\psi)}\}$, $\{\sigma_{\hat{q}+1}^{(\nu)}, \dots, \sigma_q^{(\nu)}\}$ and $\{\sigma_{\hat{p}+1}^{(\mu)}, \dots, \sigma_p^{(\mu)}\}$, i.e., we set the corresponding parts of $\mathbf{S}$ to zero. Then we have[41]

$$\|\tilde{\mathbf{G}} - \hat{\mathbf{G}}\|_F^2 \leq \sum_{i=\hat{s}+1}^{s} \sigma_i^{(\psi)} + \sum_{j=\hat{r}+1}^{r} \sigma_j^{(\phi)} + \sum_{k=\hat{q}+1}^{q} \sigma_k^{(\nu)} + \sum_{l=\hat{p}+1}^{p} \sigma_l^{(\mu)}.$$

The truncation of $\hat{\mathbf{G}} \in \mathbb{R}^{qr \times ps}$ after a basis transformation corresponding to $\mathbf{U}^{(\psi)}$, $\mathbf{U}^{(\phi)}$, $\mathbf{U}^{(\nu)}$ and $\mathbf{U}^{(\mu)}$ yields a low-dimensional representation $\bar{\mathbf{G}} \in \mathbb{R}^{\hat{q}\hat{r} \times \hat{p}\hat{s}}$.

In Figure 6 the HOSVD has been applied to a matrix $\tilde{\mathbf{G}} \in \mathbb{R}^{qs \times pr}$ for the Test case 6.2 with $p = 17$, $q = 65$ and $r = s = 64$. The first row shows the respective n-mode singular values. Underneath the first and most relevant two transformed/new basis functions $\hat{\mu}_i$, $\hat{\nu}_i$, $\hat{\phi}_i$ and $\hat{\psi}_i$, are plotted. It is not surprising that the positions of the connections between the plates can be recovered as large values of the corresponding spatial input and output basis functions.

Remark 6.1. The application of a HOSVD is useful in two ways. *First*, it delivers a low-dimensional matrix-representation of the system, which is small enough to be used for real-time feedback control design. *Second*, it allows to identify relevant input and output signals, which can be exploited in actuator and sensor design, i.e., to decide where actuators and sensors have to be placed and which resolution in time and space they should have.

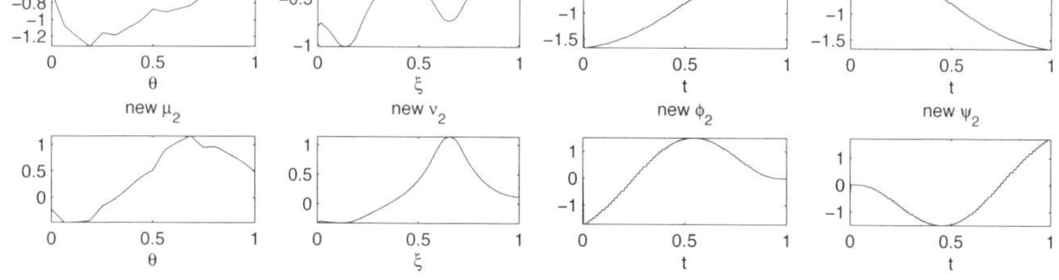

Fig. 6. HOSVD applied to the I/O map of Test case 6.2. First row: n-mode singular values in semilogarithmic scales. 2nd and 3rd row: Respective two most relevant basis functions.

6.4. *Application in optimization problems*

We investigate the use of the I/O-map approximation in optimization problems

$$\min J(u,y) \qquad \text{subject to } y = \mathbb{G}u, \quad u \in \mathcal{U}_{ad}. \tag{27}$$

Here $\mathcal{U}_{ad} \subset \mathcal{U}$ is the subset of admissible controls, $J : \mathcal{U} \times \mathcal{Y} \to \mathbb{R}$ is a quadratic cost functional $J(u,y) = \frac{1}{2}\|y - y_D\|^2_{\mathcal{Y}} + \alpha\|u\|^2_{\mathcal{U}}$, $y_D \in \mathcal{Y}$ is a desired output signal, and $\alpha > 0$ is a regularization parameter. We define the discretized cost functional

$$\bar{J}_{\mathbf{h}} : \mathbb{R}^{pr} \times \mathbb{R}^{qs} \to \mathbb{R}, \qquad \bar{J}_{\mathbf{h}}(\mathbf{u},\mathbf{y}) = \frac{1}{2}\|\mathbf{y} - \mathbf{y}_D\|^2_{qs;w} + \alpha\|\mathbf{u}\|^2_{pr;w},$$

with $\mathbf{y}_D = \kappa_{\mathcal{Y},h_2,\tau_2}\mathbb{P}_{\mathcal{Y},h_2,\tau_2}y_D$, and instead of (27) we solve

$$\min \bar{J}_{\mathbf{h}}(\mathbf{u},\mathbf{y}) \qquad \text{subject to } \mathbf{y} = \tilde{\mathbf{G}}\mathbf{u}, \quad \mathbf{u} \in \bar{U}_{ad}, \tag{28}$$

with $\bar{U}_{ad} = \{\mathbf{u} \in \mathbb{R}^{pr} : \mathbf{u} = \kappa_{\mathcal{U},h_1,\tau_1}\mathbb{P}_{\mathcal{U},h_1,\tau_1}u,\ u \in \mathcal{U}_{ad}\}$. Considering optimization problems without control constraints, i.e., $\mathcal{U}_{ad} = \mathcal{U}$ and $\bar{U}_{ad} = \mathbb{R}^{pr}$, the solution $\bar{\mathbf{u}}$ of (28) is characterized by

$$(\tilde{\mathbf{G}}^T\mathbf{M}_{\mathcal{Y}}\tilde{\mathbf{G}} + \alpha\mathbf{M}_{\mathcal{U}})\bar{\mathbf{u}} = \tilde{\mathbf{G}}^T\mathbf{M}_{\mathcal{Y}}\mathbf{y}_D. \tag{29}$$

As concrete example, we consider Test case 6.2 and choose $y_D = \mathbb{G}u_0$ to be the output for an input $u_0 \equiv 1$ which is equal to 1 on all of $[0,T] \times (0,1)$. We then try to find an *optimal* input u_* that minimizes the cost functional (27).

First we solve (29) with an approximated I/O map $\tilde{\mathbf{G}} \in \mathbb{R}^{17\cdot 64 \times 65\cdot 64}$ and $\alpha = 10^{-4}$, yielding an approximation $\bar{u} \approx u_*$.

The solution takes 0.33 seconds on a normal desktop PC. The u-norm is reduced by 27.9% and the relative deviation of $\mathbb{G}\bar{u}$ from y_D is 9.4%. In Fig. 7 the same calculations have been carried out with $\hat{\mathbf{G}} \in \mathbb{R}^{3\cdot 5 \times 3\cdot 5}$, where $\hat{\mathbf{G}}$ arises from a HOSVD-based matrix reduction of $\tilde{\mathbf{G}} \in \mathbb{R}^{17\cdot 64 \times 65\cdot 64}$, where all but the 3 most relevant spatial and the 5 most relevant temporal input and output basis functions have been truncated. Using this approximation, the norm of u is reduced by 27.4%, whereas the relative deviation of $\mathbb{G}\bar{u}$ from y_D is 9.5%. The cost functional has been reduced by 44.5%, and the calculation of $\bar{u}$ took less than 0.0004 seconds. The outputs resulting from u_0 and $\bar{u}$ have been calculated in simulations independent from the calculation of the I/O-matrix.

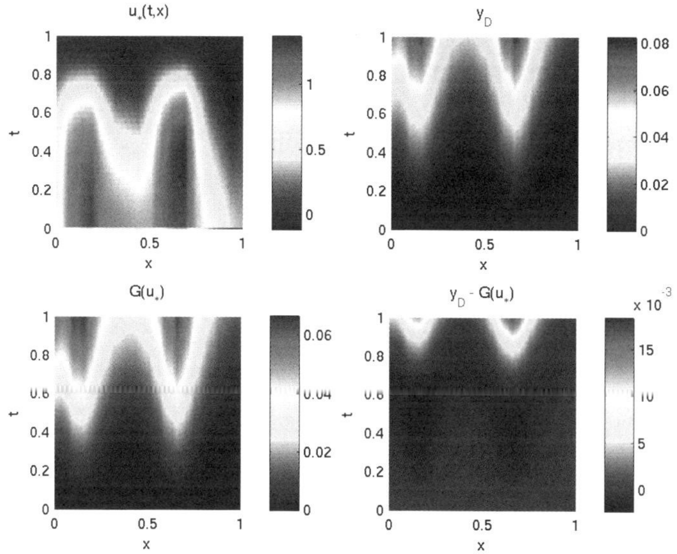

Fig. 7. Application of the SVD-reduced approximated I/O map $\hat{\mathbf{G}} \in \mathbb{R}^{3\cdot 5\times 3\cdot 5}$ in an optimization problem. From top left to bottom right: optimized control $\bar{u}$, original output $y_D = \mathbb{G}u_0$, optimized output $\mathbb{G}\bar{u}$ and their difference.

7. Final remarks and outlook

We have presented a systematic framework for the discretization of I/O-maps of linear infinite-dimensional control systems with *spatially distributed* inputs and outputs. Global error estimates have been provided, which allow to choose the involved discretization parameters in such a way that a desired overall accuracy is achieved and that the signal and the system dynamics approximation errors are balanced. Moreover, the error results are capable to take into account many practical and technical restrictions in sensor and actuator design, like limited spatial and temporal resolutions or the use of piecewise constant controls and observations due to digital devices.

The numerical costs of the approach are primarily governed by the numerical calculation of p underlying homogeneous PDEs, where p is the number of input basis functions in space. This, however can be done beforehand, and in parallel, provided there is enough storage available that allows to store these solutions. This, however, can become problematic when the spatial resolution of the input signal space has to be very accurate. In this case, code-optimization, e.g. due to parallelization and appropriate updating of mass and stiffness matrices from prior calculations, promises to have a large potential for speed-up.

The SVD-based dimension reduction for the matrix representation can be considered as an alternative model reduction approach, and the resulting reduced I/O-models proved to be successful in first numerical optimization applications. Moreover, the SVD-based reduction may be able to provide useful insight for efficient actuator and sensor design by filtering out relevant input and output signals.

References

1. A. C. Antoulas, *Approximation of large-scale dynamical systems* (Society for Industrial and Applied Mathematics (SIAM), Philadelphia, 2005).
2. P. Benner, V. Mehrmann and D. Sorensen (eds.), *Dimension Reduction of Large-Scale Systems* (Springer, 2005).
3. D. M. Luchtenburg, B. Gunter, B. R. Noack, R. King and G. Tadmor, *J. Fluid Mech.* **623**, 283 (2009).
4. B. R. Noack, M. Schlegel, B. Ahlborn, G. Mutschke, M. Morzynski, P. Comte and G. Tadmor, *J. Non-Equilib. Thermodyn.* **33**, 103 (2008).
5. M. Pastoor, L. Henning, B. R. Noack, R. King and G. Tadmor, *J. Fluid Mech.* **608**, 161 (2008).
6. G. Berkooz, P. Holmes and J. L. Lumley, The proper orthogonal decomposition in the analysis of turbulent flows, in *Annual review of fluid mechanics, Vol. 25*, (Annual Reviews Inc., Palo Alto, 1993) pp. 539–575.
7. V. Mehrmann and T. Stykel, Balanced truncation model reduction for large-scale systems in descriptor form, in *Dimension Reduction of Large-Scale Systems*, eds. P. Benner, V. Mehrmann and D. Sorensen (Springer, Heidelberg, 2005).
8. R. Becker, M. Garwon, C. Gutknecht, G. Bärwolff and R. King, *J. of Process Control* **15**, 691 (2005).
9. L. Henning, D. Kuzmin, V. Mehrmann, M. Schmidt, A. Sokolov and S. Turek, Flow control on the basis of a FEATFLOW-MATLAB coupling, in *Active Flow Control. Papers contributed to the Conference "Active Flow Control 2006", Berlin, Germany, September 27 to 29, 2006*, ed. R. King (Springer, Berlin, 2006).
10. S. Gugercin and A. C. Antoulas, *Internat. J. Control* **77**, 748 (2004).
11. R. W. Freund, *Model Reduction Methods Based on Krylov Subspaces*, tech. rep., Bell Laboratories, Lucent Technologies (2001).
12. C. W. Rowley, *Internat. J. Bifur. Chaos Appl. Sci. Engrg.* **15**, 997 (2005).
13. E. Emmrich, *Gewöhnliche und Operator-Differentialgleichungen* (Vieweg, Wiesbaden, 2004).
14. J.-L. Lions and E. Magenes, *Non-homogeneous boundary value problems and applications. Vol. II* (Springer, New York, 1972).
15. A. Pazy, *Semigroups of linear operators and applications to partial differential equations* (Springer, New York, 1983).

16. L. C. Evans, *Partial differential equations*, Graduate Studies in Mathematics, Vol. 19 (American Mathematical Society, Providence, 1998).
17. A. Lunardi, *Analytic semigroups and optimal regularity in parabolic problems* (Birkhäuser, Basel, 1995).
18. H. Sohr, *The Navier-Stokes equations* (Birkhäuser, Basel, 2001).
19. E. Emmrich and V. Mehrmann, *Analysis of a class of operator differential-algebraic equations arising in fluid mechanics. Part 1. The finite dimensional case*, tech. rep. (2010).
20. J. Heiland, V. Mehrmann and M. Schmidt, A new discretization framework for input/output maps and its application to flow control, in *Active Flow Control. Papers contributed to the Conference "Active Flow Control II 2010", Berlin, Germany, May 26 to 28, 2010*, ed. R. King (Springer, Berlin, 2006).
21. H. Yserentant, Hierarchical bases in the numerical solution of parabolic problems, in *Large scale scientific computing (Oberwolfach, 1985)*, (Birkhäuser, Boston, MA, 1987) pp. 22–36.
22. H. Yserentant, Hierarchical bases, in *ICIAM 91 (Washington, DC, 1991)*, (SIAM pp. 256–276.
23. A. Cohen, *Numerical analysis of wavelet methods*, Studies in Mathematics and its Applications, Vol. 32 (North-Holland Publishing Co., Amsterdam, 2003).
24. P. G. Ciarlet, *The finite element method for elliptic problems*, Classics in Applied Mathematics, Vol. 40 (SIAM, Philadelphia, PA, 2002).
25. E. Zeidler, *Nonlinear functional analysis and its applications. II/A* (Springer, New York, 1990).
26. W. Hackbusch, B. N. Khoromskij and E. E. Tyrtyshnikov, *J. Numer. Math.* **13**, 119 (2005).
27. C. Johnson, *Numerical solution of partial differential equations by the finite element method* (Cambridge University Press, Cambridge, 1987).
28. K. Eriksson, D. Estep, P. Hansbo and C. Johnson, Introduction to adaptive methods for differential equations, in *Acta numerica, 1995*, Acta Numer. (Cambridge University Press, Cambridge) pp. 105–158.
29. K. Eriksson and C. Johnson, *SIAM J. Numer. Anal.* **32**, 706 (1995).
30. V. Thomée, *Galerkin finite element methods for parabolic problems*, Springer Series in Computational Mathematics, Vol. 25 (Springer, Berlin, 1997).
31. I. Babuska and A. Miller, *Int. J. Numer. Methods Eng.* **20**, 2311 (1984).
32. M. Ainsworth and J. T. Oden, *A posteriori error estimation in finite element analysis* (Wiley-Interscience, New York, 2000).
33. W. Bangerth and R. Rannacher, *Adaptive finite element methods for differential equations* (Birkhäuser, Basel, 2003).
34. R. Becker, V. Heuveline and R. Rannacher, *Int. J. Numer. Methods Fluids* **40**, 105 (2002).
35. R. Becker and R. Rannacher, *East-West J. Numer. Math.* **4**, 237 (1996).
36. R. Becker and R. Rannacher, *Acta Numer.* **10**, 1 (2001).
37. V. Heuveline and R. Rannacher, *J. Numer. Math.* **11**, 95 (2003).
38. C. Johnson and R. Rannacher, On errror control in cfd, in *Numerical methods for the Navier-Stokes equations. Proceedings of the international workshop*

held, Heidelberg, Germany, October 25-28, 1993. Notes Numer. Fluid Mech. 47, 133-144 , ed. K.-F. Hebeker (Vieweg, Braunschweig, 1994)
39. C. Johnson, R. Rannacher and M. Boman, *SIAM J. Numer. Anal.* **32**, 1058 (1995).
40. W. Bangerth, R. Hartmann and G. Kanschat, `deal.II` *Differential Equations Analysis Library, Technical Reference.* http://www.dealii.org/, 5.2 edn.(September, 2005).
41. L. De Lathauwer, B. De Moor and J. Vandewalle, *SIAM J. Matrix Anal. Appl.* **21**, 1253 (2000).

VERTEX COUPLINGS IN QUANTUM GRAPHS: APPROXIMATIONS BY SCALED SCHRÖDINGER OPERATORS

PAVEL EXNER

Doppler Institute for Mathematical Physics and Applied Mathematics, Czech Technical University, Břehová 7, 11519 Prague, Czech Republic, and

Department of Theoretical Physics, Nuclear Physics Institute, Czech Academy of Sciences, 25068 Řež near Prague, Czech Republic
E-mail: exner@ujf.cas.cz

We review recent progress in understanding the physical meaning of quantum graph models through analysis of their vertex coupling approximations.

Keywords: Quantum graphs; Vertex coupling; Tube networks; Approximations.

1. Introduction

Quantum graphs attracted a lot of interest recently. There are several reasons for that. On one hand these models are useful as descriptions of various structures prepared from semiconductor wires, carbon nanotubes, and other substances. On the other hand they provide a tool to study properties of quantum dynamics in situations when the system has a nontrivial geometrical or topological structure.

Quantum graph models contain typically free parameters related to coupling of the wave functions at the graph vertices, and to get full grasp of the theory one has to understand their physical meaning. A natural approach to this question is to investigate "fat graphs", that is, systems of thin tubes built over the skeleton of a given graph, and to analyze its limit as the tube thickness tends to zero.

While simple at a glance, the problem is in fact rather difficult and its understanding is being reached through a long series of works. The aim of the present paper is to review some recent achievements in this area. We present this survey in a non-technical way referring for detailed proofs and a wider background to the literature, in particular, to our recent papers

[7,19]. Having said that it is important to stress that we will formulate the problem and the results in a fully rigorous way.

2. Quantum graphs

2.1. *A bit of history*

The quantum graph concept was born in early days of quantum mechanics being first suggested in the 1930's by Linus Pauling as a model of aromatic hydrocarbons, and worked out later by Ruedenberg and Scherr [37]. Then, as it sometimes happen in the history of science, it was happily forgotten.

In a sense it might be surprising because the idea of a quantum particle living on a graph is theoretically attractive, however, it was not enough and for three decades quantum graph models enjoyed the status of an obscure textbook example. This changed in the eighties, when the diminishing size of structures produced in solid-state-physics laboratories reached the state when the electron transport in them became dominantly ballistic and quantum graphs suddenly reemerged as a useful model.

The list of physical system to which these methods can be applied kept expanding. At the beginning it included microstructures fabricated from semiconductor or metallic materials, later carbon nanotubes were added. It is worth mentioning, however, that the same technique can be used also to investigation of electromagnetic phenomena in large network-type structures [27], at least as long stationary situation is considered.

Quantum dynamics of a particle confined to a graph can mean various things, of course. Typically one considers a nonrelativistic situation described by a *Schrödinger operator* supported by the graph. Often the motion is free but in other situations one adds potentials corresponding to external electric or magnetic fields, spin degrees of freedom, etc. Graphs can support also *Dirac operators*. Such a model, too, was for a long time regarded as a theoretician toy and attracted a limited attention only [3,4]. The situation changed dramatically two or three years ago with the discovery of graphene in which electron behave effectively as relativistic particles which triggered a wave of papers of the subject.

The literature on quantum graphs is nowadays immense; we are not going to try to give a bibliographical review and refer instead to the proceedings volume of a recent Isaac Newton Institute programme [15] where one can find an extensive guide to further reading.

2.2. *Vertex coupling*

For simplicity let us consider a graph having a single vertex in form of a *star*, i.e. n halflines with the endpoint connected. The state Hilbert space $\mathcal{H}$ of such a system is $= \bigoplus_{j=1}^{n} L^2(\mathbb{R}_+)$ and the particle Hamiltonian acts on $\mathcal{H}$ as $\psi_j \mapsto -\psi_j''$; the values of physical constants are irrelevant for our discussion and we put conventionally $\hbar = 2m = 1$.

The Hamiltonian domain consists of $W^{2,2}$ functions; in order to make it self-adjoint we need to impose suitable boundary conditions at the vertex. Since we deal with a second-order operator, the latter involve boundary value $\Psi(0) := \{\psi_j(0)\}$ and $\Psi'(0) := \{\psi_j'(0)\}$; conventionally they are written in the form

$$A\Psi(0) + B\Psi'(0) = 0 \tag{1}$$

proposed by Kostrykin and Schrader [28], where the $n \times n$ matrices A, B give rise to a self-adjoint operator *iff* they satisfy the conditions

- $\operatorname{rank}(A|B) = n$
- AB^* is self-adjoint

The obvious drawback of (1) is that the pair A, B is not unique. The common way to remove the non-uniqueness [24,26,29] is to choose

$$A = U - I\,, \quad B = i(U + I)\,, \tag{2}$$

where U is an $n \times n$ unitary matrix; there are also other unique forms more suitable for some purposes [7,8,30]; one of them we will need in Sec. 6 below. It is obvious from (2) that the coupling of n edges is characterized in general by n^2 real parameters.

There is a simple way to derive the boundary conditions with which can be traced to [23] where it was used for $n = 2$. Self-adjointness requires vanishing of the boundary form,

$$\sum_{j=1}^{n} (\bar{\psi}_j \psi_j' - \bar{\psi}_j' \psi_j)(0) = 0\,,$$

which occurs *iff* the norms $\|\Psi(0) \pm i\ell\Psi'(0)\|_{\mathbb{C}^n}$ with a fixed $\ell \neq 0$ coincide, since the difference of the squared norms is just the *lhs* of the displayed relation. Consequently, the vectors must be related by an $n \times n$ unitary matrix, which yields immediately $(U - I)\Psi(0) + i\ell(U + I)\Psi'(0) = 0$. It may seem that we have an extra parameter here, however, matrices corresponding to two different values of ℓ are related by

$$U' = \frac{(\ell + \ell')U + \ell - \ell'}{(\ell - \ell')U + \ell + \ell'}\,,$$

so it just fixes the length scale of the problem and we can put $\ell = 1$ without loss of generality. Note also that the parameter matrix is closely related to the scattering at the vertex, specifically, it coincides with the on-shell scattering matrix at the momentum $k = 1$.

2.3. *Examples of vertex coupling*

Denote by $\mathcal{J}$ the $n \times n$ matrix whose all entries are equal to one; then the unitary matrix $U = \frac{2}{n+i\alpha}\mathcal{J} - I$ corresponds to the standard δ *coupling* characterized by the conditions

$$\psi_j(0) = \psi_k(0) =: \psi(0)\,, \; j,k = 1,\ldots,n\,, \quad \sum_{j=1}^{n} \psi_j'(0) = \alpha\psi(0) \tag{3}$$

of "coupling strength" $\alpha \in \mathbb{R}$; we include also the case $\alpha = \infty$, or $U = -I$, when the edges are decoupled with Dirichlet conditions at the endpoints. Another particular case of interest is $\alpha = 0$ corresponding to the "free motion". It would be natural to call then (3) *free* boundary conditions, however, they are mostly called *Kirchhoff* in the literature[a]. Note that the δ-couplings are the only ones with wave functions continuous at the vertex.

The second example to mention is the δ_s' *coupling*, a counterpart to the above one with the roles of functions and derivatives interchanged. The corresponding unitary matrix is $U = I - \frac{2}{n-i\beta}\mathcal{J}$ giving

$$\psi_j'(0) = \psi_k'(0) =: \psi'(0)\,, \; j,k = 1,\ldots,n\,, \quad \sum_{j=1}^{n} \psi_j(0) = \beta\psi'(0) \tag{4}$$

with $\beta \in \mathbb{R}$; for $\beta = \infty$ we get decoupled edges with *Neumann* conditions.

3. Vertex understanding through approximations

3.1. *Statement of the problem*

The first question to pose is why we should be interested in quantum graph vertex couplings. There are several reasons for that:

- One is mathematical. Different couplings define different Hamiltonians which have different spectral properties. Sometimes they can be quite involved; as an example let us number theoretic properties of rectangular lattice-graph spectra [13].

[a]The name is generally accepted but unfortunate because in electricity it is associated with current conservation at the junction, and in the quantum case *any* self-adjoint coupling preserves probability current.

- On a more practical side, the conductivity of nanostructures is controlled typically by application of external fields. Understanding of vertex coupling would give us an alternative mean to this goal.
- As a specific example, the authors of Ref. 10 used the generalized point interaction on line as a model of a qubit; in a similar way star graphs with $n > 2$ edges can similarly model *qudits*.

At a glance the vertex parameters can be interpreted easily. One should replace the graph in question by a family of "fat graphs", i.e. a tube network built around the graph skeleton, with appropriate Laplacian as the Hamiltonian. Such a system has no free parameters, so it would be enough to inspect the squeezing limit with the tube diameter tends to zero and to see which graph Hamiltonian we obtain. Unfortunately, as it is often the case with simple answers, the problem is in reality rather complicated:

- The answer depends substantially on the type of the Laplacian supported by the tube network. The *Neumann* case is easier and after an effort more than a decade long an understanding was reached [17,18,22,31,34,36,38]. The drawback was that the limit gave the free (Kirchhoff) boundary conditions only.
- the *Dirichlet* case is more difficult and only recently some substantial results were obtained [1,5,12,25,32,33], nevertheless, a lot of work remains to be done

Before proceeding to our main topic, let us review briefly the existing results we have mentioned above.

3.2. *Briefly on Dirichlet networks*

The distinctive feature of the Dirichlet case is the energy blow-up associated with the fact that the transverse part of the Dirichlet Laplacian has lowest eigenvalue proportional to d^{-2} where d is the tube diameter. To get a meaningful result we have thus to use an *energy renormalization* which can be done in different ways. Molchanov and Vainberg [32] chose the en-

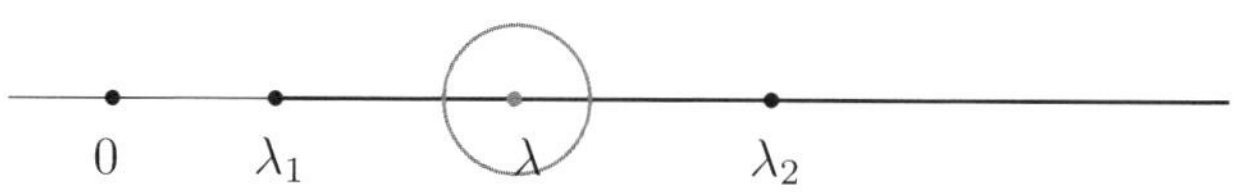

Fig. 1. Energy renormalization

ergy λ to be subtracted between the first and second transverse eigenvalue, cf. Fig. 1, and obtained a nontrivial limit determined by scattering properties of the corresponding "fat star". A drawback of this approach is that leads to energy spectrum unbounded from below which is a feature one tries to avoid in meaningful nonrelativistic models.

Most authors choose therefore the transverse threshold λ_1 as the energy to subtract. In such a case the limit is *generically trivial* giving disconnected edges with Dirichlet endpoints [32,33]. However, the limit can be nontrivial provided the tube system we start with has a *threshold resonance* [1,5,25]; a similar, closely related effect using finite star graphs was proposed in [12].

3.3. *A survey on Neumann network results*

Consider first for simplicity a finite connected graph M_0 with vertices v_k, $k \in K$ and edges $e_j \simeq I_j := [0, \ell_j]$, $j \in J$; the corresponding Hilbert space is thus $L^2(M_0) := \bigoplus_{j\in J} L^2(I_j)$. The form $u \mapsto \|u'\|^2_{M_0} := \sum_{j\in J} \|u'\|^2_{I_j}$ with $u \in W^{2,1}(M_0)$ is associated with the operator which acts as $-\Delta_{M_0} u = -u_j''$ and satisfies the *free* boundary conditions.

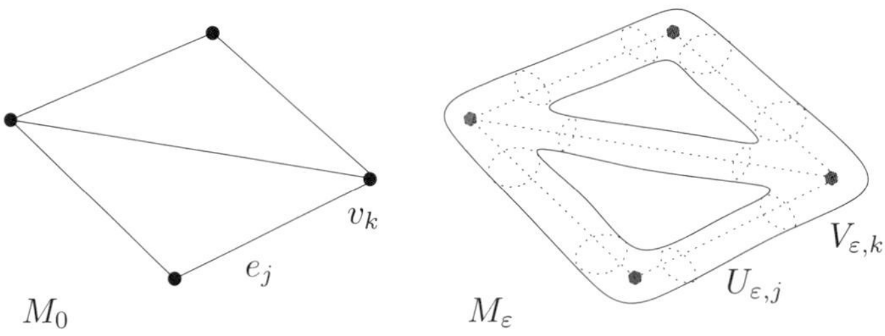

Fig. 2. Graph M_0 and fat graph M_ε

On the other hand, consider a Riemannian manifold X of dimension $d \geq 2$ and the corresponding space $L^2(X)$ w.r.t. volume $\mathrm{d}X$ equal to $(\det g)^{1/2}\mathrm{d}x$ in a fixed chart. For $u \in C^\infty_{\mathrm{comp}}(X)$ we set

$$q_X(u) := \|\mathrm{d}u\|^2_X = \int_X |\mathrm{d}u|^2 \mathrm{d}X\,, \quad |\mathrm{d}u|^2 = \sum_{i,j} g^{ij}\partial_i u\, \partial_j \overline{u}\,; \qquad (5)$$

the closure of this form is associated with the self-adjoint *Neumann* Laplacian Δ_X on the X. Let us stress that within this framework we can treat both "solid" tubes with the boundary at which Neumann condition is imposed, as well as "sleeve-type" manifolds without a boundary when the particle is supposed to live on the surface – cf. Fig. 2. This is made possible by the similarity of the transverse ground-state eigenfunction in both cases.

The "fat graphs" M_ε associated with the graph M_0 are all constructed from X by taking a suitable *ε-dependent family of metrics*. This the approach was used in [17]; in contrast to earlier work such as [31] one also need not assume that the network is embedded in a Euclidean space since only intrinsic geometrical properties are involved.

The analysis requires dissection of M_ε into a union of edge and vertex components, $U_{\varepsilon,j}$ and $V_{\varepsilon,k}$, respectively, with appropriate scaling properties,

- for edge regions we assume that $U_{\varepsilon,j}$ is diffeomorphic to $I_j \times F$ where F is a compact and connected manifold (with or without a boundary) of dimension $m := d - 1$
- for vertex regions we assume that the manifold $V_{\varepsilon,k}$ is diffeomorphic to an ε-independent manifold V_k

In this setting one can prove the following result [17]:

Theorem 3.1. *Under the stated assumptions we have eigenvalue convergence,* $\lambda_k(M_\varepsilon) \to \lambda_k(M_0)\,,\; k = 1, 2, \dots\,,$ *as* $\varepsilon \to 0$.

The shrinking limit thus leads to free boundary conditions only, but also in other respects the stated result is not particularly strong, for instance, in that it concerns the eigenvalue convergence in finite graphs only. One can do better: in Ref. 34 Olaf Post proved a norm-resolvent convergence $\Delta_{M_\varepsilon} \to \Delta_{M_0}$ as $\varepsilon \to 0+$ on generally infinite graphs under natural uniformity conditions analogous to those of used in Theorem 4.3, namely (i) existence of nontrivial bounds on vertex degrees and volumes, edge lengths, and the second Neumann eigenvalues at vertices, (ii) appropriate scaling (analogous to the described above) of the metrics at the edges and vertices. The involved operators act on different Hilbert spaces, of course, and the stated limiting relation makes sense with a suitable identification map which we will describe below.

Other extensions are possible. For graphs with semi-infinite "outer" edges, e.g., the problem typically exhibits series of *resonances*, and one may ask what happens with them if the graph is replaced by a family of "fat" graphs. Using *exterior complex scaling* in the "longitudinal" variable one can prove a convergence result for resonances as $\varepsilon \to 0$ [18]; the same is true for *embedded eigenvalues* of the graph Laplacian which may remain embedded or become resonances for $\varepsilon > 0$.

Hence we have a number of convergence results is available for squeezing limit of Neumann-type thin tube networks, however, the limiting operator corresponds always to *free* boundary conditions only. The question is whether one can do better.

4. Beyond the free coupling

4.1. *A graph inspiration*

It is obvious that one has to add a new feature to the approximating family to get more general results. Let us look how one can approximate δ coupling on graphs using families of scaled potentials. For simplicity we will consider again the n-edge star graph as in Sec. 2.2, however, we replace the Laplacian at the edges by a Schrödinger operator, $\psi_j \mapsto -\psi_j''+V_j\psi_j$. In order make the problem well-defined we have to impose requirement on the potential; we suppose that $V_j \in L^1_{\mathrm{loc}}(\mathbb{R}_+)\,,\ j=1,\dots,n$. If the boundary conditions at the vertex are (3) with the parameter $\alpha\in\mathbb{R}$ we get a self-adjoint operator which we denote as $H_\alpha(V)$. Let now the potential contain a scaled component,

$$W_{\varepsilon,j} := \frac{1}{\varepsilon}\, W_j\left(\frac{x}{\varepsilon}\right), \quad j=1,\dots,n\,, \tag{6}$$

then we have the following result [14]:

Theorem 4.1. *Suppose that the potentials* $V_j \in L^1_{\mathrm{loc}}(\mathbb{R}_+)$ *are below bounded and* $W_j\in L^1(\mathbb{R}_+)$ *for* $j=1,\dots,n$. *Then*

$$H_0(V+W_\varepsilon) \longrightarrow H_\alpha(V)$$

as $\varepsilon\to 0+$ *in the norm resolvent sense, with the coupling parameter defined as* $\alpha := \sum_{j=1}^n \int_0^\infty W_j(x)\,dx$.

Our aim is to "lift" this result to tube networks.

4.2. *Single vertex networks*

Consider first a star graph again. Let G have one vertex v and $\deg v$ adjacent edges of lengths $\ell_e \in (0,\infty]$. The corresponding Hilbert space is $L^2(G) := \bigoplus_{e\in E} L^2(I_e)$, the *decoupled* Sobolev space of order k is $W^{2,k}_{\max}(G) := \bigoplus_{e\in E} W^{2,k}(I_e)$ together with its natural norm.

Let $\underline{p}=\{p_e\}$ have components $p_e>0$ for $e\in E$; we introduce it because we want to consider squeezing limits also in the situation when the tubes have different cross sections. The Sobolev space associated with weight $\underline{p}$ is

$$W^{2,k}_{\underline{p}}(G) := \left\{ f\in W^{2,k}_{\max}(G):\ \underline{f}\in\mathbb{C}\underline{p}\right\}\,,$$

where $\underline{f} := \{f_e(0)\}$, in particular, if all the components are equal, $\underline{p} = (1, \ldots, 1)$, we arrive at the *continuous* Sobolev space $W^{2,1}(G) := W^{2,1}_{\underline{p}}(G)$.

Next we have to introduce operators on the graph. We start with the (weighted) *free* Hamiltonian Δ_G defined via the quadratic form $\mathfrak{d} = \mathfrak{d}_G$,

$$\mathfrak{d}(f) := \|f'\|^2_G = \sum_e \|e'\|^2_{I_e} \quad \text{and} \quad \operatorname{dom} \mathfrak{d} := W^{2,1}_{\underline{p}}(G)$$

for a fixed $\underline{p}$ (we drop the index $\underline{p}$); the form is a closed as related to the Sobolev norm $\|f\|^2_{W^{2,1}(G)} = \|f'\|^2_G + \|f\|^2_G$. The Hamiltonian with *$\delta$-coupling of strength q* is defined via the quadratic form $\mathfrak{h} = \mathfrak{h}_{(G,q)}$ given by

$$\mathfrak{h}(f) := \|f'\|^2_G + q(v)|f(v)|^2 \quad \text{and} \quad \operatorname{dom} \mathfrak{h} := W^{2,1}_{\underline{p}}(G)$$

Using standard Sobolev arguments one can show [19] that the δ-coupling is a "small" perturbation of Δ_G by estimating the difference $\mathfrak{h}(f) - \mathfrak{d}(f)$.

The manifold model of the "fat" graph is constructed as in the previous section. Given $\varepsilon \in (0, \varepsilon_0]$ we associate a d-dimensional manifold X_ε to the graph G in the following way: to the edge $e \in E$ and the vertex v we ascribe the Riemannian manifolds

$$X_{\varepsilon,e} := I_e \times \varepsilon Y_e \quad \text{and} \quad X_{\varepsilon,v} := \varepsilon X_v\,,$$

respectively, where εY_e is a manifold Y_e equipped with metric $h_{\varepsilon,e} := \varepsilon^2 h_e$ and $\varepsilon X_{\varepsilon,v}$ carries the metric $g_{\varepsilon,v} = \varepsilon^2 g_v$. As before, we use the ε-independent coordinates $(s, y) \in X_e = I_e \times Y_e$ and $x \in X_v$, so the radius-type parameter ε only enters via the Riemannian metric. Let us stress this includes the case of the ε-neighborhood of an embedded graph $G \subset \mathbb{R}^d$, but only up to a longitudinal error of order of ε; this problem can be dealt with again using an ε-dependence of the metric in the longitudinal direction.

The Hilbert space of the manifold model $L^2(X_\varepsilon)$ can be decomposed as

$$L^2(X_\varepsilon) = \bigoplus_e \big(L^2(I_e) \otimes L^2(\varepsilon Y_e)\big) \oplus L^2(\varepsilon X_v)$$

with the norm given accordingly by

$$\|u\|^2_{X_\varepsilon} = \sum_{e\in E} \varepsilon^{d-1} \int_{X_e} |u|^2 \mathrm{d}y_e \mathrm{d}s + \varepsilon^d \int_{X_v} |u|^2 \mathrm{d}x_v\,,$$

where $\mathrm{d}x_e = \mathrm{d}y_e\mathrm{d}s$ and $\mathrm{d}x_v$ denote the Riemannian volume measures associated to the (unscaled) manifolds $X_e = I_e \times Y_e$ and X_v, respectively. We also introduce the Sobolev space $W^{2,1}(X_\varepsilon)$ of order one defined conventionally as the completion of the space of smooth functions with compact support under the norm $\|u\|^2_{W^{2,1}(X_\varepsilon)} = \|\mathrm{d}u\|^2_{X_\varepsilon} + \|u\|^2_{X_\varepsilon}$.

Next we pass to operators on the manifold. The Laplacian Δ_{X_ε} on X_ε is given via its quadratic form $\mathfrak{d}_\varepsilon(u)$ equal to

$$\|\mathrm{d}u\|^2_{X_\varepsilon} = \sum_{e\in E} \varepsilon^{d-1} \int_{X_e} \Big(|u'(s,y)|^2 + \frac{1}{\varepsilon^2}|\mathrm{d}_{Y_e} u|^2_{h_e}\Big)\mathrm{d}y_e \mathrm{d}s + \varepsilon^{d-2} \int_{X_v} |\mathrm{d}u|^2_{g_v}\,\mathrm{d}x_v$$

where u' is the *longitudinal* derivative, $u' = \partial_s u$, and $\mathrm{d}u$ is the exterior derivative of u. Again, $\mathfrak{d}_\varepsilon$ is closed by definition. Adding a potential, we define the Hamiltonian H_ε as the operator associated with the form $\mathfrak{h}_\varepsilon = \mathfrak{h}_{(X_\varepsilon, Q_\varepsilon)}$ given by

$$\mathfrak{h}_\varepsilon = \|\mathrm{d}u\|^2_{X_\varepsilon} + \langle u, Q_\varepsilon u\rangle_{X_\varepsilon},$$

where the potential Q_ε is supported in the vertex region X_v only. Now we use graph result mentioned as an inspiration and choose

$$Q_\varepsilon(x) = \frac{1}{\varepsilon} Q(x),$$

where $Q = Q_1$ is a fixed bounded and measurable function on X_v. The reader may wonder that in comparison to (6) the factor ε^{-1} is missing in the argument, however, this is due to our choice to perform the squeezing by the change of the metric only.

One can establish the relative (form-)boundedness of H_ε with respect to the free operator Δ_{X_ε} : to a given $\eta \in (0,1)$ there is $\varepsilon_\eta > 0$ such that the form $\mathfrak{h}_\varepsilon$ is relatively form-bounded with respect to the free form $\mathfrak{d}_\varepsilon$, that is, there is $\widetilde{C}_\eta > 0$ such that

$$|\mathfrak{h}_\varepsilon(u) - \mathfrak{d}_\varepsilon(u)| \le \eta\, \mathfrak{d}_\varepsilon(u) + \widetilde{C}_\eta \|u\|^2_{X_\varepsilon}$$

whenever $0 < \varepsilon \le \varepsilon_\eta$ with explicit constants ε_η and $\widetilde{C}_\eta$. The latter are given explicitly in [19]; what is important that they are expressed in terms of the parameters of the model which we give below.

We have mentioned above that our operators acts in different spaces, namely the Hilbert spaces $\mathcal{H} = L^2(G)$ and $\widetilde{\mathcal{H}_\varepsilon} = L^2(X_\varepsilon)$ and their Sobolev counterparts, hence we need to define quasi-unitary operators to relate the graph and manifold Hamiltonians. For further purposes we denote

$$p_e := (\mathrm{vol}_{d-1} Y_e)^{1/2} \quad \text{and} \quad q(v) = \int_{X_v} Q\,\mathrm{d}x_v\,;$$

recall that we introduced the weights p_e to be able to treat situations when the tube cross sections Y_e are mutually different.

First we define the graph-to-manifold map, $J : \mathcal{H} \to \widetilde{\mathcal{H}_\varepsilon}$, by

$$Jf := \varepsilon^{-(d-1)/2} \bigoplus_{e\in E} (f_e \otimes \mathbb{1}_e) \oplus 0\,, \tag{7}$$

where $\mathbb{1}_e$ is the normalized eigenfunction of Y_e associated to the lowest (zero) eigenvalue, i.e. $\mathbb{1}_e(y) = p_e^{-1}$. Next introduce the following averaging operators

$$\fint_v u := \fint_{X_v} u \mathrm{d}x_v \quad \text{and} \quad \fint_e u(s) := \fint_{Y_e} u(s,\cdot)\mathrm{d}y_e$$

The map in the opposite direction, $J' : \widetilde{\mathcal{H}}_\varepsilon \to \mathcal{H}$, is given by the adjoint,

$$(J'u)_e(s) = \varepsilon^{(d-1)/2}\langle \mathbb{1}_e, u_e(s,\cdot)\rangle_{Y_e} = \varepsilon^{(d-1)/2} p_e \fint_e u(s)\,.$$

In an analogous way one can construct identification maps between the Sobolev spaces. They are need in the proofs but not for stating the result, hence we refer the reader to [19] for their explicit forms.

Using these notions in combination with an abstract convergence result of [34] one can then arrive at the following conclusions [19]:

Theorem 4.2. *As $\varepsilon \to 0$, we have*

$$\begin{aligned} \|J(H-z)^{-1} - (H_\varepsilon - z)^{-1}J\| &= \mathcal{O}(\varepsilon^{1/2}),\\ \|J(H-z)^{-1}J' - (H_\varepsilon - z)^{-1}\| &= \mathcal{O}(\varepsilon^{1/2}) \end{aligned}$$

for $z \notin [\lambda_0, \infty)$. Moreover, $\phi(\lambda) = (\lambda - z)^{-1}$ can be replaced by any measurable, bounded function converging to a constant as $\lambda \to \infty$ and being continuous in a neighborhood of $\sigma(H)$.

Corollary 4.1. *The spectrum of H_ε converges to $\sigma(H)$ uniformly on any finite energy interval, and the same is true for the essential spectra.*

Corollary 4.2. *For any $\lambda \in \sigma_{\mathrm{disc}}(H)$ there exists a family $\{\lambda_\varepsilon\}$ with $\lambda_\varepsilon \in \mathrm{disc}(H_\varepsilon)$ such that $\lambda_\varepsilon \to \lambda$ as $\varepsilon \to 0$, and moreover, the multiplicity is preserved. If λ is a simple eigenvalue with normalized eigenfunction ϕ, then there exists a family of simple normalized eigenfunctions $\{\phi_\varepsilon\}_\varepsilon$ of H_ε such that $\|J\phi - \phi_\varepsilon\|_{X_\varepsilon} \to 0$ holds as $\varepsilon \to 0$.*

4.3. *The general case*

So far we have talked for simplicity about the star-shaped graphs only. The same technique of "cutting" the graph and the corresponding manifold into edge and vertex regions works also in the general case. As a result of the analysis performed in Ref. 19 we get

Theorem 4.3. *Assume that G is a metric graph and X_ε the corresponding approximating manifold. If*

$$\sup_{v\in V} \frac{\operatorname{vol} X_v}{\operatorname{vol} \partial X_v} < \infty\,, \quad \sup_{v\in V} \|Q \restriction X_v\|_\infty < \infty\,, \quad \inf_{e\in E} \ell_e > 0$$

and

$$\inf_{v\in V} \lambda_2(v) > 0\,, \qquad \inf_{e\in E} \lambda_2(e) > 0\,,$$

where λ_2 *denotes the* second *Neumann eigenvalue in the appropriate manifold region, then the corresponding Hamiltonians* $H = \Delta_G + \sum_v q(v)\delta_v$ *and* $H_\varepsilon = \Delta_{X_\varepsilon} + \sum_v \varepsilon^{-1} Q_v$ *are* $\mathcal{O}(\varepsilon^{1/2})$*-close with the error depending only on the above indicated global constants.*

In this way we have managed to solve the problem for quantum graphs with δ-couplings under mild uniformity conditions.

5. Discontinuity at the vertex: the example of δ'_s

While the above results break the Kirchhoff restriction of the previous studies, they do not give a full answer; recall that the δ-couplings at a vertex v represent a one-parameter subfamily in the n^2 parameter family, $n = \deg v$, of all self-adjoint couplings. Let us now investigate the case of δ'_s as a prime example of coupling with functions discontinuous at the vertex.

5.1. *The idea of Cheon and Shigehara*

Our strategy will be the same as before, first we will construct an approximation on the graph itself and then we we will lift it to the manifold. The problem is not easy and its core is the question whether one can approximate the δ'-interaction on the line by means of (regular or singular) potentials. It was believed for a considerable time that this problem has no solution, until Cheon and Shigehara in the seminal paper [9] demonstrated a formal approximation by means of of three δ-interaction; a subsequent mathematical analysis [2,16] showed that it converges in fact in norm-resolvent sense.

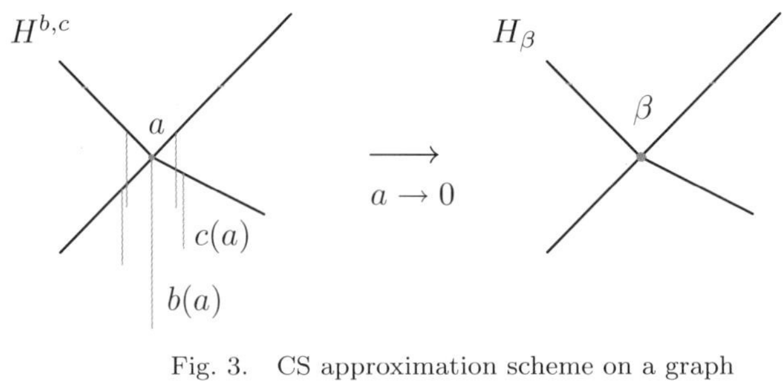

Fig. 3. CS approximation scheme on a graph

The idea can be extended to δ'_s-coupling on a graph. A scheme of the approximation is given on Fig. 3. One starts with a δ-coupling of strength $b(a)$ and adds δ-interactions of strength $c(a)$ at the graph edges; the parameter a is the distance of the additional interactions from the vertex. Core of the approximation lies in a suitable, a-dependent choice of the interaction strengths: we put

$$H^{\beta,a} := \Delta_G + b(a)\delta_{v_0} + \sum_e c(a)\delta_{v_e}, \quad b(a) = -\frac{\beta}{a^2}, \quad c(a) = -\frac{1}{a}$$

which corresponds to the quadratic form

$$\mathfrak{h}^{\beta,a}(f) := \sum_e \|f'_e\|^2 - \frac{\beta}{a^2}|f(0)|^2 - \frac{1}{a}\sum_e |f_e(a)|^2, \quad \operatorname{dom} \mathfrak{h}^a = W^{2,1}(G).$$

Then we have the following result [6]:

Theorem 5.1. $\|(H^{\beta,a} - z)^{-1} - (H^\beta - z)^{-1}\| = \mathcal{O}(a)$ *holds as* $a \to 0$ *for* $z \notin \mathbb{R}$, *where* $\|\cdot\|$ *is the operator norm on* $L^2(G)$.

Proof is by a direct computation. We note that the result is highly non-generic, both resolvents are strongly singular as $a \to 0$ but in the difference those singularities cancel.

5.2. *The convergence result*

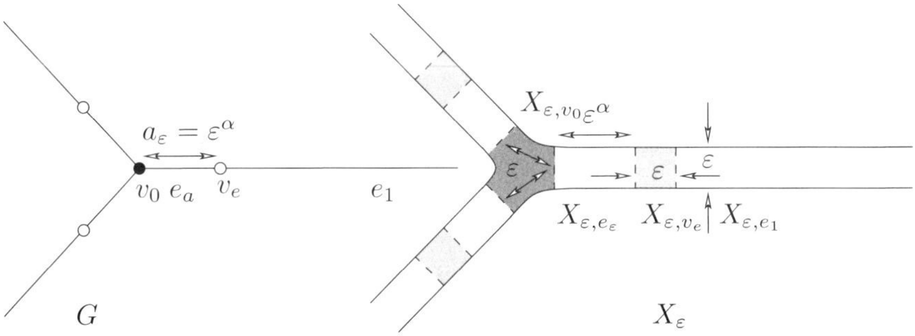

Fig. 4. Scheme of the lifting

Now we will lift the above graph approximation result to the manifold according to the scheme depicted on Fig. 4. For simplicity assume that the star graph in question is finite with all edges having the same length; without loss of generality we may put it equal to one. In contrast to the previous section we have two parameters to deal with, the tube width ε and the distance of the additional potentials; we choose $a = a_\varepsilon = \varepsilon^\alpha$ with $\alpha \in (0,1)$ to be specified later. The crucial point is the choice of the additional potentials. The simplest option is to assume that they are constant,

$$Q_{\varepsilon,v}(x) := \frac{1}{\varepsilon} \cdot \frac{q_\varepsilon(v)}{\operatorname{vol} X_v}, \qquad x \in X_v$$

so that $\int_{X_v} Q_{\varepsilon,v} \mathrm{d}x = \varepsilon^{-1} q_\varepsilon(v)$, where we put

$$q_\varepsilon(v_0) := b(\varepsilon^\alpha) = -\beta\varepsilon^{-2\alpha} \quad \text{and} \quad q_\varepsilon(v_e) := c(\varepsilon^\alpha) = -\varepsilon^{-\alpha}.$$

The corresponding manifold Hamiltonian and the respective quadratic form are then given by

$$H_\varepsilon^\beta = \Delta_{X_\varepsilon} - \varepsilon^{-1-2\alpha} \frac{\beta}{\operatorname{vol} X_{v_0}} \chi_{X_{v_0}} - \varepsilon^{-1-\alpha} \sum_{e \in E} \chi_{X_{v_e}},$$

where χ_X is the indicator function of the set X, and

$$\mathfrak{h}_\varepsilon^\beta(u) = \|\mathrm{d}u\|^2_{X_\varepsilon} - \varepsilon^{-1-2\alpha} \frac{\beta}{\operatorname{vol} X_{v_0}} \|u\|^2_{X_{\varepsilon,v_0}} - \varepsilon^{-1-\alpha} \sum_{e \in E} \|u\|^2_{X_{\varepsilon,v_e}},$$

respectively. Note that the unscaled vertex neighborhood X_{v_e} of each of the added vertices v_e has volume one by construction.

We employ again the identification operator (7). Using the same technique as in the δ case, one can prove the following result [19]:

Theorem 5.2. *Assume that* $0 < \alpha < 1/13$, *then*

$$\|(H_\varepsilon^\beta - i)^{-1} J - J(H^\beta - i)^{-1}\| \to 0$$

holds as the radius parameter $\varepsilon \to 0$.

Remark 5.1. The theorem has analogous corollaries as the δ-coupling result of the previous section, however, a caveat is due. If $\beta < 0$ the the spectrum of $H^{\beta,a}$ is uniformly bounded from below as $a \to 0$. If $\beta \geq 0$, on the other hand, the spectrum of $H^{\beta,a}$ is asymptotically unbounded from below, $\inf \sigma(H^{\beta,a}) \to -\infty$ as $a \to 0$. At the same time, for $\beta \geq 0$ the spectrum of the approximating operator H_ε^β is asymptotically unbounded from below, $\inf \sigma(H_\varepsilon^\beta) \to -\infty$ as $\varepsilon \to 0$. This fact, existence of eigenvalues which escape to $-\infty$ in the limit does not contradict the fact that the limit operator H^β is non-negative. Recall that the spectral convergence holds only for *compact* intervals $I \subset \mathbb{R}$, in particular, $\sigma(H^\beta) \cap I = \emptyset$ implies that $\sigma(H_\varepsilon^\beta) \cap I = \emptyset$ and $\sigma H^{\beta,\varepsilon} \cap I = \emptyset$ for $\varepsilon > 0$ small enough.

Remark 5.2. While it is easy to see that the parameter α in the approximation must less than one, the value $\frac{1}{13}$ is certainly not optimal.

6. Full solution on the graph level

6.1. *Going beyond δ and δ'_s*

The network approximations of the δ and δ'_s-couplings described in the two previous sections represent the present state of art in this question. One naturally asks whether one can extend the technique to other vertex couplings. Following the philosophy used here we should look first whether such approximations exist on the graph level.

The simplest extension covers the class of couplings invariant w.r.t. permutations of edges. It is a two-parameter family containing δ and δ'_s as particular cases; in the parametrization (2) its elements are characterized by matrices $U = a\mathcal{J} + bI$ with $|b| = 1$ and $|b + a \deg v| = 1$. The appropriate approximation in the spirit of Theorem 5.1 was worked out in Ref. 20; note that, as with δ and δ'_s, the problem again splits into a one-dimensional component in the subspace symmetric over the edges and its complement which is trivial from the coupling point of view.

If we relax the symmetry requirement things become more complicated. The first question is what we can achieve by modifications of the original Cheon and Shigehara idea, placing a finite number of properly scaled δ-interactions on each edge. The answer is given by the following claim [21]:

Proposition 6.1. *Let G be an n-edged star graph and $G(d)$ obtained by adding a finite number of δs at each edge, uniformly in d, at the distances $\mathcal{O}(d)$ as $d \to 0_+$. Suppose that these approximations yield conditions* (1) *with some A, B as $d \to 0$. The family which can be obtained in this way depends on $2n$ parameters if $n > 2$, and on three parameters for $n = 2$.*

It was demonstrated in Ref. 21 that a family with the maximum number of parameters given in the proposition can be indeed constructed.

In order to get a wider class one has to pass to a more general approximation. The idea put forward in Ref. 21 was *to change locally the graph topology* by adding new edges in the vicinity of the vertex whose lengths shrink to zero in the approximation. This yielded a family of couplings with $\binom{n+1}{2}$ parameters and real matrices A, B. To get a better result which will be described below one has to do two more things:

- together with adding edges in the vicinity of the vertex one has also to *remove parts of the graph* to optimize locally the approximating graph topology,
- furthermore, one has to add *local magnetic fields* described by suitable vector potentials to be able to get couplings which are not

invariant w.r.t. time reversion.

6.2. *An alternative unique parametrization*

In order to present the indicated approximation result we have to first introduce another form of the boundary conditions (1) derived in Ref. 7.

Theorem 6.1. *Consider a quantum graph vertex of degree n. If $m \le n$, $S \in \mathbb{C}^{m,m}$ is a self-adjoint matrix and $T \in \mathbb{C}^{m,n-m}$, then the relation*

$$\begin{pmatrix} I^{(m)} & T \\ 0 & 0 \end{pmatrix} \Psi' = \begin{pmatrix} S & 0 \\ -T^* & I^{(n-m)} \end{pmatrix} \Psi \tag{8}$$

expresses self-adjoint boundary conditions of the type (1)*. Conversely, for any self-adjoint vertex coupling there is an $m \le n$ and a numbering of the edges such that the coupling is described by the boundary conditions* (8) *with uniquely given matrices $T \in \mathbb{C}^{m,n-m}$ and self-adjoint $S \in \mathbb{C}^{m,m}$.*

Remark 6.1. As we have mentioned there are several unique forms of the conditions (1). Kuchment [30] splits the boundary value space using projections P, Q corresponding to Dirichlet, $P\Psi = 0$, and Neumann, $Q\Psi' = 0$, parts and the mixed conditions in the complement. It is easy to see that parts singled out correspond to eigenspaces of U corresponding to eigenvalues ∓ 1, respectively. The conditions (8) which one call the *ST-form* single out the eigenspace corresponding to -1. There is also an analogue of (8) symmetric w.r.t. the two singular parts, called *PQRS-form*, cf. Ref. 8.

6.3. *A general graph approximation*

In view of the above result one can put general self-adjoint boundary conditions into the form (8) renumbering the edges if necessary. We will now describe how those can be approximated by a family of graphs with locally changed topology and added magnetic fields. For notational purposes we adopt the following convention: the lines of the matrix T are indexed from 1 to m, the columns are indexed from $m+1$ to n.

The general vertex-coupling approximation, schematically depicted in Fig. 5, consists of the following sequence of steps:

- Take n halflines, each parametrized by $x \in \mathbb{R}_+$, with the endpoints denoted as V_j, and put a δ-coupling to the edges specified below with the parameter $v_j(d)$ at the point V_j for all $j = 1, \dots, n$.
- Some pairs V_j, V_k, $j \ne k$, of halfline endpoints are connected by edges of length $2d$, and the center of each such joining segment is

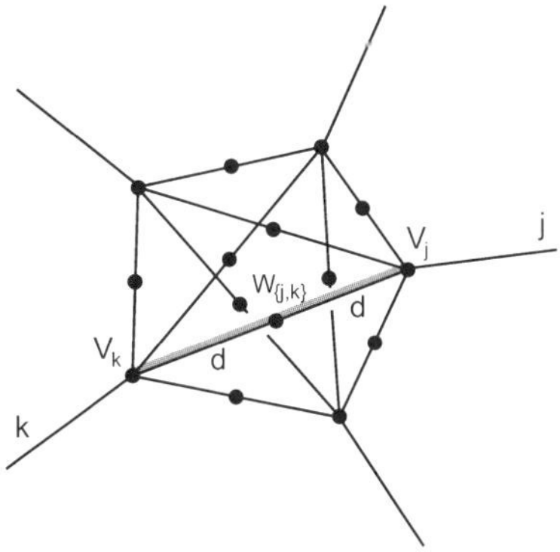

Fig. 5. The scheme of the approximation. All the inner links are of length $2d$, some may be missing. The grey line symbolizes the vector potential $A_{(j,k)}(d)$.

denoted as $W_{\{j,k\}}$. This happens if one of the following conditions is satisfied:

(a) $j = 1, \ldots, m$, $k \geq m+1$, and $T_{jk} \neq 0$ (or $j \geq m+1$, $k = 1, \ldots, m$, and $T_{kj} \neq 0$),

(b) $j, k = 1, \ldots, m$, and $S_{jk} \neq 0$ or $(\exists l \geq m+1)$ $(T_{jl} \neq 0 \wedge T_{kl} \neq 0)$.

- At each middle-segment point $W_{\{j,k\}}$ we place a δ interaction with a parameter $w_{\{j,k\}}(d)$. The connecting edges of length $2d$ are considered as consisting of two segments of length d, and on each of them the variable runs from zero at $W_{\{j,k\}}$ to d at the points V_j, V_k.
- On each connecting segment we put a vector potential of constant value between the points V_j and V_k. We denote its strength between the points $W_{\{j,k\}}$ and V_j as $A_{(j,k)}(d)$, and between the points $W_{\{j,k\}}$ and V_k as $A_{(k,j)}(d)$. It follows from the continuity that $A_{(k,j)}(d) = -A_{(j,k)}(d)$ for any pair $\{j,k\}$.

The choice of the dependence of $v_j(d)$, $w_{\{j,k\}}(d)$ and $A_{(j,k)}(d)$ on the parameter d is crucial for the approximation. In order to describe it we introduce the set $N_j \subset \{1, \ldots, n\}$ containing indices of all the edges that are joined to the j-th one by a connecting segment, i.e.

$$\begin{aligned} N_j =& \{k \leq m |\, S_{jk} \neq 0\} \cup \{k \leq m |\, (\exists l \geq m+1)(T_{jl} \neq 0 \wedge T_{kl} \neq 0)\} \\ & \cup \{k \geq m+1 |\, T_{jk} \neq 0\} \qquad \text{for } j \leq m \\ N_j =& \{k \leq m |\, T_{kj} \neq 0\} \qquad \text{for } j \geq m+1 \end{aligned} \tag{9}$$

We distinguish two cases regarding the indices involved:

Case I. First we suppose that $j = 1, \ldots, m$ and $l \in N_j \backslash \{1, \ldots, m\}$. Then the vector potential strength may be chosen as follows,

$$A_{(j,l)}(d) = \begin{cases} \frac{1}{2d} \arg T_{jl} & \text{if } \operatorname{Re} T_{jl} \geq 0\,, \\ \frac{1}{2d} (\arg T_{jl} - \pi) & \text{if } \operatorname{Re} T_{jl} < 0 \end{cases}$$

while for v_l and $w_{\{j,l\}}$ with $l \geq m+1$ we put

$$v_l(d) = \frac{1 - \#N_l + \sum_{h=1}^{m} \langle T_{hl} \rangle}{d} \qquad \forall l \geq m+1\,,$$

$$w_{\{j,l\}}(d) = \frac{1}{d} \left(-2 + \frac{1}{\langle T_{jl} \rangle} \right) \qquad \forall j, l \quad \text{indicated above}\,,$$

where the symbol $\langle \cdot \rangle$ here has the following meaning: if $c \in \mathbb{C}$, then

$$\langle c \rangle = \begin{cases} |c| & \text{if } \operatorname{Re} c \geq 0\,, \\ -|c| & \text{if } \operatorname{Re} c < 0\,. \end{cases}$$

We remark that the choice of $v_l(d)$ is not unique. This is related to the fact that for $m = \operatorname{rank} B < n$ the number of parameters of the coupling is reduced from n^2 to at most $n^2 - (n-m)^2$.

Case II. Suppose next that $j = 1, \ldots, m$ and $k \in N_j \cap \{1, \ldots, m\}$.

$$A_{(j,k)}(d) = \frac{1}{2d} \arg \left(d \cdot S_{jk} + \sum_{l=m+1}^{n} T_{jl} \overline{T_{kl}} - \mu\pi \right),$$

where $\mu = 0$ if

$$\operatorname{Re} \left(d \cdot S_{jk} + \sum_{l=m+1}^{n} T_{jl} \overline{T_{kl}} \right) \geq 0$$

and $\mu = 1$ otherwise. The functions $w_{\{j,k\}}$ are given by

$$w_{\{j,k\}} = -\frac{1}{d} \left(2 + \left\langle d \cdot S_{jk} + \sum_{l=m+1}^{n} T_{jl} \overline{T_{kl}} \right\rangle^{-1} \right)$$

and $v_j(d)$ for $j = 1, \ldots, m$ by

$$v_j(d) = S_{jj} - \frac{\#N_j}{d} - \sum_{k=1}^{m} \left\langle S_{jk} + \frac{1}{d} \sum_{l=m+1}^{n} T_{jl} \overline{T_{kl}} \right\rangle + \frac{1}{d} \sum_{l=m+1}^{n} (1 + \langle T_{jl} \rangle) \langle T_{jl} \rangle\,.$$

Having constructed the approximating graph we may now investigate how the corresponding Hamiltonian behaves in the limit $d \to 0$. We denote the Hamiltonian of the star graph G with the coupling (8) at the vertex as $H^{\rm star}$ and $H_d^{\rm approx}$ will stand for the approximating operators constructed above; the symbols $R^{\rm star}(z)$ and $R_d^{\rm approx}(z)$, respectively, will denote the corresponding resolvents. Needless to say, the operators act on different spaces: $R^{\rm star}(z)$ on $L^2(G)$, while $R_d^{\rm approx}(k^2)$ acts on $L^2(G_d)$, where G_d is the Cartesian sum $G \oplus (0,d)^{\sum_{j=1}^{n} N_j}$. To compare the resolvents, we identify $R^{\rm star}(z)$ with the orthogonal sum

$$R_d^{\rm star}(z) = R^{\rm star}(z) \oplus 0\,,$$

which acts as zero on the added edges. Comparing the resolvents is in principle a straightforward task, however, computationally rather demanding. Performing it we arrive at the following conclusion [7] which provides us with the full answer to our problem on the graph level:

Theorem 6.2. *In the described setting, the operator family $H_d^{\rm approx}$ converges to $H^{\rm star}$ in the norm-resolvent sense as $d \to 0$.*

Remark 6.2. There are various modifications of the approximation described above. In Ref. 11, for instance, the δ-interactions on the connecting segments have been replaced by varying lengths of those segments; the construction is there performed for scale-invariant vertex couplings, i.e. the conditions (8) with $S = 0$ and any T.

7. Concluding remarks

We have demonstrated how one can use scaled Schrödinger operators to approximate quantum graph Hamiltonians with different vertex couplings. We have worked out the argument for the δ and δ_s'-couplings. On the graph level we have provided a full solution of the problem.

This suggests how one could proceed further. The approximating graph of the previous section has to be replaced by a network with a fat edge width ε and the δ-couplings by constant potentials of the appropriated strength at the segment of fat edge of length ε. Similarly the Laplacian is to be replaced by magnetic Laplacian on the added edges, the halflength of which is set to be $d = \varepsilon^\alpha$. We call the resulting magnetic Schrödinger operator H_ε^ω, where ω stands now for the appropriate family of parameters, and by H^ω the corresponding limiting operator on the graph itself.

Conjecture 7.1. *If $\alpha > 0$ is sufficiently small the approximation result analogous to Theorem 5.2 is valid in the described setting for any vertex coupling* (8) *with the same identification operator J.*

Scaled potentials are not the only way how approximations of nontrivial vertex couplings can be constructed. There are other possibilities such as replacement of Neumann by suitable position dependent boundary conditions – for a survey of fresh results we recommend Ref. 35. A more difficult question is whether one can accomplish the goal by geometric means. A naive inclusion of curvature-induced potentials does not give the answer [19] a more elaborate approach has to be sought.

Let us finally comment on possible physical application of the results surveyed here. Thinking of the network as of a model of a semiconductor system, one can certainly vary the material parameters. Doping the network locally changes the Fermi energy at the spot creating effectively a potential well or barrier. From the practical point of view, however, this does not help much because our approximations need potentials which get stronger with the diminishing tube width ε.

A more promising alternative is to use external fields. In experiment with nanosystems one often adds "gates", or local electrodes, to which a voltage can be applied. In this way one can produce local potentials fitting into our approximation scheme, without material restrictions. This opens an rather intriguing possibility of creating quantum graphs with the vertex coupling controllable by an experimentalist.

Acknowledgments

The author enjoyed the pleasure of collaboration with Taksu Cheon, Olaf Post and Ondřej Turek which led to the results reviewed here. The research was supported in part by the Czech Ministry of Education, Youth and Sports within the project LC06002.

References

1. S. Albeverio, C. Cacciapuoti, D. Finco: Coupling in the singular limit of thin quantum waveguides, *J. Math. Phys.* **48** (2007), 032103.
2. S. Albeverio, L. Nizhnik: Approximation of general zero-range potentials, *Ukrainian Math. J.* **52** (2000), 664–672.
3. J. Bolte, J. Harrison: Spectral statistics for the Dirac operator on graphs, *J. Phys. A: Math. Gen.* **36** (2003), 2747–2769.
4. W. Bulla, T. Trenckler: The free Dirac operator on compact and non-compact graphs, *J. Math. Phys.* **31** (1990), 1157–1163.
5. C. Cacciapuoti, P. Exner: Nontrivial edge coupling from a Dirichlet network squeezing: the case of a bent waveguide, *J. Phys. A: Math. Theor.* **40** (2007), L511–L523.

6. T. Cheon, P. Exner: An approximation to δ' couplings on graphs, *J. Phys. A: Math. Gen.* **37** (2004), L329–L335.
7. T. Cheon, P. Exner, O. Turek: Approximation of a general singular vertex coupling in quantum graphs, *Ann. Phys.* **325** (2010), 548–578.
8. T. Cheon, P. Exner, O. Turek: Tripartite connection condition for quantum graph vertex, *Phys. Lett.* **A375** (2010), 113–118.
9. T. Cheon, T. Shigehara: Realizing discontinuous wave functions withrenormalized short-range potentials, *Phys. Lett.* **A243** (1998), 111–116.
10. T. Cheon1, I. Tsutsui, T. Fülőp: Quantum abacus, *Phys. Lett.* **A330** (2004), 338–342.
11. T. Cheon, O. Turek: Fülőp-Tsutsui interactions on quantum graphs, *Phys. Lett.* **A374** (2010), 4212-4221.
12. G.F. Dell'Antonio, E. Costa: Effective Schrödinger dynamics on ε-thin Dirichlet waveguides via quantum graphs. I. Star-shaped graphs, *J. Phys. A. Math. Theor.* **43** (2010), 474014.
13. P. Exner: Contact interactions on graph superlattices, *J. Phys. A: Math. Gen.* **29** (1996), 87–102.
14. P. Exner: Weakly coupled states on branching graphs, *Lett. Math. Phys.* **38** (1996), 313–320.
15. P. Exner, J.P. Keating, P. Kuchment, T. Sunada, A. Teplayaev (eds.): *Analysis on graphs and its applications*, Proc. Symp. Pure Math., vol. 77 (Amer. Math. Soc., Providence, R.I., 2008).
16. P. Exner, H. Neidhardt, V.A. Zagrebnov: Potential approximations to δ': an inverse Klauder phenomenon with norm-resolvent convergence, *Comm. Math. Phys.* **224** (2001), 593–612.
17. P. Exner, O. Post: Convergence of spectra of graph-like thin manifolds, *J. Geom. Phys.* **54** (2005), 77–115.
18. P. Exner, O. Post: Convergence of resonances on thin branched quantum wave guides, *J. Math. Phys.* **48** (2007), 092104.
19. P. Exner, O. Post: Approximation of quantum graph vertex couplings by scaled Schrödinger operators on thin branched manifolds, *J. Phys.* **A42** (2009), 415305.
20. P. Exner, O. Turek: Approximations of permutation-symmetric vertex couplings in quantum graphs, in *Quantum graphs and their applications*, Contemp. Math., vol. 415, pp. 109–120 (Amer. Math. Soc., Providence, RI, 2006).
21. P. Exner, O. Turek: Approximations of singular vertex couplings in quantum graphs, *Rev. Math. Phys.* **19** (2007), 571–606.
22. M.I. Freidlin, A.D. Wentzell: Diffusion processes on graphs and theaveraging principle, *Ann. Probab.* **21** (1993), 2215–2245.
23. T. Fülőp, I. Tsutsui: A free particle on a circle with point interaction, *Phys. Lett.* **A264** (2000), 366–374.
24. V.I. Gorbachuk, M.L. Gorbachuk: *Boundary value problems for operator differential equations* (Kluwer, Dordrecht, 1991).
25. D. Grieser: Spectra of graph neighborhoods and scattering, *Proc. London Math. Soc.* **97** (2008), no. 3, 718–752.
26. M. Harmer: Hermitian symplectic geometry and extension theory, *J. Phys.*

A: Math. Gen. **33** (2000), 9193–9203.
27. O. Hul et al.: Experimental simulation of quantum graphs by microwave networks, *Phys. Rev.* **E69** (2004), 056205.
28. V. Kostrykin, R. Schrader: Kirchhoff's rule for quantum wires, *J. Phys. A: Math. Gen.* **32** (1999), 595–630.
29. V. Kostrykin, R. Schrader: Kirchhoff's rule for quantum wires. II: The inverse problem with possible applications to quantum computers, *Fortschr. Phys.* **48** (2000), 703–716.
30. P. Kuchment: Quantum graphs: I. Some basic structures, *Waves in Random Media* **14** (2004), S107–S128.
31. P. Kuchment, H. Zeng: Convergence of spectra of mesoscopic systems collapsing onto a graph, *J. Math. Anal. Appl.* **258** (2001), 671–700.
32. S. Molchanov, B. Vainberg: Scattering solutions in networks of thin fibers: small diameter asymptotics, *Comm. Math. Phys.* **273** (2007), 533–559.
33. O. Post: Branched quantum wave guides with Dirichlet boundary conditions: the decoupling case, *J. Phys. A: Math. Gen.* **38** (2005), 4917–4931.
34. O. Post: Spectral convergence of quasi-one-dimensional spaces, *Ann. H. Poincaré* **7** (2006), 933–973.
35. O. Post: Convergence results for thick graps, in *Mathematical Results in Quantum Physics*, Proceedings of QMath11 Conference (P. Exner. ed.), (World Scientific, Singapore, 2011).
36. J. Rubinstein, M. Schatzman: Variational problems on multiply connected thin strips. I. Basic estimates and convergence of the Laplacian spectrum, *Arch. Rat. Mech. Anal.* **160** (2001), 271–308.
37. K. Ruedenberg, C.W. Scherr: Free-electron network model for conjugated systems, I. Theory, *J. Chem. Phys.* **21** (1953), 1565–1581.
38. Y. Saito: The limiting equation for Neumann Laplacians on shrinking domains, *Electr. J. Diff. Eq.* **31** (2000), 25 pp.

COMPLEXITY LEADS TO RANDOMNESS IN CHAOTIC SYSTEMS

RENÉ LOZI

Laboratory J. A. Dieudonné, UMR CNRS 6621, University of Nice Sophia-Antipolis, Parc Valrose, 06108 Nice Cedex 02, France

and

Institut Universitaire de Formation des Maîtres Célestin Freinet-académie de Nice, University of Nice-Sophia-Antipolis, 89 avenue George V, 06046 Nice Cedex 1, France

Abstract— Complexity of a particular coordinated system is the degree of difficulty in predicting the properties of the system if the properties of the system's correlated parts are given. The coordinated system manifests properties not carried by individual parts. The subject system can be said to emerge without any "guiding hand". In systems theory and science, emergence is the way complex systems and patterns arise out of a multiplicity of relatively simple interactions. Emergence is central to the theories of integrative levels and of complex systems. The emergent property of the ultra weak multidimensional coupling of p 1-dimensional dynamical chaotic systems for which complexity leads from chaos to randomness has been recently pointed out.

Pseudorandom or chaotic numbers are nowadays used in many areas of contemporary technology such as modern communication systems and engineering applications. Efficient Chaotic Pseudo Random Number Generators (CPRNG) have been recently introduced. They use the ultra weak multidimensional coupling of p *1*-dimensional dynamical systems which preserves the chaotic properties of the continuous models in numerical experiments. Together with chaotic sampling and mixing processes, the complexity of ultra weak coupling leads to families of CPRNG which are noteworthy. In this paper we improve again these families using a double threshold chaotic sampling instead of a single one. A window of emergence of randomness for some parameter value is numerically displayed. Moreover we emphasize that a determining property of such improved CPRNG is the high number of parameters used and the high sensitivity to the parameters value which allows choosing it as cipher-keys.

1. Introduction

Characterizing complexity is not easy and there are in science a number of approaches to do it. Many definitions tend to postulate or assume that complexity expresses a condition of numerous elements in a system and numerous forms of relationships among the elements. Some others definitions relate to the algorithmic basis for the expression of a complex phenomenon or model or mathematical expression. Warren Weaver [1] has posited that the (organized) complexity of a particular system is the degree of difficulty in predicting the properties of the system if the properties of the system's parts are given. In Weaver's view organized complexity, resides in nothing else than the non-random, or correlated, interaction between the parts. These correlated relationships create a differentiated structure which can, as a system, interact with other systems. The coordinated system manifests properties not carried by individual parts. The organized aspect of this form of complexity versus other systems than the subject system can be said to emerge without any "guiding hand". The number of parts does not have to be very large for a particular system to have emergent properties.

In systems theory and science, emergence is the way complex systems and patterns arise out of a multiplicity of relatively simple interactions. Emergence is central to the theories of integrative levels and of complex systems (M. A. Aziz-Alaoui *et al.* [2]).

In this paper we use the emergent property of the ultra weak multidimensional coupling of p 1-dimensional dynamical chaotic systems for which complexity leads from chaos to randomness. Efficient Chaotic Pseudo Random Number Generators (CPRNG) have been recently introduced (Lozi [3, 4, 5, 6]) and their properties analyzed (Hénaff *et al.* [7, 8, 9, 10]). The idea of applying discrete chaotic dynamical systems, intrinsically, exploits the property of extreme sensitivity of trajectories to small changes of initial conditions. The ultra weak multidimensional coupling of p *1*-dimensional dynamical systems preserves the chaotic properties of the continuous models in numerical experiments. The process of chaotic sampling and mixing of chaotic sequences, which is pivotal for these families, works perfectly in numerical simulation when floating point (or any multi-precision) numbers are handled by a computer.

It is noteworthy that these families of ultra weakly coupled maps are more powerful than the usual formulas used to generate chaotic sequences mainly because only additions and multiplications are used in the computation process; no division being required. Moreover the computations are done using floating point or double precision numbers, allowing the use of the powerful Floating Point Unit (FPU) of the modern microprocessors (built by both Intel and Advanced Micro Devices (AMD)). In addition, a large part of the computations can be parallelized taking advantage of the multicore microprocessors which appear on the market of laptop computers.

In this paper we improve the properties of these families using a double threshold chaotic sampling instead of a single one. The genuine map f used as one-dimensional dynamical systems to generate them is henceforth perfectly hidden. A window of emergence of randomness for some parameter value is numerically displayed.

A determining property of such improved CPRNG is the high number of parameters used ($p\times(p-1)$ for p coupled equations) which allows to choose it as cipher-keys due to the high sensitivity to the parameters values. We call these families multi-parameter chaotic pseudo-random number generators (M-p CPRNG).

Several applications can be found for these families as for example producing Gaussian noise, computing hash function or in chaotic cryptography.

In Sec. 2 we review some of the most popular chaotic mappings in low dimension in the scope of their use in numerical algorithms and PRNG.

In Sec. 3 we improve the properties of ultra weak multidimensional coupled of p 1-dimensional dynamical chaotic systems using a double threshold chaotic sampling instead of a single one.

In Sec. 4 we describe the emergence of randomness from complexity in a particular window of parameter value. We point out the parameter sensitivity in Sec. 5, with some applications of the M-p CPRNG and we give a conclusion in Sec. 6.

2. Discrete Dynamical Systems in Low Dimension

Chaotic dynamical systems in low dimension are often used since their discovery in the 70' in order to generate chaotic numbers, because they are very easy to implement in numerical algorithms [11]. However, as we point out in this section the computation of numerical approximation of their periodic orbits leads to very different results from the theoretical ones. Then they are unable to generate Pseudo Random Numbers (PRN). We review some of the most used maps in dimension from 1 to 3 in this scope.

2.1. 1-Dimensional Chaotic Dynamical Systems

2.1.1. Logistic map

The very well known logistic map $g_a : [0,1] \to [0,1]$ is simply defined as

$$g_a(x) = ax(1-x) \tag{1}$$

and generally considered for $a \in [0,4]$ (see Fig. 1). It is associated to the discrete dynamical system [12]

$$x_{n+1} = g_a(x_n) \tag{2}$$

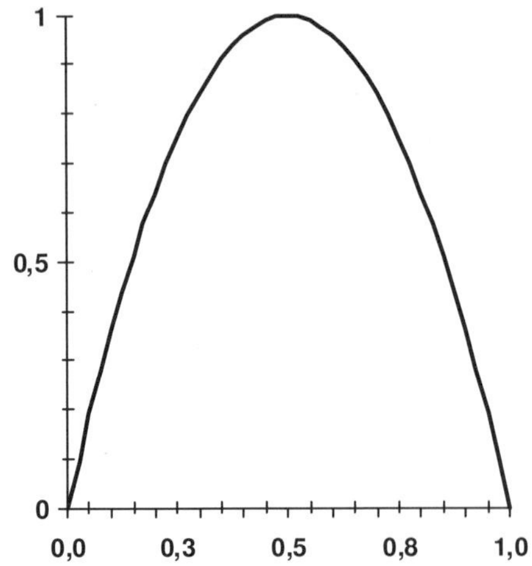

Figure 1. Graph of the logistic map for $a = 4$.

This dynamical system which has excellent ergodic properties on the real interval $[0,1]$ has been extensively studied especially by R. M. May [13], and J. Feigenbaum [14] who introduced what is now called the Feigenbaum's constant $\delta = 4.669201609102990671853203 82...$ explaining by a new theory (period doubling bifurcation) the onset of chaos.

For every value of a there exist two fixed points: $x = 0$ which is always unstable and $x = \frac{a-1}{a}$ which is stable for $a \in]1,3[$ and unstable for $a \in]0,1[\cup]1,4[$.

When $a = 4$, the system is chaotic. The set $\left\{ \frac{5-\sqrt{5}}{8}, \frac{5+\sqrt{5}}{8} \right\} = \{0.3454915028, 0.9045084972\}$ is the period-2 orbit. In fact there exist infinity of periodic orbits and infinity of periods. This dynamical system possesses an invariant measure $P(x) = \frac{1}{\pi\sqrt{x(1-x)}}$ (see Fig. 2).

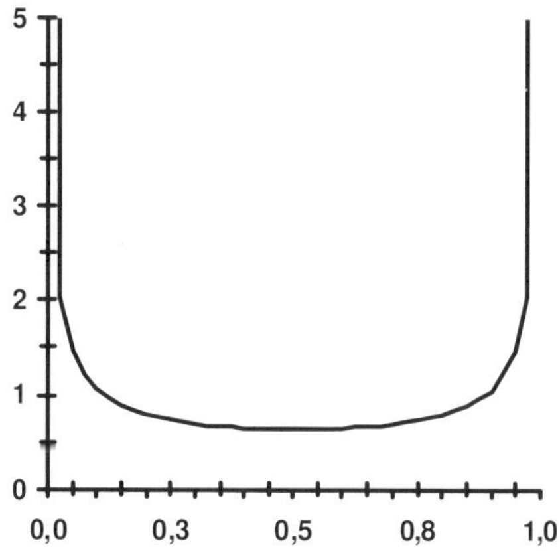

Figure 2. Graph of the invariant measure $P(x)$ of the logistic map for $a = 4$.

2.1.2. Numerical approximation of the logistic map

In order to compute longer periodic orbits the use of computer is required, as it is equivalent to find roots of polynomial equation of degree greater than 4 for which Galois theory teaches that no closed formula is available. However, numerical computation uses ordinarily double precision numbers (IEEE-754) so that the working interval contains roughly 10^{16} representable points. Doing such a computation in Eq. (2) with 1,000 randomly chosen initial guesses, 596, *i.e.*, the majority, converge to the unstable fixed point $x = 0$, and 404 converge to a cycle of period 15,784,521. (see Table 1) [15].

Table 1. Coexisting periodic orbits found using 1,000 random initial points for double precision numbers

Period	Orbit	Relative Basin size
1	$x = 0$ unstable fixed point	596 over 1,000
15,784,521	Scattered over the interval	404 over 1,000

Thus, in this case at least, the very long-term behaviour of numerical orbits is, for a substantial fraction of initial points, in flagrant disagreement with the true behaviour of typical orbits of the original smooth logistic map.

In others numerical experiments we have performed, the computer working with fixed finite precision is able to represent finitely many points in the interval in question. It is probably good, for purposes of orientation, to think of the case where the representable points are uniformly spaced in the interval. The true logistic map is then approximated by a discretized map, sending the finite set of representable points in the interval to itself.

Describing the discretized mapping exactly is usually complicated, but it is roughly the mapping obtained by applying the exact smooth mapping to each of the discrete representable points and "rounding" the result to the nearest representable point. In our experiments [16, 17], uniformly spaced points in the interval with several order of discretization (ranging from 9 to 2,001 points) are involved, the results for 2,000 and 2,001 points are displayed in Table 2. In each experiment the questions addressed are:

• how many periodic cycles are there and what are their periods ?
• how large are their respective basins of attraction, *i.e.* , for each periodic cycle, how many initial points give orbits with eventually land on the cycle in question ?

Table 2. Coexisting periodic orbits for the discretization with regular meshes of $N = 2{,}000 \text{ and } 2{,}001$ points.

N	Period	Orbit	Relative Basin Size
2,000	1	{0}	2 over 2,000
2,000	2	{1,499}	14 over 2,000
2,000	2	{691;1,808}	138 over 2,000
2,000	3	{276;1,221;1,900}	6 over 2,000
2,000	8	{3;11;43;168;615;1,703.1,008.1,998}	1,840 over 2,000
2,001	1	{0}	5 over 2,001
2,001	1	{1500}	34 over 2,001
2,001	2	{91; 1809}	92 over 2,001
2,001	8	{3;11;43;168;615;1,703;1,011;1,999}	608 over 2,001
2,001	18	{35;137;510;1,519;1,461;1,574;...}	263 over 2,001
2,001	25	{27;106;401;1,282;1,840;588;...}	1,262 over 2,001

The existence of very short periodic orbit (see Table 1), the existence of a non constant invariant measure (see Fig. 2) and the easily recognized shape of the function in the phase space (x_n, x_{n+1}) avoid the use of the logistic map as a PRN generator. However, its very simple implementation in computer program led some authors to use it as a base of cryptosytem [18, 19].

2.1.3. Symmetric tent map

Another often studied dynamical system is defined by the symmetric tent map on the interval $J = [-1, 1]$, $f_a : J \to J$

$$f_a(x) = 1 - a|x| \tag{3}$$

$$x_{n+1} = f_a(x_n) \tag{4}$$

Despite its simple shape (see Fig. 3), it has several interesting properties. First, when the parameter value $a = 2$, the system possesses chaotic orbits. Because of its piecewise-linear structure, it is easy to find those orbits explicitly. More, owing to its simple definition, the symmetric tent map's shape under iteration is very well understood. The invariant measure is the Lebesgue measure. Finally, and perhaps the most important, the tent map is conjugate to the logistic map, which in turn is conjugate to the Hénon map for small values of b [12].

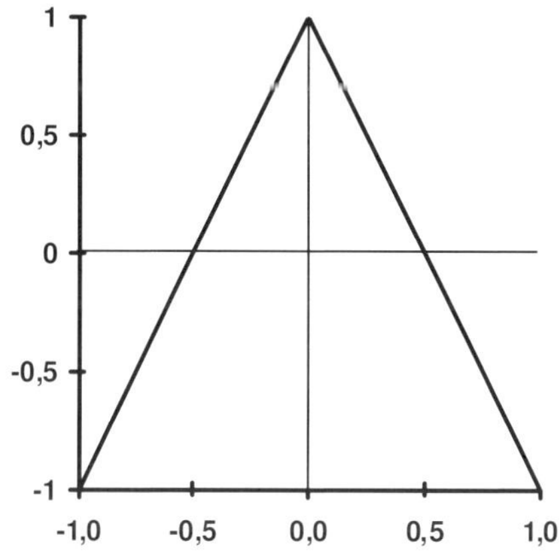

Figure 3. Graph of the symmetric tent map on J for $a = 2$.

However the symmetric tent map is dramatically numerically instable: Sharkovskiĭ's theorem applies for it [20]. When $a = 2$ there exists a period three orbit, which implies that there is infinity of periodic orbits. Nevertheless the orbit of almost every point of the interval J of the discretized tent map converges to the (unstable) fixed point $x = -1$ (this is due to the binary structure of floating points) and there is no numerical attracting periodic orbit [11].

The numerical behaviour of iterates with respect to chaos is worse than the numerical behaviour of iterates of the approximated logistic map. This is why the tent map is never used to generate numerically chaotic numbers. However in Sec. 3 we will show that it is possible to preserve its chaotic properties when several logistic maps are ultra weakly coupled.

2.2. 2-Dimensional System Chaotic Dynamical Systems

2.2.1. Hénon map

In order to study numerically the properties of the Lorenz attractor [11], M. Hénon in 1976 [21] introduced a simplified model of the Poincaré map [12] of this attractor. The Lorenz attractor being in dimension 3, the corresponding Poincaré map is a map from $\mathbb{R}^2$ to $\mathbb{R}^2$. The Hénon map is then also defined in dimension 2 as

$$F : \begin{pmatrix} x \\ y \end{pmatrix} \rightarrow \begin{pmatrix} y + 1 - ax^2 \\ bx \end{pmatrix} \tag{5}$$

It is associated to the dynamical system

$$\begin{cases} x_{n+1} = y_n + 1 - ax_n^{\ 2} \\ y_{n+1} = bx_n \end{cases} \tag{6}$$

For the parameter value $a = 1.4$, $b = 0.3$ Hénon pointed out numerically that there exists an attractor with fractal structure (see Fig. 4). This was the first example of strange attractor (previously introduced by D. Ruelle and F. Takens [22]) for a mapping defined by an analytic formula.

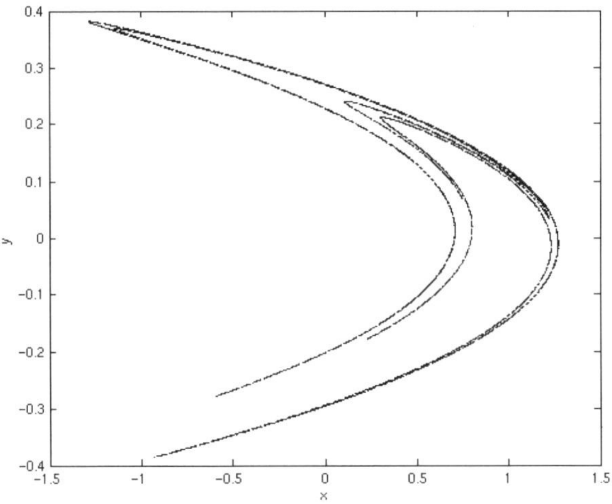

Figure 4. The strange attractor of the Hénon map for $a = 1.4$ and $b = 0.3$.

Nowadays hundreds of research papers have been published on this prototypical map in order to fully understand its innermost structure. However as in dimension 1, there is a discrepancy between the mathematical properties of this map in the plane $\mathbb{R}^2$ and the numerical computations done using (IEEE-754) double precision numbers.

If we call Megaperiodic orbits [23], those whose length of the period belongs to the interval of natural numbers $\left[10^6, 10^9\right[$ and Gigaperiodic orbits, those whose length of the period belongs to the interval $\left[10^9, 10^{12}\right[$, Hénon map possesses Gigaperiodic orbits. On a Dell computer with a Pentium IV microprocessor running at the frequency of 1.5 Gigahertz, using a Borland C compiler and computing with ordinary (IEEE-754) double precision numbers, one can find for $a = 1.4$ and $b = 0.3$ one attracting period of length 3,800,716,788 *i.e.* two hundred forty times longer than the longest period of the one-dimensional logistic map (see Table 1). This periodic orbit (we call it here Orbit 1) is numerically slowly attracting. Starting with the initial value

$(x_0, y_0)_1 = (-0.35766, 0.14722356)$ one obtains:

$$(x_{11,574,730,767}, y_{11,574,730,767})_1 = (x_{15,375,447,555}, y_{15,375,447,555})_1$$
$$= (1.27297361350955662, -0.0115735710153616837)$$

The length of the period is obtained subtracting

$$15,375,447,555 - 11,574,730,767 = 3,800,716,788.$$

However this periodic orbit is not unique: starting with the initial value $(x_0, y_0)_2 = (0.4725166, 0.25112222222356)$ the following periodic orbit (which is a Megaperiodic orbit of period 310,946,608 (Orbit 2)) is computed.

$$(x_{12,935,492,515}, y_{12,935,492,515})_2 = (x_{13,246,439,123}, y_{13,246,439,123})_2$$
$$= (1.27297361350865113, -0.0115734779561870744)$$

This orbit can be reached more rapidly starting form the other initial value

$(x_0, y_0) = (0.881877775591, 0.0000322222356)$, then

$$(x_{4,459,790,707}, y_{4,459,790,707}) = (1.27297361350865113, -0.0115734779561870744).$$

It is possible that some others periodic orbits coexist with both Orbit 1 and Orbit 2.

The comparison between Orbit 1 and Orbit 2 gives a perfect idea of the sensitive dependence on initial conditions of chaotic attractors: Orbit 1 passes through the point (**1.27297361350**955662, – **0.0115735**710153616837) and Orbit 2 passes through the point (**1.27297361350**865113, – **0.0115734**779561870744). The same digits of these points are bold printed, they are very close.

Nevertheless, as displayed in Fig. 4 the orbit are not uniformly distributed on the phase space, then it is not possible to use this map as a PRN generator.

Beside the problem of PRN generator, logistic and Hénon maps are recently used together with a secret key, by N. Pareek *et al.* [24], in order to build a chaotic block cipher which is extremely robust, due to the excellent confusion and diffusion properties of these maps. The results of the statistical analysis show that the chaotic cipher possesses all features needed for a secure system and useable for the security of communication system.

L. dos Santos Coelho *et al.* [25], introduced a chaotic particle swarm optimisation (PSO) which is a population-based swarm intelligence algorithm driven by the stimulation of a social psychological metaphor instead of the survival of the fittest individual. Based on the chaotic systems theory (using Hénon map sequences which increase its convergence rate and resulting precision) the novel PSO combined with an implicit filtering allows solving economic dispatch problems.

2.2.2. Lozi map

The Lozi map [26] is a linearized version of the Hénon map, built in order to simplify the computations, mainly because it is possible to compute explicitly any periodic orbits solving a linear system. It is defined as

$$\begin{cases} x_{n+1} = y_n + 1 - a|x_n| \\ y_{n+1} = bx_n \end{cases} \tag{7}$$

or equivalently

$$x_{n+1} = 1 - a|x_n| + bx_{n-1} \tag{8}$$

For $a = 1.7$ and $b = 0.5$ there exists a strange attractor. The particularity of this strange attractor is that it has been rigorously proved by Misiurewicz in 1980 [27].

In the same conditions of computation as for Hénon map, running the computation during nineteen hours, one can find a Gigaperiodic attracting orbit of period 436,170,188,959 more than one hundred times longer than the period of Orbit 1 found for the Hénon map.

Starting with (x_0, y_0) = (0.88187777591, 0.0000322222356) one obtains

$(x_{686,295,403,186}, y_{686,295,403,186}) = (x_{250,125,214,227}, y_{250,125,214,227})$
$= (1.34348444450739479, -2.07670041497986548.\ 10^{-7})$.

There is a transient regime before the orbit is reached. It seems that there is no periodic orbit with a smaller length. This could be due to the quasihyperbolic nature of the attractor. However, the orbit-shifted shadowing property of Lozi map (and generalized Lozi map), which is the property which ensures that pseudo-orbits of a homeomorphism f can be traceable by actual orbits even if rounding errors in computing are not inevitable has been recently proved [28].

Hence this attractor is very efficient, in order to generate chaotic numbers without repetition for standard simulation using either the first or the second component. However they are not equally distributed on the plane (see Fig. 5). The non constant density forbids its direct use as a PRN generator. Nevertheless there are some U.S. patents for "method of generating pseudo-random numbers in an electronic device, and a method of encrypting and decrypting electronic data" in which the Lozi map is involved [29, 30].

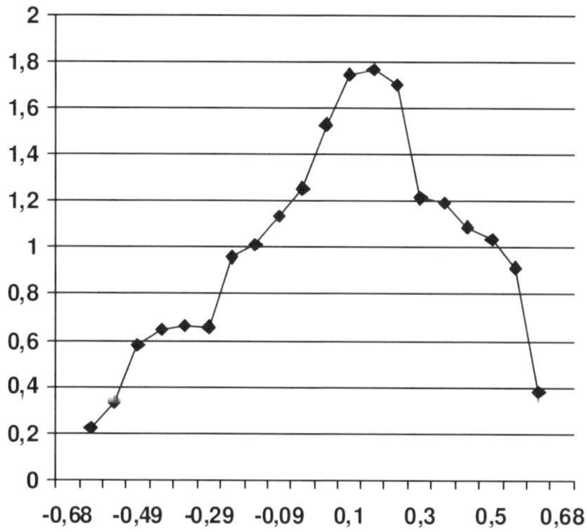

Figure 5. Invariant density of the second component y of Eq. (7) computed using 10^{10} iterations.

Nevertheless Lozi map is now widely used in chaotic optimisation which belongs to a new class of algorithms: the evolutionary algorithms (EA). In a founding paper, R. Caponetto et al. [31] propose an experimental analysis on the convergence of EA. The effect of introducing chaotic sequences instead of random ones during all the phases of the evolution process is investigated. The approach is based on the substitution of the PRNG with chaotic sequences. Several numerical examples are reported in order to compare the performance of the EA using random and chaotic generators as regards to both the results and the convergence speed. The results obtained show that chaotic sequences obtained from Lozi map are always able to increase the value of some measured algorithm-performance indexes with respect to random sequences.

Several authors following this idea use Lozi map in chaotic optimization in order to avoid local optima stagnation and embed a superior search strategy [32 – 40].

2.3. 3-Dimensional System Chaotic Dynamical Systems

In order to generalize in higher dimension the tent map, G. Manjunath *et al.* [41] introduce a three-dimensional map $F : I^3 \to I^3$ where $I = [0,1]$ (see Fig. 6) which is continuous in the Euclidian topology and prove its chaotic properties:

$$F(x,y,z)=\begin{pmatrix} \left|1-\left|2x+\frac{y+z}{2}-1\right|\right| \\ \left|1-\left|2y+\frac{x+z}{2}-1\right|\right| \\ \left|1-\left|2z+\frac{x+y}{2}-1\right|\right| \end{pmatrix} \quad (9)$$

The related dynamical system is

$$\left(x_{n+1}, y_{n+1}, z_{n+1},\right) = F\left(x_n, y_n, z_n,\right) \quad (10)$$

They emphasize that most of the well known examples of higher dimensional chaotic dynamical systems belong to the class of hyperbolic diffeomorphisms on a n-torus. These higher dimensional maps on the torus are not continuous on the standard topology of the Euclidean space since they exhibit jump discontinuities. The realization of such jump discontinuities in an electronic circuit implementation is not reliable. They prove the following theorem:

Theorem (G. Manjunath *et al.*) The map defined in (9) is topologically transitive and exhibits sensitive dependence on initial conditions with any real number $\delta \in \left(0, \frac{\sqrt{3}}{2}\right)$ as sensitivity constant.

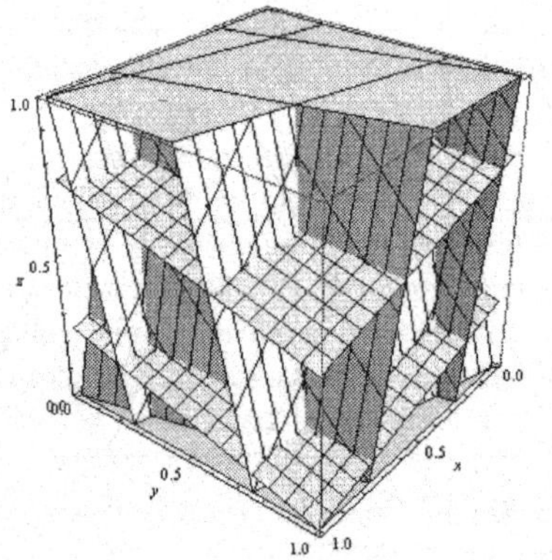

Figure 6. Tessellation of I^3 into 3^3 regions by parallel set of critical planes $\overline{T}_{y,z}(z)=0$ and 1, $\overline{T}_{x,z}(y)=0$ and 1, $\overline{T}_{x,z}(z)=0$ and 1, pertaining to the map (9).

Once again the sequence of iterated points (x_n, y_n, z_n) obtained from the dynamical system (10) is not equally distributed on the volume I^3. The invariant density of the first component x_n is displayed in Fig. 7. The relative discrepancy of this invariant density versus the uniform one lies between 4% and 5%.

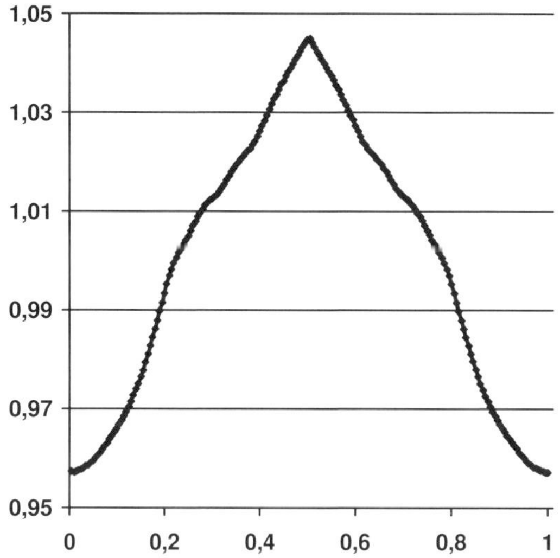

Figure 7. Invariant density of the first component x of Eq. (10) computed using 10^{11} iterations.

To allow the generation of PRN using complexity and emergence theory we consider in the next section how to generate these numbers with uniform repartition on a given interval, or on a given square of the plane or more generally in a given hypercube of $\mathbb{R}^n$ involving the ultra weak multidimensional coupling of p 1-dimensional chaotic dynamical systems.

3. Multi-parameter Chaotic Pseudo-Random Number Generator (M-p CPRNG)

As previously seen, when a dynamical system is realized on a computer using floating point or double precision numbers, the computation is of a discretization, where finite machine arithmetic replaces continuum state space. For chaotic dynamical systems, the discretization often has collapsing effects to a fixed point or to short cycles [15, 42]. In order to preserve the chaotic properties of the continuous models in numerical experiments we consider an ultra weak multidimensional coupling of p one-dimensional dynamical systems.

3.1. System of p-Coupled Symmetric Tent Map

In order to simplify the presentation of the M-p CPRNG we introduce, we use as an example the symmetric tent map previously defined (3), even though others chaotic map of the interval (as the logistic map, the baker transform, …) can be used for the same purpose (as a matter of course, the invariant measure of the chaotic map chosen is preserved).

The considered system of the p-coupled dynamical systems is described by

$$X_{n+1} = F(X_n) = A\cdot\left(\underline{f}(X_n)\right) \tag{11}$$

with

$$X_n = \begin{pmatrix} x_n^1 \\ \vdots \\ x_n^p \end{pmatrix}, \quad \underline{f}(X_n) = \begin{pmatrix} f(x_n^1) \\ \vdots \\ f(x_n^p) \end{pmatrix} \tag{12}$$

and

$$A = \begin{pmatrix} \varepsilon_{1,1}=1-\sum_{j=2}^{j=p}\varepsilon_{1,j} & \varepsilon_{1,2} & \cdots & \varepsilon_{1,p-1} & \varepsilon_{1,p} \\ \varepsilon_{2,1} & \varepsilon_{2,2}=1-\sum_{j=1,\,j\neq 2}^{j=p}\varepsilon_{2,j} & \cdots & \varepsilon_{2,p-1} & \varepsilon_{2,p} \\ \vdots & \ddots & & \vdots & \vdots \\ \vdots & & \ddots & \vdots & \vdots \\ \varepsilon_{p,1} & \cdots & \cdots & \varepsilon_{p,p-1} & \varepsilon_{p,p}=1-\sum_{j=1}^{j=p-1}\varepsilon_{p,j} \end{pmatrix} \tag{13}$$

F is a map of $J^p = [-1,1]^p \subset \mathbb{R}^p$ into itself.

Considering $\varepsilon_{i,i} = 1 - \sum_{j=1,j\neq i}^{j=p} \varepsilon_{i,j}$, the matrix A is always a stochastic matrix iff the coupling constants verify $\varepsilon_{i,j} > 0$ for every i and j.

If $\forall i,j \;\; \underset{i\neq j}{\varepsilon_{i,j}} = 0$ the maps are totally decoupled, instead they are fully crisscross coupled when for example $\underset{i\neq j}{\varepsilon_{i,j}} = \dfrac{1}{p-1}$. Generally, researchers do not consider very small values of $\varepsilon_{i,\,j}$ because it seems that the maps are quasi-decoupled with those values and no special effect of the coupling is expected. In fact it is not the case and ultra small coupling constants (as small as 10^{-7} for floating point numbers or 10^{-16} for double precision numbers), allows the construction of very long periodic orbits, leading to sterling chaotic generators. This is the way in complexity leads to randomness from chaos.

Moreover each component of these numbers belonging to $\mathbb{R}^p$ is equally distributed over the finite interval $J \subset \mathbb{R}$, when one chooses a function f with uniform invariant measure. Numerical computations (up to 10^{13} numbers) show that this distribution is obtained with a very good approximation. They have also the property that the length of the periods of the numerically observed orbits is very large [23].

3.2. Chaotic Sampling and Mixing

However chaotic numbers are not pseudo-random numbers because the plot of the couples of any component $\left(x_n^l, x_{n+1}^l\right)$ of iterated points $\left(X_n, X_{n+1}\right)$ in the corresponding phase plane reveals the map f used as one-dimensional dynamical systems to generate them *via* Eq. (11).

Nevertheless we have recently introduced a family of enhanced Chaotic Pseudo Random Number Generators (CPRNG) in order to compute very fast long series of pseudorandom numbers with desktop computer [3, 4, 5]. This family is based on the previous ultra weak coupling which is improved in order to conceal the chaotic genuine function.

In order to hide f in the phase space $\left(x_n^l, x_{n+1}^l\right)$ two mechanisms are used. The pivotal idea of the first one mechanism is to sample chaotically the sequence $\left(x_0^l, x_1^l, x_2^l, \ldots, x_n^l, x_{n+1}^l, \ldots\right)$ generated by the l-th component x^l, selecting x_n^l every time the value x_n^m of the *m-th* component x^m, is strictly greater (or smaller) than a threshold $T \in J$, with $l \neq m$, for $1 \leq l, m \leq p$.

That is to say to extract the subsequence $\left(x_{n_{(0)}}^l, x_{n_{(1)}}^l, x_{n_{(2)}}^l, \ldots, x_{n_{(q)}}^l, x_{n_{(q+1)}}^l, \ldots\right)$ denoted here $\left(\overline{x_0}, \overline{x_1}, \overline{x_2}, \cdots, \overline{x_q}, \overline{x_{q+1}}, \cdots\right)$ of the original one, in the following way

Given $1 \leq l, m \leq p, l \neq m$

$$\begin{cases} n_{(-1)} = -1 \\ \overline{x_q} = x^1_{n_{(q)}},\ with\ n_{(q)} = \underset{r\in\mathbb{N}}{Min}\left\{r > n_{(q-1)}\ \middle|\ x^m_r > T\right\} \end{cases} \tag{14}$$

The sequence $\left(\overline{x_0}, \overline{x_1}, \overline{x_2}, \cdots, \overline{x_q}, \overline{x_{q+1}}, \cdots\right)$ is then the sequence of chaotic pseudo-random numbers.

The mathematical formula (14) can be best understood in algorithmic way. The pseudo-code, for computing iterates of (14) corresponding to *N* iterates of (11) is:

$X_0 = \left(x^1_0, x^2_0, \ldots, x^{p-1}_0, x^p_0\right) = seed$

$n = 0;\ q = 0$

do $\{$ while $n < N$

do $\{$ while $\left(x^m_n \le T\right)$

compute $\left(x^1_n, x^2_n, \ldots, x^{p-1}_n, x^p_n\right)$; $n++\}$

compute $\left(x^1_n, x^2_n, \ldots, x^{p-1}_n, x^p_n\right)$;

then $n(q) = n;\ \ \overline{x_q} = x^1_{n_{(q)}};\ n++;\ q++\}$

This chaotic sampling is possible due to the independence of each component of the iterated points X_n *vs.* the others [3].

Remark 1: Albeit the number $NSampl_{iter}$ of pseudo-random numbers $\overline{x_q}$ corresponding to the computation of *N* iterates is not known *a priori*, considering that the selecting process is again linked to the uniform distribution of the iterates of the tent map on J, this number is equivalent to $\dfrac{2N}{1-T}$.

A second mechanism can improve the unpredictability of the pseudo-random sequence generated as above, using synergistically all the components of the vector X_n, instead of two. Given *p - 1* thresholds

$$T_1 < T_2 < \cdots < T_{p-1} \in J \tag{15}$$

and the corresponding partition of J

$$J = \bigcup_{k=0}^{p-1} J_k \tag{16}$$

with $J_0 = [-1, T_1]$, $J_1 =]T_1, T_2[$, $J_k = [T_k, T_{k+1}[\; for \; 1 < k < p-1$ and $J_{p-1} = [T_{p-1}, 1[$, this simple mechanism is based on the chaotic mixing of the *p-1* sequences

$$\left(x_0^1, x_1^1, x_2^1, \ldots, x_n^1, x_{n+1}^1, \ldots\right), \qquad \left(x_0^2, x_1^2, x_2^2, \ldots, x_n^2, x_{n+1}^2, \ldots\right), \ldots,$$
$$\left(x_0^{p-1}, x_1^{p-1}, x_2^{p-1}, \ldots, x_n^{p-1}, x_{n+1}^{p-1}, \ldots\right), \ldots$$

using the last one $\left(x_0^p, x_1^p, x_2^p, \ldots, x_n^p, x_{n+1}^p, \ldots\right)$ in order to distribute the iterated points with respect to this given partition defining the subsequence $\left(x_{n_{(0)}}^1, x_{n_{(1)}}^1, x_{n_{(2)}}^1, \ldots, x_{n_{(q)}}^1, x_{n_{(q+1)}}^1, \ldots\right)$ here denoted $\left(\overline{x_0}, \overline{x_1}, \overline{x_2}, \cdots, \overline{x_q}, \overline{x_{q+1}}, \cdots\right)$ by

$$\begin{cases} n_{(-1)} = -1 \\ \overline{x_q} = x_{n_{(q)}}^k, \; with \; n_{(q)} = \underset{1 \le k \le p-1}{Min} \left\{ s_k(q) = \underset{r_k \in \mathbb{N}}{Min} \left\{ r_k > n_{(q-1)} \,\middle|\, x_{r_k}^p \in J_k \right\} \right\} \end{cases} \tag{17}$$

The pseudo-code, for computing the iterates of (17) corresponding to N iterates of (11) is:

$X_0 = \left(x_0^1, x_0^2, \ldots, x_0^{p-1}, x_0^p\right) = seed$

$n = 0; \; q = 0$

do { while $n < N$

do {while $\left(x_n^p \in J_0\right)$ compute

$\left(x_n^1, x_n^2, \ldots, x_n^{p-1}, x_n^p\right)$; $n++$}

compute $\left(x_n^1, x_n^2, \ldots, x_n^{p-1}, x_n^p\right)$

let k be such that $x_n^p \in J_k$

then $n(q) = n$; $\overline{x_q} = x_{n_{(q)}}^k$; $n++$; $q++$}

Remark 2: In this case also, $NSampl_{iter}$ is not known *a priori*, however, considering that the selecting process is linked to the uniform distribution of the iterates of the tent map on J, one has $NSampl_{iter} \approx \dfrac{2N}{1-T_1}$.

Remark 3: This second mechanism is more or less linked to the whitening process [43, 44].

Remark 4: Actually, one can choose any of the components in order to sample and mix the sequence, not only the last one.

3.3. Double Threshold Chaotic Sampling

On can eventually improve the CPRNG previously introduced with respect to the infinity norm instead of the L_1 or L_2 norms because the L_∞ norm is more sensitive than the others ones to reveal the concealed function *f* [5]. For this purpose we introduce a second kind of threshold $T' \in \mathbb{N}$, together with $T_1, \cdots, T_{p-1} \in J$ such that the subsequence $\left(\overline{x_0}, \overline{x_1}, \overline{x_2}, \cdots, \overline{x_q}, \overline{x_{q+1}}, \cdots\right)$ is defined by

$$\begin{cases} n_{(-1)} = -1 \\ \overline{x_q} = x^k_{n_{(q)}},\ with\ n_{(q)} = \underset{1 \le k \le p-1}{Min}\left\{s_k(q) = \underset{r_k \in \mathbb{N}}{Min}\left\{r_k > n_{(q-1)} + T' \ \middle|\ x^p_{r_k} \in J_k\right\}\right\} \end{cases} \tag{18}$$

In pseudo-code Eq. (18) is then:

$X_0 = \left(x^1_0, x^2_0, \ldots, x^{p-1}_0, x^p_0\right) = seed$

$n = 0, q = 0$

do $\{$ while $n < N$

do {while $\left(n \le n_{(q-1)} + T' \quad and \quad x^p_n \in J_0\right)$

compute $\left(x^1_n, x^2_n, \ldots, x^{p-1}_n, x^p_n\right)$; $n++$}

compute $\left(x^1_n, x^2_n, \ldots, x^{p-1}_n, x^p_n\right)$

let *k* be such that $x^p_n \in J_k$

then $n(q) = n$; $\overline{x_q} = x^k_{n_{(q)}}$; $n++$; $q++\}$

Remark 5: In this case also, $NSampl_{iter}$ is not known a priori, it is more complicated to give an equivalent to it. However, considering that the selecting process is linked to the uniform distribution of the iterates of the tent map on J, and to the second threshold T', it comes that $NSampl_{iter} \leq Min\left\{\frac{2N}{1-T_1}, \frac{N}{T'}\right\}$.

Remark 6: the second kind of threshold T' can also be used with only the chaotic sampling, without the chaotic mixing.

4. Emergence of Randomness

Numerical results about chaotic numbers produced by (11) — (17) show that they are equally distributed over the interval J with a very good precision [3, 4].

In this section we emphasize that when the parameters $\varepsilon_{i,j}$ belong to a special window (called the window of emergence) the M-p CPRNG defined above behaves well.

4.1. Approximated Invariant Measures

In order to perform numerical computation, we have to define some numerical tools: the approximated invariant measures.

First we define an approximation $P_{M,N}(x)$ of the invariant measure also called the probability distribution function linked to the 1-dimensional map f when computed with floating numbers (or numbers in double precision). In this scope we consider a regular partition of M small intervals (boxes) r_i of J defined by:

$$s_i = -1 + \frac{2i}{M}, \quad i = 0, M$$

$$r_i = [s_i, s_{i+1}[, \; i = 0, M-2$$

$$r_{M-1} = [s_{M-1}, 1]$$

$$J = \bigcup_{0}^{M-1} r_i$$

the length of each box is

$$s_{i+1} - s_i = \frac{2}{M}$$

(note that this regular partition of J is different from the previous one linked to the threshold values T_i (16)).

All iterates $f^{(n)}(x)$ belonging to these boxes are collected (after a transient regime of Q iterations decided *a priori*, *i.e.* the first Q iterates are neglected). Once the computation of $N + Q$ iterates is completed, the relative number of iterates with respect to N/M in each box r_i represents the value $P_N(s_i)$. The approximated $P_N(x)$ defined in this article is then a step function, with M steps. As M may vary, we define

$$P_{M,N}(s_i) = \frac{1}{2}\frac{M}{N}(\# r_i)$$

where $\# r_i$ is the number of iterates belonging to the interval r_i and the constant 1/2 allows the normalisation of $P_{M,N}(x)$ on the interval J.

$$P_{M,N}(x) = P_{M,N}(s_i) \quad \forall x \in r_i$$

In the case of p-coupled maps, we are more interested by the distribution of each component $\left(x^1, x^2, x_2^1, \ldots, x^p\right)$ of X rather than the distribution of the variable X itself in J^p. We then consider the approximated probability distribution function $P_{M,N}(x^j)$ associated to one among several components of $F(X)$ defined by (11) which are one-dimensional maps. In this paper we use equally N_{disc} for M and N_{iter} for N when they are more explicit.

The discrepancies E_1 (in norm L_1), E_2 (in norm L_2) and E_∞ (in norm L_∞) between $P_{N_{disc}, N_{iter}}(x)$ and the Lebesgue measure which is the invariant measure associated to the symmetric tent map, are defined by

$$E_{1,N_{disc},N_{iter}}(x) = \left\| P_{N_{disc}, N_{iter}}(x) - 0.5 \right\|_{L_1}$$

$$E_{2,N_{disc},N_{iter}}(x) = \left\| P_{N_{disc}, N_{iter}}(x) - 0.5 \right\|_{L_2}$$

$$E_{\infty,N_{disc},N_{iter}}(x) = \left\| P_{N_{disc}, N_{iter}}(x) - 0.5 \right\|_{L_\infty}$$

In the same way an approximation of the correlation distribution function $C_{M,N}(x, y)$ is obtained numerically building a regular partition of M^2 small squares (boxes) of J^2 imbedded in the phase subspace (x^l, x^m)

$$s_i = -1 + \frac{2i}{M}, \quad t_j = -1 + \frac{2j}{M}, \quad i, j = 0, M$$

$$r_{i,j} = \left[s_i, s_{i+1}\right[\times \left[t_j, t_{j+1}\right[, \; i, j = 0, M\text{-}2$$

$$r_{i,M-1} = \left[s_i, s_{i+1}\right[\times \left[t_{M-1}, 1\right], \; j = 0, M\text{-}2$$

$$r_{M-1,M-1} = [s_{M-1}, 1]\times[t_{M-1}, 1]$$

the measure of the area of each box is

$$(s_{i+1}-s_i)\times(t_{i+1}-t_i)=\left(\frac{2}{M}\right)^2$$

Once $N+Q$ iterated points $\left(x_n^l, x_n^m\right)$ belonging to these boxes are collected the relative number of iterates with respect to N/M^2 in each box $r_{i,j}$ represents the value C_N (s_i, t_j). The approximated probability distribution function C_N (x, y) defined here is then a 2-dimensional step function, with M^2 steps. As M can take several values in the next sections, we define

$$C_{M,N}(s_i, t_j) = \frac{1}{4}\frac{M^2}{N}\left(\# r_{i,j}\right) \tag{19}$$

where $\# r_{i,j}$ is the number of iterates belonging to the square $r_{i,j}$ and the constant 1/4 allows the normalisation of $C_{M,N}(x, y)$ on the square J^2

$$C_{M,N}(x,y) = C_{M,N}(s_i,t_j) \quad \forall (x,y)\in r_{i,j} \tag{20}$$

The discrepancies E_{C_1} (in norm L_1), E_{C_2} (in norm L_2) and E_{C_∞} (in norm L_∞) between $C_{N_{disc}, N_{iter}}(x, y)$ and the uniform distribution on the square, are defined by

$$E_{C_1,N_{disc},N_{iter}}(x,y) = \left\| C_{N_{disc},N_{iter}}(x,y) - 0.25 \right\|_{L_1}$$

$$E_{C_2,N_{disc},N_{iter}}(x,y) = \left\| C_{N_{disc},N_{iter}}(x,y) - 0.25 \right\|_{L_2}$$

$$E_{C_\infty,N_{disc},N_{iter}}(x,y) = \left\| C_{N_{disc},N_{iter}}(x,y) - 0.25 \right\|_{L_\infty}$$

Finally let $AC_{N_{disc}, N_{iter}}(x, y)$ be the autocorrelation distribution function which is the correlation function $C_{N_{disc}, N_{iter}}(x, y)$ of (20) defined in the phase space $\left(x_n^l, x_{n+1}^l\right)$ instead of the phase space $\left(x^l, x^m\right)$. In order to control that the enhanced chaotic numbers $\left(\overline{x_0}, \overline{x_1}, \overline{x_2}, \cdots, \overline{x_q}, \overline{x_{q+1}}, \cdots\right)$ are uncorrelated, we plot them in the phase subspace $\left(\overline{x_q}, \overline{x_{q+1}}\right)$ and we check if they are uniformly distributed in the square J^2 and if f is concealed (*i.e.* $E_{AC_1,N_{disc},N_{iter}}(\overline{x}_q,\overline{x}_{q+1})$, $E_{AC_2,N_{disc},N_{iter}}(\overline{x}_q,\overline{x}_{q+1})$, $E_{AC\infty,N_{disc},N_{iter}}(\overline{x}_q,\overline{x}_{q+1})$ vanish).

4.2. A window of Emergence of Randomness

In order to point out the usefulness of the double threshold chaotic sampling with simply consider the case of only 4 coupled equations, and such that $\varepsilon_{i,j}=\varepsilon_i \ \forall i \neq j$ and $\varepsilon_{i,i}= 1\text{-}3\varepsilon_i$. Eq. (11) becomes

$$\begin{cases} x_{n+1}^1 = (1-3\varepsilon_1)f(x_n^1) + \varepsilon_1 f(x_n^2) + \varepsilon_1 f(x_n^3) + \varepsilon_1 f(x_n^4) \\ x_{n+1}^2 = \varepsilon_2 f(x_n^1) + (1-3\varepsilon_2)f(x_n^2) + \varepsilon_2 f(x_n^3) + \varepsilon_2 f(x_n^4) \\ x_{n+1}^3 = \varepsilon_3 f(x_n^1) + \varepsilon_3 f(x_n^2) + (1-3\varepsilon_3)f(x_n^3) + \varepsilon_3 f(x_n^4) \\ x_{n+1}^4 = \varepsilon_4 f(x_n^1) + \varepsilon_4 f(x_n^2) + \varepsilon_4 f(x_n^3) + (1-3\varepsilon_4)f(x_n^4) \end{cases} \quad (21)$$

Moreover we assume that $\varepsilon_i = i\,\varepsilon_1$

For the shake of simplicity we consider only the chaotic sampling method (*i.e.* we use only one threshold T), without the chaotic mixing. We then compute $E_{1,N_{disc},N_{iter}}(\overline{x})$, $E_{2,N_{disc},N_{iter}}(\overline{x})$, $E_{\infty,N_{disc},N_{iter}}(\overline{x})$ and $E_{AC\infty,N_{disc},N_{iter}}(\overline{x}_q,\overline{x}_{q+1})$ $E_{AC_1,N_{disc},N_{iter}}(\overline{x}_q,\overline{x}_{q+1})$, $E_{AC_2,N_{disc},N_{iter}}(\overline{x}_q,\overline{x}_{q+1})$, for $N_{disc} = 1{,}024$ and $N_{iter} = 10^{11}$. We choose, $T = 0.9$ and $T' = 20$. We display on Fig. 8 the values of the six computed error when $\varepsilon_1 \in \left[10^{-17}, 10^{-1}\right]$. The seed (initial values) being

$$x_0^1 = 0.330000,\ x_0^2 = 0.338756,\ x_0^3 = 0.504923,\ x_0^4 = 0.324082.$$

A window of emergence comes clearly into sight for the values $\varepsilon_1 \in \left[10^{-15}, 10^{-7}\right]$ if one considers all together the six errors.

The errors $E_{\infty,N_{disc},N_{iter}}(\overline{x})$, $E_{AC\infty,N_{disc},N_{iter}}(\overline{x}_q,\overline{x}_{q+1})$ narrowing this window in which $340{,}753{,}095 \leq NSampl_{iter} \leq 340{,}768{,}513$ out of $N_{iter} = 10^{11}$.

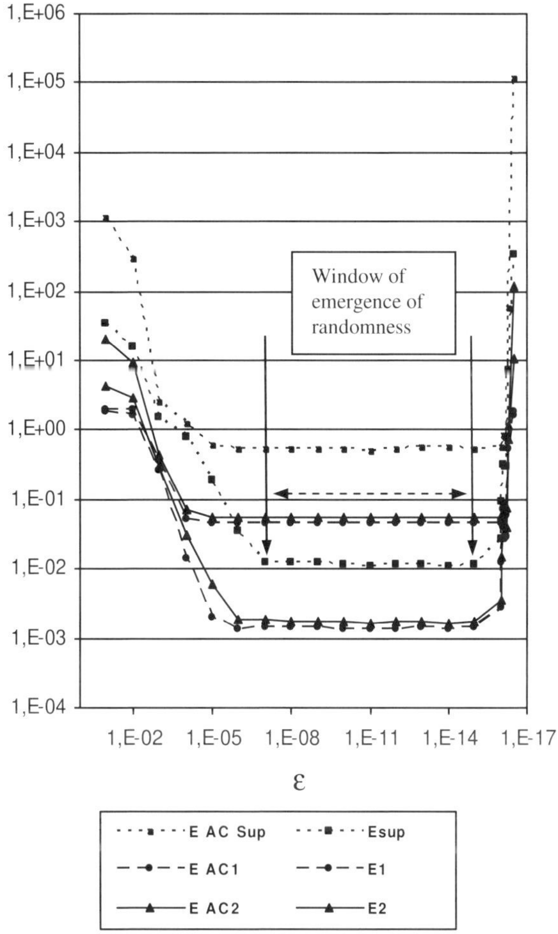

Figure 8. The window of emergence of randomness

4.3. The Underneath of Randomness

The double threshold chaotic sampling is very efficient because its aim is mainly to conceal f in the most drastic way. In order to understand the underneath mechanism consider first that in the phase space $\left(x_n^l, x_{n+1}^l\right)$ the graph of the chaotically sampled chaotic numbers is a mix of the graphs of the $f^{(r)}$ for all $r \in \mathbb{N}$ (see Fig. 9).

It is obvious as showed on Fig. 10 that for $r = 1$ if $M = 1$ or 2, $AC_{M,N}(x, y)$ is constant and normalized on the square hence $E_{AC_1,N_{disc},N_{iter}}(x, y) = E_{AC_2,N_{disc},N_{iter}}(x, y) = E_{AC\infty,N_{disc},N_{iter}}(x, y) = 0$

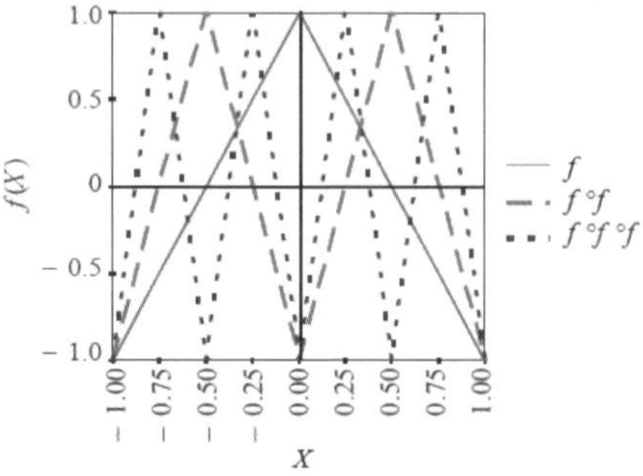

Figure 9. Graphs of the symmetric tent map $f, f^{(2)}$ and $f^{(3)}$ on the interval [-1, 1].

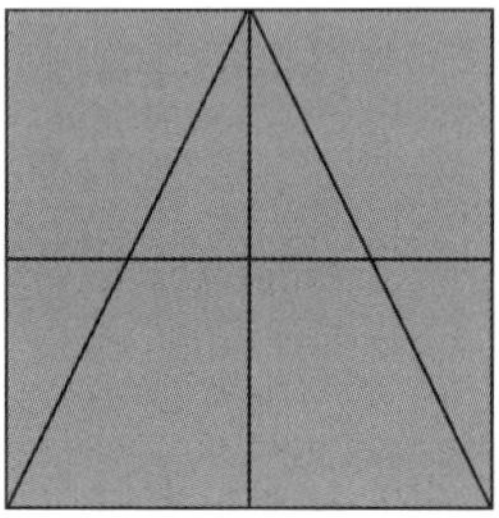

Figure 10. In shaded regions the autocorrelation distribution $AC_{M,N}(x, y)$ is constant for the symmetric tent map f on the interval [-1, 1] for $M = 1$ or 2.

The autocorrelation function is different from zero only if $M > 2$ (see Fig. 11).

In the same way as displayed on Fig. 12, 13 and 14, $E_{AC_1,N_{disc},N_{iter}}(x, y) = E_{AC_2,N_{disc},N_{iter}}(x, y) = E_{AC\infty,N_{disc},N_{iter}}(x, y) = 0$ for $f^{(i)}$ iff $M < 2^i$. Hence for a given M, if we cancel the contribution of all the $f^{(i)}$ for $2^i < M$, it is not possible to identify the genuine function f.

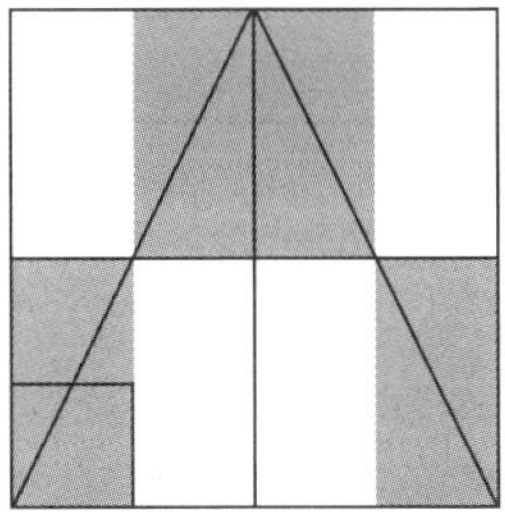

Figure 11. Regions where the autocorrelation distribution $AC_{M,N}(x, y)$ is constant for the symmetric tent map f are shaded, for $M = 4$. (The square on the bottom left hand side of the graph shows the size of the $r_{i,j}$ box). $AC_{M,N}(x, y)$ vanishes on the white regions.

Figure 12. In shaded regions the autocorrelation distribution $AC_{M,N}(x, y)$ is constant for the symmetric tent map $f^{(2)}$ on the interval [-1, 1] for $M = 1$, 2 and 4.

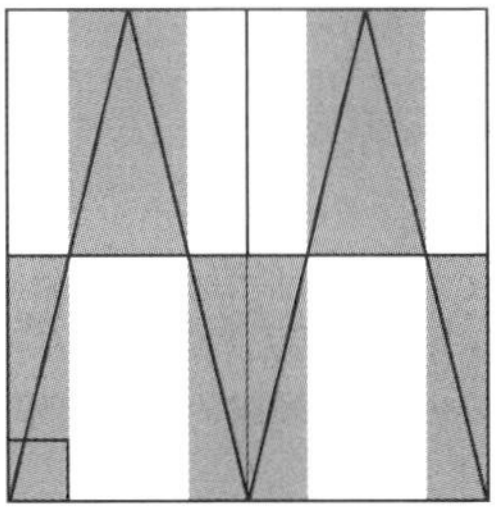

Figure 13. Regions where the autocorrelation distribution $AC_{M,N}(x, y)$ is constant for the symmetric tent map $f^{(2)}$ are shaded for $M = 8$.

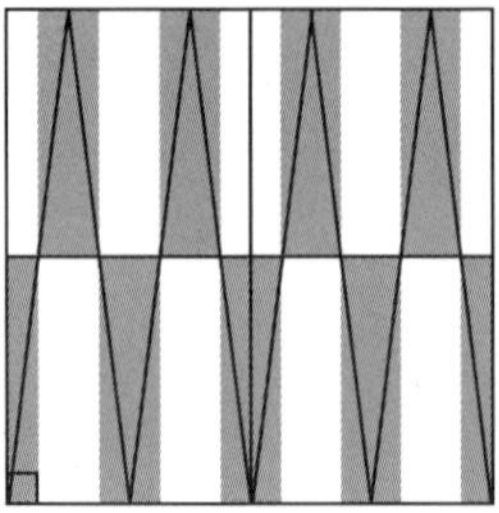

Figure 14. Regions where the autocorrelation distribution $AC_{M,N}(x,y)$ is constant for the symmetric tent map $f^{(3)}$ are shaded for $M = 16$.

4.4. Testing the Randomness

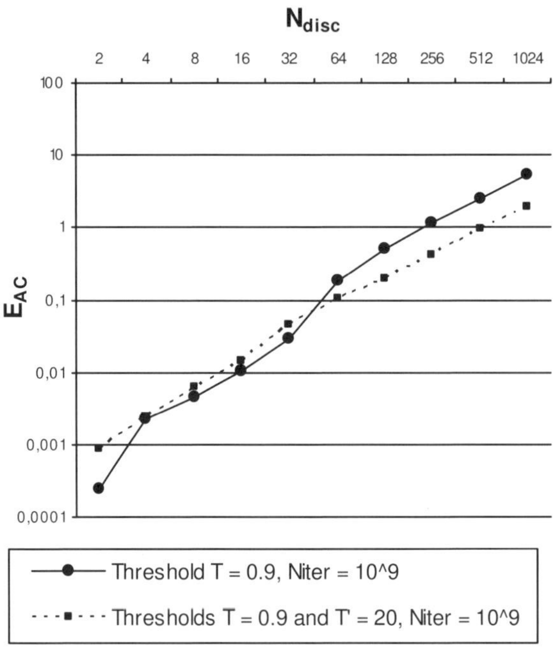

Figure 15. Error of $E_{AC\infty,N_{disc},N_{iter}}(\overline{x}_q,\overline{x}_{q+1})$, $N_{disc} = 2^1\ to\ 2^{10}$, $N_{iter} = 10^9$, thresholds $T = 0.9$ and $T' = 20$, $\varepsilon_i = i\,\varepsilon_1$, $\varepsilon_i = 10^{-14}$. Computations are done using double precision numbers (~14-15 digits).

As shown previously [3] the errors in L_1 or L_2 norms decrease with the number of chaotic points (as in the law of large numbers) and conversely increase with the number M of boxes used to define $AC_{M,N}(x, y)$. It is the same for the error in L_∞ norm. Fig. 15 shows that when M is greater than 2^5, the sequence defined by (18) behaves better than the one defined by (14) or (17) when applied to Eq. (21).

Fig. 16 shows that when the number of chaotic points increases the error $E_{AC\infty,N_{disc},N_{iter}}(\overline{x}_q,\overline{x}_{q+1})$ decreases drastically. If for example $T' > 100$, it is necessary to use a huge grid of 2^{100}x2^{100} boxes splitting the square J^2 in order to find a trace of the genuine function *f*. This is numerically impossible with double precision numbers. Then the chaotic numbers emerge as random numbers.

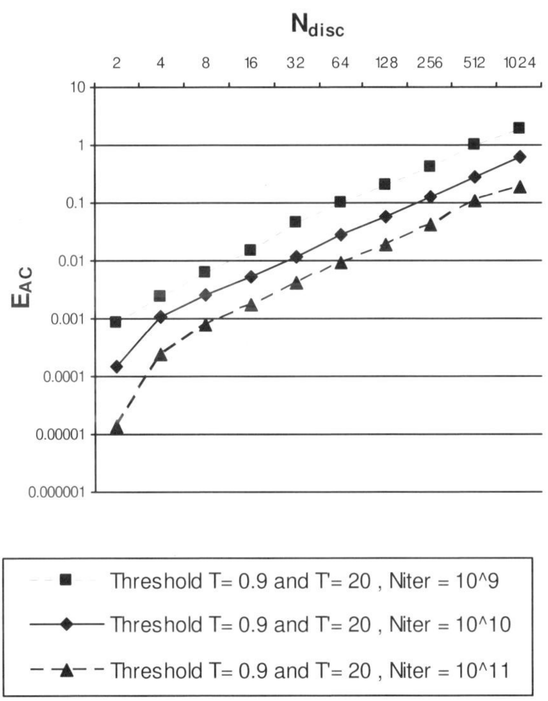

Figure 16. Error of $E_{AC\infty,N_{disc},N_{iter}}(\overline{x}_q,\overline{x}_{q+1})$ $N_{disc} = 2^1\ to\ 2^{10}$, $N_{iter} = 10^9\ to\ 10^{11}$, thresholds $T = 0.9$ and $T' = 20$, $\varepsilon_i = i\,\varepsilon_l$, $\varepsilon_i = 10^{-14}$. Computations are done using double precision numbers (~14-15 digits).

5. Applications

Generation of random or pseudorandom numbers, nowadays, is a key feature of industrial mathematics. Pseudorandom or chaotic numbers are used in many areas of contemporary technology such as modern communication systems and engineering applications.

More and more European or US patents using discrete mappings for this purpose are obtained by researchers of discrete dynamical systems [29, 30].

When an efficient M-p CPRNG is defined, there exists a huge number of applications for the pseudo-random numbers it can generate, as for example chaotic masking, chaotic modulation or chaotic shift keying in the fields of secure communications [7, 8, 9, 10].

5.1. Parameter sensitivity

A determining property of the M-p CPRNG we have improved in this paper *via* Eq. (21) and double threshold chaotic sampling (18) is the high number of parameters used ($p\times(p-1)$ for p coupled equations) which allows to choose it as cipher-keys however this achievement is possible only if there is a high sensitivity to the parameters values.

In order to point up this sensitivity, it is enough to consider the simplest case of 2-coupled equations with two sets of slightly different parameters $(\varepsilon_1, \varepsilon_2)$ and $(\varepsilon_1^*, \varepsilon_2)$ $\varepsilon_1 = 0.000,001$, $\varepsilon_1^* = 0.000,001,000,000,000,000,3$, and $\varepsilon_2 = 0.000,002$.

$$\begin{cases} x_{n+1}^1 = (1-\varepsilon_1) f(x_n^1) + \varepsilon_1 f(x_n^2) \\ x_{n+1}^2 = \varepsilon_2 f(x_n^1) + (1-\varepsilon_2) f(x_n^2) \end{cases} \tag{22}$$

$$\begin{cases} x_{n+1}^{*1} = (1-\varepsilon_1) f(x_n^{*1}) + \varepsilon_1^* f(x_n^{*2}) \\ x_{n+1}^{*2} = \varepsilon_2 f(x_n^{*1}) + (1-\varepsilon_2) f(x_n^{*2}) \end{cases} \tag{23}$$

The double threshold sampling is done using $T = 0.9$ and $T' = 20$ and the same seed is taken

$$X_0 = (x_0^1, x_0^2) = X_0^* = (x_0^{*1}, x_0^{*2})$$

Despite the fact that the difference between ε_1 and ε_1^* is tiny: $\frac{\left|\varepsilon_1 - \varepsilon_1^*\right|}{\varepsilon_1} = 3\times 10^{-13}$ the sequences $\left(\overline{x_0}, \overline{x_1}, \overline{x_2}, \cdots, \overline{x_q}, \overline{x_{q+1}}, \cdots\right)$ and $\left(\overline{x_0^*}, \overline{x_1^*}, \overline{x_2^*}, \cdots, \overline{x_q^*}, \overline{x_{q+1}^*}, \cdots\right)$ differ completely as displayed in Table 3 (In fact all the components $\left(x^1_{n_{(q)}}, x^2_{n_{(q)}}\right)$ and $\left(x^{*1}_{n_{(q)}}, x^{*2}_{n_{(q)}}\right)$ are different).

Then rather than a unique CPRNG which is introduced here, there is a quasi-infinite family of CPRNG that the M-p CPRNG define allowing several possibilities of applications.

5.2. Gaussian Noise

As an example of such application, the generation of Gaussian noise from the sequences $\left(\overline{x_0}, \overline{x_1}, \overline{x_2}, \cdots, \overline{x_q}, \overline{x_{q+1}}, \cdots\right)$ is very easy when a Box-Muller transform is applied.

A Box–Muller transform [45] is a method of generating pairs of independent standard normally distributed (zero expectation, unit variance) random numbers, given a source of uniformly distributed random numbers. The polar form [46] of such a transform takes two samples from a different interval, $\left[-1,1\right]$ and maps them to two normally distributed samples without the use of sine or cosine functions. This form of the polar transform is widely used, in part due to its inclusion in Numerical Recipes.

ε_1	$0.000,001$	ε_1^*	$0.000,001,000,$ $000,000,000,3$
x_0^1	$0.330,000,013,$ $113,021,851$	x_0^{*1}	$0.330,000,013,$ $113,021,851$
$x^1_{n_{(0)}}$	$-0.959,214,817,$ $207,605,153$	$x^{*1}_{n_{(0)}}$	-0.058,536,729, 173,974,455,5
$x^1_{n_{(1)}}$	0.657,775,688, 600,752,417	$x^{*1}_{n_{(1)}}$	0.386,129,403, 866,398,935
$x^1_{n_{(2)}}$	-0.784,600,935, 471,051,031	$x^{*1}_{n_{(2)}}$	0.471,824,729, 381,262,631

ε_1	0.000,001	ε_1^*	0.000,001,000, 000,000,000,3
x_0^2	0.338,756,413, 113,021,848	x_0^{*2}	0.338,756,413, 113,021,848
$x_{n_{(0)}}^2$	0.914,472,270, 898,123,885	$x_{n_{(0)}}^{*2}$	-0.646,249,812, 458,326,023
$x_{n_{(1)}}^2$	0.915,684,412, 995,676,6	$x_{n_{(1)}}^{*2}$	0.894,262,910, 879,751,405
$x_{n_{(2)}}^3$	0.910,813,705, 361,448,345	$x_{n_{(2)}}^{*2}$	0.820,811,987, 022,524,114

Table 3. Sequences $\left(x_{n_{(q)}}^1, x_{n_{(q)}}^{*1}\right)$ and $\left(x_{n_{(q)}}^2, x_{n_{(q)}}^{*2}\right)$ of Eq. (22) and (23) with $\varepsilon_1 = 0.000,001$, $\varepsilon_1^* = 0.000,001,000,000,000,000,3$ and $\varepsilon_2 = 0.000,002$. $X_0 = \left(x_0^1, x_0^2\right) = X_0^* = \left(x_0^{*1}, x_0^{*2}\right)$

As the sequences $\left(\overline{x_0}, \overline{x_1}, \overline{x_2}, \cdots, \overline{x_q}, \overline{x_{q+1}}, \cdots\right)$ are uniformly distributed in $J = [-1,1] \subset \mathbb{R}$, the application is straightforward.

5.3. Hash Function

Another example of application could be the computation of hash function. A hash function is any well-defined procedure or mathematical function that converts a large, possibly variable-sized amount of data into a small one. The values returned by a hash function are called hash values, hash codes, hash sums, checksums or simply hashes.

Hash functions are mostly used to speed up table lookup or data comparison tasks — such as finding items in a database, detecting duplicated or similar records in a large file, finding similar stretches in DNA sequences, and so on.

A hash function may map two or more keys to the same hash value. In many applications, it is desirable to minimize the occurrence of such collisions, which means that the hash function must map the keys to the hash values as evenly as possible. Depending on the application, other properties may be required as well. Although the idea was conceived in the 1950s, the design of good hash functions is still a topic of active research.

Although hash function generally involve integers, on can consider that the application which maps the initial seed $X_0 = \left(x_0^1, x_0^2, \ldots, x_0^{p-1}, x_0^p\right)$ into any predetermined term of the sequence $\left(\overline{x_0}, \overline{x_1}, \overline{x_2}, \cdots, \overline{x_q}, \overline{x_{q+1}}, \cdots\right)$ is a hash function working on floating point numbers.

We will explore this application in a forthcoming paper.

Others applications show the high-potency of such M-p CPRNG. Due to limitation of this article, they will be published elsewhere.

6. Conclusion

Using a double threshold in order to sample a chaotic sequence, we have improved with respect to the infinity norm the M-p CPRNG previously introduced. When the value of the second threshold T' is greater than 100, it is impossible to find the genuine function used to generate the chaotic numbers. The new M-p CPRNG family is robust versus the choice of the weak parameter of the system for $10^{-14} < \varepsilon < 10^{-5}$, allowing the use of this family in several applications as for example producing Gaussian noise, computing hash function or in chaotic cryptography.

References

1. W. Weaver, *American Scientist* **36,** (4), 536 (1948).
2. M. A. Aziz-Alaoui and C. Bertelle, *From System Complexity to Emergent Properties (Understanding Complex Systems)* Springer-Verlag, Berlin (2009).
3. R. Lozi, *Indian Journal of Industrial and Applied Mathematics* Vol.1, n° 1, 1 (2008).
4. R. Lozi, *Proceedings of 6th EUROMECH Non Linear Dynamics Conference, Saint-Petersburg, ENOC 2008, IPACS Open Access Electronic Library,* 1715 (2008).
5. R. Lozi, *Conference Proceedings of ICCSA 2009* 20 (2009).
6. R. Lozi, *Intern. J. Bifurcation & Chaos* to appear (2011).
7. S. Hénaff, I. Taralova and R. Lozi, *Conference Proceedings of ICCSA 2009* 47 (2009).
8. S. Hénaff, I. Taralova and R. Lozi, *Indian Journal of Industrial and Applied Mathematics* Vol.2, No 2, 1 (2009).
9. S. Hénaff, I. Taralova and R. Lozi, *Proceedings of the Physics and Control Conference, Catania 2009, IPACS Open Access Electronic Library*, 1939 (2009).

10. S. Hénaff, I. Taralova and R. Lozi, *Journal of Nonlinear Systems and Applications* Vol.1, (3-4), 87 (2010).
11. J. C. Sprott, *Chaos and Time-Series Analysis* Oxford University Press, Oxford, UK (2003).
12. K. T. Alligood, T. D. Sauer and J. A. Yorke, *Chaos. An introduction to dynamical systems* Springer, Textbooks in mathematical sciences, New-York (1996).
13. R. M. May, *Science, New Series* Vol. 186, No 4164, 645 (1974).
14. M. J. Feigenbaum, *J. Stat. Phys* 21, 669 (1979).
15. O. E. Lanford III, *Experimental Mathematics* Vol. 7, **4**, 317 (1998).
16. R. Lozi and C. Fiol, *Grazer Math. Bericht* Nr 354, 112 (2009).
17. R. Lozi and C. Fiol, *Conference Proceedings A.I.P.* **1146**, 303 (2009).
18. M. S. Baptista, *Phys. Lett. A* 240, 50 (1998).
19. M. R. K. Ariffin and M. S. M. Noorani, *Phys. Lett. A* 372, 5427 (2008).
20. A. N. Sharkovskiĭ, *Intern. J. Bifurcation & Chaos* Vol. 5, **5**, 1263 (1995).
21. M. Hénon, *Comm. Math. Phys.* **50**, 69 (1976).
22. D. Ruelle and F. Takens, *Comm. Math. Phys.* **20**, 167 (1971).
23. R. Lozi, in *Modern Mathematical Models, Methods and Algorithms for Real World Systems*, eds. A. H. Siddiqi, I. S. Duff and O. Christensen, Anamaya Publishers, New Delhi, India, 80 (2006).
24. N. K. Pareek, V. Patidar and K. K. Sud, *Int. J. Information and Communication Technology* Vol. 2, n°3, 244 (2010).
25. L. dos Santos Coelho, *Chaos, Solitons and Fractals* **39**, 510 (2009).
26. R. Lozi, *J. Phys. Colloques* **39**, C5 (1978).
27. M. Misiurewicz, *Ann. N. Y. Acad. Sci.* **375**, 348, (1980).
28. A. Sakurai, *Taiwanese J. of Math.* Vol. 14, N° 4, 1609 (2010).
29. M. V. Petersen & H. M. Sorensen, *United States Patent* 7170997 (2007).
30. D. Ruggiero, D. Mascolo, I. Pedaci and P. Amato, *United States Patent Application* 20060251250 (2006).
31. R. Caponetto, L. Fortuna, S. Fazzino and M. G. Xibilia, *IEEE Transactions on Evolutionary Computation* Vol. 7, Iss. 3, 289 (2003).
32. L. dos Santos Coelho, *Chaos, Solitons and Fractals* **39**, 1504 (2009).
33. L. dos Santos Coelho, *Chaos, Solitons and Fractals* **41**, 594 (2009).
34. L. dos Santos Coelho and D. L. de Andrade Bernet, *Chaos, Solitons and Fractals* **42**, 634 (2009).
35. S. Jalilzadeh, H. Shayeghi, A. Safari and E. Aliabadi, *Proceedings of ECTI-CON 2009, 6th International Conference on Electrical Engineering/Electronics, Computer, Telecommunications and Information Technology* 24 (2009).
36. H. Shayeghi, S. Jalilzadeh, H. A. Shayanfar and A. Safari, *Proceedings of ECTI-CON 2009, 6th International Conference on Electrical Engineering/Electronics, Computer, Telecommunications and Information Technology* 40 (2009).

37. H. Shayeghi, H. A. Shayanfar, S. Jalilzadeh and A. Safari, *Energy Conversion and Management* **51**, No 7, 1572 (2010).
38. H. Shayeghi, H. A. Shayanfar, S. Jalilzadeh and A. Safari, *Energy Conversion and Management* **51**, No 12, 2678 (2010).
39. A. Safari, H. Shayeghi and H. A. Shayanfar, *International Journal on Technical and Physical Problems of Engineering* Vol. 1, Number 3, Issue 4, 44 (2010).
40. D. Davendra, I. Zelinka and R. Senkerik, *Computers and Mathematics with Applications* 60, 1088 (2010).
41. G. Manjunath, D. Fournier-Prunaret and A.-K. Taha, *Grazer Math. Bericht* Nr 354, 145 (2009).
42. P. Gora, A. Boyarsky, Md. S. Islam and W. Bahsoun, *SIAM J. Appl. Dyn. Syst. (electronic)* 5:1, 81 (2006).
43. J. Viega, *Proceedings of 19th Annual Annual Computer Security Applications Conference*, 129 (2003).
44. J. Viega, and M. Messier, *Secure programming cook book for C and C++* O'Reilly, Sebastopol CA (2003).
45. G. E. P. Box and M. E. Muller, *Ann. Math. Statist.* Vol. 29, No 2, 610 (1958).
46. R. Knop, *Comm. of ACM* Vol. 12, Issue 5, 28 (1969).

MATHEMATICAL MODELING FOR UNIFYING DIFFERENT BRANCHES OF SCIENCE, ENGINEERING AND TECHNOLOGY

N. RUDRAIAH*

National Research Institute for Applied Mathematics (NRIAM)
No. 492/G, 7th Cross, 7th Block (West), Jayanagar Bangalore - 560 070.
and
UGC-CAS in Fluid Mechanics,
Department of Mathematics, Central College Campus,
Bangalore University, Bangalore - 560 001.
**E-mail: rudraiahn@hotmail.com*

This paper reviews the basic concepts of Mathematical Modeling, different analytical techniques to solve nonlinear differential equations and their advantages before resorting to find numerical solutions. It explains how mathematics unifies different branches of science, engineering and technology. This paper also provides a mathematical model to show that a problem in science, engineering and technology belongs to one of the physical phenomena of the oscillations, diffusion and potential governed, by hyperbolic, parabolic and elliptic partial differential equations respectively.

1. Introduction

There are many distinguished Applied Mathematicians in the country and my appeal to them is, in addition to their own work, try to propose suitable and efficient Mathematical Models to find solutions to the following ten challenging problems faced by the people of our country and become one of the contributors in improving the quality of living of people in rural area on par with those living in an urban area.

1. Health services for every citizen of our country which are economically within their reach.
2. Technologies to take care of health problems of mothers in our country. (According to a report released by UNO and published through the columns of the press reveals that every day about 1000 pregnant mothers die during their delivery.)
3. Efficient and potential solar technologies.

4. Methods to save electricity in the prevailing system.
5. Adequate pure and clean potable water.
6. Quality education system.
7. Methods to eliminate different types of waste materials.
8. Suitable information technology to ensure internal security.
9. To protect and preserve food.
10. Efficient and transparency in the administration for rural development.

Suitable solutions to these challenging problems may improve the quality of living of people in rural area in our country. If Q is the quality of life, N the natural resource, E the energy, I the ingenuity of an individual, P the population and n is the index, then the quality of living can be quantified through the following formula

$$Q = \frac{NEI}{P^n}$$

The index n depends on the population of a country. For a thickly populated countries like India, China and so on, $n = 1$ and for a sparsely populated countries $n < 1$ on suitable indexes for the conditions prevailing in their countries. It is belived that the quality of life is one of the factors to access a country to be a developed country. At our Center, we motivate our teachers and research students to think and work hard to provide suitable solutions for the above ten challenging problems.
At our Center, we motivate our teachers and research students to think and work hard to provide suitable solutions for the above ten challenging problems

(1) A possible solution to the problem 1 listed above: One of our PhD students who worked on biomedical engineering problems developed a mathematical model to design artificial organs, with maximum efficiency and minimum weight, like endothelium in coronary artery diseases (CAD), cartilages in synovial joints (SJ), angiograph and so on using a Smart material of Nano structure, as explained below.

(2) A mathematical model to design a Smart material of Nano structure: We consider poorly conducting alloys like Nickel-Titanium (Ni-Ti), Aluminum Oxides and so on which have very poor electrical conductivity $\sigma \ll 1$, increasing with temperature like

$$\sigma = \sigma_0(1 + \alpha_h(T - T_0)) \qquad (1)$$

where α_h is the volumetric expansion co-efficient for conductivity, T_0 is the room temperature and σ_0 is that of σ at $T = T_0$. Since we deal with a poorly conducting fluid, any fluctuation on it is negligible compared to its basic state. If $T = T' + T_b$, where T' is a perturbation on and T_b is the conduction temperature obtained by solving pure conduction equation

$$\frac{d^2T_b}{dy^2} = 0 \tag{2}$$

satisfying the conductions

$$T_b = T_1 \quad at \quad y = h \quad and \quad T_b = T_0 \quad at \quad y = 0 \tag{3}$$

Then the solution of Eqn. (2), satisfying the conditions (3) is

$$T_b = \frac{y}{h}\triangle T + T_0 \tag{4}$$

Equation (1), with $T = T_b$ and using Eqn. (4), becomes

$$\sigma = \sigma_0 \tag{5}$$

Making it dimensionless using $\sigma^* = \dfrac{\sigma}{\sigma_0}, \alpha^* = \alpha_h \triangle T, y^* = \dfrac{y}{h}$ and for simplicity neglecting the asterisks $(*)$ we get

$$\sigma = (1 + \alpha y) = e^{\alpha y} \tag{6}$$

1.1. *Nano Materials:*

Nano comes from the Greek word "nanos" means dwarf. Scientist, Engineer and Technologists use prefix 10^{-9} or one- billionth of a meter to indicate it. The emergence of nano technology in atomic assembly was first announced in public in 1959, by a physicist Rechard Feryman. The fundamentals of nano technology lies in the fact that properties of substances drastically change when their size is reduced to nano meter range (for details see [1]). For example, ceramics which are normally brittle can easily be made deformable when their grain size is reduced to the nano meter range.

1.2. *Smart Materials:*

Smart materials are those materials which have the properties of sensing as well as actuation. We have developed a smart material of nano structure by solidifying the Ni-Ti alloy by cooling from below and heating from above. The difference in temperature produces the difference in electrical conductivity, $\triangle\sigma$. This $\triangle\sigma$ releases the charges from the nuclei forming the distribution of charge density, ρ_e. These charges produce an electric field, $\vec{E}_i$, called the induced electric field. If high strength of electric field is needed we can generate it by embedding the electrodes of different potentials. This difference in potential produces an electric field,$\vec{E}_a$, called the applied electric field. The total electric field $\vec{E} = \vec{E}_a + \vec{E}_i$ produces the current density, $\vec{J}$, according to the Ohm's law,$\vec{J} = \sigma\vec{E}$. This electric field,$\vec{E}$, together withρ_e, produces a force $\rho_e\vec{E}$. This current acts as sensor and force acts as actuation. These are the properties to make a material to be a smart material. In addition, the solidification process explained above produces a mushy layer also called dendrites which are the mixture of solids and poorly conducting fluid. The solid particles have the structure of nano crystals (see Srirammurthy and Arunachalam [2]). Therefore, this solidification process in a poorly conducting fluid produces a smart material of nano structure.

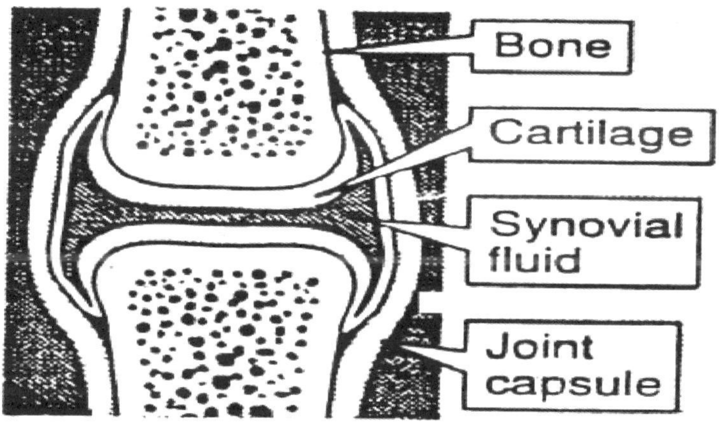

Fig. 1. The Synovial Joints

One of our PhD students at UGC-CAS in Fluid Mechanics used this smart material of nano structure to synthesize cartilage in synolial joints as explained below.

In joints, there are three types, namely freely movable joints called diorthoidal joints and also called Synovial joints (SJ), slowly movable joints called Amphorthroidal and immovable joints called synorthroidal. Synovial joints are important because they involve lubrication and have two important parts - Synovial fluid (SF) and articulate cartilages (AC). Synovial fluid in the cavity between the two bones (see Fig.1) has high viscosity of about 1000 times higher than that of water. This high viscosity is due to Hyaluronic acid (HA) in SF. The quantity of synovial fluid gradually reduces with age. The SF can be synthesized in the laboratory because HA is easily available and can be injected into the joints if need be.

The cartilage is a sponge type of porous material covering the ends of the bones (see Fig.1) through which nutrients and other substances required for the survival of the joints will be transported. If the cartilage is damaged then it can not be recouped by any kind of medicine and surgery. In that case, it has to be replaced by artificial cartilages. At present they are made up of metals. This type of artificial cartilage made up of metal lacks biocompatibility resulting in either rough or smooth surfaces. Both of them are dangerous because they produce stresses which in turn generate a force. This force drives the RBC's to a particular region. The accumulation of RBC's in that region will be bursted, letting loose the haemoglobins causing loss of haemoglobin. It is a disease known as 'Haemolysis.' We have suggested [3] a mechanism, using electrohydrodynamic aspect, to synthesize a cartilage using a Smart material of Nano structure. The artificial cartilages made up of metals are expensive in addition to causing side effects of Haemolysis. We have shown [3] that if a cartilage is made up of smart material of nano structure it will not only reduces the side effect of haemolysis but also reduces the cost. This is within the means of a common man and hence may be a possible solution for the problem 1 out of ten cited above. A possible solution to the problem 3 mentioned above. In a residential area, we often come across the flat plate solar concentrators on the roof top. Its efficiency is believed to be about 59%. In contrast to this, we [4] have designed a solar concentrator (see Fig.2) of the shape in between parabolic and elliptic using geometry and found a suitable focal point using optimization technique and showed mathematically that it has 82% efficiency. This theoretical efficiency has been confirmed by conducting experiments and considered to be one of the possible solutions for the problem 3 out of 10 problems mentioned above.

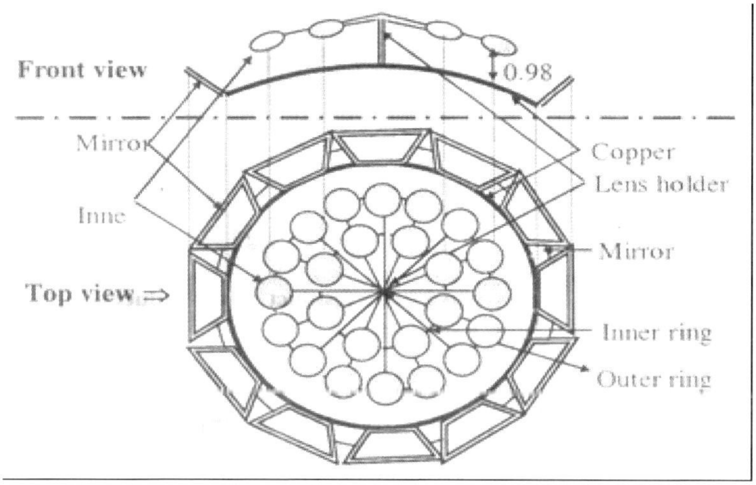

Fig. 2. Solar Concentrator

1.3. *Solar Technology:*

The depletion of fossil fuel and vagaries of monsoon have tremendous strain on providing required power supply in the country, in general, and in our state, in particular. To provide adequate power supply to the exponential growth of population, industries and for the overall development of our country, we have to resort to alternative unconventional methods of power generation. One of the effective, efficient and everlasting sources is the solar energy. Although, most of the days in a year we have blue sky with bright sun, it is not economically feasible to generate solar energy, because of the large scale. Therefore, it is our opinion that it is economically feasible to combine solar energy with magnetohydrodynamics (MHD) principles because of the following regions:

(1) Our model, called poor man generator [see Fig.3], works at a place where it is needed and save the cost and loss involved in transmiting electricity.
(2) Environmental friendly, because we use a new solar concentrator explained in section 1.2.
(3) It has no mechanical moving part and hence works even at higher altitudes.
(4) Our model works as a generator or a pump depending on the value of load factor.

Our generator is based on the combined solar energy and MHD principles. The usual MHD generator uses either ionized gas or liquid metal. Both of them are not economically feasible for rural areas in our country. However, in our model, we develop an indigenous and economically feasible model which suits conditions of rural area. For this purpose, we use a stratified fluid like alkali content bore well water mixed with municipal waste to achieve higher conversion efficiency by increasing the stratification factor β. We have found that the efficiency, η_g, of our generator as

$$\eta_g = [1 - \frac{a^* K}{(f(\beta) - K)}] \tag{7}$$

where $K = \dfrac{2Q_w}{UB - 0h}$ is the load factor, Q_w is the flux, $a^* = \dfrac{ah}{l}$ is the aspect ratio and $a = \dfrac{2}{\pi} log 2$ the end losses in the channel and $f(\beta) = \dfrac{2}{\beta h}\left(e^{\frac{\beta h}{2}} - 1\right)$, h is the width and l is the length of the generator. From this we have found that

$$(\eta_g)_{max} = f(\beta)\frac{\sqrt{1+a^*} - \sqrt{a^*}}{\sqrt{1+a^*} + \sqrt{a^*}} \tag{8}$$

This $(\eta_g)_{max}$ is computed for different values of, K, β, a and found that $(\eta_g)_{max}$ lies above 0.6 even for small values of aspect ratio a^*. The computed values are experimentally verified by designing a suitable model as shown in the following Fig.3. Further, our model works as a generator for $K < 1$ and as a pump for $K > 1$.

We have also achieved a desired stratification by mixing in proper proportion of alkaline water and municipal water to achieve the higher efficiency of the generator. We have taken patent for it. This is one of the possible solutions for the problem 3 out of 10 mentioned above.

1.4. *To save electricity in the existing system:*

We have developed [3] a mathematical model based on numerical analysis to isolate the eddy current losses from the stray losses in a transformer. We have shown that a suitable tank material and proper dimensions of the tank will reduce the eddy current losses considerably. This is one of the possible solutions for the problem 4 in the above list of ten problems.

Fig. 3. The Poorman Generator

1.5. *Potable water:*

At present, considerable amount of waste water from municipalities, hospitals, industries and so on have been discharged into drainage. This waste water involves organic and inorganic substances dissolving in water making it a poorly conducting liquid. In this, the conductivity increases with temperature and concentration of substance. In the process of solidifying this waste water by heating from above and cooling from below, the electrical conductivity varies with difference in temperature. This variation of conductivity releases the charges from the nuclei forming the distribution of density of charges, ρ_e. These charges produce an electric field called induced electric field, $\vec{E}_i$. If high strength of electric field is needed we can generate it, called applied electric field, $\vec{E}_a$, by embedding electrodes of different potentials at the bottom and top surfaces. This electric field, $\vec{E} = \vec{E}_a + \vec{E}_i$ not only produces a current according to Ohm's law but also produces an electric force. This current acts as sensing and force acts as actuation which are the two required properties to form a material as a smart material. We [5,6] have shown that, this property of smart material purifies waste water more effectively than any other methods. This method can be used as an effective method to purify waste water and make it potable water. This is one of the possible solutions for the problem 5 out of ten listed above.

1.6. *Possible solution to control waste materials:*

In these days of liberalization, modernization and arbitrary use of automobiles produce waste material when released to the atmosphere called suspended particulate matter (SPM). If these SPMs are retained in the atmosphere, they are called aerosols. The coagulation in the atmosphere hit these aerosols and sometimes sticks to each other forming aerosol of larger sizes. Sometimes the coagulation splits aerosols forming tiny particles. We [7] have shown that large size aerosols are advantages for the formation of clouds and tiny aerosols are disadvantageous when inhaled through the nasals blocking the trachea (wind pipe) causing breathing problem. The process of forming large size aerosols using waste materials is advantageous in the formation of clouds. This is one of the possible solution to get rid of waste materials posed under problem 7 out of ten listed above.

2. Mathematical Modeling

To solve some of the problems posed above we need Mathematical models. In this section we briefly explain it and its importance in science, engineering and technology.

2.1. *Need for Mathematical Modeling*

A single experiment covering all aspects of a problem is not possible to find all the parameters involved in it. Only the behavior of certain aspects can be directly observed. Also, it is difficult to measure, simultaneously through an experiment, more than one aspect of the several problems. A mathematical model in which all the parameters can be controlled may be a valuable tool for a better understanding of the problems to be investigated. This is because if one aspect of the parameters in a mathematical model is measured and verified it experimentally, then the other remaining parameters in the modeling can easily be computed from the mathematical model and can be used if needed in the practical problem.

2.2. *Flow Chart for Mathematical Modeling*

The processes involved in a mathematical modeling are:

Step1: State the problem to be solved in such a way that even a layman can understand it. We should also explain why we need it, where to apply, when to apply and how to apply.

Step2: Develop the equations required using the physical phenomena. In civil and mechanical engineering, problems like vibrations, bending of beams and so on, one usually uses Newton's second law to derive the required equations. The electrical and electronics engineers dealing with circuit theory use Kirchoff's second law. One has to stress that Kirchoff's second law is analogous to Newton's second law.

Step3: Solve these equations using mathematical tools. It is known that nonlinear equations cannot be solved analytically in all cases except in some trivial cases. Therefore, often engineers resort to a numerical solution. A numerical method gives only numbers and fails to give a physical insight. Therefore, a scientist or an engineer, before resorting to a numerical method it is always beneficial to use the following analytical techniques to find the solutions to understand the physics of the problem.

i) Method of characteristics
ii) Truncated representation of Fourier series.
iii) Lyapanov technique
iv) Energy method
v) Moment method
vi) Galerkin method
vii) Regular and singular perturbation techniques.
viii) Computer extended series solution (CESS).

In the remaining part of this section, I will discuss the following two biomedical problems to illustrate the use of mathematical models.

2.3. *Spreading of a Contagious Disease - AIDS*

Human immunodeficiency virus (HIV) is the causative agent for Acquired Immune Deficiency Syndrome (AIDS), which is proving fatal to millions of Indians. HIV depends on many specific proteins from the virus as well as the host. Many of the drugs prescribed for treatment of this disease are inhibitors of the viral enzymes. At present, the following two approaches to control this disease are proposed:

1. Development of vaccines
2. Development of antiviral drugs

In addition to these we have shown that awareness factor of this disease prevents spreading of the disease as explained below.

Consider a residential area having the population n, out of which let P(t), be the number of people who are prone to this disease and Q(t) be the number of people, who are not prone to this disease. It is reasonable, from phenomenological point of view, to assume that the rate of P at which the disease will spread is given by

$$\frac{dp}{dt} = kPQ \tag{9}$$

where k is the constant of proportionality attributed to awareness factor of AIDS. Let one infected person be introduced into a fixed population of n people, then P and Q are related by

$$P + Q = n + 1 \tag{10}$$

Eliminating Q in (9), using (10), we get,

$$\frac{dp}{dt} = kP(n + 1 - P) \tag{11}$$

This is a non-linear differential equation of Riccati type which has to be solved using the initial condition

$$P = 1 \quad at \quad t = 0 \tag{12}$$

Analytical solution of this equation is possible using a mathematical model based on a suitable transformation. For example, consider a transformation

$$P = \frac{1}{V} \tag{13}$$

Then

$$\frac{dp}{dt} = \frac{-1}{V^2}\frac{dV}{dt} \tag{14}$$

Substituting Eqn. (14) into Eqn.(11), we get

$$\frac{-1}{V^2}\frac{dV}{dt} = \frac{k}{V}(n + 1 - \frac{1}{V})$$

That is,

$$\frac{dV}{dt} + k(n + 1)V = k \tag{15}$$

The advantage of the transformation (13) is to transform the nonlinear ordinary differential Eqn. (11) to a linear Eqn. (15). Its integrating factor is $e^{k(n+1)t}$. Then the above equation can be written as

$$\frac{d\left(e^{k(n+1)t}V\right)}{dt} = ke^{k(n+1)t} \tag{16}$$

Integrating it and rearranging it, we get

$$V = \frac{1}{n+1} + Ae^{-k(n+1)t} \tag{17}$$

where A is a constant of integration to be determined using the initial condition (12). This condition yields

$$A = \frac{n}{n+1} \tag{18}$$

Then,

$$V = \frac{1}{n+1} + \frac{n}{n+1}e^{-k(n+1)t} \tag{19}$$

From Eqn. (13), using Eqn. (18), we get

$$P = \frac{n+1}{1+ne^{-k(n+1)t}} \tag{20}$$

This equation predicts the spread of the disease over the time, t, for a particular population. This solution reveals that if the awareness factor k increases the spreading of this disease decreases. Hence we conclude that the increase in awareness is one of the possible solutions in the control of spreading of AID.

2.4. *A Couple Stress Model Of Blood Flow In Microcirculation*

Microcirculation is the study of blood flow in small blood vessels, particularly in the capillaries. In physiology the most important functions of the circulation of blood through capillaries are to supply nutrients to every living cell of the organisms and also to remove various waste products from every cell. The capillaries are bounded by endothelial cells which have ultra microscopic pores through which substances of various molecular size can penetrate the surrounding tissue and also the capillary. One of the important features of the capillary geometry which distinguishes from the arteries is the permeability of the wall. The deposition of the cholesterol is believed to increase the permeability of the wall. Such, an increase in permeability also result from dilated damaged or inflamed capillary walls. Thus, it is worthwhile to study the effect of wall permeability of the blood vessel from a Fluid Mechanics point of view. Most of the available works on blood flow are concerned with the assumption of Newtonian Fluid. The available experimental works (see Dulal Pal et al [8]) on blood flows indicate that under certain flow conditions, blood flow exhibit strong deviation from Newtonian flow behavior. This deviation mainly occurs, in the form of non-parabolic velocity profile, for flow through tubes of small diameter. This is because of the presence of non-uniform red blood cells (RBC) in the blood and hence spins with certain angular velocity $\vec{\Omega}$. When this angular velocity balances with the natural vorticity of the blood producing a couple stress in the blood which is called a Couple Stress Fluid (see Stokes [9]). The study of blood flow with couple stress may play a vital role in understanding Rheological anomalies associated with the blood flows. In this section, we study the combined effect of the couple stress and the exchange of fluids across the capillary walls on the flow of blood in microcirculation. For this purpose we consider a simplified model for a capillary flow and the blood flow and solve the basic equations using Starling's hypothesis of fluid exchange which states that the difference in hydraulic pressure between the blood and tissue fluid is not only responsible for the process of filtration but also depends on the difference in colloidal pressure between the blood and tissue fluid.

2.4.1. *Formulation of the Problem*

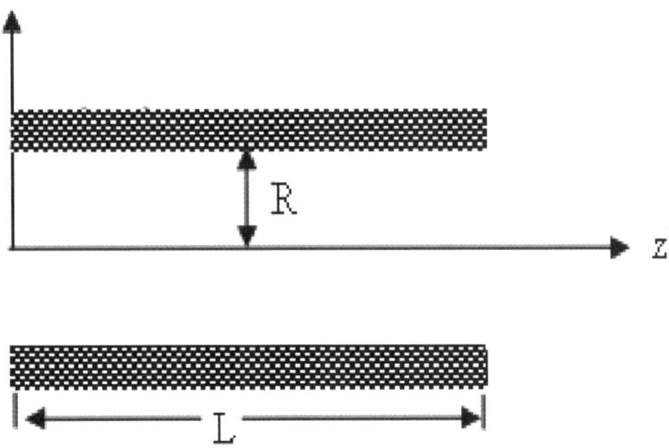

Fig. 4. The Synovial Joints

We consider a physical configuration as shown in Fig.4. Following Lighthill [10], we use the assumption that the effect of curvature can be neglected in the case of creeping flow. This implies that the capillary between an arteriole and a venule is a tube of uniform circular cross-section with permeable wall as shown in Fig.4. In addition, we assume the steady flow of couple stress fluid in the capillary tube. The cylindrical co-ordinates (r, θ, z) are chosen with z-axis coinciding with the axis of the tube. The flow is assumed to be axisymmetric implying all the physical quantities are independent of θ $\left(i.e. \frac{\partial}{\partial\theta} = 0\right)$. The permeability of the wall is governed by Starling's law, defined above, which is a modification of Fick's law and states that the net filtration pressure is given by the difference between hydrostatic and osmotic pressure between the blood and tissue fluid. The filtered water which passes into the tissue is either reabsorbed into the capillary blood or returned to the blood via the Lymphatic systems. Starlings hypothesis is usually expressed in the form

$$M = k(P_c - P_i - \Pi_c + \Pi_i) \tag{21}$$

where M is the flow rate per unit area of wall surface, the constant k is the measure of the permeability of the capillary wall to water and is called the filtration constant, P_c the hydrostatic capillary blood pressure, P_i the interstitial fluid pressure, Π_c the osmotic pressure of the plasma and Π_i is the pressure of the proteins in the interstitial fluid. Here M is positive when hydrostatic pressure difference is greater than the osmotic pressure difference and it implies the filtration of fluid out of the capillary. If M is negative implies reabsorption of fluid from the interstitial space into the capillary takes place. If M is as defined above then P_c cannot vary along the length of the capillary and it has to be replaced by an appropriate value.

2.4.2. *Equations of motion, Constitutive Equations and Boundary Conditions*

The constitutive equations and equations of motion for a couple stress fluid flows, in the absence of body moment and body couple, are

$$\tau_{ij,j} = \rho \frac{dv_i}{dt} \tag{22}$$

$$e_{ijk} T^A{}_{jk} + M_{ji,i} = 0 \tag{23}$$

$$\tau_{ij} = -p\delta_{ij} + 2\mu\alpha_{ij} \tag{24}$$

$$\mu_{ij} = 4\eta\omega_{j,i} + 4\eta'\omega_{i,j} \tag{25}$$

where ρ is the density,τ_{ij} and $T^A{}_{ij}$ are respectively the symmetrical and antisymmetrical parts of the stress tensor, v_i is the velocity vector, M_{ij} the couple stress tensor, μ_{ij} the deviotoric part of M_{ij}, ω_i is the vorticity vector, δ_{ij} is the symmetric part of velocity gradient, η and η' are constants associated with the couple stress, p the pressure and other terms have their usual meanings in tensor analysis. Following Skalak [11], we neglect the inertial term in the basic equations because the rate of inertia is not so significant in microcirculation. These basic equations of motion (22) and (23), using (24) and (25) and the assumption stated above take the form

$$\nabla p = \nabla^2 \left(\mu \vec{q} - \eta \nabla^2 \vec{q}\right) \tag{26}$$

where

$$\nabla^2 = \frac{1}{r}\frac{\partial}{\partial r}\left(r\frac{\partial}{\partial r}\right) \tag{27}$$

The two-dimensional form of equation (26), using $\vec{q} = (u, v)$, is The conservation of momentum:

$$\begin{aligned}\frac{\partial p}{\partial r} =&\mu\left(\frac{\partial^2 v}{\partial r^2} + \frac{1}{r}\frac{\partial v}{\partial r} - \frac{v}{r^2} + \frac{\partial^2 v}{\partial z^2}\right) - l^2\mu\left(\frac{\partial^4 v}{\partial r^4} + \frac{\partial^4 v}{\partial z^4} + 2\frac{\partial^4 v}{\partial r^2 \partial z^2}\right)\\ &- l^2\mu\left(\frac{2}{r}\frac{\partial^3 v}{\partial r^3} + \frac{2}{r}\frac{\partial^3 v}{\partial r \partial z^2} - \frac{3}{r^2}\frac{\partial^2 v}{\partial r^2} + \frac{3}{r^3}\frac{\partial v}{\partial r} - \frac{3}{r^4}v\right)\end{aligned} \tag{28}$$

$$\begin{aligned}\frac{\partial p}{\partial z} =&\mu\left(\frac{\partial^2 u}{\partial r^2} + \frac{1}{r}\frac{\partial u}{\partial r} - \frac{u}{r^2} + \frac{\partial^2 u}{\partial z^2}\right) - l^2\mu\left(\frac{\partial^4 u}{\partial r^4} + \frac{\partial^4 u}{\partial z^4}(2\frac{\partial^4 u}{\partial r^2 \partial z^2}\right)\\ &- l^2\mu\left(\frac{2}{r}\frac{\partial^3 u}{\partial r^3} + \frac{2}{r}\frac{\partial^3 u}{\partial r \partial z^2} - \frac{3}{r^2}\frac{\partial^2 u}{\partial r^2} + \frac{3}{r^3}\frac{\partial u}{\partial r} - \frac{3}{r^4}u\right)\end{aligned} \tag{29}$$

The conservation of mass for incompressible fluid :

$$\frac{1}{r}\frac{\partial}{\partial r}(rv) + \frac{\partial u}{\partial z} = 0 \tag{30}$$

where μ is the coefficient of viscosity of the fluid and η is the couple stress parameter of the fluid and both μ and η have the dimensions of MLT and $l^2 = \frac{\eta}{\mu}$ has the dimensions of length squared. The required boundary conditions are

$$v = 0, \frac{\partial u}{\partial r} \quad at \quad r = 0 \tag{31}$$

$$v = k\,(p - \alpha) \qquad u = 0 \quad at \quad r = R \tag{32}$$

$$\overline{p} = p_a \quad at \quad z = 0 \tag{33}$$

$$\overline{p} = p_v \quad at \quad z = L \tag{34}$$

where R is the radius and L is the length of the capillary, $\overline{p}$ is the average of the pressure p over the cross-section of the capillary, p_a and p_v are respectively the arterial and venous end pressures, which are taken to be the constants. The coefficient k in equation (31) is a measure of the permeability of the wall and $\alpha = \pi_c + p_i - \pi_i$ is assumed to be a constant. The boundary condition (31) is the no slip condition. Note that this no slip condition, instead of slip condition proposed by Beavers and Joseph [12] or by Rudraiah [13] at the interface between fluids in a capillary and the porous boundary is assumed to be valid because the permeability of the porous wall of the capillary is very small. Further, in a couple stress fluid both a yield stress and shear dependent viscosity exist so that the fluid element in contact with the boundary adheres to it and hence has the same velocity as the boundary. The boundary condition (31) is Starling's hypothesis which takes care of the smooth transfer of mass across the porous wall.

2.4.3. *Mathematical Model for the Solution of Flow Field*

To solve for the flow field, we introduce the dimensionless parameter

$$\varepsilon = \frac{\mu k}{R} \tag{35}$$

where μ is the coefficient of viscosity and the filtration constant, k, varies widely for capillaries. The average value of muscle capillaries of dog and cat is $2.5 \times 10^{-}8cm/(sec.cmH_2o)$. If we take $\eta = kp$ and $R = 5\mu m$ then we get$\varepsilon = 1 \times 10^{-}7$. Hence ε can be regarded as very small. The order of u, v and p may be estimated respectively as $1, \varepsilon, \varepsilon^2$. The derivatives and with respect to z will be of higher order of than that of the derivatives with respect to r.

In view of the mathematical model of the order analysis, a rough estimate of the orders of magnitude of the various terms in equations (28)-(30) is given below:

$$\begin{aligned}
&O(1): \frac{\partial p}{\partial z}, \frac{\partial u}{\partial r}, \frac{\partial^2 u}{\partial r^2}, \frac{\partial^3}{u}\partial r^3, \frac{\partial^4 u}{\partial r^4} \\
&O(\varepsilon): \frac{\partial p}{\partial r}, \frac{\partial u}{\partial z}, \frac{\partial^2 u}{\partial r^2}, \frac{\partial v}{\partial r}, \frac{\partial^2 v}{\partial r^2}, \frac{\partial^3 v}{\partial r^3}, \frac{\partial^4 v}{\partial r^4} \\
&\frac{\partial^4 u}{\partial r^4}, \frac{\partial^4 u}{\partial r^2 \partial z^2}, \frac{\partial^4 u}{\partial r \partial z^3}, \frac{\partial v}{\partial r \partial z^2}, \frac{1}{r}\frac{\partial (rv)}{\partial r}, \frac{v}{r^2} \\
&O(\varepsilon^2): \frac{\partial^2 v}{\partial z^2}, \frac{\partial^4 u}{\partial r^2 \partial z^2}, \frac{\partial^4 v}{\partial z^4}
\end{aligned} \tag{36}$$

A mathematical model to find solutions for the basic equations (28)-(30) is to split into two parts, the first and second order in the form

$$\begin{aligned} u(r,z) &= u_1(r,z) + u_2(r,z) \\ v(r,z) &= v_1(r,z) + v_2(r,z) \\ p_1(r,z) &+ p_2(r,z) \end{aligned} \tag{37}$$

where u_1,p_1 $\sim$ $O(1)$,u_2,p_2,v_1 $\sim$ $O(\varepsilon)$,v_2 $\sim$ $O(\varepsilon^2)$. Substituting (37) into (28)-(30), using the order of analysis of the terms, we obtain the following simplified version for the first and second approximation. (i) First approximation:

$$\frac{\partial p_1}{\partial r} = 0 \tag{38}$$

$$\frac{\partial p_1}{\partial z} = \mu D(1 - l^2 D)u_1 \tag{39}$$

$$\frac{1}{r}\frac{\partial(rv_1)}{\partial r} + \frac{\partial u_1}{\partial z} = 0 \tag{40}$$

where

$$D = \frac{1}{r}\frac{\partial}{\partial r}\left(r\frac{\partial}{\partial r}\right) \tag{41}$$

with the boundary conditions

$$\frac{\partial u_1}{\partial r} = 0 \quad \textit{and} \quad v_1 = 0 \quad \textit{at} \quad r = 0 \tag{42}$$

$$u_1 = 0 \quad \textit{and} \quad v_1 = k(p_1 - \alpha) \quad \textit{at} \quad r = R \tag{43}$$

$$\overline{p_1} = p_u \quad \textit{at} \quad z = 0 \tag{44}$$

$$\overline{p_1} = p_v \quad \textit{at} \quad z = L \tag{45}$$

(ii) Second approximation:

$$\frac{\partial p_z}{\partial r} = \mu\left(D - l^2 D_1^2 - \frac{1}{r^2}\frac{\partial^2}{\partial r^2} - \frac{3}{r^3}\frac{\partial}{\partial r}\right)v_2 \tag{46}$$

$$\frac{\partial p_z}{\partial z} = \mu\left[\frac{\partial^2 u_2}{\partial r^2} + \frac{1}{r}\frac{\partial u_2}{\partial r} + \frac{\partial^2 u_1}{\partial z^2} - l^2\left(D_2 u_1 + D_3 u_2\right)\right] \tag{47}$$

$$\frac{1}{r}\frac{\partial(rv_2)}{\partial r} + \frac{\partial u_2}{\partial z} = 0 \tag{48}$$

where

$$D_1 = \frac{\partial^2}{\partial r^2} + \frac{1}{r}\frac{\partial}{\partial r} - \frac{1}{r^2} \tag{49}$$

$$D_2 = \frac{\partial^4}{\partial r^4} + 2\frac{\partial^4}{\partial r^2 \partial z^2} + \frac{2}{r}\frac{\partial^3}{\partial r \partial z^2} \tag{50}$$

$$D_3 = \frac{\partial^4}{\partial r^4} + \frac{2}{r}\frac{\partial^3}{\partial r^3} - \frac{1}{r^2}\frac{\partial^2}{\partial r^2} + \frac{1}{r^3}\frac{\partial}{\partial r} \tag{51}$$

with the boundary conditions

$$\frac{\partial u_2}{\partial r} = 0 \quad and \quad v_2 = 0 \quad at \quad r = 0 \tag{52}$$

$$u_2 = 0 \quad and \quad v_2 = kp_2 \quad at \quad r = R \tag{53}$$

$$\overline{p_2} = 0 \quad at \quad z = 0 \tag{54}$$

$$\overline{p_2} = 0 \quad at \quad z = L \tag{55}$$

2.4.4. ***Solution for small permeability***

According to the physiological data (see Dulal pal et al[8]) the permeability of the porous capillary is very small, of the order of 10^{-6}. Hence, without giving the details (interested reader may refer to the work of Oka and Murata [14] for details), we directly write, as given below, the solutions for first and second order velocity and pressure for small permeability, obtained for first and second approximations:

$$\begin{aligned} u_1(\zeta,\xi) =& \frac{R^2}{4\mu}\left(\frac{\Delta p}{L}\right)\left[(1-\zeta^2) + \frac{4}{a^2}(g_0(\zeta)-1)\right] \\ & \left[1 - \frac{8\varepsilon}{3\beta^2\lambda_0}\left(1 - 3\frac{\Delta\alpha}{\Delta p} + 6\frac{\Delta\alpha}{\Delta p}\xi - 3\xi^2\right)\right] \end{aligned} \tag{56}$$

$$v_1(\zeta,\xi) = \frac{\varepsilon R \Delta p}{\mu\lambda_0}\left(\frac{\Delta\alpha}{\Delta p} - \xi\right)\left[2\zeta - \zeta^3 + \frac{16}{a^3}(g_1(\zeta) - \frac{1}{2}\zeta a)\right] \tag{57}$$

$$p_1(\zeta,\xi) = p_a - \xi\Delta p + \frac{8\varepsilon\Delta p}{3\lambda_0\beta^2}\left[\left(1 - 3\frac{\Delta\alpha}{\Delta p}\right)\xi\frac{\Delta\alpha}{\Delta p} + 3\xi^2 - \xi^3\right] \tag{58}$$

$$
\begin{aligned}
u_2(\zeta,\xi) =& \frac{8\varepsilon\Delta pR^2}{4\lambda_0\mu L}\left[\left(1-\frac{8\varepsilon}{a^2\lambda_0}\right)\zeta^4+\frac{8}{a^2}\left(2s_1+\frac{\varepsilon}{\lambda_0}\right)g_0(\zeta)\right] \\
&-\frac{8\varepsilon\Delta pR^2}{4\lambda_0\mu L}\left[\left(16\frac{s_1}{a^2}+\frac{m_1}{3\beta_1}+\beta_2\right)\zeta^2\right] \\
&-\frac{8\varepsilon\Delta pR^2}{4\lambda_0\mu L}\left[\frac{4}{a^2}\left(\frac{m_1}{3\beta_1}+\beta_2\right)(1-g_0(\zeta))\right] \\
&+\frac{8\varepsilon\Delta pR^2}{4\lambda_0\mu L}\left[-\frac{64\delta_7(\zeta)}{a^4}+\frac{m_1\beta_2}{3\beta_1}-1\right]
\end{aligned} \tag{59}
$$

$$
v_2(\zeta,\xi) = 0 \tag{60}
$$

$$
p_2(\zeta,\xi) = -\frac{4\varepsilon\Delta p}{\lambda_0}\left(\frac{\Delta\alpha}{\Delta p}-\xi\right)\left[\xi^2-\frac{\beta_2(1+2\beta_1)}{6I_0(a)}h_1(\zeta)+h_2(\zeta)+\beta_3\right] \tag{61}
$$

where $\xi = \dfrac{z}{L}$ is the normalized axial distance from the arteriolar end of the capillary,$\zeta = \dfrac{r}{R}$, $a = \dfrac{R}{l}$ is the couple stress parameter,$\delta p = p_0 - p_v$, $\delta\alpha = p_a - \alpha$, $\beta = \dfrac{R}{L}$

$$
\lambda_0 = 1+\frac{16}{a^3}\left[\frac{I_1(a)}{I_0(a)}-\frac{a}{2}\right] \tag{62}
$$

where $g_0(\zeta) = \dfrac{I_0(\zeta a)}{I_0(a)}$, $g_1(\zeta) = \dfrac{I_1(\zeta a)}{I_0(a)}$, I is the modified Bessel function and other constants appearing in the equations (59) - (61) are defined in the appendix.

Finally, combining the first and second approximations for the velocity and pressure fields we obtain the solution in the form

$$
\begin{aligned}
U(\zeta,\xi) &= \frac{u(\zeta,\xi)}{(\frac{R^2\Delta p}{4\mu L})} \\
&= \left[(1-\zeta^2)+\frac{4(g_0(\zeta)-1)}{a^2}\right][1+\varepsilon f_1(0,\xi)]+\varepsilon f_2(\zeta,0)
\end{aligned} \tag{63}
$$

$$
\begin{aligned}
V(\zeta,\xi) &= \frac{v(\zeta,\xi)}{\frac{R^2\Delta p}{4\mu L}} \\
&= \frac{4\varepsilon}{\beta\lambda_0}\left(\frac{\Delta\alpha}{\Delta p}-\xi\right)\left[2\zeta-\zeta^3+\frac{16\left(g_0(\zeta)-\frac{\zeta a}{2}\right)}{a^3}\right]
\end{aligned} \tag{64}
$$

$$
P(\zeta,\xi) = p_a - \xi\Delta p + \varepsilon g_2(\zeta,\xi)\Delta p \tag{65}
$$

where $f_0(\zeta,\xi)$, $f_1(0,\xi)$ and $g_2(\zeta,\xi)$ are given in the appendix. To know the accuracy of the solutions given by the equation (63) to (65) we can reduce them to the available solutions in the literature for Poiseuille flow in the limit of $\varepsilon \to 0$and a $\to \infty$ (Newtonian flow). In these limits equations (63) to (65) reduce to the solutions

$$u = \frac{R^2 \Delta p(1-\zeta^2)}{4\mu L}, v = 0, p = p_a - \xi \Delta p \tag{66}$$

which is available in the literature. We also note that our solutions (63)-(65) reduce to those given by Oka and Murata [13] as $a \to \infty$ (Newtonian fluid).

2.4.5. *Solutions for Streamlines*

To understand the nature of the streamlines, they are determined using the equation

$$\frac{dr}{v} = \frac{dz}{u} \tag{67}$$

Integrating Eqn. (67), using Eqns. (63) and (64), we get

$$\delta_1(\zeta)\left[1 + \varepsilon f_1(0,\xi)\right] + \delta_2(\zeta) = C \tag{68}$$

where $\delta_1(\zeta)$ and $\delta_2(\zeta)$ are given in the appendix and C is an arbitrary constant.
Volume of flow Q per Unit Time
It is of practical importance to calculate the volume of flow of fluid per unit time across the cross-section. It is at a point, z,is given by

$$Q^* = \int_0^R 2\pi u r \, dr \tag{69}$$

Using Eqn. (63), Q is obtained from Eqn. (69) as

$$Q = \frac{Q^*(\xi)}{\left(\frac{R^4 \Delta p}{4\mu L}\right)} = \frac{\pi \lambda_0}{2}\left[(1 + \varepsilon f_1(0,\xi)) + \frac{4\varepsilon A_0}{\lambda_0^2}\right] \tag{70}$$

where λ_0 is given in Eqn. (62) and A_0 is given in the appendix.
The net outflow, M, of water into the tissue per unit time across the capillary wall can be calculated using

$$M = Q^*(0) - Q^*(1) \tag{71}$$

This, using Eqn. (70), becomes

$$M = \frac{2\pi\varepsilon R^4}{\mu\beta^2}\frac{\Delta p}{L}\left(\Delta\alpha\Delta p - \frac{1}{2}\right) = ks\left(\frac{\Delta\alpha}{\Delta p} - \frac{1}{2}\right) \tag{72}$$

$$\therefore m' = \frac{M}{S} = k\left(\frac{\Delta\alpha}{\Delta p} - \frac{1}{2}\right) = k(p_m - \alpha) \tag{73}$$

$$or\ m' = k(p_m - p_i - \pi_c + \pi_i) \tag{74}$$

where p_m is the arithmetical mean of p_a and p_v. It is to noted that p_c in equation (1) is replaced by p_m.
From Eqn. (72) it is clear that $M = 0$ when $\frac{\Delta\alpha}{\Delta p} = \frac{1}{2}$ that is outflow and inflow are balanced across the wall. Only outflow *i.e.*, $M > 0$ if $\frac{\Delta\alpha}{\Delta p} > \frac{1}{2}$ and there is no outflow *i.e.*, $M < 0$ for the case where $\frac{\Delta\alpha}{\Delta p} < \frac{1}{2}$.

2.4.6. *Discussion and Conclusions*

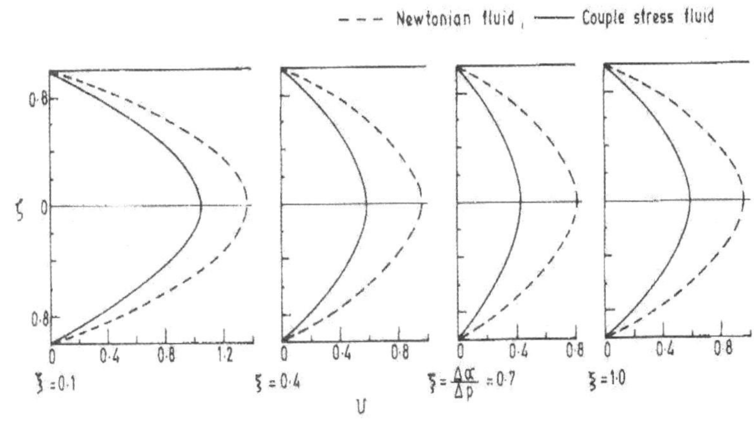

Fig. 5. Profiles of longitudinal velocity component u for Newtonian and couple stress fluids for $a = \frac{R}{l} = 3.0, \varepsilon = 4.9 \times 10^{-}6, \frac{\Delta\alpha}{\Delta p} = 0.7$.

A mathematical model describing blood flow through a capillary with permeability of the vessel has been investigated and analytical solutions have

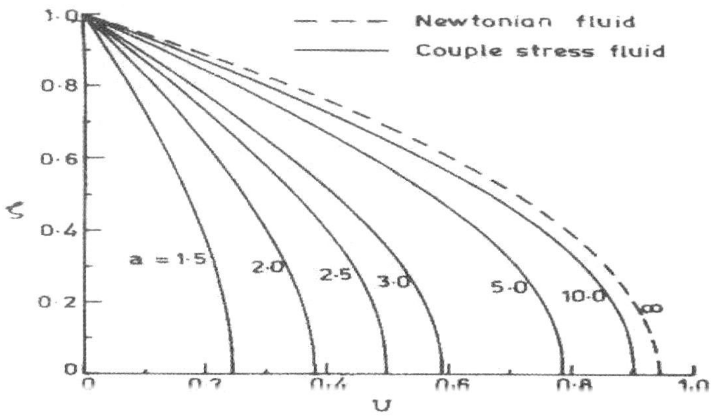

Fig. 6. Comparison of profiles of longitudinal velocity u for Newtonian and couple stress fluids for $\dfrac{\Delta\alpha}{\Delta p} = 0.7, \varepsilon = 4.9 \times 10^{-6}$.

been obtained. Although several assumptions have been made in our mathematical model the results obtained agree fairly well with the experimental results obtained by Oka and Murata [14]. Both the longitudinal and radial velocities, stream functions and the volume of fluid Q are numerically computed for Newtonian as well as couple stress fluids and the results are depicted graphically in Fig.5-Fig.9. From Fig5 it is clear that the axial velocity attains a minimum value at the point $\xi = \dfrac{\Delta\alpha}{\Delta p}$ and it decreases or increases in the region $\xi > \dfrac{\Delta\alpha}{\Delta p}$ or $\xi < \dfrac{\Delta\alpha}{\Delta p}$ respectively. From this we conclude that the axial velocity has a decreasing tendency in blood flow as compared to a Newtonian fluid. On the other hand the axial velocity increases with an increase in $a = \dfrac{R}{L}$ and coincides with a Newtonian profile for very large a (Fig.6). The stream line patterns shown in Fig.7 reveals that some of them are clustered along the central part while others are directed towards the wall. From this we conclude that the red cells in the blood accumulate near the axis of the capillary in conformity with the real situation. Similarly the plasma tends towards the wall.

The radial velocity is shown in Fig.8 which shows that it is zero at $\xi = \dfrac{\Delta\alpha}{\Delta p}$ whereas it is positive or negative depending on $\xi < \dfrac{\Delta\alpha}{\Delta p}$ or $\xi > \dfrac{\Delta\alpha}{\Delta p}$ or respectively. Further we see that there appears outflow and inflow at the wall in the region $\xi < \frac{\Delta\alpha}{\Delta p}$ and $\xi < \frac{\Delta\alpha}{\Delta p}$. Further, we observe that radial velocity decreases with an increase in a and coincides with the Newtonian profile for large a. We also note that as in the axial velocity, the radial velocity also decreases in the blood flow. The volumetric rate Q is depicted in the Figure 9. This Figure reveals that Q is minimum at $\xi = \frac{\Delta\alpha}{\Delta p}$ and the effect of couple stress is to decrease Q. It is also clear that an increase in a increases Q and finally tends to the constant value for a Newtonian fluid. Physically a decrease in Q with an increase in ξ in the region $0 < \xi < \frac{\Delta\alpha}{\Delta p}$ is attributed to filtration. Further, an increase in Q with ξ in the region $\frac{\Delta\alpha}{\Delta p} < \xi < 1$ is due to absorption. The net flow will cause the edema. Finally we conclude that lymphatic will pay a role in protecting the tissues against edema.

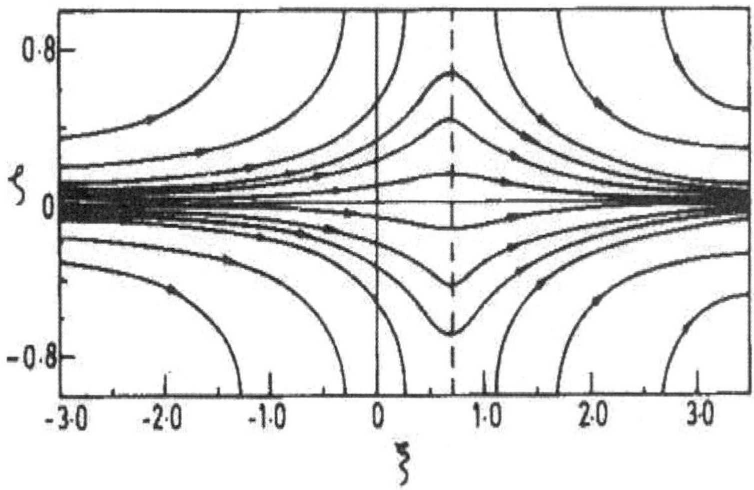

Fig. 7. Streamlines for couple stress fluid.

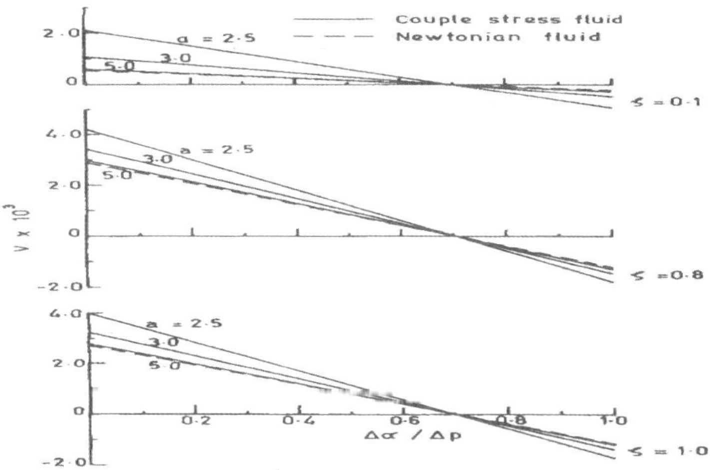

Fig. 8. Radial velocity profiles for Newtonian and couple stress fluids for $\dfrac{\Delta\alpha}{\Delta p} = 0.7$

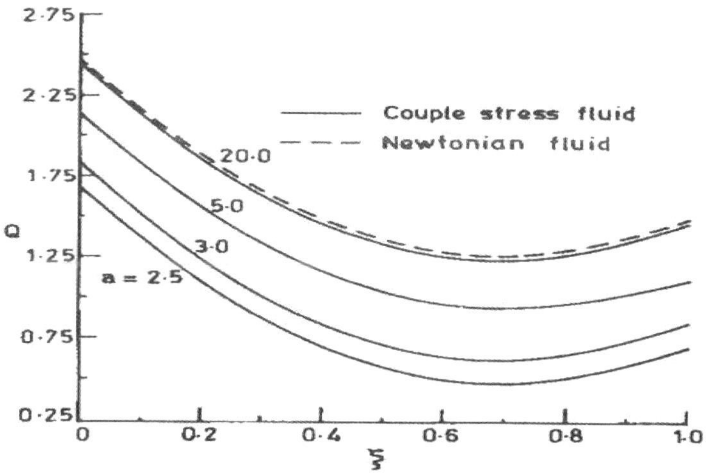

Fig. 9. Relationship between Q and ξ for Newtonian and couple stress fluids for $\dfrac{\Delta\alpha}{\Delta p} = 4.9 \times 10^{-6}$.

3. Mathematics concerned with ODE unifies Different Branches of Science, Engineering and Technology

This is illustrated, first, by considering practical problems in section (3.1) to (3.4), dealing with ordinary differential equations and then explains it in section 3.5 by considering partial differential equation. In ordinary differential equation (ODE) we usually teach students an equation of the form

$$a\frac{d^2y}{dt^2} + b\frac{dy}{dt} + cy = f(t) \tag{75}$$

where y is a dependent variable (i.e. an unknown) t is an independent variable (i.e. known) $f(t)$ is a known function, a, b, and c are the coefficients of ODE. Here a, b and c may be constants, function of t only and they may be functions of y and t. In the first two cases Eqn. (75) is a linear ODE and in the third case it is a nonlinear ODE. Equation of the type (75) appears in many branches of science, engineering and technology, as explained below.

3.1. *Mechanical Oscillations*

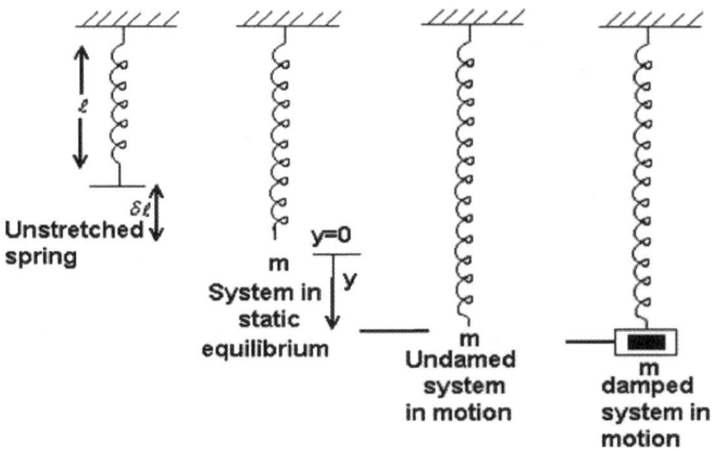

Fig. 10. Displacement in a Spring

Consider a spring of length l suspended vertically from a fixed support with a body of mass m attached to the lower end of the spring as shown in Fig.10. In deriving the required equation, we make use of the following assumptions:
m is assumed to be so large that the mass of the spring can be neglected compared to m. Let δl be the elongation when a mass m is attached to

it which is small compared to l and leads to Static equilibrium, namely where k is the Young's modulus obtained by applying Hook's law to the spring and g is the acceleration due to gravity. Let $y(t)$, measured positive downwards, be the displacement of the mass from the equilibrium position. $y(t)$ is related to forces acting on the system through Newton's second law. According to convention, the force in the down ward direction is taken as positive while upward direction as negative. The following forces act on the mass m:

(1) Force F_1 due to gravity, acting downwards, is

$$F_1 = mg \tag{76}$$

(2) The force F_2, exerted by the spring when it is stretched, acting upwards, according to Hooke's Law, is

$$F_2 = -k\delta l \tag{77}$$

where k is a spring constant, called Young's modulus.

(3) The force F_3, due to spring which according to Hooke's law acts upward is

$$F_3 = -ky \tag{78}$$

(4) Let F_4 be the damping force due to dash pot which acts upwards is

$$F_4 = -c\frac{dy}{dt} \tag{79}$$

where $(c > 0)$ is the damping constant.

(5) Let $f(t)$ be the external force acting on the system. Then according to Newton's second law

$$m\frac{d^2y}{dt^2} = mg - k\delta l - ky - c\frac{dy}{dt} + f(t) \tag{80}$$

This equation, using Eqns. (76) to (80) for F_1, F_2, F_3 and F_4, becomes

$$m\frac{d^2y}{dt^2} = F_1 + F_2 + F_3 + F_4 = f(t) \tag{81}$$

In the equilibrium state,$F_1 + F_2 = 0$ i.e.,

$$mg - k\delta l = 0 \tag{82}$$

Then Eqn. (81), using Eqn. (82) becomes

$$m\frac{d^2y}{dt^2} + c\frac{dy}{dt} + ky = f(t) \tag{83}$$

If the damping is small (*i.e.*, $c \to 0$) then Eqn. (83) becomes

$$m\frac{d^2y}{dt^2} + ky = f(t) \tag{84}$$

In the absence of external force $f(t)$, Eqn. (84) becomes

$$m\frac{d^2y}{dt^2} + ky = 0 \tag{85}$$

This is the equation for simple harmonic oscillations, with internal frequency $w_e = \sqrt{\frac{k}{m}}$

3.2. *Electrical Circuits*

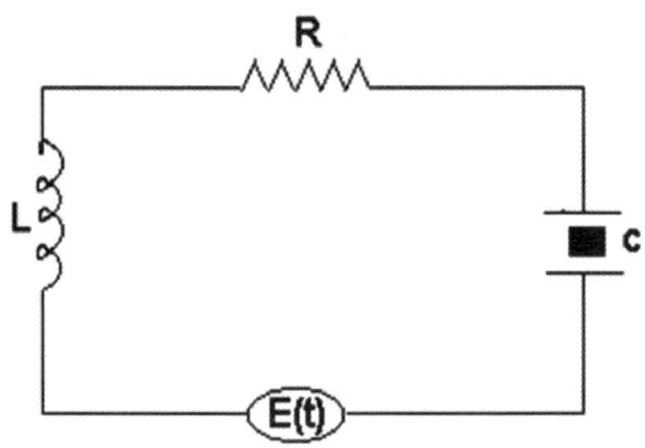

Fig. 11. *RLC* Circuit

Consider a RLC circuit as shown in Figure 11. In this electrical circuit, R is the resistance (Ohms), L is the inductance (Henries), and c is the capacitance (Farad), and $E(t)$ be an applied voltage (Volts). Let I (Amps) be the current in the circuit. To derive the required equation for the current I or charge Q, we use the $Kirchoff's$ second law, which states that, the sum of the voltage drops in the circuit is equal to the applied voltage in a closed circuit. This law is analogous to Newton's second law. The following voltage drops exist.

The voltage drop across a resistance, E_R, is

$$E_R = RI \tag{86}$$

The voltage drop across the capacitor, E_c, is

$$E_c = \frac{Q}{c} = \frac{1}{c}\int I\,dt \tag{87}$$

where Q is the charge in the capacitor such that

$$I = \frac{dQ}{dt} \tag{88}$$

the Voltage drop across an inductor, E_L, is

$$E_L = L\frac{dI}{dt} \tag{89}$$

Then from Kirchoff's second law we have

$$L\frac{dI}{dt} + RI + \frac{1}{c}\int I\,dt = E(t) \tag{90}$$

Differentiating this w.r.t t we get

$$L\frac{d^2I}{dt^2} + R\frac{dI}{dt} + \frac{I}{c} = \frac{dE}{dt} \tag{91}$$

Equation (90), interms of Q, using Eqn. (88), becomes

$$L\frac{d^2Q}{dt^2} + R\frac{dQ}{dt} + \frac{Q}{c} = E(t) \tag{92}$$

Note that Eqn. (83) for mechanical vibration and Eqn. (91) or (92) for RLC circuit are analogous to the general Eqn. (76) of this section, although they belong to entirely different physical entities. This analysis shows that mathematical model plays an important role in unifying various phenomena of entirely different physical nature.

3.3. *Deflection of a Beam*

Consider the deflection, $y(x)$ as shown in Fig.12 of a beam of rectangular cross-section which is subjected to uniform loading, while the ends of the beam are supported so that they under go no reflection. This situation is governed by the ODE of the form.

$$EI\frac{d^2y}{dx^2} - Sy = \frac{qx(x-l)}{2} \tag{93}$$

Satisfying the conditions

$$y(0) = y(l) = 0 \tag{94}$$

where $y = y(x)$ is the deflection, l the length of the beam, q the uniform load, E the modulus of elasticity, S the stress at the end points, I the

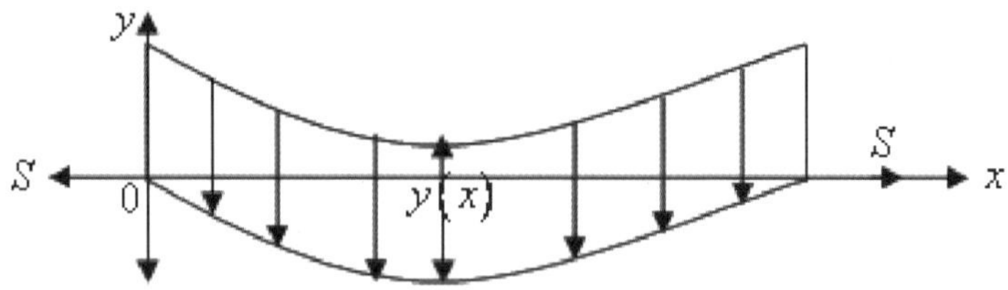

Fig. 12. Deflection of a Beam

central moment of inertia. No deflection occurs at the ends of the beam, implies, $y(0) = y(l) = 0$. When the beam is of uniform thickness, the product EI will be a constant and the exact solutions can be obtained. In many applications, however, the thickness is not uniform, so the moment of inertia I is a function of x only and approximate techniques are required to find analytical solution.

3.4. *Simple Pendulum*

In some situations one can proceed part of the way with the analytical solution and the final answer needs a numerical method. For example, consider the case of a simple pendulum. The governing differential equation is

$$\frac{d^2\theta}{dt^2} + \frac{gsin\theta}{l} = 0 \tag{95}$$

where g is the angular displacement and θ is the deflection.

Multiplying this by $2\frac{d\theta}{dt}$ we get $2\frac{d\theta}{dt}\frac{d^2\theta}{dt^2} + \frac{2gsin\theta}{l}\frac{d\theta}{dt} = 0$

Integrating this once and using $\frac{d\theta}{dt} = \theta$ when $\theta = \alpha$, the maximum displacement, we get $dt = \frac{d\theta}{\sqrt{\frac{2g(cos\theta - cos\alpha)}{l}}}$

This has to be integrated numerically.

3.5. *Mathematics, Concerned with PDE, Unifies Different Branches of Science, Engineering and Technology*

The model, discussed in the section 3.4, is concerned with ODE where the unknown is a function of only one known variable t. In the case of partial differential equations (PDE) the unknown θ is a function of more than one variable. The problems in Science, Engineering and Technology belong to one of the following phenomena

(1) Oscillatory Phenomenon

(2) Diffusion Phenomenon

(3) Potential Phenomenon

The oscillatory phenomenon is governed by the wave equation of the form

$$\frac{\partial^2 u}{\partial t^2} = c^2 \nabla^2 u \tag{96}$$

where c is the velocity of the wave and u is a physical quantity like velocity of propagation. In the case of electric or magnetic fields, Eqn. (96) is called electro magnetic wave equation. This can be obtained using the Maxwell's equations, in a general media, given by

$$\nabla \cdot \overrightarrow{E} = \frac{\rho_e}{\varepsilon} \tag{97}$$

$$\nabla \times \overrightarrow{E} = -\mu \frac{\partial \overrightarrow{H}}{\partial t} \tag{98}$$

$$\nabla \times \overrightarrow{H} = \overrightarrow{J} + \varepsilon \frac{\partial \overrightarrow{E}}{\partial t} \tag{99}$$

$$\nabla \cdot \overrightarrow{H} = 0 \tag{100}$$

$$\overrightarrow{J} = \sigma \overrightarrow{E} \tag{101}$$

where $\overrightarrow{E}$ is the electric field, $\overrightarrow{H}$ the magnetic field, $\overrightarrow{J}$ the conduction current density, $\varepsilon \frac{\partial \vec{E}}{\partial t}$ the displacement current, ρ_e the distribution of charge density, μ the magnetic permeability, ε the dielectric constant and σ is the electrical conductivity. Eqn. (97) is called Gauss law, Eqn. (98) the Faradays law, Eqn. (99) the Amperes law, Eqn. (100) the solenoidal property of magnetic field and Eqn. (101) is the Ohm's law. Operating Curl on Eqn. (98), we get

$$\nabla \times \nabla \times \overrightarrow{E} = -\mu \frac{\partial (\nabla \times \overrightarrow{H})}{\partial t} \tag{102}$$

But, from vector identity, we have

$$\nabla \times \nabla \times \overrightarrow{E} = \nabla(\nabla \cdot \overrightarrow{E}) - \nabla^2 \overrightarrow{E} = \frac{\nabla \rho}{\varepsilon} - \nabla^2 \overrightarrow{E}$$

using Eqn. (97) From Eqn. (99), using (101), we get

$$\nabla \times \overrightarrow{H} = \sigma \overrightarrow{E} + \varepsilon \frac{\partial \vec{E}}{\partial t}$$

Differentiating this with respect to t, we get

$$\frac{\partial}{\partial t}(\nabla \times \overrightarrow{H}) = \sigma \overrightarrow{E} + \varepsilon \frac{\partial^2 \overrightarrow{E}}{\partial t^2}$$

Then using above Eqns becomes

$$\frac{\partial^2 \overrightarrow{E}}{\partial t^2} + R \frac{\partial \overrightarrow{E}}{\partial t} = c^2 \nabla^2 \overrightarrow{E} - \frac{c^2 \nabla \rho}{\varepsilon} \tag{103}$$

where $c = \frac{1}{\sqrt{\mu\varepsilon}}$ is the velocity of light, $R = \frac{\sigma}{\varepsilon}$ is the Relaxation frequency and the last term on the R.H.S of Eqn. (103) is the source term. Eqn. (103) is called the non-homogeneous telegraphic equation. In free space $\rho = 0$ and if $\sigma \to 0$ $i.e., R = 0$ we have

$$\frac{\partial^2 \overrightarrow{E}}{\partial t^2} = c^2 \nabla^2 \overrightarrow{E} \tag{104}$$

This is the equation for electromagnetic wave in free space and is analogous to Eqn. (96). In mathematics, this wave equation is called hyperbolic partial differential equation. In a conducting material the displacement current $\varepsilon \frac{\partial \overrightarrow{E}}{\partial t}$, in Eqn. (99), is negligible compared to conduction current, $\sigma \overrightarrow{E}$, and we have

$\nabla \times \overrightarrow{H} = \overrightarrow{J} = \sigma \overrightarrow{E}$

Operating curl on this, we get

$\nabla \times \nabla \times \overrightarrow{H} = \sigma \nabla \times \overrightarrow{E}$

This, using Eqn.(98), becomes

$\nabla \times \nabla \times \overrightarrow{H} = -\mu\sigma \frac{\partial \overrightarrow{H}}{\partial t}$

From the Vector Identities, $\nabla \times \nabla \times = \nabla(\nabla \cdot \overrightarrow{H}) - \nabla^2 \overrightarrow{H}$

This, using Eqn. (100), becomes

$\nabla \times \nabla \times \vec{H} = -\nabla^2 \overrightarrow{H}$

Then,

$$-\nabla^2 \overrightarrow{H} = -\mu\sigma \frac{\partial \overrightarrow{H}}{\partial t} \tag{105}$$

$$\frac{\partial \overrightarrow{H}}{\partial t} = \frac{\nabla^2 \overrightarrow{H}}{\mu\sigma} = \gamma_m \nabla^2 \overrightarrow{H} \tag{106}$$

where $\gamma_m = \frac{1}{\mu_a}$ is the magnetic viscosity. This is a diffusion equation in electromagnetic fields. In the conduction of heat, the energy equation, in the absence of heat source and radiation, is

$$\frac{\partial T}{\partial t} = k\nabla^2 T \tag{107}$$

where T is the temperature, k is the thermal diffusivity and

$$\nabla^2 = \frac{\partial^2}{\partial y^2} + \frac{\partial^2}{\partial y^2} + \frac{\partial^2}{\partial z^2} \tag{108}$$

equations (105) and (106), represent diffusion of magnetic field and heat respectively. In mathematics, this diffusion equation, is called parabolic PDE. The potential phenomena is governed by the equation of the form

$$\nabla^2 u = 0 \tag{109}$$

This equatuion can be obtained from Eqns. (96), (104), (105) or (106) for steady flow $\left(\frac{\partial}{\partial t} = 0\right)$ Eqn. (109) in mathematics is called the elliptic PDE. In other words a conservative field in science, engineering and technology represents potential phenomena and is governed by the elliptic partial differential equation. A conservative field means the work done by that field is independent of the path and represents the form given Eqn. (109). The electrical engineers usually deal with electro magnetic waves given by Eqn. (104). The mechanical engineers deal with elastic and acoustic waves, civil engineers deal with water waves. However, for a mathematician, all these waves are governed by one single hyperbolic PDE of the form given by Eqn. (96). This implies that even PDEs in mathematics unify different branches of Science, Engineering and Technology.

Acknowledgements

This work is supported by ISRO under the research projects no $ISRO/RES/2/338/2007-08$ and $ISRO/RES/2/335/2007-08$. ISRO's financial support to carry out this work is gratefully acknowledged.

Appendix

$$
\begin{aligned}
a_1 =& R^2 \left(1 + \frac{4}{a^2}\right)\left(1 + \frac{3}{16 I_0(a)}\right) \\
&+ R^2 \left[\frac{(1+\lambda^2 l^2)(a^2+8)}{2a^2 I_0(a)} \frac{\lambda^2 R^2 (a^4 + 16a^2 + 64)}{32a^4}\right]
\end{aligned}
$$

$$a_2 = \frac{1+\lambda^2 l^2}{2I_0(a)}$$

$$a_3 = \frac{3l^2}{4I_0(a)}$$

$$b_1 = \frac{a_1 l^2}{I_0(a)}$$

$$b_2 = \frac{l^2(16I_0(a)(1+a_2)+3)}{16I_0(a)}$$

$$b_3 = l^2(4l^2 - 2\lambda^2 l^2 + a_3)$$

$$b_4 = \frac{2(a_1 + 2l^2(1+\lambda^2 l^2))}{I_0(a)}$$

$$b_5 = \frac{l^2(11 + 8\lambda^2 l^2 + 8I_0(a)(\lambda^2 l^2 + 1))}{I_0(a)}$$

$$b_6 = \frac{I_1(a)}{I_0(a)}$$

$$d_1 = 1 - \frac{12b_2}{R^2} + \frac{\lambda^2 l^2 (a^2+24)}{4a^2} - \frac{24b_3}{R^4} + 48l^2 a_1 b_6$$

$$d_2 = 1 - \frac{8}{a^2} + \frac{16b_6}{a^3}$$

$$
\begin{aligned}
d_3 =& 1 - \frac{b_1(2aI_0()a)(a^2+1) - I_1(a)(a^2+4)}{2R^2 l^2 a^3} \\
&- \frac{2b_2}{R^2 a} - \frac{b_4 I_0(a)}{2R^2} + \frac{b_5}{2R^2} - \frac{\lambda^2 l^2}{2}
\end{aligned}
$$

$$d_4 = \frac{2(a^2+1)}{a^2} - \frac{b_6(a^2+4)}{a^3}$$

$$\beta_1 = \sqrt{\left(\frac{d_4\lambda_0}{d_2}\right)}$$

$$\beta_2 = \frac{3\lambda_0 d_3 - d_1}{\beta_1^2 d2}$$

$$\beta_3 = 1 - \frac{\beta_2}{6} - \frac{\beta_1\beta_2}{3}$$

$$\beta_4 = \frac{l^2(I_1(a)(a^2+2) - aI_{0(a)})}{a}$$

$$\beta_5 = \frac{R^2}{2}\left[b_5 - R^2\lambda^2 l^2 \frac{2b_4 I_1(a)}{a}\right] + \frac{2b_1\beta_4}{l^2}$$

$$\beta_6 = 2\beta_3 - \frac{2\beta_2\beta_4(1+2\beta_1)}{3R^2 I_0(a)} + \frac{4\beta_5}{R^4}$$

$$\beta_7 = \frac{2\beta_4}{R^2 I_0(a)}$$

$$\beta_8 = \beta_1\beta 2\beta 7$$

$$g_1 = \frac{I_1(\zeta a)}{I_0(a)}$$

$$m_1 = \frac{\beta_1(1+\beta_6)}{1+\beta_7}$$

$$m_2 = \frac{\beta_1^2\beta_2(1+\beta_8)}{1+\beta_7}$$

$$m_3 = 1 + \frac{11}{16 I_0(a)} - \frac{8\varepsilon}{a^2\lambda_0\left(1 - \frac{1}{I_0(a)}\right)}$$

$$\lambda_1 = b_6\frac{a}{2}$$

$$A_0 = \frac{16\varepsilon}{a^3\lambda_0} + \frac{32}{a^3}\left[m_3\left(b_6 - \frac{a}{4}\right) + \frac{8\lambda_1}{a^2}\left(1 - \frac{8\varepsilon}{a^2\lambda_0} + \frac{3}{16I_0(a)}\right)\right]$$
$$+ \frac{1}{2a^3}\left(\frac{m_1}{\beta_1} - \frac{\beta_2}{3}\right) + (16\lambda_1 + a^3) - \left(b_6 - \frac{a}{5} + \frac{2}{3}\right)$$

$$h_1(\zeta) = I_0(\zeta a)\left(1 + \frac{2}{\zeta^2 a^2} - \frac{I_1(\zeta a)(\zeta^2 a^2 - 4)}{\zeta^3 a^3}\right)$$

$$h_2(\zeta) = \frac{2b_1 h_1}{l^2} - \frac{4b_2}{\zeta^2 a^2} - \frac{I_0(\zeta a)(b_4 l^2 - 2b_1)}{l^2} - \lambda^2 l^2 R^2 + b_5$$

$$\delta_1(\zeta) = \frac{g_1(\zeta)}{\zeta a} - \frac{1}{2}$$

$$\delta_2 = \frac{g_1(\zeta)}{\zeta a} - \frac{\zeta^3}{5}$$

$$\delta_3 = \left(1 - \frac{8\varepsilon}{a^2\lambda_0} + \frac{3}{16I_0(a)}\right)(1 - g_0(\zeta))$$

$$f_0(\zeta, 0) = \frac{4(g_1(\zeta) - 1}{a^2}$$

$$f_1(0, \xi) = \frac{8}{\lambda_0\beta^2}\left(\frac{\Delta\alpha}{\Delta p} - \frac{2\Delta\alpha\xi}{\Delta p} + \xi^2 - \frac{1}{3}\right)$$

$$f_3(\zeta, 0) = \frac{1}{\lambda_0}\left[\frac{16\varepsilon(g_0(\zeta)) - \zeta^4}{a^2\lambda_0} - 2(1 - \zeta^4) + \frac{32(g_0(\zeta) - \zeta^2)m_3}{a^2}\right]$$
$$- \frac{1}{\lambda_0}\left[\frac{256\delta_3(\zeta)}{a^4} + \frac{2\beta_2(4(1 - g_0(\zeta)) - a^2(1 - \zeta^2))}{3a^2}\right]$$
$$- \frac{1}{\lambda_0}\left[\frac{8m_1(1 - g_0(\zeta) - \frac{a^2(1-\zeta^2)}{4})}{\beta_1 a^2}\right]$$

$$g_2(\zeta, 0) = \frac{8}{\lambda_0\beta^2}\left[\left(\frac{1}{3} - \frac{\Delta\alpha}{\Delta p}\right)\xi + \frac{\Delta\alpha}{\Delta p}\xi^2 - \frac{\xi^3}{3}\right]$$
$$- \frac{8}{\lambda_0\beta^2}\left[\frac{\beta^2}{2}\left(\frac{\Delta\alpha}{\Delta p} - \xi\right)(\xi^2 + \beta_3 + h_2(\xi))\right]$$
$$- \frac{8}{\lambda_0\beta^2}\left[\frac{\beta^2}{2}\left(\frac{\Delta\alpha}{\Delta p} - \xi\right)\left(-\frac{\beta_2(1 + 2\beta_1)h_1(\xi)}{6I_0(a)}\right)\right]$$

References

1. N. Rudraiah, *Modelling of Nano and Smart Materials.* (A Book Published by Book Paradise, 8th E Main NIIT Towers, 4th Block Jayanagar, Bangalore-560011, 2003).
2. A. M. Srirammurthy and K. S. Arunachalam, *Electric Growth In Space, Material Science Bulletin*, (Proc. Indian Academy of Scis, **4**, 1982).
3. C. O. Ng, N. Rudraiah, C. Nagaraj and H. N. Nagaraj, *Electrohydrodynamic Dispersion Of Macromolecular Components In Biological Bearing*, (J. Engg. Heat and Mass Transfer, **28**, 261-280, 2006).
4. N. Rudraiah and V. Venkatesh, *Solar-MHD Power Generator*, (Proc. of Int. Conf. on Frontiers in Fluid Mechanics (ICFFM-06), Oct 26-28, published by Sapna Book House, 676-696, 2006).
5. N. Rudraiah and G. Ranganna, *Studies on Nonlinear Transport of Poorly Conducting Polluted Ground Water Through Vadose zone porous media in the presence of electric field*, (IAHR Int. Ground Water Symposium, Istan Bul, Turkey, 2008).
6. N. Rudraiah, G. Ranganna and C.O. Ng, *Effects of Electric Field Couple Stress and Permeability of Poorly Conducting Fluid Through Composite Materials*, (7th Asian-Australian Conference on Composite Materials (ACCM-7/15-18, November 2010), Taipei, Taiwan. Accepted for oral presentation and to include in the proceedings, 2010).
7. N. Rudraiah, *Advantages And Disadvantages of Aerosols*, (Proc.of Ind. Sci. Congress held at Annamalai University, 2007).
8. Dulal Pal, N. Rudraiah and Rathna Devanathan, *A Couple Stress Model Of Blood Flow In The Microcirculation*, (Bull. Math.Biology, **50**,212-227, 1988).
9. V. K. Stokes,*Couple Stress In Fluids*, (Phys. Fluids, **9**,1709-1715, 1966).
10. M. J. Lighthill, *Motion In Narrow Capillaries From The Stand Point Of Lubrication Theory*, (In Circulatory and Respiratory Mass Transport, G.E.N. Wolstenholme and J. Knight (Eds),London, Churchill, 85-96, 1969).
11. S. Oka and T. Murata, *A Theoretical Study Of Flow Of Blood In A Capillary With Permeable Wall*, (Jap. J. Appl. Phys, **9**,345-352, 1970).
12. G. S.Beavers, and D. D.Joseph, *Boundary conditions at a naturally permeable wall*, (J. Fluid. Mech, **30**, pp 197-207, 1967).
13. N. Rudraiah, *Coupled Parallel Flows In A Channel And Bounding Porous Medium Of Finite Thickness*, (ASME, J.of Fluid Eng, **322**(107), 153-160, 1985).

ON EQUIVALENCE TRANSFORMATIONS AND EXACT SOLUTIONS OF A HELMHOLTZ TYPE EQUATION

O.P.BHUTANI

Honorary Scientist(INSA),B-1/1057,Vasant Kunj, New Delhi-110070,India

LIPIKA ROY CHOWDHURY

B1K,136#101-40,Vishun Ring Road, Singapore-760136

We have herein utilized equivalence transformation approach to seek similarity reduction and exact solutions of the entitled class of equations and deduce as special cases the solutions to Klein-Gordon and Liouville equations in n-dimensional Euclidean space. Further, the sine-Gordon and Poisson-Boltzmann equations have been reduced, for the n-dimensional Euclidean space, to nlodes which for the case n=3 assume the form of Painleve third transcendent. Beside two more examples are provided wherein f depends on new independent variable s_1, s_2(that are symmetric functions of n Euclidean coordinates $x_1, x_2, \ldots x_n$), φ_{s_1} and φ_{s_2}.

1. Introduction

Differential equations as mathematical models for a wide spectrum of natural phenomena involve parameters, known as the arbitrary elements in the literature, which are determined experimentally or simplified under certain assumptions. Akhatov, Gazizov and Ibragimov [1], Ibragimov, Torrisi and Valenti[2], Torrissi, Tracina and Valenti[3], used these arbitrary elements for the classification of equations into equivalent classes, by observing that the results obtained through experimental determination of these can be achieved by requiring the corresponding differential equation to satisfy some additional symmetry groups, so that the equations from class under consideration permute amongst each other under an equivalence transformation. In this paper, we use the equivalence transformation approach for the similarity reduction[4] and hence exact solutions of the following Helmholtz type equation in n-Euclidean dimension:

$$\Box_n \Phi + f(x_1, x_2, \ldots x_n, \Phi, \Phi_{x_1}, \Phi_{x_2}, \ldots, \Phi_{x_n}) = 0 \tag{1}$$

where h_n denotes the n-dimensional Laplace operator and is given by

$$\Box_n = \sum_{i=1}^{n} \frac{\partial^2}{\partial x_i^2}$$

Notable examples of Eq.(1) are those of Sine-Gordon, Liouville, Poisson-Botzmann equations etc. which, via the dependent variable transformation, assume a particular form of (1).

The paper is planned as follows. Following Akhtov et al[1], we treat the arbitrary function f in Eq(1) as a dependent variable to calculate the equivalence generators in section II. In section III, we tabulate the similarity transformations and the corresponding reductions of partial differential equation under consideration to ordinary differential equations. The main result of investigations is a theorem which when utilized can yield(i) the form of f for a given transformation and (ii) the transformation when f is known. The significance of these results is illustrated through handling of certain well known equations in n-dimensional Euclidean space in section iv. More specifically, it is shown through examples, how for a given f, the corresponding infinitesimal transformation can be found, it may be remarked that the study carried out here is, in some sense, a sequel to the one dealt in[5] via the iso vector approach.

Introducing the following elementary symmetric functions of the n-Euclidean coordinates $x_1, x_2, \ldots, x_n$ as new independent variables[6]:

$$s_1 = x_1 + x_2 + \ldots + x_n$$

$$s_2 = x_1 x_2 + x_1 x_3 + \ldots + x_{n1} x_n$$

$$s_3 = x_1 x_2 \ldots x_n \tag{2}$$

equation (1) can be expressed as:

$$\sum_{i=1}^{n}[\sum_{k=1}^{n}(s_{k-1})_{x_i=0}\{\sum_{j=1}^{n}(s_{j-1})_{x_i=0}\frac{\partial^2}{\partial s_j \partial s_k}\}]\Phi + f(s, \Phi, \Phi_s) = 0 \tag{3}$$

where $s_0 = 1$.

2. The Equivalence Generator

Considering φ to be a function of s_1 and s_2 only, Eq.(3) assume the following form:

$$n\Phi_{s_1 s_2}+2(n-1)s_1\Phi_{s_1 s_2}+\{(n-1)s_1^2-2s_2\}\Phi_{s_1 s_2}+f(s_1,s_2,\Phi,\Phi_{s_1},\Phi_{s_2})=0 \tag{4}$$

Following[3] we assume the generator of the equivalence group in the form

$$V=\xi\frac{\partial}{\partial s_1}+\tau\frac{\partial}{\partial s_2}+\psi\frac{\partial}{\partial \phi}+\mu\frac{\partial}{\partial f}+\zeta_1\frac{\partial}{\partial \phi_{s_1}}+\zeta_2\frac{\partial}{\partial \phi_{s_2}} \tag{5}$$

where the coordinates ξ,τ,ψ are the functions of s_1,s_2 and ϕ while the coordinate μ depends on all the six variables $s_1,s_2,\phi,\phi s_1,\phi s_2$ and f. Further, ζ_1 and ζ_2, are given by,

$$\zeta_1=D_{s_1}(\psi)-\phi_{s_1}D_{s_1}(\xi)-\phi_{s_2}D_{s_1}(\tau)$$

$$\zeta_2=D_{s_2}(\psi)-\phi_{s_1}D_{s_2}(\xi)-\phi_{s_2}D_{s_2}(\tau) \tag{6}$$

where D_{s_1} and D_{s_2} denote the total derivative with respect to s_1 and s_2 and are computed using the following formulae:

$$D_{s_1}=\frac{\partial}{\partial s_1}+\varphi_{s_1}\frac{\partial}{\partial \varphi}+\varphi_{s_1,s_1}\frac{\partial}{\partial \varphi_{s_1}}+\varphi_{s_1 s_2}\frac{\partial}{\partial \varphi_{s_2}}$$

$$D_{s_2}=\frac{\partial}{\partial s_2}+\varphi_{s_2}\frac{\partial}{\partial \varphi}+\varphi_{s_1,s_2}\frac{\partial}{\partial \varphi_{s_1}}+\varphi_{s_1 s_2}\frac{\partial}{\partial \varphi_{s_2}}$$

The prolongation of operator (5) that we need for the second order pde (4) is:

$$PrV=V+\psi^{11}\frac{\partial}{\partial \phi_{s_1 s_1}}+\psi^{12}\frac{\partial}{\partial \phi_{s_1 s_2}}+\psi^{22}\frac{\partial}{\partial \phi_{s_2 s_2}} \tag{7}$$

where

$$\phi^{11}=D_{s_1}(\zeta_1)-\phi_{s_1 s_1}D_{s_1}(\xi)-\phi_{s_1 s_2}D_{s_1}(\tau)$$

$$\phi^{12}=D_{s_1}(\zeta_2)-\phi_{s_1 s_2}D_{s_1}(\xi)-\phi_{s_2 s_2}D_{s_1}(\tau)$$

$$\phi^{22}=D_{s_2}(\zeta_2)-\phi_{s_1 s_2}D_{s_2}(\xi)-\phi_{s_2 s_2}D_{s_2}(\tau) \tag{8}$$

Applying operator (7) to Eq.(4), we get,

$$n\psi^{11} + 2(n-1)s_1\psi^{12} + \{(n-1)s_1^2 - 2s_2\}\phi^{22} + \mu$$
$$+ 2(n-1)(\phi_{s_1s_2} + s_1\phi_{s_2s_2})\xi - 2\tau\phi_{s_2s_2} = 0 \tag{9}$$

$$n[\psi_{s_1s_1} + (2\psi_{s_1\phi} - \xi_{s_1s_1})\phi_{s_1} - \tau_{s_1s_1}\phi_{s_2} + (\psi_{\phi\phi} - 2\xi_{s_1\phi})\phi_{s_1}^2 -$$

$$2\tau_{s_1\phi}\phi_{s_1}\phi_{s_2} - \xi_{\phi\phi}\phi_{s_1}^3 - \tau_{\phi\phi}\phi_{s_1}^2\phi_{s_2} + (\psi_\phi - 2\xi_{s_1})\phi_{s_1s_1} -$$

$$2\tau_{s_1}\phi_{s_1s_2} - 3\xi_\phi\phi_{s_1}\phi_{s_1s_1} - \tau_\phi\phi_{s_2}\phi_{s_1s_1} - 2\tau\phi\phi_{s_1}\phi_{s_1s_2}] +$$

$$2(n-1)s_1[\psi_{s_1s_2} + (\psi_{\phi s_2} - \xi_{s_1s_2})\phi_{s_1} + (\psi_{s_1\phi} - \tau_{s_1s_2})\phi_{s_2} - \xi_{\phi s_2}\phi_{s_1}^2 +$$

$$(\psi_{\phi\phi} - \xi_{s_1\phi} - \tau_{\phi s_2})\phi_{s_1}\phi_{s_2} - \tau_{s_1\phi}\phi_{s_2}^2 - \xi_{\phi\phi}\phi_{s_1}^2\phi_{s_2} - \tau_{\phi\phi}\phi_{s_1}\phi_{s_2}^2 - \xi_{s_2}\phi_{s_1s_1} +$$

$$(\psi_\phi - \xi_{s_1} - \tau_{s_2})\phi_{s_1s_2} - \tau_{s_1}\phi_{s_2s_2} - \xi_\phi\phi_{s_2}\phi_{s_1s_1} - 2\tau_\phi\phi_{s_2}\phi_{s_1s_2} - 2\xi_\phi\phi_{s_1}\phi_{s_1s_2} -$$

$$\tau_\phi\phi_{s_1}\phi_{s_2s_2} + \{(n-1)s_1^2 - 2s_2\}\psi_{s_2s_2} + (2\psi_{s_2\phi} - \tau_{s_2s_2})\phi_{s_2} - \xi_{s_2s_2}\phi_{s_1} - 2\xi_{\phi s_2}\phi_{s_1}\phi_{s_2} +$$

$$(\psi_{\phi\phi} - 2\tau_{\phi s_2})\phi_{s_2}^2 - \xi_{\phi\phi}\phi_{s_1}\phi_{s_2}^2 - \tau_{\phi\phi}\phi_{s_2}^3 - 2\xi_{s_2}\phi_{s_1s_2} + (\psi_\phi - 2\tau_{s_2})\phi_{s_2s_2} -$$

$$2\xi_\phi\phi_{s_2}\phi_{s_1s_2} - \xi_\phi\phi_{s_1}\phi_{s_2s_2} - 3\tau_\phi\phi_{s_2}\phi_{s_2s_2} +$$

$$\mu + 2(n-1)\xi(\phi_{s_1s_2} + s_1\phi_{s_2s_2}) - 2\tau\phi_{s_2s_2} = 0 \tag{10}$$

Using Eq.(4) in (10) and then collecting coefficients of various second order derivative terms, we arrive at the following system of pde determining ξ,τ and μ:

$$(n-1)s_1\xi_\phi - n\tau_\phi = 0 \tag{11}$$

$$\{\frac{(n-1)(n-2)}{n}s_1^2 + 2s_2\}\xi_\phi - (n-1)s_1\tau_\phi = 0 \tag{12}$$

$$(n-1)\xi + (n-1)s_1\xi_{s_1} + \{\frac{(n-1)(n-2)}{n}s_1^2 + 2s_2\}\xi_{s_2} - n\tau_{s_1} - (n-1)s_1\tau_{s_2} = 0 \tag{13}$$

$$(n-1)s_1\xi + \{(n-1)s_1^2 - 2s_2\}\xi_{s_1} + \frac{(n-1)}{n}s_1\{(n-1)s_1^2\}\xi_{s_2}$$

$$= \tau + (n-1)s_1\tau_{s_1} + \{(n-1)s_1^2 - 2s_2\}\tau_{s_2} \tag{14}$$

$$\mu = -n\psi_{s_1s_1} + (2\psi_{s_1\phi} - \xi_{s_1s_1})\phi_{s_1} - \tau_{s_1s_1}\phi_{s_2} + \psi_{\phi\phi}\phi_{s_1}^2 -$$

$$2(n-1)s_1\psi_{s_1s_2} + (\psi_{s_2\phi} - \xi_{s_1s_2})\phi_{s_1} + (\psi_{s_1\phi} - \tau_{s_1s_2})\phi_{s_2} + \psi_{\phi\phi}\phi_{s_1}\phi_{s_2} -$$

$$\{(n-1)s_1^2 - 2s_2\}\psi_{s_2s_2} + (2\psi_{s_2\phi} - \tau_{s_2s_2})\phi_{s_2} - \xi_{s_2s_2}\phi_{s_1} + \psi_{\phi\phi}\phi_{s_2}^2 +$$

$$\frac{1}{n}f\{n(\psi_\phi - 2\xi_{s_1}) - 2(n-1)s_1\xi_{s_2}\} \quad (15)$$

Eqs.(11) imply,

$$\xi_\phi = 0 = \tau_\phi$$

Assuming

$$\xi_{s_2} = 0$$

and

$$\tau = \xi + a_0 s_1 + b_0 s_2 + c_0, \quad (16)$$

Eqs.(13)-(14), yield

$$\xi = as_1 + \frac{n}{n-1}b \quad (17)$$

$$\tau = bs_1 + 2as_2 \quad (18)$$

where $2a = b_0$ and$b = a_0 + a$
Using Eqs.(17) and (18) in Eq.(16), we get,

$$\mu = -n[\psi_{s_1s_1} + 2\psi_{s_1\phi}\phi_{s_1} + \psi_{\phi\phi}\phi_{s_1}^2] - 2(n-1)s_1[\psi_{s_1s_2} + \psi_{s_2\phi}\phi_{s_1} + \psi_{s_1\phi}\phi_{s_2} +$$

$$\psi_{\phi\phi}\phi_{s_1}\phi_{s_2}] - \{(n-1)s_1^2 - 2s_2\}[\psi_{s_2s_2} + 2\psi_{s_2\phi}\phi_{s_2} + \psi_{\phi\phi}\phi_{s_2}^2] + f(\psi_\phi - 2a) \quad (19)$$

In expression (19) ψ is an arbitrary function. This implies that the operator (5) depends upon two arbitrary constant a and b and one arbitrary function ψ. However, to find out an equivalent transformation when ξ and τ are as in Eqs.(17)-(18), the function ψ can depend upon s_1 and s_2 in a particular form which is consistent with the relation obtained by solving.

$$\frac{ds_1}{as_1 + \frac{n}{n-1}b} = \frac{ds_2}{bs_1 + 2as_2} \quad (20)$$

The following three possibilities arise due to the arbitrary nature of the constants a and b:
i) $a = 0, b \neq 0$
The solution of Eq.(20) is

$$q_{01} = \frac{n-1}{2}s_1^2 - ns_2 \quad (21)$$

ii) $a \neq 0, b = 0$
In this case, the solution of Eq.(20) is

$$q_{02} = s_1^2 s_2^{-1} \tag{22}$$

and finally, for the case
iii) $a \neq 0, b \neq 0$
the solution of Eqn.(20) is,

$$q_{03} = \{s_2 + \frac{b}{a}s_1 + \frac{nb^2}{2(n-1)a^2}\}(s_1 + \frac{n}{n-1}\frac{b}{a})^{-2} \tag{23}$$

Hence,

$$\psi = \psi(\phi, q_{0i}), \qquad i = 1, 2, 3 \tag{24}$$

Even though, we know the form of ψ from Eq.(24) we still cannot derive an equivalence transformation as the solutions to the auxiliary equations(for i=1,2,3)

$$\frac{ds_1}{as_1 + \frac{n}{n-1}b} = \frac{ds_2}{bs_1 + 2as_2} = \frac{d\phi}{\psi(\phi, q_{0i})} \tag{25}$$

still maintain a certain degree of arbitrariness.Consequently, the need to put some restriction(s) on ψ arises. But before choosing ψ we find it essential to summarize the results obtained above in the form of the following theorem.

Theorem A
The following equivalence generator of Eq.(4) act as the symmetry operator of this equation:

$$\overline{V_1} = \frac{n}{n-1}b\frac{\partial}{\partial s_1} + bs_1\frac{\partial}{\partial s_2} + \psi(\phi, \frac{n-1}{2}s_1^2 - ns_2)\frac{\partial}{\partial \phi} + \mu\frac{\partial}{\partial f} + \zeta_1\frac{\partial}{\partial \phi s_1} + \zeta_2\frac{\partial}{\partial \phi s_2} \tag{26}$$

$$\overline{V_2} = as_1\frac{\partial}{\partial s_1} + 2as_2\frac{\partial}{\partial s_2} + \varphi(\phi, \frac{s_1^2}{s_2})\frac{\partial}{\partial \phi} + \mu\frac{\partial}{\partial f} + \zeta_1\frac{\partial}{\partial \phi_{s_1}} + \zeta_2\frac{\partial}{\partial \phi_{s_2}} \tag{27}$$

$$\overline{V_3} = (as_1 + \frac{n}{n-1}b)\frac{\partial}{\partial s_1} + (bs_1 + 2as_2)\frac{\partial}{\partial s_2} +$$

$$\psi(\phi,(s_2+\frac{b}{a}s_1+\frac{b^2n}{a^2 2(n-1)})(s_1+\frac{nb}{(n-1)a})^{-2})\frac{\partial}{\partial\phi}+\mu\frac{\partial}{\partial f}+\zeta_1\frac{\partial}{\partial\phi_{s_1}}+\zeta_2\frac{\partial}{\partial\phi_{s_2}} \tag{28}$$

where the explicit expressions of μ, ζ_1 and ζ_2 are to be calculated from Eqs.(19), (6) and (7), using the values of ξ, τ and ψ.

3. Application of Theorem A

We can apply the theorem given above to determine the form of that f assumes in different cases and eventually arrive at the equivalence transformation which reduces the pde to an ordinary differential equation.
If $\hat{f}$ is the form that f assumes, then

$$\overline{V_i}(f-f_i)|_{f=\hat{f}_i}=0 \qquad\qquad i=1,2,3 \tag{29}$$

where $\overline{V_i}'s$ are defined in Theorem A

Case 1: Determination of $\hat{f}_i$.

For the case under consideration Eq.(29) assumes the following form (dropping the subscript 1):

$$\frac{n}{n-1}b\frac{\partial\hat{f}}{\partial s_1}+bs_1\frac{\partial\hat{f}}{\partial s_2}+\psi(\phi,\frac{n-1}{2}s_1^2-ns_2)\frac{\partial\hat{f}}{\partial\phi}+\zeta_1\frac{\partial\hat{f}}{\partial\phi_{s_1}}+C\frac{\partial\hat{f}}{\partial\phi_{s_2}}=\mu \tag{30}$$

On account of the arbitrary nature of ψ the expression forζ_1, ζ_2 and hence, for μ are still too general. As a particular form of ψ shall serve the purpose we, therefore, choose

$$\psi=k\phi+\delta(\frac{n-1}{2}s_2^2-ns_2)+\gamma \tag{31}$$

In Eq.(31), k, δ and γ are arbitrary constants.
Eqs.(6),(19) and (31) when combined yield:

$$\zeta_1=(n-1)\delta s_1+k\phi_{s_1}-b\phi_{s_2}$$

$$\zeta_2=-n\delta+k\phi_{s_2}$$

$$\mu=-n(n-1)\delta+k\hat{f} \tag{32}$$

Next, we consider two special cases of Eq.(31) giving zero or nonzero values to the arbitrary constants.

Subcase 1: k=0=γ

We need to solve the following equation to find $\hat{f}$:

$$\frac{ds_1}{\frac{n}{n-1}b} = \frac{ds_2}{bs_1} = \frac{d\phi}{\delta(\frac{(n-1)}{2}s_1^2 - ns_2)}$$

$$= \frac{d\phi_{s_1}}{\delta(n-1)s_1 - b\phi_{s_2}} = \frac{d\phi_{s_2}}{-n\delta} = \frac{d\hat{f}}{-n(n-1)\delta} \tag{33}$$

The first and second terms of the above set has the integral

$$y_1 = \frac{(n-1)}{2}s_1^2 - ns_2. \tag{34}$$

Similarly, the second and fourth terms of the set (33) have the internal

$$y_2 = \frac{-(n-1)}{n}\frac{\delta}{b}s_1 y_1 + \phi, \tag{35}$$

and the first and fifth terms of the set (33) yield

$$y_3 = \phi_{s_2} + \frac{\delta(n-1)}{b}s_1 \tag{36}$$

Further, the combination of fourth and first terms with the fifth term of the set yield respectively, the following solutions:

$$y_4 = \phi_{s_1} + \frac{(n-1)}{n}s_1\phi_{s_2} \tag{37}$$

and

$$\hat{f} = \frac{-(n-1)^2\delta}{b}s_1 + F(y_1, y_2, y_3, y_4) \tag{38}$$

Nothing that

$$y_3 = -ny_2'(y_1) \tag{39}$$

$$y_4 = \frac{(n-1)\delta}{nb}y_1 \tag{40}$$

we can write (38) in the following form (putting back subscript 1, for $\hat{f}$):

$$\hat{f}_1 = \frac{-(n-1)^2}{b}\delta s_1 + F(y_1, y_2, y_2') \tag{41}$$

We take the following as the similarity transformation

$$y = y_1$$

$$Y(y) = y_2 \tag{42}$$

Consequently Eqs.(4), (41) and (42) together yield

$$2nyY'' + n(n-1)Y' + F(y, Y, Y') = 0 \tag{43}$$

Subcase 2: $k \neq 0, \gamma = 0.\delta = 1$

For the case under consideration Eq.(30) takes the form:

$$\frac{ds_1}{\frac{n}{(n-1)}b} = \frac{ds_2}{bs_1} = \frac{d\phi}{k\phi + \frac{n-1}{2}{s_1}^2 - ns_2} = \frac{d\phi_{s_1}}{k\phi_{s_1} - b\phi_{s_2} + (n-1)s_1}$$

$$= \frac{d\phi_{s_2}}{k\phi_{s_2} - n} = \frac{d\hat{f}}{k\hat{f} - n(n-1)} \tag{44}$$

Solving Eqs.(44), we get the form of $\hat{f}$ (on putting back the subscript 1,$\hat{f}$) as:

$$\hat{f} = \frac{n(n-1)}{k} + F(y_1, y_2, y_3, y_4) \exp\left(\frac{(n-1)}{n}\frac{k}{b}s_1\right) \tag{45}$$

where

$$y_1 = \frac{n-1}{2}s_1^2 - ns_2 \tag{46}$$

$$y_2 = (\phi + \frac{1}{k}y_1) \exp\left(\frac{-(n-1)}{n}\frac{k}{b}s_1\right. \tag{47}$$

$$y_3 = (\phi_{s_2} + \frac{n}{k}) \exp\left(\frac{-(n-1)}{n}\frac{k}{b}s_1\right. \tag{48}$$

$$y_4 = \{\phi_{s_1} + (\frac{n-1}{n})s_1\phi_{s_2}\} \exp\left(\frac{-(n-1)}{n}\frac{k}{b}s_1\right. \tag{49}$$

The similarity transformation is taken to be

$$y = y_1$$

$$Y(y) = y_2 \tag{50}$$

Using Eqs.(45) and (50), Eq.(4) transforms to

$$2nyY'' + n(n-1)Y' + \frac{k^2(n-1)^2}{nb^2}Y + F(y, Y, Y') = 0 \tag{51}$$

Case 2:Determination of $\hat{f}_2$

For this case Eq.(29) with (28) assumes the following form:

$$as_1\frac{\partial \hat{f}_2}{\partial s_1} + 2as_2\frac{\partial \hat{f}_2}{\partial s_2} + \psi(\phi, \frac{s_1^2}{s_2})\frac{\partial \hat{f}_2}{\partial \phi} + \zeta_1\frac{\partial \hat{f}}{\partial \phi_{s_1}} + \zeta_2\frac{\partial \hat{f}}{\partial \phi_{s_2}} = \mu \tag{52}$$

Choosing ψ in the following form

$$\psi = k\phi + \gamma \tag{53}$$

Eqs.(6) and (19) yield

$$\zeta_1 = (k-a)\phi_{s_1},$$

$$\zeta_2 = (k-2a)\phi_{s_2}$$

$$\mu = (k-2a)\hat{f}_2 \tag{54}$$

Solving Eq.(52), we get

$$\hat{f}_2 = s_1{}^{\frac{k}{a}-2}F(y_1, y_2, y_3, y_4) \tag{55}$$

where

$$y_1 = s_1^2 s_2^{-1},$$

$$y_2 = s_1{}^{\frac{-k}{a}}(\phi + \frac{\gamma}{k}),$$

$$y_3 = s_1{}^{(1-\frac{k}{a})}\phi_{s_1},$$

$$y_4 = s_1{}^{(2-\frac{k}{a})}\phi_{s_2} \tag{56}$$

Thus, the similarity transformation is taken as

$$y = y_1,$$

$$Y(y) = y_2 \tag{57}$$

implying

$$y_3 = \frac{k}{a}Y + 2yY',$$

$$y_4 = -y^2 Y'. \tag{58}$$

Using this transformation pde (4) reduces to

$$[(n-1)y^4 + 2(1-2n)y^3 + 4ny^2]Y'' + [2(n-1)y^3 - 2\{(n-1)(\frac{k}{a}+2)+2\}y^2 +$$

$$2n(2\frac{k}{a}+1)y]Y' + \frac{nk}{a}(\frac{k}{a}-1)Y + F(y, Y, Y') = 0 \tag{59}$$

Case 3: Determination of $\hat{f}_3$

In this case, Eq.(29) yields the following expression for $\hat{f}_3$

$$\hat{f}_3 = (s_1 + \frac{n}{n-1}\frac{b}{a})^{\frac{k}{a}-2} F(y_1, y_2, y_3, y_4), \tag{60}$$

where

$$y_1 = (s_2 + \frac{b}{a}s_1 + \frac{b^2}{2a^2}\frac{n}{(n-1)})(s_1 + \frac{n}{(n-1)}\frac{b}{a})^{-2} \tag{61}$$

$$y_2 = (\phi + \frac{\gamma}{k})(s_1 + \frac{n}{n-1}\frac{b}{a})^{\frac{-k}{a}} \tag{62}$$

$$y_3 = \phi_{s_2}(s_1 + \frac{n}{n-1}\frac{b}{a})^{2-\frac{k}{a}} \tag{63}$$

$$y_4 = (\phi_{s_1} - \frac{b}{a}\phi_{s_2})^{\frac{-k-a}{k-2a}}\phi_{s_2} \tag{64}$$

As in Case 2, ψ is chosen to be

$$\psi = k\phi + \gamma \tag{65}$$

The similarity transformation is taken as

$$y = y_1$$

$$Y(y) = y_2 \tag{66}$$

which implies

$$y_3 = Y'$$

$$y_4 = (\frac{k}{a}Y - 2yY')^{\frac{-k-a}{k-2a}} Y'$$

Consequently Eq.(4) assumes the following form:

$$[4ny^2 - 2(2n-1)y + (n-1)]Y'' - 2[n(\frac{2k}{a} - 3)y - (n-1)(\frac{k}{a} - 2)]Y' +$$

$$n(\frac{k}{a} - 1)\frac{K}{a}Y + F(y, Y, Y') = 0 \tag{67}$$

It may be mentioned here that these are only few of the possible ode reductions with the chosen pair of ξ and τ. For some other choices of ψ one can arrive at other possibilities. Further, some other possible choices of ξ and τ can give rise to alternative choices of similarity transformations.
In the following section, we confine our attention to the determination of exact solutions of the odes obtained above for particular choices of $F(y, Y, Y')$ occuring in them.

4. Examples

As mentioned in section I, the emphasis here is on the dependency of the arbitrary function f occurring in Eq.(1), on first order partial derivatives of the dependent variable. But the results for the case when f depends upon the independent and dependent variables only, should also be deducible as special case of this more general case, for this purpose, we have first of all taken up the Klein-gordon type equation in[5] and have shown that similar results can still be obtained. Next we give some examples in which f depends upon the first order partial derivatives too. To this effect we deduced the results obtained for the Liouvile type equation in[5]. In addition to these, the Sine-Gordon equation in n-Euclidean dimensions and the n-dimensional Poisson-Boltzmann equations are reduced to nlpdes. Two more equations with explicit occurrence of the first order partial derivatives are solved to get exact solutions.

Example 4.1: The Klein-Gordon Type Equation
The equation under considerations

$$L\phi + f(s_1, s_2, \phi) = 0 \tag{68}$$

where

$$L = n\frac{\partial^2}{\partial {s_1}^2} + 2(n-1)s_1\frac{\partial^2}{\partial s_1 \partial s_2} + \{(n-1)s_1^2 - 2s_2\}\frac{\partial^2}{\partial {s_2}^2} \tag{69}$$

Eq.(68) is a special case of Eq.(4).

Torrisi et al(1992) treated a similar case by adding the following two equations to Eq.(68).

Transformation	Form 'f'	Reduced ode
$y=\left(s_2+\frac{b}{a}s_1+\frac{b^2n}{2a^2(n-1)}\right)\left(s_1+\frac{n}{n-1}\frac{b}{a}\right)^{-2}$ $Y(y)=(\phi+\frac{y}{k})\left(s_1+\frac{n}{n-1}\frac{b}{a}\right)^{-k/a}$	$\left(s_1+\frac{n}{n-1}\frac{b}{a}\right)^{\frac{k}{a}-2}F(y,Y)$	$[4ny^2-2(2n-1)y+(n-1)]Y''$ $-2\left[n\left(\frac{2k}{a}-3\right)y-(n-1)\left(\frac{k}{a}-2\right)\right]$ $+n\left(\frac{k}{a}-1\right)\frac{k}{a}Y+F(y,Y)=0$
$y=\frac{1}{2}(n-1)s_1^2-ns_2$ $Y(y)=\phi-\frac{(n-1)}{nb}s_1y$	$\frac{-(n-1)^2}{b}s_1+F(y,Y)$	$2nyY''+n(n-1)Y'+F(y,Y)=0$
$y=s_1^2s_2^{-1}$ $Y(y)=s_1^{\frac{-k}{a}}\left(\phi+\frac{y}{k}\right)$	$s_1^{\frac{k}{a}-2}F(y,Y)$	$[(n-1)y^4+2(1-2n)y^3+4ny^2]Y''$ $\left[2(n-1)y^3-2\left\{(n-1)\left(\frac{k}{a}+2\right)+2\right\}y^2\right.$ $\left.+2n\left(\frac{2k}{a}+1\right)y\right]Y'+\frac{nk}{a}\left(\frac{k}{a}-1\right)Y$ $+F(y,Y)=0$
$y=\frac{n-1}{2}s_1^2-ns_2$ $Y(y)=\left(\phi+\frac{1}{k}y\right)e^{-\frac{(n-1)}{n}\frac{k}{b}s_1}$	$\frac{n(n-1)}{k}+F(y,Y)e^{\frac{(n-1)}{n}\frac{k}{b}s_1}$	$2nY''+n(n-1)Y'+\frac{k^2(n-1)^2}{nb^2}Y$ $+F(y,Y)=0$

Table 3.1

$$\frac{\partial f}{\partial \phi_{s_1}} = 0, \qquad \frac{\partial f}{\partial \phi_{s_2}} = 0 \tag{70}$$

and calculated the equivalence algebra under these conditions which required additional processing. Here, we simply add the following conditions

$$\frac{\partial \hat{f}_i}{\partial \phi_{s_j}} = 0, \qquad i = 1, 2, 3; j = 1, 2 \tag{71}$$

to Eq.(29), which rather simplifies the process of determination of $\hat{f_{i's}}$. The three cases, studied in Section III, give similar results. These are listed in Table 3.1.

Choosing $F(y, Y)$, we seek solutions of the nlode by known techniques, which are then transformed via the equivalence transformations(in reverse direction i.e.) to yield solutions of the pdes. Further, the forms of $'f'$ also get specified. More specifically, we would like to mention here that the results listed in Table 2.1 of[5] are obtained for row-1 of Table 3.1. As mentioned in[5], other odes listed in Table 3.1 can be solved using this or other known techniques. As shown in[5], Matsuno's results can easily be deduced here as special cases.

Example 4.2: The Liouville Type Equation

The equation under consideration is

$$\Box_n g + h(x_i) \exp(\lambda g) = 0 \tag{72}$$

Substituting $g = \frac{1}{\lambda} ln\phi$, we get

$$\phi \Box_n \phi - \sum_{i=1}^{n} (\phi_{x_i})^2 + \lambda h(x_i)\phi^3 = 0 \tag{73}$$

On switching over to $\{s_i\}$ coordinates, Eq.(32) is transformed to

$$L\phi - \frac{1}{\phi}[n\phi_{s_1}^2 + 2(n-1)s_1\phi_{s_1}\phi_{s_2} + \{(n-1)s_1^2 - 2s_2\}\phi_{s_2}^2 + \lambda\phi^2 h(s_1, s_2) = 0 \tag{74}$$

Knowing the form of $'f'$ our next step is to find out what constraints it puts on the arbitrary constants or functions, involved in the equivalence generator. For this purpose, we take up Case 3 of Section III. The function

$$f = \frac{-1}{\phi}[n\phi_{s_1}^2 + 2(n-1)s_1\phi_{s_1}\phi_{s_2} + \{(n-1)s_1^2 - 2s_2\}\phi_{s_2}^2] + \lambda\phi^2 h(s_1, s_2) \tag{75}$$

must satisfy Eq.(60) for $\hat{f}_3$. With a particular choice of ψ taken as

$$\psi = k\phi + \gamma \tag{76}$$

we have

$$\zeta_1 = (k-a)\phi_{s_1} - b\phi_{s_2}$$

$$\zeta_2 = (k-2a)\phi_{s_2}$$

$$\mu = (k-2a)f \tag{77}$$

Eqs.(75)(77) and Eq.(60) when combined, yield

$$\gamma = 0 \tag{78}$$

and an equation for $h(s_1, S_2)$

$$(as_1 + \frac{n}{(n-1)}b)h_{s_1} + (bs_1 + 2as_2)h_{s_2} = -(k+2a)h \tag{79}$$

whose solution, can be expressed as

$$h(s_1, s_2) = (s_1 + \frac{n}{n-1}\frac{b}{a})^{\frac{k}{a}+2}F(y) \tag{80}$$

where

$$y = (s_2 + \frac{b}{a}s_1 + \frac{n}{2(n-1)}\frac{b^2}{a^2})(s_1 + \frac{n}{n-1}\frac{b}{a})^2 \tag{81}$$

Eq.(81) along with

$$Y(y) = (s_1 + \frac{n}{n-1}\frac{b}{a})^{\frac{-k}{a}}\phi \tag{82}$$

reduces the pde (74) to the following nlode:

$$[4ny^2 - 2(2n-1)y + (n-1)][YY'' - Y'^2] + [6ny - 4(n-1)]YY' - \frac{nk}{a}Y^2 +$$

$$\lambda F(Y)Y^3(y) = 0 \tag{83}$$

Solution to Eq.(83) for a particular form of $F(y)$ were listed in Table 2.2[5]. Solution for the pde are obtained from the solutions of the ode using Eqs.(81)-(82). As mentioned there, Matsuno's (1987) results can be recovered easily.

It is worth mentioning here that other such reductions to nlodes are possible via the transformations given in cases 1 and 2 in Section 3.3.

Example 4.3: The Sine-Gordon Type Equations

Here we consider the equation

$$\Box_n g = h(x) sin g \tag{84}$$

which is the generalization of the Sine-Gordon equation in n-Euclidean dimensions.The sine-Gordon equation has many applications in physics, a particular one is the description of the movement of the vectors of the magnetization at the weakly excited states of exchanged ferromagnetic[Martinov and Vitanov(1992)]. The classical sine-Gordon equation has been studied in connection with the transformation of surfaces of constants negative curvature.

Putting

$$\phi = \tan \frac{g}{4}, \tag{85}$$

Eq.(84) is transformed to

$$\sum_{i=1}^{n} \phi_{x_i x_i} = \frac{2\phi}{1+\phi^2} \sum_{i=1}^{n} \phi_{x_i}^2 + h(x) \frac{\phi(1-\phi^2)}{(1+\phi^2)^2} \tag{86}$$

Proceeding as in previous examples we find that this equation is equivalent to

$$L\phi - \frac{2\phi}{1+\phi^2}[n\phi_{s_1}^2 + 2(n-1)s_1\phi_{s_1}\phi_{s_2} + \{(n-1)s_1^2 - 2s_1\}\phi_{s_2}^2$$

$$-h(s_1, s_2)\frac{\phi(1-\phi^2)}{(1+\phi^2)} = 0, \tag{87}$$

and $h(s_1, s_2)$ satisfies the following pde:

$$[as_1 + \frac{n}{n-1}b]h_{s_1} + (bs_1 + 2as_2)h_{s_2} = 2ah \tag{88}$$

On solving Eq.(88), we get

$$h(s_1, s_2) = (s_1 + \frac{n}{n-1}\frac{b}{a})^2 H(y) \tag{89}$$

where

$$y = (s_2 + \frac{b}{a}s_1 + \frac{n}{n-1}\frac{b^2}{a^2})(s_1 + \frac{n}{n-1}\frac{b}{a})^2 \tag{90}$$

The transformation

$$\phi(s_1, s_2) = \phi(y) \tag{91}$$

reduces the pde (87) to the following ode:

$$[4ny^2 - 2(2n-1)y + (n-1)](\phi'' - \frac{2\phi}{1+\phi^2}\phi'^2) + [4(n-1) + 6ny]\phi'$$

$$-H(y)\frac{\phi(1+\phi^2)}{(1+\phi^2)} = 0 \tag{92}$$

Alternative Reduction

On using the transformation

$$\phi = \exp ig \tag{93}$$

Eq.(84) reduces to

$$\Box_n - \frac{1}{\phi}\sum_{i=1}^{n}\phi_{x_i}^2 + h(x)\frac{1-\phi^2}{2} = 0, \tag{94}$$

which under the transformation of coordinates $\{x_i\}$ to $\{s_i\}$ is equivalent to

$$L\phi - \frac{1}{\phi}[n\phi_{s_1}^2 + 2(n-1)s_1\phi_{s_1}\phi_{s_2} + \{(n-1){s_2}^2 - 2s_2\}\phi_{s_2}^2]$$

$$-\frac{1}{2}h(s)(\phi^2 - 1) = 0 \tag{95}$$

Following the same procedure as for the previous transformation this equation can be reduced via (92) to

$$[4ny^2 - 2(2n-1)y + (n-1)][\phi'' - \frac{\phi'^2}{\phi}] = [6ny + 4(n-1)]\phi' - \frac{1}{2}H(y)(\phi^2 - 1) = 0 \tag{96}$$

For the particular case

$$h(x) = 1, \tag{97}$$

Eq.(88) implies $a = 0$. Hence, with the transformation

$$\xi = (n-1)s_1^2 - 2ns_2$$

$$\phi(s_1, s_2) = \phi\xi \tag{98}$$

Eq.(95) with $h(x) = 1$ transforms to

$$4n\xi\phi\phi'' - 4n\xi\phi'^2 - 2n(n - 1\phi\phi' + \frac{1}{2}(\phi - \phi^3) = 0 \tag{99}$$

For the case $n = 3$ this is Painleve III transcendent which suggests complete integrability.

Example 4.4: The Generalized n-Dimensional Poisson-Boltzmann Equation

The equation under consideration here is

$$\Box_n g = h(x)\sinh g \tag{100}$$

The solution of th two dimensional Poisson-Boltzmann equation describe the plane distribution of the particles in two component Coulomb gas[Martinov and Vitanov(1992)].

On replaying the transformation (85) and (93) respectively by

$$\phi = \tanh\frac{g}{4},$$

and

$$\phi = \exp g \tag{101}$$

We find that Eq.(100) is transformed to Eq.(86) and (94) of Example 4.3. Consequently, the results turn out to ne identical with that of the previous example. However, the difference lies in the solution to Eq.(84) and(100) which are being handled via different transformations for ode-reductions.

Example 4.5

Let us suppose

$$f = \frac{n(n-1)}{k} + (2m+n-3)(\phi_{s_2} - \frac{n}{k}) - \frac{k^2(n-1)^2}{nb^2}(\phi + \frac{n-1}{2k}s_1^2 - \frac{n}{k}s_2) \tag{102}$$

This is special case of $\hat{f}_1$ in Eq.(45). Using the transformation (50) the pde (4) with f given by Eq.(102) transform to

$$2nY'' + n(3-n-2m)Y' = 0 \tag{103}$$

A general solution of Eq.(4) can be expressed as

$$\phi = \frac{-1}{k}(\frac{n-1}{2}s_1^2 - ns_2) + u_0(\frac{n-1}{2}s_1^2 - ns_2)^m \exp\frac{k(n-1)}{bn}s_1 \tag{104}$$

where u_0 is an arbitrary constant.

Example 4.6

A similar result can be obtained via Case 3 of Section 3.3 for

$$f = A(\phi + \frac{\gamma}{k})(s_1+c)^{-2} + B\phi_{s_2} + C(\phi_{s_1} - \frac{b}{a}\phi_{s_2})(s_1+c)^{\frac{k}{a}-2}\phi_{s_2}{}^{\frac{a-k}{k-2a}} \tag{105}$$

where c,A,B and C are given by

$$c = \frac{n}{n-1}\frac{b}{a}$$

$$A = -n[\frac{8(3k-5a)(k-2a)}{(k-a)^2} - \frac{2(2k-3a)(3k-5a)}{a(k-a)} + \frac{(k-a)k}{a^2}]$$

$$B = 4(2n-1)(\frac{k-2a}{k-a}) - 2(n-1)(\frac{k-2a}{a})$$

$$C = -n(n-1)u_0{}^{\frac{k-a}{k-2a}} \frac{(k-a)^{\frac{3a-k}{k-2a}}(3k-5a)^{\frac{3k-3a}{k-2a}}(k-2a)}{\frac{k}{a} - \frac{2(3k-5a)}{k-a}} \tag{106}$$

It can be easily seen that f given in Eq.(105) is equivalent to

$$F = AY + BY' + C(\frac{k}{a}Y - 2yY')(Y')^{\frac{-(k-a)}{k-2a}} \tag{107}$$

which yield the following solution of the equation under consideration:

$$\phi = \frac{-\gamma}{a} + u_0(s_2 + \frac{b}{a}s_1 + \frac{n}{2(n-1)}\frac{b^2}{a^2})^{\frac{3k-5a}{k-a}}(s_1 + \frac{nb}{(n-1)a})^{\frac{k}{a} - \frac{2(3k-5a)}{(k-a)}} \tag{108}$$

Concluding Remarks

In this paper we have been able to determine an equivalence transformation that reduced a class of nlpdes to nlodes. Some physically interesting special cases that have been taken up are of Klein-Gordon, Liouville, sine-Gordon and Poisson-Boltzmann type equations in n-Euclidean dimensions. Exact solutions are reported for some special choices. It may be worth mentioning here that the results are quite general, but are not exhaustive. That it, to say, for different choices of ξ and τ and hence other quantities depending on these two, other such transformations can be found.

Further, following the procedure given in Ibraginov et al[2] and Torrisi et al[3] different equivalence algebras can be calculated from the different equivalent generators given in Theorem A. Preliminary group classification is also possible using these. As shown in these papers, the principal Lie Algebra can be extended.

Acknowledgments

The author gratefully acknowledge the financial support provided by INSA and CSIR(India) through the research project (03)(0815)/95/EMR-II.

References

1. Akhatov, I. Sh.,Gazizov, R.K., and Ibragimov,N.H.;Mod. Probl. Math.,34, VINITI,3(1989) and J.Soviet Math.,55,1401(1991).
2. Ibragimov,N.H., Torrisi M., and Valenti, A.; J.Math. Phys.,32,2988(1991).
3. Torrisi M.,Tarcina, R., and Valenti, A; *"Modern Group Analysis:Advanced Analytical and Computational Methods in Mathematical Physics"*, Kluwer Acad. Publishers,367(1992)
4. Ovsiannikov, L.V.; *"Group Analysis of Differential Equations"*, Academic Press, New York(1982)
5. Bhutani, O.P. And Bhattacharya, Lipika; J.Math.Phys.36(7).3759(1995).
6. Matsuno, Y.;J.Math. Phys.28(10),2317(1987).
7. Martinov, N. and Vitanov, N,;J.Phys.A:Math. Gen.,25,L51(1992).

COGNITIVE RADIO: STATE-OF-THE-ART AND MATHEMATICAL CHALLENGES

TASKEEN NADKAR, VINAY THUMAR, AAQIB PATEL,
MD. ZAFAR ALI KHAN*, U. B. DESAI* and S. N. MERCHANT

Department of Electrical Engineering, Indian Institute of Technology Bombay, Mumbai 400076, Maharashtra, India
E-mail: {taskeenn, vinay_thumar, aaqib, merchant}@ee.iitb.ac.in

**Indian Institute of Technology Hyderabad, Ordnance Factory Estate, Yeddumailaram 502205, Andhra Pradesh, India*
E-mail: {zafar, ubdesai}@iith.ac.in

In the last few years, a new paradigm has emerged in the field of wireless communication, called *Cognitive Radio (CR)*. CR attempts to alleviate the imbalance created by a fixed spectrum allocation policy and the irregular usage of the frequency bands. Since its inception, as an *intelligent* and *reconfigurable* radio, CR has evolved into a multi-disciplinary technology, and has invoked the interest of engineers, economists, scientists and mathematicians alike. A vast amount of literature has stemmed from this revolutionary paradigm, with varied focuses and perspectives. The goal of this paper is two-fold: (i) First, we catalogue the myriad aspects which encompass the CR technology, and on most issues, we highlight the open research problems and challenges faced. Though cursorily, the multiple facets of CR have been addressed, to invoke the interest of readers across disciplines. (ii) An interesting approach to document the plethora of research activities pertaining to CR, is to cast them in the framework of enabling tools, viz. optimization, game theory, fuzzy logic, genetic algorithms and neural networks, and to provide a state-of-the-art of the research contributions therein.

An important research area in CR is power allocation; the optimum power allocation problem for an Orthogonal Frequency Division Multiplexing-based CR is investigated, and relevant simulation results are provided to demonstrate the effectiveness of the technique.

Keywords: Cognitive Radio; Optimization; Game theory, Fuzzy logic, Genetic algorithms; Neural networks

1. Introduction

Extensive measurements of the radio spectrum usage has revealed its under-utilization; a fact, which is attributed to a fixed spectrum assignment policy [1]. This discovery has triggered exciting activities in the engineering, economics and regulation fraternities in searching for better spectrum management policies. To alleviate the imbalance between spectrum allocation and its use, opportunistic spectrum access has been proposed; many schools of thought have emerged from this innovative concept, all of which manifest themselves in the technology solution called *Dynamic Spectrum Access (DSA)*, also referred to as *Cognitive Radio (CR)*. Opportunistic spectrum access entails the temporary usage of unused portions of the spectrum (*spectrum holes or white spaces*), owned by the licensed users *(Primary Users or PUs)* to be accessed by unlicensed users *(Secondary Users or SUs)*. CR is characterized by an adaptive, multi-dimensionally aware, autonomous radio system empowered by advanced intelligent functionality, which interacts with its operating environment and *learns* from its experiences to *reason*, *plan*, and *decide* future actions to meet various needs. This approach can lead to a significant increase in spectrum efficiency, networking efficiency as well as energy efficiency [2]. Due to the phenomenal advances in digital signal processing, computer software and hardware, networking, and machine learning, the implementation of this far-reaching combination of *cognition* and *reconfigurability* has become practically achievable [3].

Since its introduction in the seminal paper by Joseph Mitola (1999) in the context of *radio knowledge representation language* for flexible personal wireless services [4], CR has evolved into a multidisciplinary research topic. The CR technology is projected to make a significant and lasting impact on wireless communication, and form the basis of *Next-Generation* networks. The visionary work of Mitola, substantiated by Federal Communications Commission's (FCC's) report (2002) on spectrum utilization [1], was built upon by two more comprehensive papers by Haykin [3] (2005) and Akyldiz [5] (2006).

Standardization efforts of the working groups IEEE 802.22 [6] (wireless regional area network for secondary use); 802.11h [7] (dynamic frequency selection for wireless local area networks); P1900/SCC41 [8] (technologies and techniques for *Next-Generation* radio and advanced spectrum management); and IEEE 802.11y [7] (shared 802.11 operation with other users), evince the great interest surrounding the CR paradigm.

To achieve a two-fold objective of cataloguing the multiple facets of the CR technology, and providing a state-of-the-art of the research contributions within the framework of some generic enabling tools, the rest of the paper is organized as follows: Section 2 describes the CR technology. Section 3 emphasizes the role of CR in *Next-Generation* networks, and the associated functionalities. Section 4 explains cross-layer design for CR. Section 5 outlines the application of cooperative technology in CR. Section 6 is dedicated to the tools which enable the various functionalities of CR, viz. optimization, game theory, fuzzy logic, genetic algorithms and neural networks. Section 7 presents the fundamental research in CR from an information theoretic perspective. In Section 8, we detail a specific problem, that of power allocation for CR. Section 9 concludes the paper.

2. Cognitive Radio Technology

With the primary objective of efficient opportunistic utilization of the radio spectrum while achieving highly reliable communications, CR has the potential for making a significant difference to the way in which the radio spectrum can be accessed. The *cognitive* capability of a CR node refers to its ability to interact with its environment in real-time, to determine appropriate communication parameters, and dynamically adapt to the radio environment [3,5] . In this section, we describe the key concepts of CR, and the physical architecture which makes the aforementioned tasks possible.

2.1. *The Cognitive Cycle*

The *cognitive* process starts with the sensing of Radio Frequency (RF) stimuli and culminates with action. Each cognitive transceiver (the SU) needs to implement a *cognitive cycle* [3] , which is depicted in Figure 1. The three main steps of a *cognitive cycle* are as follows:

- Radio scene analysis, which encompasses estimation of the *interference temperature*, and detection of *spectrum holes*.

  The *interference temperature* has been recommended by the FCC as a measure of the acceptable level of RF interference in the frequency band of interest. It serves as a threshold on the potential RF energy that can be introduced in the PU band by the unlicensed users.

- Channel identification, which involves estimation of the channel state information (CSI), and analyzing the characteristics of the *spectrum holes* detected. It may also involve prediction of the channel and PU traffic.

The goal of traffic prediction is to forecast future traffic variations, as well as the idle times of the PU, as precisely as possible, based on the measurement history [9] . This knowledge will enable the SUs in a CR system, to efficiently utilize the spectrum opportunity. Each PU channel should be sampled to determine its ON/OFF state and additional traffic information such as periodicity, distribution of idle and busy times and the utilization percentage of the channel. Traffic patterns may change over time, and thus, a limited timescale, in terms of some kind of moving time window, should be used for measurement and estimation.

- Spectrum decision and adaptive communication, in which the CR node chooses a spectrum band for usage and reconfigures its transmission parameters such as operating frequency, modulation type and transmit power, based on the input from the radio-scene analysis and channel identification modules.

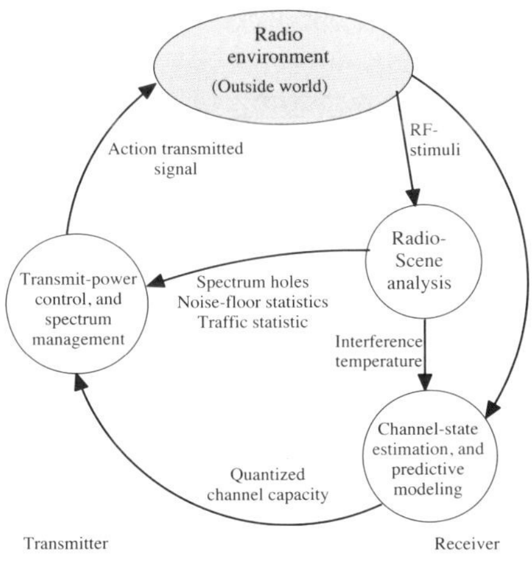

Fig. 1. Cognitive cycle (adapted from [3])

2.2. *Physical Architecture*

The SU should have the capability of detecting the temporarily unused spectrum of the PU, and rapidly jumping in and out of it, without interfering with the transmission of the licensed user. A generic physical architecture, that makes this possible, is depicted in Figure 2. It mainly consists of two chains: the transceiver chain, and the monitoring chain [10]. The transceiver chain comprises the data antenna, the RF front-end (where the signal is amplified, mixed, analog-to-digital converted) and baseband processing (where modulation/demodulation and encoding/decoding takes place) on a *software-defined radio (SDR)* platform. The SDR allows the operating parameters such as frequency range, modulation type or output power to be reconfigured in software, without making any alteration in the hardware [5]. The monitoring chain includes a sensing antenna to detect the presence of a PU in a specific frequency range and a spectrum database which is continuously updated. Connecting the two chains is a spectrum switching unit; when the PU occurs, the data transceiver discontinues transmission on the current working spectrum and searches for an available channel from the spectrum database. This process can guarantee an efficient utilization of the PU's idle time. The main challenge in the hardware implementation is in developing the wide-band RF front-end for spectrum sensing, which should be able to detect the weak signals from the licensed users over a wide spectrum range.

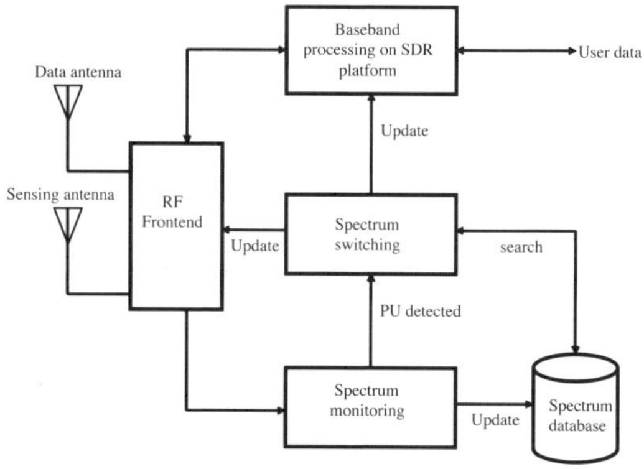

Fig. 2. Physical architecture

3. Cognitive Radio and Next-generation networks

It is anticipated that the *Next-Generation* (xG) communication networks will be based on CR [5] . These networks will provide high bandwidth to mobile users via heterogenous wireless architectures and dynamic spectrum access techniques. The main functions for CR nodes to meet the challenges of spectrum aware xG networks, such as wide spectrum range, Quality-of-Service (QoS) requirements of diverse applications, user mobility across wireless architectures, are summarized as under:

- Spectrum sensing: Detecting unused spectrum so that communication is achieved without causing harmful interference to the licensed user
- Spectrum allocation: Capturing the best available spectrum to meet user communication requirements.
- Spectrum utilization: Exploiting the allocated spectrum in the most efficient manner within the limited resources and QoS constraints.
- Spectrum sharing: Providing fair scheduling among contending users.
- Spectrum mobility: Maintaining seamless communication during transition to a better spectrum.

In the subsequent sub-sections, each of these functions are elaborated. Most of the details of spectrum allocation, spectrum sharing and spectrum mobility have been compiled from the work of Akyildiz et al. [5] .

3.1. *Spectrum sensing*

Spectrum sensing is a very challenging task for cognitive users, and has been pursued as an active research area over recent years. Quite a few sensing techniques have been proposed in literature, viz. energy detection (ED), matched filtering detection (MD), cyclostationary detection (CSD), eigenvalue-based sensing, wavelet-based sensing, classic likelihood ratio test (LRT), covariance-based sensing and blindly combined energy detection. These techniques can be classified into three general categories based on different requirements for their implementation: (i) Techniques requiring both source signal and noise power information; (ii) Techniques requiring only noise power information; and (iii) Techniques requiring no information of source signal or noise power (blind detection) [11] .

From among the above techniques, LRT, MD and CSD belong to category 1; ED and wavelet-based belong to category 2, while blindly combined energy detection belongs to category 3. In this section we describe some of the aforementioned spectrum sensing techniques.

3.1.1. *Energy Detection*

Energy detection is a popular technique to detect the presence of the PU because of its low computational and implementation complexity [12,13] . In this technique, the energy of an unknown signal is measured in the presence of noise.

Let us assume that the received signal has the following form:

$$y(n) = x(n) + w(n) \tag{1}$$

where $x(n)$ is the signal to be detected, $w(n)$ is the additive white Gaussian noise (AWGN), and n is the sample index. The decision matrix for this detector can be represented as

$$M = \sum_{n=0}^{N-1} |y(n)|^2, \tag{2}$$

where N is the size of the observation vector. The decision matrix M is compared with a fixed threshold λ to decide the PU signal's presence. The energy detector is a simple and efficient technique, however, it does not work well when detecting spread spectrum signals.

3.1.2. *Matched Filter Detection*

In matched filter-based detection prior knowledge of the PU's signal ($x(n)$) is necessary. The decision metric for a matched filter-based sensing can be represented as [12,13]

$$M = \Re\left[\sum_{n=0}^{N-1} y(n)x^*(n)\right], \tag{3}$$

where $*$ denotes conjugation operation, and $\Re$ denotes the real part of the signal. In the presence of the PU, the decision metric M can be calculated as

$$M = \sum_{n=0}^{N-1} |x(n)|^2 + \Re\left[\sum_{n=0}^{N-1} w(n)x^*(n)\right]. \tag{4}$$

Similarly, in the absence of the PU the decision metric M can be calculated as (the second term of Eqn. 4)

$$M = \Re\left[\sum_{n=0}^{N-1} w(n)x^*(n)\right]. \tag{5}$$

Similar to the ED technique, the decision metric M is compared with a fixed threshold λ to decide the PU signal's presence.

The main advantage of matched filter-based detection is that it requires less time for detection because of coherency. However, a significant drawback is the need for a dedicated receiver for every signal it may have to detect [13].

3.1.3. *Cyclostationary Detection*

Cyclostationary detection technique uses periodicity property or the moments (mean, autocorrelation,etc.) of the PU's transmitted signal [11,13]. In this approach, noise is differentiated from the PU's signal because noise is wide sense stationary (WSS), with no correlation with the PU signal.

If $x(t)$ is a zero-mean cyclostationary signal, the autocorrelation function of $x(t)$ is periodic in time t, and is given by

$$R_x(t,\tau) = E(x(t)x(t+\tau)) = R_x(t+T,\tau), \tag{6}$$

where T represents the cyclic period. Due to its periodicity, $R_x(t,\tau)$ can be represented as a Fourier series

$$R_x(t,\tau) = \sum_{\alpha} R_x^{\alpha}(\tau)e^{j2\pi\alpha t} \tag{7}$$

where α is called the cyclic frequency, and $R_x^{\alpha}(\tau)$ is the cyclic autocorrelation, defined as

$$R_x^{\alpha}(\tau) = \frac{1}{T}\sum_{t=-\frac{T}{2}}^{\frac{T}{2}} R_x(t,\tau)e^{-j2\pi\alpha t}, \tag{8}$$

The cyclic features of the signal may, alternatively, be represented by the cyclic power spectral density (PSD) function $S_x^{\alpha}(f)$ which is the Fourier transform of $R_x^{\alpha}(\tau)$. The presence or absence of the PU signal is determined by evaluating the spectral component corresponding to the cyclic frequency

α, as follows:

$$S_x^\alpha(f) = \begin{cases} S_n^0(f), & \alpha = 0, \text{ signal absent} \\ |H(f)|^2 S_x^0(f) + S_n^0(f), & \alpha = 0, \text{ signal present} \\ 0, & \alpha \neq 0, \text{ signal absent} \\ H(f+\frac{\alpha}{2})H^*(f-\frac{\alpha}{2})S_x^\alpha(f), & \alpha \neq 0, \text{ signal present} \end{cases}$$

where $H(f)$ is the channel gain and $S_n^0(f)$ is noise PSD at $\alpha = 0$. A PSD threshold detection may be used to determine the presence of the PU. The major advantage of this technique is its robustness to the uncertainty noise power and channel conditions[11] . However, it also has some disadvantages: it needs very high sampling rate and high computational complexity; sampling time error and frequency offset may affect the cyclic frequencies[11,13] .

3.1.4. *Wavelet-based Detection*

Wavelet transform techniques are well known for signal discontinuity detection[14] . In wavelet-based sensing, the *spectrum hole* identification depends on the detection of edges of the received PSD. Generally, PSD is smooth within each sub-band but exhibits discontinuities at its edges. A spectrum sensing technique has been proposed based on this fact[14,15] .

If $\phi(f)$ is a wavelet smoothing function, then its dilation by a scale factor s is given by

$$\phi_s(f) = \frac{1}{s}\phi\left(\frac{f}{s}\right) \tag{9}$$

If the Continuous Wavelet Transform (CWT) of the received PSD is expressed as $Sy(f)$, then $W_s S_y(f)$ represents a measure of the correlation between the dilated wavelet function at one specific scale s, and $S_y(f)$, and is given by

$$W_s S_y(f) = S_y \star \phi_s(f) \tag{10}$$

where ($\star$) denotes the convolution operation. At fine scales $W_s S_y(f)$ provides a localized information of $S_y(f)$, and the edges and irregularities in $S_y(f)$ are obtained by using derivatives.

The advantage of wavelet-based sensing is high-speed sensing over a wide spectrum bandwidth, with low power consumption. Also, by virtue of the scalable feature of the wavelet transform, multi-resolution is achieved without any additional hardware burden[16] .

3.1.5. *Cooperative Detection*

If there are multiple users located in different geographical locations, then it is possible for them to cooperate to achieve higher spectrum sensing reliability [11,17–20]. Each user sends its observed data or processed data to a centralized controller to make a final decision about the presence of the PU. The decision-making can be of two types, viz. data fusion and decision fusion.

Data Fusion

In this method each user sends its observed data to a specific user or centralized controller, where it is jointly processed to make a final decision about spectrum sensing. In energy-based cooperative sensing, each user computes its received signal energy in the presence of noise and sends it to centralized controller [11,18]. The centralized controller sums the collected energy values using a linear combiner (LC) to obtain the following test statistic:

$$E_{LC} = \sum_{i=1}^{M} g_i E_{ED,i} \tag{11}$$

where $E_{ED,i}$ is the ith user's detected energy, and g_i (≥ 0) is the combining coefficient, such that $\sum_{i=1}^{M} g_i = 1$. If signal power received by each user is known, then optimal combining coefficients can be obtained as

$$g_i = \frac{\mu_i^2}{\sum_{k=1}^{M} \mu_k^2}, \tag{12}$$

where μ_i^2 is the received signal power (excluding the noise) for user i.

Decision Fusion

In this method each user processes its observed data independently and sends its decision to the centralized controller. The centralized controller takes the final decision based on the deployed fusion rules, some of which are as follows [11,19,20]:

(1) "*OR*" fusion Rule: If one of the decisions is "1", then the final decision declares the PU's presence. The probability of detection P_d and probability of false detection P_{fd} are given by Eqns. (13) and (14) respectively, where $P_{d,i}$ and $P_{fd,i}$ represent the probability of detection and probability of false detection of the ith user among a set of M users.

$$P_d = 1 - \prod_{i=1}^{M} (1 - P_{d,i}) \tag{13}$$

$$P_{fd} = 1 - \prod_{i=1}^{M} (1 - P_{fd,i}) \tag{14}$$

(2) "AND" fusion rule: If and only if all the decisions are "1", then the final decision declares the PU's presence. The probability of detection P_d and probability of false detection P_{fd} of the final decision are given by Eqns. (15) and (16) respectively.

$$P_d = \prod_{i=1}^{M} P_{d,i} \tag{15}$$

$$P_{fd} = \prod_{i=1}^{M} P_{fd,i} \tag{16}$$

(3) "K out of M" rule: If and only if, K or more decisions are "1"s, then the final decision declares the PU's presence. The probability of detection P_d and probability of false detection P_{fd} of the final decision are given by Eqns. (17) and (18) respectively.

$$P_d = \sum_{i=0}^{M-K} \binom{M}{K+i} (P_{d,i})^{M-K-i} \times (1 - P_{d,i})^{K+i} \tag{17}$$

$$P_{fd} = \sum_{i=0}^{M-K} \binom{M}{K+i} (P_{fd,i})^{M-K-i} \times (1 - P_{fd,i})^{K+i} \tag{18}$$

For higher reliability, the centralized controller requires more information (multiple-bit decision) from the SUs, at the cost of increased communication overheads.

3.1.6. *Other Sensing Techniques*

Other spectrum sensing techniques include covariance detector, eigenvalue-based spectrum sensing and blind spectrum sensing. Zeng et al. [21] and Kim et al. [22] have proposed a spectral covariance-based sensing algorithm that exploits different statistical correlations of the PU signal and noise in the frequency domain. Sensing methods based on the eigenvalues of the covariance matrix of signals received at the SUs have been proposed [23,24] . Zayen et al. have suggested a blind detection technique, which exploits model selection tools like Akaike information criterion and Akaike weights [15] .

Some of the open research challenges in spectrum sensing include detection of the PU in a very short time, and the complexity that arises in sensing in a multi-user environment. Another crucial issue in sensing the presence of the PU is the hidden terminal problem [25] that arises because of shadowing; the SU cannot reliably detect the presence of the PU if it is shadowed from the PU transmitter due to some physical obstacle opaque to radio signals (Figure 3).

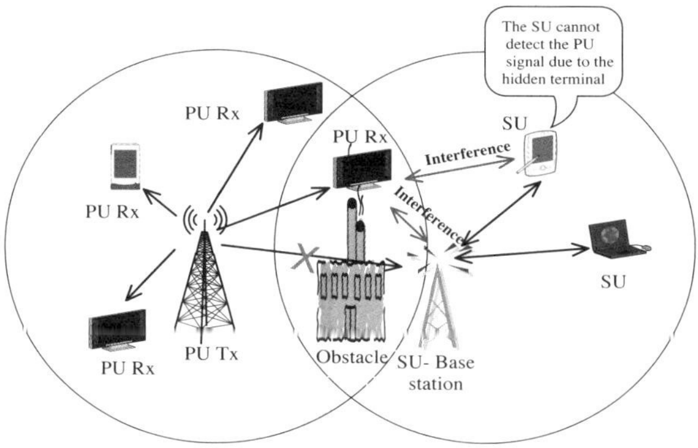

Fig. 3. Hidden terminal problem

3.2. *Spectrum allocation*

This function entails analyzing the quality of the spectrum for its suitability of usage by the SUs. The spectrum band has to be characterized considering the time-varying nature of the channel, activity of the PU, and also its parameters such as operating frequency, bandwidth and channel state. Moreover, the band selected need not be contiguous; CR allows data to be sent over multiple non-contiguous spectrum bands. Transmission over non-contiguous spectrum shows less performance degradation as compared to conventional transmission on a single band during *spectrum hand-off* [5] (to be defined later).

The main challenge in spectrum allocation is to arrive at an appropriate model, which will combine the various decision parameters to effectively characterize the quality of the spectrum.

3.3. *Spectrum utilization*

The spectrum allocated to an SU should be optimally used to meet its requirements, without degrading the performance of the licensed user. This entails efficient *Radio Resource Management (RRM)* implemented at the Media Access Control (MAC) sub-layer of the network protocol stack [26] . The aim of RRM is to evaluate the available resources (power, time slots, bandwidth, etc) and assign them to meet the QoS objectives of the SU, within some constraints on factors (typically interference) which limit the performance of the PU.

Furthermore, for optimum spectrum utilization it is necessary to be adaptive to, one or more, time-varying characteristics of the system, such as the wireless channel state, number of users, QoS requirements, etc. *Adaptive Radio Resource Management (ARRM)* is an important feature of the spectrum utilization task, which seeks to harmonize two contradictory concepts of limited resources and strict QoS requirements, depending on the instantaneous state of the system, and suitably reconfigure after having detected the new state. *Adaptiveness* may be achieved in the transmission scheme through a variation of coding schemes, constellation size, power level, symbol transmission rate, etc., or any combination of these parameters. However, ARRM requires awareness and continuous monitoring of the operating environment, and information exchange between the receiver and transmitter, which increases system complexity and overheads.

Orthogonal Frequency Division Multiplexing (OFDM) is the most popular choice of communication technology for CR. OFDM presents a promising solution to enable opportunistic spectrum access in CR networks by dynamically nulling those sub-carriers where the PU claims its spectrum. This variant of OFDM is called dis-contiguous OFDM (D-OFDM). Besides its ability to handle multi-path fading and inter-symbol interference, it offers flexibility of resource allocation on its individual sub-carriers. When OFDM is used in CR transmission, the power, modulation and bandwidth of the sub-carrier, are parameters which may be reconfigured to improve the performance and achieve the desired system goals [27] . Power allocation may be done with various objectives: to maximize the overall throughput within a power budget; to minimize the total energy consumption while transmitting a fixed number of bits per symbol; to transmit at the lowest possible bit error rate (BER). In OFDM-based CR, the side-lobe interference may hamper the PU communication; thereby posing an additional important constraint in the power allocation problem (discussed in Section 8).

The complexity of RRM, for efficient spectrum utilization in CR, increases with the number of constraints and their stringency, and is exacerbated in a multi-user scenario. The challenge is in developing low-complexity algorithms for practical deployment of RRM in a channel-adaptive manner.

3.4. *Spectrum sharing*

Spectrum sharing is analogous to MAC scheduling in conventional (non-cognitive) wireless systems. Since multiple SUs are trying to access the same spectrum, the access should be coordinated to prevent collisions. Besides, the sharing technique should ensure fairness in meeting the individual users' QoS requirements. If the users forward their measurements and information (global information) to a central entity, which is responsible for executing the sharing algorithm, it is known as centralized spectrum sharing. On the other hand, if the construction of such an infrastructure is not feasible, and each user decides on its own spectrum allocation based on local information, it is called distributed spectrum sharing. Distributed techniques introduce a tradeoff between efficient spectrum sharing and practical solutions having minimal communication overheads.

Spectrum sharing techniques are classified based on the access technology as

- Overlay spectrum sharing: It refers to a Frequency Division Multiplexed (FDM) spectrum access technique, in which each user uses that portion of the spectrum which is not being used by any other user. This approach is used to share spectrum between the PU and SU, and also among the multiple contending SUs.
- Underlay spectrum sharing: This method exploits the spread spectrum techniques [28] used in wireless communication. Every user occupies the complete available bandwidth, but transmits at a power level that is regarded as noise by the other users.

The challenge is in executing efficient spectrum sharing among multiple SUs given the dis-contiguity of the available spectrum and the heterogeneity of the wireless environment. Besides, an open research area can be identified as the implementation of a suitable control channel. However, this challenge is not exclusive to spectrum sharing; the allocation of a control channel is necessitated whenever there is an exchange of transmitter-receiver handshaking signals, sensing information exchange, cooperation among multiple users, as well as communication with a central entity.

3.5. *Spectrum mobility*

Spectrum mobility is defined as the process of the SU changing its frequency of operation. The need for it arises when current channel conditions worsen or a PU appears. Spectrum mobility leads to a new type of hand-off called a *spectrum hand-off*. The information about the change of spectrum should percolate through the layers of the protocol stack with minimum latency. The *Spectrum Mobility Management (SMM)* function should ensure that transition to a new spectrum is smooth, and with almost imperceivable performance degradation of the application running on the SU.

Spectrum hand-off may also be triggered due to reasons other than the detection of a PU; mobility of the SU, due to which the available bands may change, is termed as an *inter-cell hand-off*, and in case of transition between different networks, it is known as *vertical hand-off*.

The main challenge in SMM is in ensuring minimum delays and loss, and the consequent performance degradation, in the event a hand-off occurs.

4. Cross-layer design in Cognitive Radio

Computer networks follow a layered protocol approach, so network functionalities are designed in isolation from each other. Each layer uses the services provided by the layer below it and provides services to the layer above it. Inter-layer communication happens only between adjacent layers and is limited to procedure calls and responses, as seen in the Open Systems Interconnection (OSI) and the TCP/IP models. In CR networks, there is a need for greater interaction between the different layers of the protocol stack in order to achieve the end-to-end goals and desired performance [29]. Cross-layer design refers to the protocol design done by actively exploiting the dependence between the protocol layers to obtain performance gains.

To substantiate the need for a cross-layer design in CR networks, the following examples can be cited [5]:

- The dynamic nature of the underlying spectrum in CR networks necessitates communication protocols to adapt to the wireless channel parameters. Moreover, the behavior of each protocol affects the performance of other protocols. For example, when re-routing is done because of link failures arising from spectrum mobility, the round trip time (RTT) and error probability in the communication change accordingly. The change in error probability also affects the performance of the MAC protocols. Consequently, all these changes affect the overall quality of the SU application. These interdependencies among functionalities of the communication stack, and their close coupling with the PHY (physical) layer necessitate a cross-layer spectrum management function which considers medium access, routing, transport, and application requirements, in addition to the available spectrum, in the selection of the operating spectrum.

- Spectrum hand-off results in latency, which affects the performance of the communication protocols. Moreover, during spectrum hand-off, the channel parameters such as path loss, interference, wireless link error, and link layer delay are influenced by the dynamic use of the spectrum. On the other hand, the changes in the PHY and MAC channel parameters can initiate spectrum hand-off. In order to estimate the effect of the spectrum hand-off latency, information about the link layer and sensing delays are required. Transport and application layer should also be aware of the latency to reduce the abrupt quality degradation. In addition, the routing information is also important for the route recovery using spectrum hand-off. For these reasons, the spectrum hand-off is closely related to the operations in all communication layers.
- Spectrum sensing is primarily a PHY layer function. However, in the case of cooperative detection, the multiple SUs exchange sensing information. Cooperative techniques require transmitters to consider their interference to other users, and the interference at their receivers from other users. Such a collaboration increases the communication overhead and may lead to overall system performance degradation in terms of effective channel capacity or energy consumption. The other challenge about spectrum sensing is the huge range of spectrum that has to be sensed, and the amount of time that is required for it. Since sensing consumes energy, this process has to be carefully scheduled and requires cross-layer interaction between the PHY and the upper layers.

Motivated by the aforementioned needs, Slavik et al. have proposed a cross-layer design for CR that has two main data flows in the network stack [30]. The first, like any network stack, is the flow of user data vertically (Figure 4). This data originates at one application, flows down through its node's network stack, up through a receiving node's network stack, and terminates at the receiving node's application. The second data flow is that of control data, which represents the current state of the network stack. This flow originates in the layer(s) of the stack, flows to the controller, is processed by it, and then flows into the control points of the layer(s) in the stack. The data processing at the controller may involve optimization, and application of artificial intelligence (AI) to make important decisions which enable the vertical user data flow to achieve some performance objective. The controller may also be referred to as the *cognitive engine*. Such a cross-layer design requires that existing stack components be modified to interact with the controller.

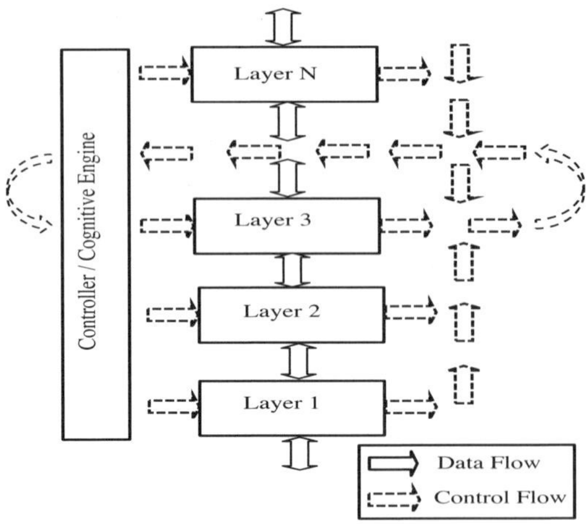

Fig. 4. Cross-layer design (adapted from [30])

A vast amount of literature is available on implementation of specific cross-layer problems for CR, and these have been described in Section 6, in the context of the tools (optimization, fuzzy logic, genetic algorithms, neural networks, etc.) used by them to obtain the solution.

Architecture violations introduced by cross-layer design, clearly undermine the significance of the original layered architecture, and can have a detrimental impact on system longevity. Besides, unintended cross-layer interactions can have undesirable consequences on overall system performance [31].

5. Cooperative Technology and Cognitive Radio

To satiate the ever increasing demand for high data rate wireless services, cooperative communication has evolved. Parallel to the developments in the field of CR, research in cooperative technologies has progressed significantly. In its basic definition, cooperative transmission refers to the information theoretic model of a three terminal relay channel in which the relay forwards the transmission from the source towards the destination. The processing at the relay may simply involve forwarding an amplified version of the received signal to the destination i.e. amplify-and-forward (AF), or decoding the received signal completely and re-encoding it to forward it to the destination i.e. decode-and-forward (DF) [32]. Performance advantages achievable from collaborating relays arise in two forms, both of which translate into enhancing overall network capacity: (i) power gains which can be achieved if the relay is suitable located, typically half way between the source and destination; (ii) diversity gains which arise due to the multiple paths taken by the signal to reach the destination [33] .

The incorporation of cooperative communication in CR networks i.e. cognitive cooperative communication, creates a promising solution to efficient radio resource utilization. Varied forms of cooperation have been investigated to suit different topologies and to meet various requirements of the CR network. Strategies that provide efficient usage of the spectrum opportunity and guarantee QoS constraints for both PU and SU have been suggested [34–36]. Cooperation in terms of a symbiotic architecture between the PU and SU has also also been investigated [37,38] . Cooperative sensing is an important area where cooperative technology can be deployed for robust detection of spectrum holes and estimation of channel conditions[11]. Cooperation among the PUs and SUs can be achieved in the following formats: (i) No Relay Aided- in which the SUs simply relay the CSI among themselves for improved spectrum sharing; (ii) Secondary Relay Aided- in which SUs assist each other in forwarding data to the intended destination; and (iii) Primary Relay Aided- the SUs relay the PU's data to reduce its power consumption [39] .

Multiple-input multiple-output (MIMO) wireless communication systems employ multiple antennas at the transmitter and the receiver. Introducing MIMO in CR communication, increases the scope for cooperation and enhances the system capacity by mitigating the impairments of a wireless environment such as multi-path fading, delay-spread and co-channel interference [40,41] . In this regard, CR MIMO channel capacity under different assumptions regarding the knowledge of CSI at the transmitter and receiver has been analyzed [40] , which brings out the benefit of MIMO CR. Cooperative spectrum sensing in OFDM-based MIMO CR sensor networks demonstrates that it gives a better detection performance compared to spectrum sensing without MIMO [42] .

6. Enabling Tools for Cognitive Radio

After having provided a detailed account of the CR technology, its fundamental components, and essential functionalities in the context of *Next-Generation* networks, in this section we review CR from the perspective of its enabling tools, and provide an extensive literature survey in each category.

6.1. *Optimization*

The main goal of an optimization problem is "to find the best solution among the available set of solutions under limited resources" [43] . A well-known example to visualize this definition is the "0-1 knapsack problem", according to which a hiker wants to put several items into his bag, but has a constraint on the weight, *maxwt*, he can carry. He needs to find the "best combination" of items which weigh as close as possible to the *maxwt*, according to the relative value of each item determined by himself.

The formal model of any optimization problem depends on its *variables*, *objective function* and *constraints*. In general , it can be written as [44]

Find x^* *which*
minimizes $f(x)$
subject to $c_i(x) \leq 0,\ i=1,2,....,r$
with $m_j(x) = 0,\ j=1,2,....,h$

where $x^* = [x_1, x_2, ..., x_n]^T$, $(.)^T$ denotes transpose operation. f represents the objective function, c_i and m_j denote the constraints.

If the optimization objective, the inequality constraints, and the equality constraints are all linear in the parameter function, the problem is called a *linear program.* If either the optimization goal or the constraint functions are nonlinear, the problem is called *nonlinear program.* One special kind of nonlinear program is the *convex optimization* problem, where the feasible set, the optimization goal, and the inequality constraints are all convex, while the equality constraints are affine [44] . In this case, any locally optimal point is also globally optimal. If all of the optimization variables are integers, the problem is called an *integer program.* If there is a combination of real and integer variables, it is called a *mixed integer program.* To find closed-form solutions, one of the most important methods for constrained optimization is the Lagrangian method [44] . However, in case of non-linear and non-convex constraints and optimization goal, the *Lagrangian* multiplier function is difficult to handle and the optimal points are hard to obtain [44,26] . Convex optimization problems can be solved globally and efficiently through the interior point primal dual method, with polynomial running times that are $O(\sqrt{N})$, where N is the size of the problem [43] . Most of the time, integer/combinatorial optimization problems are Non-deterministic-Polynomial-hard (NP-hard) and unsolvable in polynomial time. In practice, many parameters can have only integer values, like modulation level, route selection, etc. Moreover, even for some continuous parameters such as transmission power, the real implementation has finite granularity, leading to limited integer values as choices. To solve integer/combinatorial problems one can resort to *relaxation and decomposition*, *brand-and-bound*, and *cutting-plane* techniques [43] . The performance of these is largely problem-oriented.

Adaptive Radio Resource Management (ARRM) problems in CR can be formulated as constrained optimization problems, from a network or individual viewpoint. The primary objective of these problems is maximizing the SU throughput, where the transmit power, constellation size, and bandwidth should be dynamically allocated, and/or the antenna beam be adjusted, within constraints on battery level, interference temperature of the PU, target BER, etc., assuming that the CSI of the PU and SU networks is available. Some of the ARRM problems that have been cast in the framework of optimization in a CR scenario are as follows: power allocation for an OFDM-based single SU transceiver [45–47] , power allocation for a multi-user case [48–50] , adaptive modulation [51–54] , and beam-forming, in which the direction of the antenna beam is adaptively adjust to maximize the capacity of the cognitive link [55,56] . Optimization framework is also useful in spectrum sharing (sharing of sub-carriers among frequency division multiplexed SUs [48–50]). Another application of optimization is mitigating interference to the PU, created by the side-lobes of OFDM-based SUs, by using techniques like sub-carrier weighting [57] and introduction of cancelation carriers [58] .

Cross-layer optimization manifests itself in a more complicated way for CR networks. It considers the optimization variables across multiple layers of the network protocol stack to achieve one, or most of the times, multiple objectives. A multi-objective optimization, in general, is represented as [44]

$$\begin{aligned}&\textit{Minimize } F(x) = [F_1(x), F_2(x), ..., F_n(x)]^T\\&\textit{subject to } c_i(x) \leq 0,\ i\text{=}1,2,....,r\\&\textit{with } m_j(x) = 0,\ j\text{=}1,2,....,h\end{aligned}$$

In multi-objective optimization it is extremely difficult to find a solution that can maximize (or minimize) each objective; instead, the term solution corresponds to a set which represents a trade-off between objective functions.

Wang et al. have proposed a joint cross-layer scheduling and sensing design, and studied its performance advantages over the traditional decoupled approaches [59] . Cross-layer optimization problems for routing, MAC scheduling and power allocation in a multi-hop multi-channel network are formulated as *mixed integer nonlinear programs*, and solved by *branch-and-bound* techniques [60,61] .

6.2. *Game Theory*

Game theory is an analytical tool designed to help us understand the phenomena that is observed when decision-makers (players) interact [62] . The fundamental assumptions that govern the use of game theory are,

- The players are *rational* i.e. they pursue well-defined exogenous objectives
- The players *reason strategically* i.e. they take into account their knowledge or expectations of other players

Definition 6.1. (Game theory) Game theory is a formal way to analyze interactions among a group of rational decision-makers called players who reason and act strategically.

A game requires the specification of the three parameters: the set of *players* in the game, the exhaustive *strategy set* available to each player, and the *payoffs* or utility functions associated with any strategy combination of each player. Equipped with the knowledge of these parameters, any decision-making problem can be solved using the well-established concepts of game theory, provided the solution exists. We present some definitions and basic theorems in game theory and use an illustrative example to better understand these concepts. Some commonly used notations in game theory are tabulated below (Table 1).

6.2.1. *Fundamental Lessons*

The ideas in game theory have a very interesting flow. We build these leading towards the more general situations that occur in any competition.

Definition 6.2. (Dominated Strategy) The strategy $s_i^{'}$ of player i is strictly *dominated* by the strategy s_i of player i if,

$$u_i(s_i, s_{-i}) \geq u_i(s_i^{'}, s_{-i}) \ \forall \quad s_{-i} \in S_{-i}$$

The idea here is that if a game has a dominated strategy it is *never to be played*, as doing so will always yield a lower payoff. However, in most games there are seldom any dominated strategies, and hence, there is very little

Table 1. Standard Notations in Game theory

No	Entity	Notation
1	Player	i
2	Strategy of player i	s_i
3	Set of all strategies of player i	S_i
4	Strategies of all players except player i	s_{-i}
5	Strategy profile of a game wrt player i	(s_i, s_{-i})
6	Payoff of player i for strategy s_i against strategies s_{-i} of other players	$u_i(s_i, s_{-i})$

use of this idea. A more general concept is to look at one particular strategy of the opponent and then decide what is the *best response (BR)* strategy that one could play.

Definition 6.3. (Best Response) The strategy $\hat{s_i}$ of the player i is *best response* to the strategy s_{-i} of the other players if

$$u_i(\hat{s_i}, s_{-i}) \geq u_i(s_i^{'}, s_{-i}) \;\forall\; s_i^{'} \in \{S_i - \hat{s_i}\} \text{ and } s_{-i} \in S_{-i}$$

Analysis of the definition reveals that a strategy which is *never a best response* should be avoided. However, what strategy is to be played when one is oblivious to the strategy set of the opponent is yet unknown. Further analysis shows that when these strategies are deleted from the strategy set, there will be new strategies that *now* come in the category of being *never a best response*. This when done continuously for all players leads us to what is known as the famous *Nash Equilibrium (NE)*.

Definition 6.4. (Nash Equilibrium) In an N player game, a strategy profile $(s_1^\star, s_2^\star, \ldots, s_N^\star)$ is a *Nash Equilibrium*, if for each player i the choice of the strategy $s_i^\star$ is a best response to the opponents choices $s_{-i}^\star$.

A game may or may not have a NE. Moreover, if it has a NE then it may not be unique. We look at an example to understand the above ideas.

The Investment Game

Consider a game in which two people (players 1 and 2) run a company, share equal profits, and the strategy of these players (S_1, S_2) is to put an effort on a scale of $[0, 4]$ (a real number which is proportional to both the quality and quantity of the work the players put). The payoffs are defined as follows:

$$U_1(s_1, s_2) = s_1 + s_2 + bs_1s_2 - s_1^2 \tag{19}$$

$$U_2(s_2, s_1) = s_1 + s_2 + bs_1s_2 - s_2^2 \tag{20}$$

where $s_1 \in S_1$, $s_2 \in S_2$, $b \in [0, 1/4]$, and the factor bs_1s_2 represents the profit obtained when working in a team (*synergy*). The negative term contribution is due to the cost incurred by each player due the effort they put.

The best response curves of player 1 and 2 are obtained by differentiating each of the above two functions to get $\frac{\partial U_1(s_1,s_2)}{\partial s_1}$ and $\frac{\partial U_2(s_2,s_1)}{\partial s_2}$. We equate these to zero and get BR of player 1 as a function of s_2 and BR of player 2 as a function of s_1. We have

$$BR_1(s_2) = 1 + bs_2 \tag{21}$$

$$BR_2(s_1) = 1 + bs_1 \tag{22}$$

From Figure 5 it is easy to see that those strategies that are *never a best response* are not considered (shaded). Once we remove those strategies that are *never a best response* we can repeat the process over and over again. In the limiting case, we can see that we are left with only one strategy profile for the entire game, and such a strategy profile is the NE of the game.

An extension of this 2-player game is the N-player game in which the Nash Equilibrium is the intersection of N curves, each of which is a function of the other $N - 1$ opponents' strategies in the N-dimensional space. We note the following two important points regarding *existence* and *uniqueness* of NE, for strategy sets continuous over a finite interval:

- the *NE* exists if all these curves intersect at-least at one common point
- and it is unique if they intersect in one and only one point

The following theorem formally states the conditions in which the NE exists uniquely.

Theorem 6.1. [63] *A game modeled with finite number of players N, an exhaustive strategy set S_i and a well-defined utility function $u_i(s_i, s_{-i})$ for each player $i \in N$, has a unique Nash Equilibrium iff there exist one and only one strategy profile $S^\star = (s_1^\star, s_2^\star, \ldots, s_N^\star)$ such that $u_i(s_i^\star, s_{-i}^\star) \geq u_i(s_i, s_{-i}^\star) \quad \forall s_i \in \{S_i - s_i^\star\}$ and $i \in N$.*

This strategy profile can be viewed as being the only N dimensional strategy profile present in each of the $N - 1$ dimensional best response curves. In

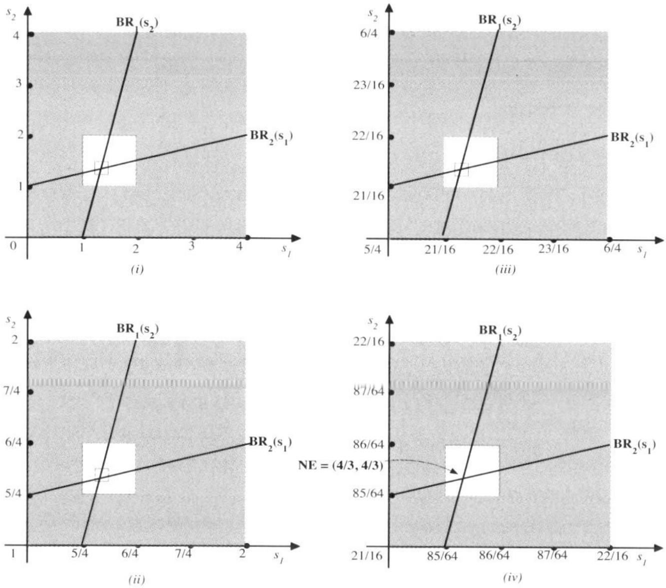

Fig. 5. Nash Equilibrium

some cases, the utility function will be defined such that the best response functions will be a linear combination of the opponents strategies. In such situations, one can conveniently represent these linear equations in the form of a matrix equation, and use the well-established concepts of linear algebra to get the solution. The matrix equation can be written as follows

$$AS = B \tag{23}$$

where A is a square matrix of size $N \times N$ in which the diagonal elements are equal. The vector S, with N elements is the strategy profile vector. B is a constant vector. We state a corollary of the above theorem for the special case of linear best response functions.

Corollary 6.1. *The necessary and sufficient condition for a unique Nash equilibrium to exist when the best response functions are linear in the opponents strategies is that the representing matrix A has only the 0_V (the zero vector) in its kernel or null space.*

This implies that for the Nash Equilibrium to exist uniquely, the matrix A is necessarily a full rank matrix, i.e. the matrix A^{-1} exists and is well-defined. We also note that there are infinite Nash equilibria when the system of equations in Eqn. 23 is consistent but incomplete. Moreover, the Nash equilibrium does not exist if this system of equations is inconsistent.

6.2.2. *Game theory for Cognitive Radio*

CR instills completion not only between the PU and the SU, but also amongst the SUs themselves. As a result, game theory is a natural tool in resolving the conflicting interests. Game theory is applied to CR networks to find optimum strategies of the PU or SU or both, given a set of network assumptions, such as communication technology, CSI, power levels etc.

A well-established game model is the *Stackelberg* game, wherein a *leader* first chooses a strategy, and the *follower's* strategy depends on the *leader's* strategy. Such models are adopted for CR, where in general, the PU enjoys more privileges than the SU, and hence is deemed as a leader. Some other game theory models like the *Bertrand* and *Cournot* are also used where the players have equal rights [64] .

Stanojev et al. have proposed a symbiotic cooperative architecture for CR, whereby a PU leases its bandwidth for a fraction of time to a network of SUs in exchange for cooperation [38] . On one hand, the PU decides whether to exploit the cooperation from the SUs in order to enhance its own transmission rate, while the SUs decide to participate only if compensated with a large enough fraction of time for their own transmission, with the objective of maximizing their rate discounted by the overall cost of transmitted power. The problem is cast in the framework of *Stackelberg* games. Another example of symbiotic cooperative architecture is: when the PU infrastructure (bandwidth and relay nodes) is not utilized, it is leased to the SUs, in return for payments made by SUs for the service they receive [65] . The interactions between the PU and SU are considered as a buyer/seller model to determine the price which maximizes the PU's revenue. Further, within the SU network, given the specified price, the users compete to access the PU infrastructure, which is modeled as a non-cooperative game. In a similar architecture, albeit with multiple PUs competing with each other in the pricing process to get higher revenues, the game has been modeled as a *Bertrand* game [66] .

A *static* game is one in which all players make decisions simultaneously, without knowledge of the strategies of other players. In *dynamic* games, however, there is an explicit time-schedule that describes when the individual players make their decisions [64] . In a Time Division Multiple Access (TDMA) CR system model, which schedules one SU per spectrum hole in each time slot according to a predefined decentralized scheduling policy, the transmission rate adaptation problem for each SU is formulated as a *dynamic* game with a latency constraint [67] . A *stochastic* game is a *dynamic* game with probabilistic transitions played by one or more players [64] . Such games have been widely used in addressing security issues in CR. *Malicious SUs* and *PU emulation* are some of the attacks that can take place in a CR system. In a *PU emulation* attack, the attacker sends signals having the same features as the PU, during spectrum sensing, and can claim the whole frequency band or interrupt the operation of SUs. It is similar to *jamming* in traditional wireless communication systems. A *stochastic zero-sum* (one participant's gains result only from another's equivalent losses) game model is used to study the strategies of an attacker and an SU, in a jamming and anti-jamming scenario [68] . The PUs dictate the system states and their transitions, while the SUs and jammers behave non-cooperatively to achieve their goals independently. The results indicate that the SUs should enhance their security levels, or increase their pay-offs by either improving their sensing capabilities or choosing to communicate under states where the available channels are less prone to jamming.

6.3. *Fuzzy logic*

Fuzzy logic is a framework, which aims at modeling the vagueness that exists in real world problems, which cannot be efficiently expressed by means of mathematical models. According to fuzzy set theory [69,70,26] when A is a fuzzy set and x a relevant object, the proposition x is a member of A is not necessarily *true* or *false*, but it may be *true* or *false* only to some degree. Fuzzy logic provides an inference morphology enabling approximate reasoning capabilities applicable to knowledge based systems. To implement decision making processes, it makes use of *Fuzzy Logic Controllers (FLCs)*, whose essential part is a set of linguistic control rules based on expert knowledge, in the form:
IF (a set of conditions are satisfied) THEN (a set of consequences can be inferred).

An FLC operates by repeating a cycle of the following steps: First, measurements are taken of all variables that represent relevant conditions of the process. Next, these measurements are converted into appropriate fuzzy sets (fuzzification). The fuzzified measurements are used by the inference engine to evaluate control rules stored in the fuzzy rule base. The result of this evaluation is a fuzzy set(s) defined on the universe of discourse of possible actions. This fuzzy set is finally converted into a crisp value(s) (defuzzification).

In essence, the FLC provides an algorithm which can convert the linguistic control strategy based on expert knowledge into an automatic control strategy; this appears very useful when processes are too complex for analysis through conventional quantitative techniques, or when the available sources of information are interpreted qualitatively, inexactly or uncertainly, which may be the case in a CR network. The main weakness of FLCs is the dependability of their decisions on the way the membership functions and fuzzy inference rules are formulated. To overcome this limit, fuzzy logic is often combined with learning algorithms based on neural networks or genetic algorithms [26] .

A fuzzy-based spectrum hand-off is proposed for the SU, so that it is able to make the hand-off decision in a decentralized fashion without causing any interference to the PU receiver [71] . Considering the uncertainty that arises due to the characteristics of the wireless channel, the interference caused by multiple SUs, the decentralization of the decision etc., the estimations are realized by means of FLCs. Two FLCs are implemented: (i) the first is used to qualitatively determine the distance between a SU and a PU, which depends on the transmission power at which the SU should transmit without interfering with the PU (P_{SU}), the signal-to-noise ratio at the PU (SNR_{PU}), and the signal strength at the SU from the PU (SS_{PU}). Each input linguistic variable is characterized by a set T of three attributes, as follows:

$$T(SS_{PU}) = T(SNR_{PU}) = T(P_{SU}) = \{Low \quad Medium \quad High\} \quad (24)$$

(ii) the second FLC has been designed to qualitatively determine whether a spectrum hand-off should be realized or not. It consists of three input linguistic variables: the SU transmission power (P_{SU}), the signal strength received at the SU (SS_{PU}), the bit rate of the SU (R_{SU}); which are also characterized as follows:

$$T(SS_{PU}) = T(R_{SU}) = T(P_{SU}) = \{Low \quad Medium \quad High\} \quad (25)$$

There are two output linguistic variables: HO decides whether the hand-off has to be realized or not, MOD_{PSU} indicates whether the SU transmission power should be modified and how.

The fuzzy-based approach when compared with a fixed-strategy hand-off indicates a reduced hand-off rate, thus guaranteeing a better perception of the service and a reduced signalling cost for the SUs, and a higher percentage reduction of interference temperature at the PU receiver.

Baldo et al. have proposed fuzzy cross-layering for the enhancement of TCP performance over 802.11 [72] . The FLC is either embedded into a layer or implemented as a centralized cognitive engine. In 802.11, the link reliability is represented by the fuzzy variable *linkErrors*, which is characterized by two attributes *low and high.* In TCP, the fuzzy variable used is *networkCongestion* having attributes *low, mid and high.* Moreover, two fuzzy output variables *cwndIncrement* and *cwndReduction* are defined, each having the attributes *small, average and strong. cwnd* represents the congestion window, which determines the number of bytes that can be outstanding at any time, and limits throughput in the face of loss. *cwnd* gets larger after every new acknowledgement *(ACK)* and *cwnd* get smaller when loss is detected. It is shown that fuzzy logic based cross layering provides significant performance gains, while being much more modular and reusable than traditional cross-layer solutions.

A fuzzy logic system can also give solutions for the opportunistic spectrum access problem [73] . The most suitable SU, having the rights to access the spectrum, is chosen based on three descriptors: spectrum usage efficiency of the SU, its degree of mobility, and its distance to the PU. The linguistic variables used to represent the spectrum utilization efficiency and degree of mobility are divided into three levels: *low, moderate, and high*; while three levels: *near, moderate, and far* are used to represent the distance. The consequence, i.e., the possibility that the SU is chosen to access the spectrum is divided into five levels which are *very low, low, medium, high and very high.* From simulation results, it is proved that the spectrum access decision is a trade-off among the three descriptors chosen to design the FLC.

6.4. *Genetic Algorithms*

Genetic Algorithms (GA) are search startegies based on the principal of evolution [74] . They are based on the following main principles:

- *Selection*: Selection is a procedure for selecting individuals for reproduction according to their fitness function.
- *Crossover*: Crossover is a recombination operator that combines the genetic information of two individuals.
- *Mutation*: Mutation is an operator that introduces variations into the the genetic material.
- *Sampling*: Sampling is a procedure which computes a new generation from the previous one and its off-springs.

The basic structure of a GA is given by [74] :

Algorithm 6.1.
t:=0
Compute the initial population X_o*;*
WHILE stopping condition not fulfilled DO
BEGIN
Select individual for reproduction;
Create offspring's by crossing individuals;
Eventually mutate some individuals;
Compute new generation
END

In general, the solution to a problem set is represented by binary strings. These strings are allowed to grow in a genetic manner. Strings which are considered good, combine with other good strings and form a new solution, while poor strings are discarded from the solution set. This decision is taken by the fitness function based on an output score for specific input parameters.

Newman et al. have derived fitness functions for power consumption minimization, BER minimization and throughput maximization [75] . For an optimal decision, the individual fitness functions have been combined using the weighted sum approach. The combined weighted fitness function is given by

$$f = w_1 f_{min\_power} + w_2 f_{min\_BER} + w_3 f_{max\_throughput} \tag{26}$$

where f_{min_power}, f_{min_BER} and $f_{max\_throughput}$ are the individual fitness functions corresponding to power consumption, BER and throughput respectively, while w_1, w_2 and w_3 are the weights which depend on the importance of each of these performance objectives, respectively.

Chen et al. have proposed a cross-layer design, with a GA-based optimization engine, to simultaneously minimize the BER, minimize the out-of-band interference and maximize the throughput for an OFDM-based CR [76,77] . Yuan et al. have suggested a fast GA for bit-allocation in OFDM sub-carriers in a CR system [78] . The bit-allocation is modeled as a binary sequences search problem, which executes iteratively, and terminates when the power consumed is equal to the power budget or the interference to the PU is greater then a specified threshold. Gelenbe et al. have proposed a routing algorithm for CR packet networks [79] . A GA is implemented at the source routers of each connection, to compose new routes from the existing ones and to select which routes to use, based on their fitness functions, which are estimated from the QoS.

GAs are designed to search highly non-linear spaces for global optima, whilst traditional optimization techniques are likely to converge to local optima once they are in its vicinity.

6.5. *Neural Networks*

Biological neural networks are made up of biological neurons that are physically connected and functionally-related to the human brain. *Artificial Neural Networks (ANN)* on the other hand, are made up of artificial neurons interconnected to each other to form a programming structure that mimics the behavior and processing (organization and learning) of biological neurons, and can be applied for solving problems with an associative or cognitive tinge [80] . In ANN, neurons are arranged in three layers: neurons which receive data from outside are organized in the input layer; neurons which send data out of the ANN comprise the output layer; and neurons whose input and output signals remain within the ANN, form hidden layer(s). Neurons communicate with each other by sending signals over a large number of weighted connections, thus creating a network with a high degree of interconnection. Generally each connection (j to k) is defined by a weight, w_{jk}, which determines the effect that the neuron j has on neuron k. Every neuron has a state of activation, which is equivalent to the output of the neuron. During processing, each neuron receives input from neighbors belonging to different layers, as well as from external sources, and uses them to compute the updated level of activation. For this purpose activation functions (such as the step function, sigmoid function, piecewise-linear function) are used [80] .

The topology of a ANN plays an important role for its achievable performance. Depending on the pattern of connections that an ANN uses to propagate data among the neurons, it can be classified into two basic categories: (i) Feed-forward ANNs, where data enters at the inputs and passes through the network, layer by layer, until it arrives at the outputs; (ii) Recurrent ANNs that contain feedback connections, which are connections extending from outputs of neurons to inputs of neurons in the same layer or previous layers. In contrast with feed-forward networks, the recurrent network has a sense of history.

Many problems in CR can easily accommodate ANNs. In particular, ANNs can be used in any of the phases of the cognition cycle (described in

Section 2.1). For instance, during the radio-scene analysis, temporal statistics of a radio environment can be used to isolate its distinct characteristics, which in turn correspond to different modulations. These statistics can then feed an ANN in order to classify a signal's modulation type [81] .

The work of Liu et al. focuses on the channel estimation and predictive modeling phases of the cognition cycle; the proposed ANN, implemented in the learning module of the cognitive engine, assists in the optimum decision regarding the radio-configuration settings (mainly PHY and MAC layer) that will provide the best QoS for the given problem and user/application needs [82] . The ANN is designed to learn from the information measured by the CR node during the radio-scene analysis, and to provide the data rate in the output, that is most anticipated to be obtained per radio configuration, thus behaving as a predictor of the next expected data rate.

Spectrum sensing is an important function for the SUs to determine the availability of a frequency band in the PUs spectrum. However, it consumes considerable energy, which can be reduced by employing predictive methods for discovering spectrum holes. Tumuluru et al. have proposed the deployment of an ANN-based prediction scheme, by which the SUs will sense only those frequency bands which are predicted to be idle [83] . Reliable prediction will significantly improve the spectrum utilization, besides conserving energy. Since the traffic characteristics of most PU systems encountered in real life are not known apriori, ANNs give an advantage over statistical models.

Zhang et al. have proposed a cognitive engine based on ANN [84] . The decision of the engine considers changeable factors (bandwidth, data rate, BER), as well as unchangeable factors (infrastructure cost, licensed user) to make the best decision, viz. adjusting its resources and performing the appropriate signaling.

General Remarks

A CR should be adaptive to the channels time-varying nature. In a fully reconfigurable real-time CR, methods such as optimization and game theory are not the best solutions for resource allocation due to high complexity and the associated high computational time and cost. So, algorithm design should be based on heuristic methods for more practical solutions in realistic scenarios. Besides, the mathematical complexity of optimization increases with the number of parameters used to characterize the system. However, optimal solutions are still useful as a benchmark. Simplified suboptimal algorithms may compete with heuristic algorithms. One cannot refute the usefulness of game theoretic approaches in pricing and revenue models for the PU and SU. Most such models can be conveniently cast in the framework of existing games, and well-established principles therein can be applied to resolve the conflicting interests of the players.

Perceiving and *learning* from the RF stimuli are the two essential modules of the cognitive engine, which enable enhanced responses and adaptation to a changing environment. *Learning* algorithms such as neural networks and genetic algorithms have significant applicability in these modules, for increased reliability and robustness of the decisions of the engine. Besides, they are effective techniques for predictive estimations of the channel and PU traffic, which can dramatically improve the efficiency of the overall system. Genetic algorithms have established themselves as vital tools for cross layer optimizations, specially those which are multi-objective in nature. Fuzzy logic is a convenient tool for decision-making in complex processes where there is vagueness in the available information, but its accuracy is largely dominated by the manner in which the membership functions and fuzzy inference rules are formulated. They are reasonable when qualitative results suffice, rather than precise numerical values.

A CR system is a juxtaposition of various functionalities, each of which may demand a specific tool for its optimum performance. Based on our analysis of the various enabling tools provided above, we conjecture that many a times, it may be a good idea to use diverse tools in different modules of the same CR system. However, caution should be exercised, in that all the modules should be inter-operable, and should have no conflict-of-interest, so that it results in an overall optimized system performance.

7. Fundamental Research in CR: An information theoretic perspective

How cognitive radios may be best employed for efficient spectrum usage by the SUs can be investigated from a number of aspects, one of them being information theory. An information theoretic framework is ideal for analyzing the fundamental limits on capacity, rate regions achieved in a network, and scaling laws for a network.

An example that offers some insights, is the Gaussian cognitive channel, which is a 2 transmitter 2 receiver channel. *Tx1, Rx1* represent the PU transceiver, while *Tx2, Rx2* represent the SU transceiver (Figure 6). The SU transmitter has full knowledge of the message that the PU is transmitting, but not vice versa. This is called asymmetric noncausal message knowledge, and it has been used to obtain better achievable rates than what time sharing schemes for SU spectrum access yield [85] . The reasons for choosing Gaussian channels is that they are the most commonly considered continuous alphabet channels and are often used to model noisy channels; more importantly, their analysis is computationally tractable. As an additional point, the Shannon's Channel Capacity theorem [86] assumes that the noise is an additive white Gaussian process. An illustration of how asymmetric noncausal message knowledge obtains better achievable rates is given by Dirty Paper Coding.

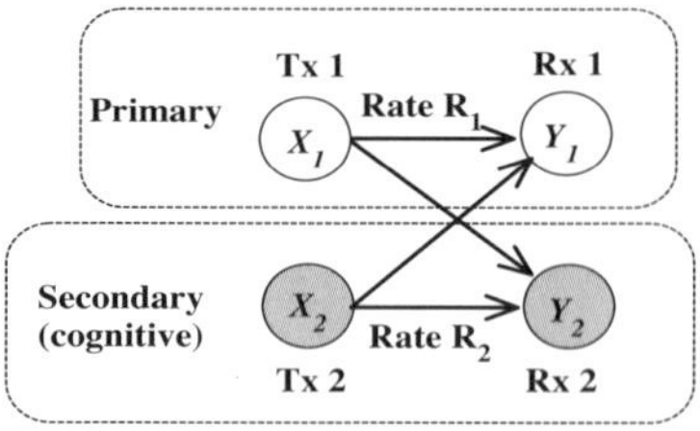

Fig. 6. Cognitive Channel

Dirty Paper Coding was first considered by Costa [87] . It is motivated by the following theorem:

Theorem 7.1. *Consider the following channel:*

$$Y = X + S + N, \quad E\left[|X|\right]^2 \leq P, \quad N \sim \mathbb{N}(0, Q)$$

Where N is the noise distributed as a Gaussian random variable, Q is the power/variance of the noise. Input X is subject to a power constraint as indicated, S is an additive interference whose value is known to the transmitter but not the receiver. Y is the received output of the channel. This channel has the same capacity as that of an interference free channel and the capacity is given by:

$$C = \frac{1}{2}\log_2\left(1 + \frac{P}{Q}\right)$$

According to this theorem, even in the presence of interference, provided it is known at the transmitter (the SU transmitter, *Tx2*, in the CR scenario), the channel behaves like an interference free channel, and is capable of achieving high rates.

7.1. *Multiplexing gains of Cognitive channels*

It has been shown that when two interfering point-to-point links employ asymmetric noncausal side information, it allows for higher spectral efficiency. It helps to analyze the performance of cognitive channels when noise power is minimal, or is not an obstacle. For Gaussian noise channels, multiplexing gain(MUX_g) is defined as [85]

$$MUX_g = \lim_{SNR \to \infty} \frac{R(SNR)}{\log(SNR)} \tag{27}$$

where R is the maximal achieved sum rate. Multiplexing gain is a measure of how well a MIMO channel can avoid self-interference. When multiple transmitters and receivers wish to share the same medium, they may be thought of as a number of parallel point-to-point links captured by MIMO channels. In this context, the following propositions are stated [85] :

Proposition 7.1. *For a single user point-to-point MIMO channel with M_T transmit and N_R receive antennas, the maximum multiplexing gain is* $\min(M_T, N_R)$.

Proposition 7.2. *For a two user MIMO multiple access channel with M_{T_1} transmit antennas for user 1, M_{T_2} transmit antennas for user 2 and N_R receive antennas; the maximum multiplexing gain is given by* $\min(M_{T_1} + M_{T_2}, N_R)$.

Proposition 7.3. *As a dual result, for a two user MIMO broadcast channel with N_{R_1} receive antennas for user 1, N_{R_2} receive antennas for user 2 and M_T transmit antennas; the maximum multiplexing gain is given by* $\min(N_{R_1} + N_{R_2}, M_T)$.

From these propositions, it is deduced that when joint signal processing is available at either transmit or receive sides, the multiplexing gain is significant.

7.2. *Scaling Laws in Cognitive Radio*

In this section, it is discussed how the total throughput, achieved by cognitive users, scales with the number of users. The communication scenario is as follows [85] : Let the PU transmitter have fixed power P_0 and minimum distance R_0 from any cognitive receiver. Assume that each cognitive user transmits with the same power P. A lower bound on the network sum capacity can be derived by upper bounding the interference to a cognitive receiver. An interference upper bound is obtained by, first, filling the primary exclusive region with cognitive users. The primary exclusive region is that in which the outage probability is constrained. Next, consider a uniform network of n cognitive users. The worst-case interference, then, is to the user with the receiver at the center of the network. Let R_c be the radius of the circle centered at the considered receiver that covers all other cognitive transmitters. With constant user density (λ users per unit area), R_c^2 grows linearly with n. Furthermore, any interfering cognitive transmitter must be at least a distance ϵ away from the interfered receiver, for some $\epsilon > 0$, and α is the path loss exponent.

Proposition 7.4. *The average worst-case interference caused by* $n = \lambda\pi(R_c^2 - \epsilon^2)$ *cognitive users is given by* [85]

$$I_{avg,n} = \frac{2\pi\lambda P}{\alpha - 1}\left(\frac{1}{\epsilon^{\alpha-2}} - \frac{1}{R_c^{\alpha-2}}\right) \tag{28}$$

From Eqn. (28), as $n \to \infty$ and for $\alpha > 2$, the following relation is obtained:

$$I_{\infty} = \frac{2\pi\lambda P}{(\alpha - 1)\epsilon^{\alpha-2}} \tag{29}$$

I_{∞} is a constant. This is used to show the following result wherein the expected capacity of each user C_i is bounded by a constant as $n \to \infty$.

$$\mathbb{E}[C_i] \geq \log\left(1 + \frac{P_{r,min}}{\sigma_{0,max}^2 + I_{\infty}}\right) = \bar{C}_1 \tag{30}$$

where $P_{r,min} = P/D_{max}^{\alpha}$ and $\sigma_{0,max}^2 = \sigma_n^2 + P_0/R_0^{\alpha}$, D_{max} being the maximum distance between a cognitive transmitter-receiver pair, and σ_n^2 being the thermal noise power of each.

Thus $\bar{C}_1$ is the achievable average rate of a single cognitive user under constant noise and interference power.

For the upper bound, the interference from all other cognitive users is removed. Assuming that the capacity of a single cognitive user under noise alone is bounded by a constant, then the total network capacity grows at most linearly with the number of users. From these lower and upper bounds, it is concluded that the sum capacity of the cognitive network grows linearly in the number of users [85] , i.e.

$$\mathbb{E}[C_n] = nK\bar{C}_1 \tag{31}$$

For some constant K.

8. Power allocation in OFDM-based CR

In this section, we present a power allocation algorithm for an OFDM-based CR [47] , in the context of *Radio Resource Management (RRM)*. The objective of the algorithm is to maximize the throughput of the SU, under a power budget, while ensuring that the side-lobe interference to the PU band is under a specified threshold. However, for interference mitigation, a threshold is defined for each PU sub-band, which is more practical and effective.

The system model is described as follows (Figure 7):
An SU transceiver is considered, and a PU exists in its radio range. OFDM is the communication technology of the PU as well as the SU, the use of which divides the available bandwidth into N frequency-flat sub-carriers. When the PU claims a portion of the spectrum, the SU nulls the corresponding sub-carriers. If the PU requires bandwidth equivalent to N_p sub-carriers, then there are $N_s=N-N_p$ active sub carriers for the SU. The spectrum hole is detected by the SU in the spectrum sensing phase of its cognitive cycle. The channel power gain of the ith sub-carrier on the link between the secondary source (S) and destination (D) is denoted by h_i, and that of the jth sub-carrier on the S to PU link is given by g_j. It is assumed that the channel state information (CSI) is available with the source.

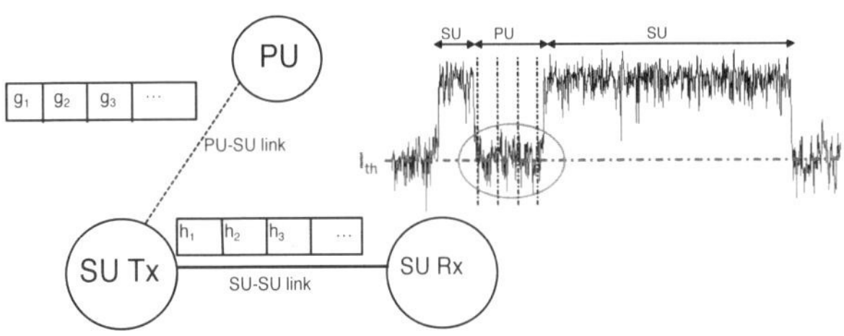

Fig. 7. System model

The maximum achievable throughput of the SU is given by

$$C = \sum_{i=1}^{N_s} \log_2(1 + \frac{P_i h_i}{\sigma^2}) \tag{32}$$

where P_i is the power allocated to the ith SU sub-carrier and σ^2 is the AWGN variance. We have neglected the interference that the PU may cause to the SU, as well as the effect of inter-carrier interference (ICI).

The interference on the jth PU sub-band is formulated as

$$I_j = g_j \sum_{i=1}^{N_s} P_i \int_{j^{th} PUband} (\frac{\sin \pi f T_s}{\pi f T_s})^2 \tag{33}$$

Our objective is to maximize the SU throughput under a total node power constraint P_t, in such a way that the interference to each PU sub-band is less than a threshold I_{th}.

The optimization problem can stated as

$$obj = max \sum_{i=1}^{N_s} \log_2(1 + \frac{P_i h_i}{\sigma^2}) \tag{34}$$

subject to

$$I_j \leq I_{th} \quad \forall j \tag{35}$$

and

$$\sum_{i=1}^{N_s} P_i \leq P_t \tag{36}$$

$$P_i \geq 0 \tag{37}$$

The Lagrangian for the above is formulated as

$$L(P_i, \lambda_j, \mu) = \sum_{i=1}^{N_s} \log_2(1 + \frac{P_i h_i}{\sigma^2}) - \sum_{j=1}^{N_p} \lambda_j (I_j - I_{th}) - \mu(\sum_{i=1}^{N_s} P_i - P_t) + \sum_{i=1}^{N_s} \gamma_i P_i$$

where λ_j, μ and γ_i are the Lagrangian multipliers. The problem is a convex optimization problem, and KKT conditions are applied for the optimum solution. The optimum power allocation is given by

$$P_i^* = max(\frac{1}{\sum_{j=1}^{N_p} \lambda_j g_j Q_{j,i} + \mu} - \frac{\sigma^2}{h_i}, 0) \tag{38}$$

where

$$Q_{j,i} = \int_{j^{th} PUband} (\frac{\sin \pi f T_s}{\pi f T_s})^2 \tag{39}$$

and

$$\lambda_j \geq 0, \mu \geq 0, \gamma_i \geq 0 \tag{40}$$

Though the above solution looks like water-filling, it is different from the conventional water-filling technique in the fact that each SU sub-carrier has a different water level.

As the problem is convex optimization with linear constraints, at the optimum point some constraints are binding, while the others are non-binding. If the power budget of the SU (P_t) is too small, then that will be a binding constraint and all interference constraints are non-binding; the corresponding Lagrange multipliers (λ_j) are zero and the solution looks like that of conventional water-filling with a constant water level:

$$P_i^* = max(\frac{1}{\mu} - \frac{\sigma^2}{h_i}, 0) \tag{41}$$

If the power budget is very high, then only the interference constraint will be binding. Generally, the PU sub-carrier which receives the maximum interference will be responsible for the binding constraint; and the solution looks like

$$P_i^* = max(\frac{1}{\lambda_j g_j Q_{j,i}} - \frac{\sigma^2}{h_i}, 0) \tag{42}$$

To make it a general water-filling solution with a constant water-level, we can multiply by $Q_{j,i}$, to get

$$\vartheta_i = max(\frac{1}{\lambda_j g_j} - \frac{Q_{j,i}\sigma^2}{h_i}, 0) \tag{43}$$

and the power allocation is

$$P_i^* = \frac{\vartheta_i}{Q_{j,i}} \tag{44}$$

If we consider the above solution as the peak power on each SU sub-carrier i.e. P_i^{max}, under the PU interference constraint, and then execute water-filling, it is referred to as *cap-limited* water-filling [49] . The solution takes the form

$$P_i^* = min(max(\frac{1}{\mu} - \frac{\sigma^2}{h_i}, 0), P_i^{max}) \tag{45}$$

If the power budget is neither too high nor too low, the solution will take the form given by Eqn. 38. On substituting P_i^* in the constraint of Eqn. 35, we get N_p equations of the form

$$g_j \sum_{k=1}^{N_p} P_i^* Q_{j,i} = I_{th} \tag{46}$$

The solution to the above cannot be obtained directly, and we have proposed an iterative algorithm to achieve the objective of Eqn. 34, for which we would like to refer the readers to our previous work [47] .

8.1. *Simulation Results and Discussion*

An SU transceiver is considered which uses 32 sub-carrier OFDM for communication. The PU in its radio range occupies a bandwidth corresponding to 8 sub-carriers, leaving 24 active sub-carriers for the SU. The channels undergo Rayleigh multi-path fading, defined in the time domain by

$\sum_{l=0}^{L-1} h_l \delta(t - lT)$ where h_l is the complex amplitude of path l and L is the number of channel taps. The lth channel coefficient is distributed as N(0,σ^2), and the frequency domain channel is given by its Fourier Transform. $\sigma^2 =$ 1e-4 and T_s=1.

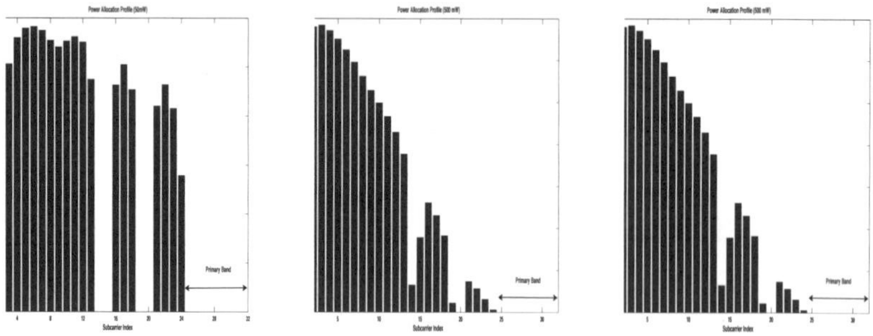

(a) (b) (c)

Fig. 8. Power Allocation Profile : (a) P_t=50 mW (b) P_t=100 mW (c) P_t=500 mW

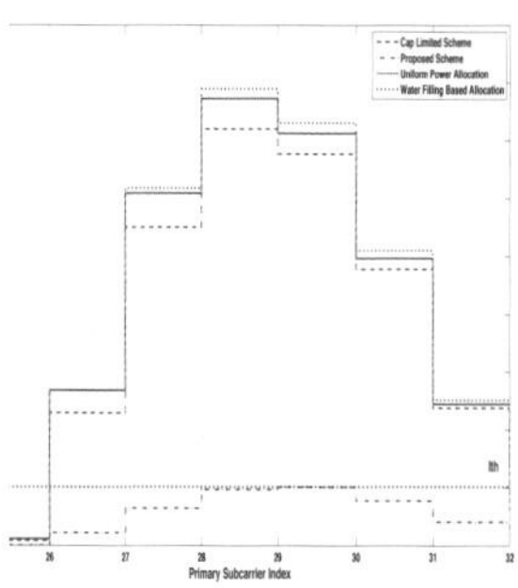

Fig. 9 Interference Profile of the PU

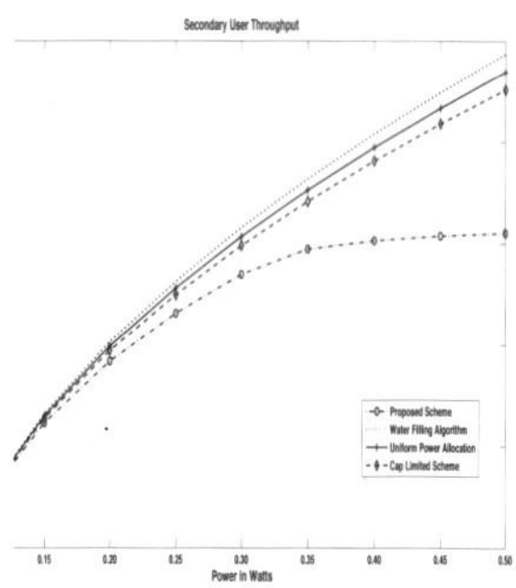

Fig. 10 SU Throughput vs. Power budget

The results of the SU power allocation, for the PU occupying sub-carrier numbers 24-32, are shown in Figure 8. We set I_{th}= 1e-5 W for each PU sub-band. For the result of Figure 8(a), the power budget P_t=50 mW. This value being very small, the interference constraints are non-binding, and it is observed (though the channel gains have not been plotted) that the solution closely resembles that of conventional water-filling: better channels are allocated higher powers as compared to the poorer ones.

As the power budget is increased (Figures 8(b) and 8(c)), the interference constraint becomes binding. Lesser power is allocated in the SU sub-carriers closer to the PU as they contribute more interference to it, and hence the graph tapers towards the PU band for P_t=100 mW, and even more steeply for P_t=500 mW.

In Figure 9, we have plotted the interference profile to the 8 PU sub-bands on execution of the various power allocation schemes. The proposed algorithm maintains the interference to each PU sub-band under the threshold. For comparison, a uniform power allocation and conventional water-filling are also executed on the SU sub-carriers, these techniques do nothing to mitigate the interference to the PU. We also include for comparison, the cap-limited water-filling that has been proposed in literature [49] . Its performance is better than uniform allocation and water-filling, but poorer than the proposed algorithm, since it is successful in keeping only the interference to the closest PU sub-band under the threshold. These results are reported for a power budget P_t=500 mW.

The SU throughput vs. power budget is plotted in Figure 10. Conventional water-filling gives the highest SU throughput, since it is unconstrained by the PU interference threshold. It is closely followed by the uniform power allocation. The cap-limited water-filling, is only partially interference constrained by the closest PU sub-band, on the other hand the proposed scheme considers the interference threshold to each PU sub-band. The SU throughput achieved is the optimum result with the given power budget and interference constraints. Furthermore, after a certain power budget (P_t=350 mW), the throughput hardly increases, since only the interference constraint is now binding. Any further increase in the power budget, cannot increase the SU sub-carrier power allocation without violating the interference constraint.

9. Conclusion

The irrefutable verity of the fact that *Next-Generation* wireless networks will be *spectrum-aware*, has kindled the interest of researchers from diverse disciplines. Consequently, a vast amount of literature has emerged on the CR paradigm. In this piece of work, we have provided an account of the intrinsic functionalities of a CR node, such as spectrum sensing, spectrum allocation, spectrum utilization, spectrum sharing and spectrum mobility, highlighting the main open issues therein. Another, interesting feature of this document is the survey of the technologies or tools that enable the CR system to perceive, think, decide, learn and adapt to the changing environmental conditions, in order to achieve reliable and efficient opportunistic access to the radio spectrum. In the end, we have also presented the optimum power allocation problem for an OFDM-based CR, which aims at maximizing the SU throughput, while mitigating interference to the PU spectrum.

Acknowledgment

The authors would like to acknowledge the support received from the Department of Information Technology (DIT), Ministry of Information and Communication Technology (MCIT), Government of India, for the the project on Cognitive Radio at IIT Hyderabad and IIT Bombay.

References

1. FCC spectrum policy task force: Report of the spectrum efficiency working group, (2002), `http://www.hraunfoss.fcc.gov/edocs\_public/attachmatch/DOC\-228542A1.pdf`.
2. H. Zhang, Cognitive radio networking for green communications and green spectrum, (1993), `http://www.comnets.org/keynote.html`.
3. S. Haykin, Cognitive radio: brain-empowered wireless communications, *IEEE Journal on Selected Areas in Communications,* 2005.
4. J. Mitola III, Cognitive radio for flexible mobile multimedia communications, *Mobile Networks and Applications, Springer,* **Vol. 6,** 2001.
5. I. F. Akyildiz, W. Y. Lee, M. C.Vuran and S. Mohanty, Next generation dynamic spectrum access cognitive radio wireless networks: A survey, *Journal of Computer Networks,* **Vol. 50,** 2006.
6. IEEE 802.22 WRAN, `http://www.ieee802.org/22/`.
7. IEEE 802.11 WLAN, `http://www.ieee802.org/11/`.
8. IEEE SCC41, `http://www.scc41.org/`.
9. M. Matinmikko, M. Hyhty, M. Mustonen, H. Sarvanko, A. Hekkala, M. Katz, A. Mmmel, M. Kiviranta and A. Kautio, *Cognitive radio: An intelligent wireless communication system*, research report, VTT Technical Research Centre of Finland (Kaitovyl 1, PL 1100, FI-90571 Oulu, Finland, 2008).
10. F. Ge, Q. Chen, Y. Wang, T. Bostian, C.W.Rondeau and B. Le, Cognitive radio: From spectrum sharing to adaptive learning and reconfiguration, in *IEEE Aerospace Conference'08*, 2008.
11. Y. Zeng, Y. C. Liang, A. T. Hoang and R. Zhang, A review on spectrum sensing for cognitive radio: challenges and solutions, *EURASIP J. Adv. Signal Process,* **Vol. 2010,** (Hindawi Publishing Corp., New York, NY, United States, Jan., 2010).
12. T. Yucek and H. Arslan, A survey of spectrum sensing algorithms for cognitive radio applications, *IEEE Communications Surveys Tutorials,* **Vol. 11,** 2009.

13. D. Noguet, *Sensing techniques for Cognitive Radio - State of the art and trends*, white paper, CEA-LETI (France, 2009).
14. D. Almeida, D. Carvalho, Cordeiro and Vieira, Experimental study of a wavelet-based spectrum sensing technique, in *42nd Asilomar Conference on Signals, Systems and Computers'08*, 2008.
15. B. Zayen, Hayar and D. Nussbaum, Blind spectrum sensing for cognitive radio based on model selection, in *3rd International Conference on Cognitive Radio Oriented Wireless Networks and Communications (CrownCom'08)*, 2008.
16. V. Stoianovici and V. P. M. Murroni, A survey on spectrum sensing techniques for cognitive radio, `http://vega.unitbv.ro/~popescu/university20bulletin%202008_final_VI.pdf`.
17. Y.-C. Liang, Y. Zeng, E. Peh and A. T. Hoang, Sensing-throughput tradeoff for cognitive radio networks, in *IEEE International Conference on Communications (ICC'07)*, 2007.
18. Z. Quan, S. Cui and A. Sayed, Optimal linear cooperation for spectrum sensing in cognitive radio networks, *IEEE Journal of Selected Topics in Signal Processing,* **Vol.2,** 2008.
19. J. Unnikrishnan and V. Veeravalli, Cooperative sensing for primary detection in cognitive radio, *IEEE Journal of Selected Topics in Signal Processing,* **Vol.2,** 2008.
20. E. Peh, Y.-C. Liang and Y. L. Guan, Optimization of cooperative sensing in cognitive radio networks: A sensing-throughput tradeoff view, in *IEEE International Conference on Communications (ICC '09)*, 2009.
21. Y. Zeng and Y.-C. Liang, Spectrum-sensing algorithms for cognitive radio based on statistical covariances, *IEEE Transactions on Vehicular Technology,* 2009.
22. J. Kim and J. Andrews, Spectral covariance for spectrum sensing, with application to IEEE 808.22, in *IEEE International Conference on Acoustics Speech and Signal Processing (ICASSP'10)*, 2010.
23. Y. Zeng, Y. Chang and Liang, Eigenvalue-based spectrum sensing algorithms for cognitive radio, *IEEE Transactions on Communications,* **Vol. 57,** 2009.
24. S. Xu, Y. Shang and H. Wang, Eigenvalue based spectrum sensing against untrusted users in cognitive radio networks, in *4th International Conference on Cognitive Radio Oriented Wireless Networks and Communications (CROWNCOM'09)*, 2009.
25. Y. Peng, F. H. Xiang, H. Long and J. Peng, The research of cross-layer architecture design and security for cognitive radio network, in *International Symposium on Information Engineering and Electronic Commerce (IEE'09)*, May 2009.
26. V. Corvino, L. Giupponi, A. Perez Neira, V. Tralli and R. Verdone, Cross-layer radio resource allocation: The journey so far and the road ahead, in *Second International Workshop on Cross Layer Design,(IWCLD'09)*, 2009.
27. R. Rakesh, Wyglinski, A. M and G. J. Minden, An efficient implementation of NC-OFDM transceivers for cognitive radios, in *1st International Conference on Cognitive Radio Oriented Wireless Networks and Communications'06*, 2006.

28. V. Chakravarthy, X. Li, R. Zhou, Z. Wu and M. Temple, A novel hybrid overlay/underlay cognitive radio waveform in frequency selective fading channels, in *4th International Conference on Cognitive Radio Oriented Wireless Networks and Communications (CROWNCOM'09)*, 2009.
29. Y. Peng, J. Peng and X. Zheng, The research of cross-layer architecture design for MCM-based cognitive radio network, in *First International Workshop on Education Technology and Computer Science (ETCS'09)*, 2009.
30. M. Slavik, I. Mahgoub and A. Badi, Cross-layer design for wireless networks with cognitive controllers, in *Wireless Telecommunications Symposium (WTS'08)*, 2008.
31. V. Kawadia and P. Kumar, A cautionary perspective on cross-layer design, *IEEE Wireless Communications,* **Vol. 12,** 2005.
32. I. Mariac, Cooperative strategies for wireless relay networks, PhD thesis, Rutgers State University, (New Jersey, 2006).
33. Y. Zang, H. Chen and M. Guizani, *Cooperative wireless communications* (CRC Press, NewYork, 2009).
34. J. Jia, J. Zhang and Q. Zhang, Relay-assisted routing in cognitive radio networks, in *IEEE International Conference on Communications,(ICC'09)*, 2009.
35. L. Fulu, A networking perspective of cooperative spectrum sharing in wireless networks: Analysis and experiments, in *Wireless Telecommunications Symposium,(WTS'08)*, 2008.
36. X. Gong, W. Yuan, W. Liu, W. Cheng and S. Wang, A cooperative relay scheme for secondary communication in cognitive radio networks, in *IEEE Global Telecommunications Conference,(GLOBECOM'08)*, 2008.
37. T. Nadkar, V. M. Thumar, U. B. Desai and S. N. Merchant, Symbiotic cooperative relaying in cognitive radio networks with time and frequency incentive, *Springer Telecommunication Systems Journal, Special Issue on Mobile Computing and Networking Technologies,* 2011, accepted.
38. I. Stanojev, O. Simeone, Y. Bar-Ness and T. Yu, Spectrum leasing via distributed cooperation in cognitive radio, in *IEEE International Conference on Communications (ICC'08)*, 2008.
39. H. Luo, Z. Zhang and G. Yu, Cognitive cooperative relaying, in *11th IEEE Singapore International Conference on Communication Systems (ICCS'08)*, 2008.
40. M. Ibnkahla, *Adaptive signal processing in wireless comunications* (CRC Press, NewYork, 2008).
41. K. Liu, A. K. Sadek, W. Su and A. Kwasinski, *Cooperative communications and networking* (Cambridge university Press, Cambridge, 2009).
42. W. Ma, S. Z. Hu, Y. C. Wang and L. Zhue, Cooperative spectrum sensing in OFDM based on MIMO cognitive radio sensor networks, in *5th International Conference on Wireless Communications, Networking and Mobile Computing (WiCom '09)*, 2009.
43. H.Arslan, *Cognitive radio, software defined radio, and adaptive wireless systems* (Springer, Netherlands, 2007).
44. S. Boyd and L. Vandenberghe, *Convex Optimization* (Cambridge University

Press, Cambridge, 2004).
45. G. Bansal, M. Hossain and V. Bhargava, Adaptive power loading for OFDM-based cognitive radio systems, in *IEEE International Conference on Communications (ICC '07)*, 2007.
46. P. Wang, M. Zhao, L. Xiao, S. Zhou and J. Wang, Power allocation in OFDM-based cognitive radio systems, in *IEEE Global Telecommunications Conference (GLOBECOM'07)*, 2007.
47. V. M. Thumar, T. Nadkar, U. B. Desai and S. N. Merchant, Adaptive power allocation for secondary users in cognitive radio networks, in *2nd UK-India-IDRC International Workshop on Cognitive (UKIWCWS'10)*, 2010.
48. G. Bansal, Z. Hasan, M. Hossain and V. Bhargava, Subcarrier and power adaptation for multiuser OFDM-based cognitive radio systems, in *National Conference on Communications (NCC'10)*, 2010.
49. M. Shaat and F. Bader, Computationally efficient power allocation algorithm in multicarrier-based cognitive radio networks: OFDM and FBMC systems, *EURASIP Journal on Advances in Signal Processing,* **Vol. 2010,** 2010.
50. G. Tej, T. Nadkar, V. M. Thumar, U. B. Desai and S. N. Merchant, Adaptive power allocation for secondary users in cognitive radio networks, in *IEEE Wireless Communications and Networking Conference (WCNC'11)*, 2010.
51. C. Shilun, Z. Yang and H. Zhang, Adaptive modulation and power control for throughput enhancement in cognitive radios, *Journal of Electronics (China),* **Vol. 25** (Science Press, co-published with Springer-Verlag GmbH, 2008).
52. D. Huang, Z. Shen, C. Miao and C. Leung, Resource allocation in MU-OFDM cognitive radio systems with partial channel state information, *EURASIP Journal on Wireless Communications and Networking,* **Vol. 2010** (Hindawi Publishing Corporation, 2010).
53. T. Nadkar, V. M. Thumar, U. B. Desai and S. N. Merchant, Optimum bit loading for cognitive relaying, in *IEEE Wireless Communications and Networking Conference (WCNC'10)*, 2010.
54. T. Nadkar, V. M. Thumar, U. B. Desai and S. N. Merchant, Adaptive constellation sizing for OFDM-based cognitive radio networks, in *IEEE Radio and Wireless symposium (RWS'09)*, 2009.
55. W. Xi, C. Yin, D. Liu and G. Yue, Adaptive beamforming based on subspace theory in cognitive networks, in *International Conference on Wireless Communications, Networking and Mobile Computing(WiCom'09)*, 2009.
56. X. Lian, H. Nikookar and L. P. Ligthart, Adaptive OFDM beamformer with constrained weights for cognitive radio, in *IEEE Vehicular Technology Conference (VTC Spring'09)*, 2009.
57. I. Cosovic, S. Brandes and M. Schnell, A technique for sidelobe suppression in OFDM systems, in *IEEE Global Telecommunications Conference (GLOBECOM'05)*, 2005.
58. S. Brandes, I. Cosovic and M. Schnell, Sidelobe suppression in OFDM systems by insertion of cancellation carriers, in *IEEE 62nd Vehicular Technology Conference (VTC'05-Fall)*, 2005.
59. R. Wang, V. Lau, L. Lv and B. Chen, Joint cross-layer scheduling and spectrum sensing for OFDMA cognitive radio systems, *IEEE Transactions on*

Wireless Communications, **Vol. 8,** May 2009.
60. Y. Shi and Y. Hou, A distributed optimization algorithm for multi-hop cognitive radio networks, in *IEEE INFOCOM'08.*, 2008.
61. M. Ma and D. Tsang, Joint spectrum sharing and fair routing in cognitive radio networks, in *5th IEEE Consumer Communications and Networking Conference(CCNC'08)*, 2008.
62. M. J. Osborne and A. Rubenstien, *A Course in Game Theory* (The MIT Press, Cambridge, Massachusetts, 1998).
63. B. Polak, Video lectures on game theory, (2010), `http://oyc.yale.edu`.
64. D. Fudenberg and J. Tirole, *Game Theory* (The MIT Press, Cambridge, Massachusetts, 1991).
65. S. Ren and M. van der Schaar, Revenue maximization and distributed power allocation in cognitive radio networks, in *Proceedings of the 2009 ACM workshop on Cognitive radio networks*, CoRoNet '09 (ACM, New York, NY, USA, 2009).
66. Y. Li, X. Wang and M. Guizani, Resource pricing with primary service guarantees in cognitive radio networks: A stackelberg game approach, in *IEEE Global Telecommunications Conference (GLOBECOM'09)*, 2009.
67. J. W. Huang and V. Krishnamurthy, Game theoretic issues in cognitive radio systems (invited paper), *Journal of Communications,* **Vol. 4,** 2009.
68. Q. Zhu, H. Li, Z. Han and T. Bas andar, A stochastic game model for jamming in multi-channel cognitive radio systems, in *IEEE International Conference on Communications (ICC'10)*, 2010.
69. A. Kaufmann, *Introduction to Theory of Fuzzy Subsets* (Academic Press, NewYork, 1975).
70. Zimmermann and Hans, *Fuzzy sets, decision making and expert systems* (Kluwer, B.V., Deventer, The Netherlands, 1986).
71. L. Giupponi and A. Perez Neira, Fuzzy-based spectrum handoff in cognitive radio networks, in *3rd International Conference on Cognitive Radio Oriented Wireless Networks and Communications (CrownCom '08)*, 2008.
72. N. Baldo and M. Zorzi, Fuzzy logic for cross-layer optimization in cognitive radio networks, *IEEE Communications Magazine,* **Vol. 46,** 2008.
73. H. Le and H. Ly, Opportunistic spectrum access using fuzzy logic for cognitive radio networks, in *Second International Conference on Communications and Electronics,(ICCE'08)*, 2008.
74. D. M. Mukhopadhyay, M. O. Balitanas, A. F. A., S. H. Jeon and D. Bhattacharyya, Genetic algorithm: A tutorial review, *International Journal of of Grid and Distributed Computing,* **Vol. 2,** 2009.
75. T. R. Newman, R. Rajbanshi, A. M. Wyglinski, J. B. Evans and G. J. Minden, Population adaptation for genetic algorithm-based cognitive radios, in *2nd International Conference on Cognitive Radio Oriented Wireless Networks and Communications (CrownCom'07)*, 2007.
76. S. Chen, T. Newman, J. Evans and A. Wyglinski, Genetic algorithm-based optimization for cognitive radio networks, in *IEEE Sarnoff Symposium'10*, 2010.
77. S. Chen and A. M. Wyglinski, Efficient spectrum utilization via cross-layer

optimization in distributed cognitive radio networks *Comput. Commun.*, **Vol. 32,** (Butterworth-Heinemann, Newton, MA, USA, Dec. 2009).

78. M. Yuan, S. Wang and S. Du, Fast genetic algorithm for bit allocation in OFDM based cognitive radio systems, in *19th Annual Wireless and Optical Communications Conference (WOCC'10)*, 2010.
79. E. Gelenbe, P. Liu and J. Laine, Genetic algorithms for route discovery, *IEEE Transactions on Systems, Man, and Cybernetics, Part B: Cybernetics,* **Vol. 36,** 2006.
80. K. Tsagkaris, A. Katidiotis and P. Demestichas, Neural network-based learning schemes for cognitive radio systems, *Computer Communications,* **Vol. 31,** 2008.
81. B. Le, T. W. Rondeau, D. Maldonado and C. W. Bostian, Modulation identification using neural network for cognitive radios, in *Proceeding of the SDR 05 Technical Conference and Product Exposition*, 2005.
82. Y. Liu, B. Tamma, B. Manoj and R. Rao, Traffic prediction for cognitive networking in multi-channel wireless networks, in *IEEE Conference on Computer Communications Workshops (INFOCOM'10)*, 2010.
83. V. Tumuluru, P. Wang and D. Niyato, A neural network based spectrum prediction scheme for cognitive radio, in *IEEE International Conference on Communications (ICC'10)*, 2010.
84. Z. Zhang and X. Xie, Intelligent cognitive radio: Research on learning and evaluation of cr based on neural network, in *ITI 5th International Conference on Information and Communications Technology (ICICT'07)*, 2007.
85. N. Devroye, M. Vu and V. Tarokh, Achievable rates and scaling laws for cognitive radio channels, *EURASIP J. Wirel. Commun. Netw.*, **Vol. 2008,** (Hindawi Publishing Corp., New York, NY, United States, Jan. 2008).
86. T. M. Cover and J. A. Thomas, *Elements of information theory* (Wiley-Interscience, New York, NY, USA, 1991).
87. M. Costa, Writing on dirty paper, *IEEE Transactions on Information Theory,* **Vol. 29,** 1983.

INVERSE PROBLEMS OF PARAMETER IDENTIFICATION IN PARTIAL DIFFERENTIAL EQUATIONS

B. JADAMBA and A. A. KHAN

School of Mathematical Sciences
Rochester Institute of Technology
Rochester, New York, 14623, USA.
E-mail: {bxjsma,aaksma}@rit.edu

M. SAMA

Departamento de Matematica Aplicada
Universidad Nacional de Educacion a Distancia
Madrid, Spain.
E-mail: msama@ind.uned.es

Inverse problems is a vibrant and expanding branch of mathematics that has found numerous applications. In this work, we will survey some of the main techniques available in the literature to solve the inverse problem of identifying variable parameters in partial differential equations. Besides carefully defining the problem, we give simple examples to depict some of the difficulties associated with the study of this inverse problem. We will discuss in sufficient detail twelve different methods available to solve this particular class of inverse problems. We also analyze some of their most exciting applications. We will also point out some of the research directions which can be pursued in this fascinating branch of applied and industrial mathematics.

Keywords: Parameter identification, inverse problems, ill-posed problems, error estimates, output least squares, Lagrange multipliers, equation error approach, elasticity imaging, total variational regularization, Tikhonov regularization, asymptotic regularization, variational and quasi-variational inequalities.

1. Inverse problems

In inverse problems the objective is to seek unknown causes from the observation of their effects. On the other hand, for the associated direct problem, one seeks effects based on sufficient knowledge concerning the causes.

It is often required to relate physical parameters x that characterizes a model to acquired observations making up some set of data y. Having a clear understanding of the underlying model, an operator can be specified relating x to y through the equation

$$F(x) = y,$$

formulated in some appropriate vector space setting. The problem of estimating x from a measurement of y is a prototype of an inverse problem.

If the operator F is linear, the inverse problem is termed to be linear; otherwise it is a nonlinear inverse problem. It turns out that nonlinear inverse problems are considerably harder to solve than the linear ones. One crucial aspect of data collection is that the data y is often corrupted by some amount of noise.

Due to their special characteristics most inverse problems are *ill-posed.* In 1932 Hadamard attempted to describe the generic properties of problems. In the sense of Hadamard, a mathematical problem is termed to be *well-posed* if it has the following features:

1. **Existence:** For a suitable data set, the problem has a solution.
2. **Uniqueness:** The solution is unique.
3. **Stability:** The solution depends continuously on the observations.

A problem is ill-posed if it fails to respect any of the above three conditions. The main concern in the study of inverse problems is the violation of the third condition, that is, the case in which the solution does not depend continuously on the data.

In this survey we focus our attention to the study of inverse problems of determining variable parameters which appear in various partial differential equations. These inverse problems are also known as distributed parameter identification problems. For more details on the fast and rapidly growing field of inverse problems the reader is referred to excellent books[6,7,25].

2. Ill-posed is real: A few examples

One characteristic of ill-posed problems is that the error between the exact and the noisy solution can be arbitrarily large even for the case when the error in the data can be kept arbitrarily small. The following example taken from[13] depicts this phenomenon.

Example 1. Differentiation of noisy functions. Assume that we wish to compute the derivative of a noisy function. That is, instead of a function f, the noisy function f_δ is available and we wish to compute the derivative

$\frac{df_\delta}{dx}$. Assume that

$$\begin{aligned} f_\delta(x) &= f(x) + e_\delta(x), \quad x \in S := [0,1] \\ f_\delta(0) &= f(0) = 0 \\ f_\delta(1) &= f(1) = 0 \end{aligned}$$

where $e_\delta(x)$ represents the data noise.
We choose

$$e_\delta = \sqrt{2}\delta \sin(2\pi n x)$$

for $n \in \mathbb{N}$. Notice that, for all n,

$$\int_0^1 |e_\delta(x)|^2 dx = \delta^2.$$

Moreover,

$$\frac{df_\delta}{dx}(x) = \frac{df}{dx}(x) + 2\sqrt{2}\pi\delta n \cos(2n\pi x).$$

Consequently, the $L_2(S)$ and the $L_\infty(S)$ errors, given by

$$\left\| \frac{df_\delta}{dx}(x) - \frac{df}{dx}(x) \right\|_{L_2(S)} = 2\pi\delta n,$$

$$\left\| \frac{df_\delta}{dx}(x) - \frac{df}{dx}(x) \right\|_{L_\infty(S)} = 2\sqrt{2}\pi\delta n,$$

can be made arbitrarily large by taking n lare enough. □

A similar behavior is shown by the numerical differentiation in the following example.

Example 2. Numerical differentiation. Consider the following boundary value problem: Find $y : [0,1] \to \mathbb{R}$ to solve

$$\begin{aligned} \frac{d^2y}{dx^2} &= f(x), \quad 0 < x < 1, \\ y(0) &= 0, \\ y(1) &= 0. \end{aligned}$$

The inverse problem we are interested in is:

Given $y : [0,1] \to \mathbb{R}$ such that $y(0) = y(1) = 1$, compute $f(x) = -y''(x)$.

To solve the above inverse problem, we discretize the above BVP by using a finite difference scheme. We establish a regular grid on the interval $[0,1]$ by defining $x_i = i\Delta x$, $i = 0, 1, \ldots, n$, $\Delta x = 1/n$. Then, restricting the differential equation $-y'' = f(x)$ to the grid points, we obtain

$$-y''(x_i) = \frac{-y(x_i - \Delta x) + 2y(x_i) - y(x_i + \Delta x)}{\Delta x^2} + O(\Delta x^2),\ i = 1, \ldots, n-1.$$

In the above, we have employed the central difference scheme and assumed that $y \in C^4[0,1]$.

The above equations can also be represented in a matrix-vector form,

$$Ly = f,$$

where

$$L = \frac{1}{\Delta x^2} \begin{bmatrix} 2 & -1 & & & & \\ -1 & 2 & -1 & & & \\ & -1 & 2 & -1 & & \\ & \ddots & \ddots & \ddots & & \\ & & & -1 & 2 & -1 \\ & & & & -1 & 2 \end{bmatrix}$$

$$y = \begin{bmatrix} y_1 \\ y_2 \\ \vdots \\ y_{n-1} \end{bmatrix} \in \mathbb{R}^{n-1}, \quad f = \begin{bmatrix} f_1 \\ f_2 \\ \vdots \\ f_{n-1} \end{bmatrix} \in \mathbb{R}^{n-1}.$$

To depict the ill-posedness, we choose (discrete versions of)

$$y(x) = x\cos\left(\frac{\pi x}{2}\right)$$
$$f(x) = \pi \sin\left(\frac{\pi x}{2}\right) + \frac{\pi^2}{4} x \cos\left(\frac{\pi x}{2}\right).$$

For $n = 50$, we solve both the forward and the inverse problems. In each case, we add normally and independently distributed errors to the components of the data, (amounting to 0.5% in the Euclidean norm) and then compute the solution. The results are shown in Figure 1. As the graphs show, the noisy data does not affect the computed solution to the forward problem in any great degree. On the other hand, the computed solution to the inverse problem is essentially useless. □

We conclude this section with the following example taken from[66] showing the ill-posedness of the inverse problem of identifying parameters in

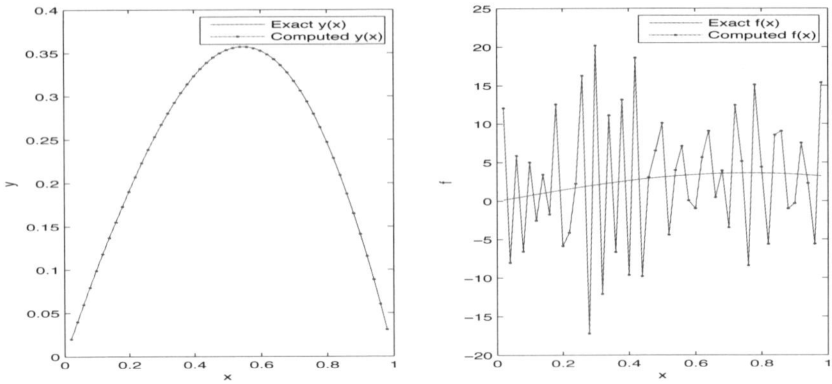

Fig. 1. Exact and the computed solution for forward problem (left) and the inverse problem (right).

boundary value problems. This example also introduces the type of problems this review will be focusing on.

Example 3. Coefficient identification in BVPs. Consider the following one-dimensional BVP: Find $u(x)$ such that

$$
\begin{aligned}
-\frac{d}{dx}\left(a\frac{du}{dx}\right) &= f \quad \text{on } (0,1),\\
a(0)\frac{du}{dx}(0) &= g_0,\\
a(1)\frac{du}{dx}(1) &= g_1.
\end{aligned}
$$

For $a \in H^1(0,1)$ with $k_0 \leq a \leq k_1$, $f \in L^2(0,1)$, a solution to the above direct problem exists if the compatibility condition $\int_0^1 f\,dx = g_0 - g_1$ holds. We are interested in the following inverse problem:
Given z, a measurement of u, find a which together with z makes the BVP true.

Assume that $z = u + \delta$ where $\delta(x) \in L^2(0,1)$ is a noisy data. We choose $f = 0$, $g_0 = g_1 = 1$. Then for the noise-free data $u(x) = x + 1$, the unique solution is $a(x) = 1$. Assume that the noisy data are given by

$$z_n(x) = u(x) + \epsilon\frac{\cos(2n\pi x)}{2n\pi}, \quad 0 < \epsilon < 1.$$

Then the solution which corresponds to this data is

$$a_n = \frac{1}{1 - \epsilon\sin(2n\pi x)}.$$

Clearly, z_n converges uniformly to u, whereas, we have

$$\|a_n - a\|_{L_\infty(S)} = \frac{\epsilon}{1-\epsilon}$$

$$\|a_n - a\|^2_{L_2(S)} = 1 - \frac{1-2\epsilon^2}{(1-\epsilon^2)\sqrt{1-\epsilon^2}}.$$

In conclusion, $z_n \to u$ but $a_n \not\to a$ and hence a is not a continuous function of the data with respect to the L_∞ or L_2 norm. □

3. Prototypical elliptic inverse problems

3.1. *Scalar elliptic BVP*

Consider the following BVP

$$-\nabla \cdot (a\nabla u) = f, \quad \text{in } \Omega \tag{1}$$

where $\Omega \subset \mathbb{R}^2$ is an open bounded domain. Given a and f, the direct problem in this setting is to find u provided that the suitable boundary conditions augment the BVP. On the other hand the corresponding inverse problem is to find a, given some measurement of u.

This problem is a particular case of the more general following BVP

$$\nabla \cdot (a\nabla u(x,t)) = B(x)\frac{\partial u}{\partial t} + C(x,t)$$

which models a confined inhomogeneous isotropic aquifer. Here u represents the piezometric head, a the transmissivity, $C(x,t)$ the recharge, and $B(x)$ the stroativity of the acquifer. It is commonly observed that aquifers tend to be *thin* relative to their horizontal extent and thus a natural simplification is the assumption that the transmissivity varies little with depth, so that the ground water flow in these cases can be viewed as essentially two dimensional, and we can take $x = (x_1, x_2)$ in a two dimensional space. If the flow of the water has reached a steady state and we assume for simplicity that $C = 0$, then one recovers (1).

Numerous researchers have focused on above inverse problem. An interesting survey article by Yeh[72] presents an overview of the approaches used in ground water modeling. As early as in 1972, Pinder and Frind[61] weighed the efficiency of Galerkin method against the efficiency of finite difference scheme for a variant of (1).

3.2. *Isotropic linear elasticity*

Let $\Omega \subset \mathbb{R}^2$ be a bounded open set representing the area occupied by an elastic membrane. Let $\partial\Omega$, boundary of Ω, be partitioned into two parts

Γ_1 and Γ_2, with the measure of Γ_1 being strictly positive. Assume that the body is fixed along Γ_1. Assume that a body force $f = (f_1, f_2)$, acts on the body and that a surface force $g = (g_1, g_2)$ acts on Γ_2. Let $u = (u_1, u_2)$ be the displacement vector. The linear isotropic elasticity system consists of the following BVP:

$$-\nabla \cdot \sigma = f \text{ in } \Omega, \tag{2a}$$

$$\sigma = 2\mu\epsilon_u + \lambda \mathrm{tr}\,(\epsilon_u)\, I, \tag{2b}$$

$$u = 0 \text{ on } \Gamma_1, \tag{2c}$$

$$\sigma n = h \text{ on } \Gamma_2. \tag{2d}$$

In the above, the strain tensor ϵ_u is given by

$$\epsilon_u = \frac{1}{2}\left(\nabla u + \nabla u^T\right),$$

σ is the corresponding stress tensor and n is the outward-pointing unit normal to $\partial\Omega$.

The inverse problem in this setting is to estimate μ and λ, the Lamé moduli. In recent years, this problem has been studied in the context of applications in biomedical imaging. The technical setup of these problems and a brief account of the most relevant works will be presented in the later part of this paper.

3.3. *Helmholtz equations*

Consider the inhomogeneous Helmholtz equation

$$(\Delta + k^2 n^2)u = f \text{ in } \Omega,$$

where Ω is a domain in $\mathbb{R}^3$ and $k > 0$ is given.
The inverse problem here is to identify the coefficient $n(x)$, the index of refraction. This inverse problem has been discussed in.[5]

A similar model BVP is the following

$$-\nabla \cdot (a\nabla u) + ku = f \quad \text{in } \Omega,$$

where Ω is open bounded domain and the constant K is known. One instance of the appearance of the above BVP is the standing waves on a bounded shallow body of water with

$$k = \frac{4\pi^2}{gT^2},$$

where a is the water depth at the quiescent state, u is the elevation of the free surface above the quiescent level, g acceleration of gravity and T period of oscillation. The above equation then holds under some simplifications and Neumann boundary conditions augment it. The direct problem in this setting is to find u. On the other hand, the corresponding inverse problem is to find a, given some measurement of u. From a computational stand point this inverse problem has been studied in[44] .

3.4. *Other classes of BVPs*

Other elliptic inverse problems arise when more complicated physical models are assumed. For example, the scalar a in the BVP could be replaced by a matrix A, introducing anisotropy in the model. Similarly, an anisotropic stress-strain law could be assumed for the elasticity problem. However, these more complicated problems have received relatively little attention in the literature.

4. Inverse problems with boundary measurements

In this work we are dealing with elliptic inverse problems where the measurements of u is available in the whole interior of the underlying domain. There is another type of inverse problem in which only the boundary data is available. A well-known example is of impedance tomography. The technical setup of this inverse problem is as follows: At the boundary of an object, say Ω, different electrical voltages are applied, and the arising electrical currents are measured on the boundary. The inverse problem is then to reconstruct the conductivity as a function of space, which gives information about different materials inside the object. A simple version of this model can be represented by the BVP

$$\begin{aligned} -\nabla \cdot (a\nabla u) &= 0 \quad \text{in } \Omega, \\ u &= g \quad \text{on } \partial\Omega, \end{aligned}$$

The measured currents over the boundary for a specific voltage g are given by

$$f_g = a\frac{\partial u}{\partial n} \text{ on } \partial\Omega.$$

In the present setting the data consists of Dirichlet-to-Neumann-map P_a : $g \mapsto f_g$ which in this case is a linear operator. The inverse problem of impedance tomography or inverse conductivity problem of reconstructing the conductivity a from a measurement of the Dirichlet-to-Neumann map P_a.

5. An overview of existing methodologies

In recent years, the field of inverse problems has certainly been one of the fastest growing fields in applied mathematics (see[7,25]). This growth has largely been driven by ever-growing applications in engineering and industry. New development in solution methods for optimization problems or partial differential equations has also motivated researchers to improve already-existing methods.

In the following we will give a brief overview of the main techniques being considered for solving inverse problems of identifying parameters in partial differential equations. Our discussion of the available methods is by no means complete. For more details the interested reader can refer to the relatively large bibliography, which also does not claim of any completeness.

Most of the work on elliptic inverse problems has been in the context of the scalar elliptic problem

$$-\nabla \cdot (a\nabla u) = f \text{ in } \Omega, \tag{3a}$$

$$u = 0 \text{ on } \partial\Omega, \tag{3b}$$

where Ω is a bounded domain with boundary $\partial\Omega$. The Dirichlet boundary condition is only for simplicity, either Neumann boundary condition or a combination of both can be made. Here the inverse problem is to identify the parameter a.

In the context of (3), there have been two primary approaches for attacking the corresponding inverse problem of finding a. The first approach reformulates the inverse problem as an optimization problem and then employs some suitable method for the solution. The second approach treats (3) as a hyperbolic partial differential equation in a.

As is well-known, the inverse problem is ill-posed, and some type of regularization is necessary. Since this is more easily accomplished in the optimization setting, this class of methods has been the subject of most of the research, and will be the focus of this work. However, viewing (3) as a hyperbolic PDE in a does produce certain insights, such as uniqueness results. Moreover, the hyperbolic viewpoint is the basis for some analysis even when the problem is approached via optimization.

There are two basic approaches to posing the inverse problem as an optimization problem. In one approach, the PDE is treated as an explicit constraint and constrained optimization algorithms are employed. In such an approach, both a and u are regarded as independent variables and they may not satisfy the PDE (although they should in the limit, as the optimization algorithm converges to a solution). In the other approach, the

PDE implicitly defines u as a function of a, so that a is the only variable and unconstrained optimization techniques may be used (although constrained optimization methods may be needed to treat other constraints, such as bounds on a).

In the following we briefly outline some of the main approaches available in the literature for the inverse problems.

5.1. *Output least-squares*

A common approach for solving parameter identification problems is the output least-squares (OLS) approach. Applied to the elliptic inverse problem of finding a in (3), the OLS approach minimizes the functional

$$J_1(a) = \|u(a) - z\|^2, \tag{4}$$

where z is the data (the measurement of u), $\|\cdot\|$ is a suitable norm and u solves the BVP or its variational form: Find $u \in H_0^1(\Omega)$ such that

$$\int_\Omega a\nabla u \nabla v = \int_\Omega fv \quad \text{for all} \quad v \in H_0^1(\Omega). \tag{5}$$

In order to handle the ill-posedness of the inverse problems, a regularized analogue of J_1 needs to be considered. More precisely, one minimizes the regularized OLS functional

$$J_1^\epsilon(a) = \|u(a) - z\|^2 + \epsilon\|u\|^2. \tag{6}$$

Here $\epsilon > 0$ is the regularization parameter and the term $\|u\|^2$ is the regularizing functional. In the literature, a wide variety of norms and semi-norms have been used as a regularizer.

Falk[27] considered (3a) with Neumann boundary conditions and analyzed the situation in which the $L^2(\Omega)$ norm is used in (4). In Falk's approach, the coefficient a is approximated by piecewise polynomials of degree r defined on a family $\{\mathcal{T}_h\}$ of meshes, and the solution u by degree $r+1$ piecewise polynomials u_h on the same meshes. He then proved that any minimizer a_h of (the discrete version of) J_1 satisfies

$$\|a - a_h\|_{L^2(\Omega)} \leq C\left[h^r + \frac{\|u - z_h\|_{L^2(\Omega)}}{h^2}\right],$$

where a is the true coefficient, u is the solution of (5) corresponding to a, z_h is the measurement of u, and C is a constant independent of h. This result proves convergence of a_h to a if, for example,

$$\|u - z_h\|_{L^2(\Omega)} = O(h^3).$$

In particular, if z_h is the degree $r+1$ piecewise polynomial interpolant of u, then

$$\|a - a_h\|_{L^2(\Omega)} = O(h^r).$$

The key assumptions needed for Falk's analysis are

(1) $a \in H^{r+1}(\Omega)$; and
(2) there exists a constant vector $\vec{\nu}$ and a constant $\sigma > 0$ such that $\nabla u \cdot \nu \geq \sigma$ throughout Ω.

The second assumption is a physical hypothesis stating that there is always some flow in the $\vec{\nu}$ direction. It implies, in particular, that the PDE (3) is truly hyperbolic with respect to a, a point that makes Falk's analysis possible. Under sufficient smoothness on u, it is shown that there is a unique coefficient a which satisfies the variational problem.

Recently, by using an OLS formulation Gockenbach[28] obtained useful error estimates for the system of isotropic elasticity (2) (with $f = 0$, $\Gamma_1 = \varnothing$ and hence $\Gamma_1 = \Omega$).

The OLS functional considered in Gockenbach's approach reads as follows

$$J_h(m) = \|u_h(m) - z\|^2_{L^2(\Omega)} + h^3\|\sigma(m), u_h(m) - h\|^2_{L^2(\partial\Omega)}, \tag{7}$$

where $m = (\mu, \sigma)$ with $\sigma = \mu + \lambda$ and the discretization was made over a triangulation $\mathcal{T}_h$. To have more control over the term $h^3\|\sigma(m), u_h(m) - h\|^2_{L(\partial\Omega)}$ it was necessary to have it incorporated in the objective functional.

In Gockenbach's approach, the Lamé moluli are approximated by piecewise polynomials of degree r defined on a family $\mathcal{T}_h$ of meshes, and each component of the solution u by degree $r+1$ piecewise polynomials u_h on the same meshes. Gockenbach[28] showed that that there exists a constant C which is independent of h such that, with $z = u^*$

$$\inf_{m \in K_h} J_m(m) \leq Ch^{2r+4}.$$

Here K_h is the discretized set of admissible coefficients.

The above estimate, which holds for $z = u^*$, is easily turned into an estimate for inexact data. In fact, for a minimizer m_h, it holds:

$$\|u_h(m_h) - z\|_{L^2(\Omega)} \leq C(h^{r+2} + \|z - u^*\|_{L^2(\Omega)})$$
$$\|\rho(m_h, u_h(m_h))n - h\|_{L^2(\partial\Omega)} \leq C\left(h^{r+1/2} + \frac{\|z - u^*\|_{L^2(\Omega)}}{h^{3/2}}\right).$$

The reader is referred to the bibliography of this paper for numerous papers which rely on the OLS approach.

5.2. *Modified or energy-norm based output least-squares*

A variation on the OLS approach was proposed independently by Tucciarelli and Ahlfeld,[68] Zou[75] and Knowles;[50] in their approach, the L^2 norm is replaced by the coefficient-dependent energy norm:

$$J_2(a) = \frac{1}{2} \int_\Omega a\nabla \left(u(a) - z\right) \cdot \nabla \left(u(a) - z\right). \tag{8}$$

Zou showed how to combine the OLS approach with total variation regularization to allow the estimation of discontinuous coefficients, a point that will be discussed further below. On the other hand, although Knowles[50] also aimed to relax the smoothness hypothesis on a, the main contribution was an elegant proof of the convexity of the OLS functional, a rare property in a nonlinear inverse problems. Knowles and his coworkers extended the convexity proof and its usefulness to a variety of problems (see[48,49]). Although Tucciarelli and Ahlfeld[68] also proved the convexity, their work was confined only to the discrete version of (8). None of the authors gave an error analysis comparable to Falk's results.

In a recent contribution by Jadamba and Khan,[42] an error analysis of the OLS approach based on the functional (8) has been carried out. Under hypotheses similar to those assumed by Falk, it is shown that any minimizer a_h of (the discrete version) of (8) satisfies

$$\|a - a_h\|_{L^2(\Omega)} \leq C \left[h^r + \frac{\|u - z_h\|_a}{h} \right], \tag{9}$$

where

$$\|v\|_a = \sqrt{\int_\Omega a\nabla v \cdot \nabla v}.$$

This result is entirely comparable to Falk's; a stronger norm is used to measure the error in the data, but this is compensated by one less factor of h in the denominator. Once again, if z_h is the degree $r+1$ piecewise polynomial interpolant of the exact data u, then convergence is guaranteed. Preliminary numerical experiments suggest that (8) can be minimized much more efficiently than the OLS functional based on the L^2 norm, presumably because of the convexity of (8).

Incidentally, the error estimates of[27] and[42] suggest that it may be preferable to use different finite element spaces for a and u, such as piecewise linear functions for a and piecewise quadratic functions for u. This is in contrast to Zou[75] and Knowles,[50] which suggest the use of piecewise linear functions for both spaces.

An extension of the above function from an abstract point of view was given by Gockenbach and Khan[33]. In this work, the author also proved the convexity of the abstract functional and applied it to the identification of Lame parameters in the system of isotropic elasticity[43].

5.3. *Equation error approach*

A shortcoming of OLS approaches is that each iteration of the optimization algorithm requires at least one solution of (5), often several, making these methods relatively expensive. Another general technique for parameter identification problems is the method of equation error, in which the unknown parameter is chosen to minimize the residual error.

To describe this method, we will use the approach presented by Acar.[1] In his work the BVP (3a) was augmented by Neumann type boundary condition of the type

$$a\frac{\partial u}{\partial n} = g$$

where n is an outward normal derivative and g is a given function. The underlying variational form of the BVP is then: Find $u \in H^1(\Omega)$ such that

$$\int_\Omega a\nabla u \nabla v = \int_\Omega fv + \int_{\partial\Omega} gv \quad \text{for all } v \in H^1(\Omega).$$

For a fixed $(a, u) \in V \times H^1(\Omega)$, ($V$ is the parameter space) the functional

$$v \mapsto -\int_\Omega a\nabla u \nabla v + \int_\Omega fv + \int_{\partial\Omega} gv$$

is linear and continuous and hence the Riesz representation theorem ensures that there exists $e = e(u, a) \in H^1(\Omega)$ such that

$$\langle e, v\rangle_{H^1(\Omega)} = -\int_\Omega a\nabla u \nabla v + \int_\Omega fv + \int_{\partial\Omega} gv \quad \text{for all } v \in H^1(\Omega).$$

If z is a measurement of u, then the equation error approach chooses a to minimize

$$J_3(a) = \|e(z, a)\|^2_{H^1(\Omega)}.$$

The functional J_3 is quadratic in a, so that (after discretization) minimizing J_3 reduces to solving a symmetric positive (semi)definite linear system. Acar[1] has regularized the above functional to obtain a unique and stable solution. Under suitable conditions, he has shown that a solution $a_{\epsilon,\delta}$ of the regularized problem converges to a solution a, if the regularization parameter ϵ and the error δ go to zero. He also demands that the regularization parameter ϵ should not go to 0 faster than $\sqrt{\delta}$. Kärkkäinen[44] also considers an equation error approach, similar to,[1] and derives an error estimate that appears to be stronger than those given by Falk,[27] Jadamba and Khan,[42] and also by Kohn and Lowe[51] (see below). In an interesting paper,[67] Tautenhahn and Muhs presented a regularized equation error method with data smoothing and gave general stability estimates.

The equation error approach from an abstract point of view is proposed by Gockenbach, Jadamba and Khan[32] .

5.4. *Kohn and Lowe's approach*

Kohn and Lowe[51] proposed a variant of the equation error method, in which the constitutive and balance laws are separated in the physical model. The PDE (3a) is written as the pair of equations

$$\begin{aligned} -\nabla \cdot \sigma &= f, \\ \sigma &= a\nabla u. \end{aligned}$$

The following objective functional is then minimized over both σ and a:

$$J_4(a,\sigma) = \int_\Omega |\sigma - a\nabla z|^2 + \gamma_1 \int_\Omega (\nabla\cdot\sigma + f)^2 + \gamma_2 \int_{\partial\Omega} (\sigma\cdot n - g)^2 .$$

Here the boundary conditions are

$$a\frac{\partial u}{\partial n} = g \text{ on } \partial\Omega$$

and γ_1, γ_2 are positive weights. The functional J_4 is a convex quadratic in (σ, a) and, as in the equation error method, a minimizer can be found by solving a symmetric positive (semi)definite system. Assuming the forcing function f and the Neumann data g are known exactly, then Kohn and Lowe prove that any minimizer of J_4 satisfies

$$\|a - a_h\|_{L^2(\Omega)} \le C\left[h + \frac{\|u - z\|_{H^1(\Omega)}}{h}\right].$$

This estimate assumes that piecewise linear functions are used to represent the coefficient a_h and piecewise quadratic functions for u_h. Moreover, these results are derived under the assumption

$$\min_{\overline{\Omega}} \max\{|\nabla u|, \Delta u\} > 0,$$

which is considerably weaker than the assumption used in[27] and.[42] Another detailed treatment of the above approach was given by Lin and Ramirez[56] where the authors considered a one dimensional BVP augmented with mixed boundary conditions and gave error estimates and detailed numerical results.

5.5. *Lagrange multiplier methods*

The PDE governing the inverse problem is a constraint relating the coefficient(s) and the solution of the PDE. In the OLS method, this constraint is used implicitly to define the solution as a function of the coefficient. A different approach is to regard the coefficient and the solution as independent variables which are to be chosen to jointly satisfy the constraint. Since the PDE appears as an explicit constraint in the optimization problem, this approach requires the use of a constrained optimization algorithm, and therefore involves Lagrange multipliers, at least indirectly. The following discussion will refer to the scalar problem (3). Ito and Kunisch[41] applied the augmented Lagrangian algorithm to minimize

$$J_5(a, u) = \frac{1}{2}\|u - z\|^2 + \frac{1}{2}\rho\|a\|^2$$

subject to (3a) as the constraint. Here ρ is the regularization parameter. They proved convergence of the algorithm in its infinite dimensional version and present numerical results for the discretized problem.

The work of Ito and Kunisch was generalized by Chen and Zou[21] to allow the use of total variation regularization together with an objective functional similar to J_2 (but with a and u as independent variables). They proved existence of a solution to the constrained optimization problem and convergence of the augmented Lagrangian method; since the total variation functional is not differentiable, their results are novel. However, neither Ito and Kunisch or Chen and Zou have developed error estimates comparable to those described above for other methods. The main results of Chen and Zou were extended in[31] .

5.6. *Luce and Perez's approach*

Luce and Perez[57] proposed to minimize

$$\frac{1}{2}\|a\|^2_{H^1(\Omega)}$$

subject to the constraints

$$\begin{aligned} -\nabla \cdot (a\nabla z) &= f, \\ a &\geq \gamma > 0, \\ \|u - z\|_{L^2(\Omega)} &\leq \delta\|z\|, \end{aligned}$$

where $\delta > 0$ is an estimate of the error in the data. This is an alternative to the usual Tikhonov regularization, in which one minimizes

$$\frac{1}{2}\|u - z\|^2 + \frac{\rho}{2}\|a\|^2_{H^1(\Omega)}.$$

In the Tikhonov approach, it is necessary to choose the regularization weight ρ; in the Luce-Perez approach, the level of error δ must be chosen instead. Since it may be possible to choose δ based on scientific grounds, the Luce-Perez approach is an attractive alternative.

Luce and Perez employed a penalty method to treat the constraints on the sign of a and the error in $u = u(a)$ and treat the PDE constraint implicitly, as in the OLS method. To estimate discontinuous coefficients, they replace the H^1 norm in their objective function with the total variation seminorm.

5.7. *Variational inequality approach*

Variational inequality theory is a powerful tool of the current mathematical technology and has applications in many branches of science, engineering, optimization, economics, equilibrium theory, etc. We briefly discuss its formulation before relating it to inverse problems. Let H be a Hilbert space, H^* be the topological dual of H and let $K \subset H$ be a nonempty closed and convex subset. Given a mapping $T : H \to H^*$, the variational inequality seeks $u \in K$ such that

$$\langle Tu, v - u\rangle \geq 0, \quad \forall v \in K.$$

If T is a potential operator, that is, there is a functional $g : H \to \mathbb{R}$ such that $Dg(\cdot) = T(\cdot)$, where $Dg(\cdot)$ is the Gateaux derivative of g, then the variational inequality is a necessary condition to the minimization problem: Find $u \in K$ such that

$$g(u) \leq g(v) \quad \forall\, v \in K.$$

In particular, if g is convex then the above two problems are equivalent.

In view of this observation, Kluge[47] considered the output least-squares functional

$$J_1(a) = \frac{1}{2}\|u(a) - z\|^2$$

and studied the variational inequality involving the directional derivative of J_1. In this setting the set K consists of all admissible coefficients. He proposed the following fixed point type algorithm to compute a numerical solution: Find $a_{n+1} \in K$ such that

$$a_{n+1} = P_K[a_n + \alpha(DJ_1(a_n) + \rho R(a_n))]$$

where P_K is the projection operator on K, R is a regularizing operator and ρ is a regularization parameter. Kluge[47] proved that a sequence $\{a_n\}$ generated by the above algorithm converges strongly to an optimal coefficient. Since then this approach has been followed by many authors. One main disadvantage of Kluge's approach is that the output-least squares functional is in general nonconvex and for the strong convergence of $\{a_n\}$ the regularization parameter cannot be chosen arbitrary small. An extension of Kluge approach, using modified OLS, is reported in[29] .

5.8. *Asymptotic regularization*

Asymptotic regularization is an interesting approach for parameter identification in which the original set of equations is embedded into a sequence of regularizing equations. Hoffmann and Sprekels[39] give a very detailed treatment of the approach to identify parameters in multi-valued variational inequalities. In the following, we discuss the basic ideas of asymptotic regularization and some of its extensions where we follow the work of Ang and Vy.[5]

Let Ω be a sufficiently smooth domain with $\partial\Omega = \Gamma_0 \cap \Gamma_1$. The objective is to identify the coefficient a in the following BVP:

$$\Delta u + au = f_1 \quad \text{in} \quad \Omega, \tag{10a}$$

$$u = \varphi \quad \text{on} \quad \Gamma_0 \tag{10b}$$

$$\frac{\partial u}{\partial n} = g \quad \text{on} \quad \Gamma_1. \tag{10c}$$

Let V be a finite dimensional subspace of $H^1(\Omega)$ and let $K := \{v \in H^1(\Omega) : v = \varphi \quad \text{on } \Gamma_0\}$. The following finite dimensional variational inequality can be considered as a discretized analogue of (10):

$$\int_\Omega \nabla\bar{u}\nabla(v - \bar{u}) - \int_\Omega a(v - \bar{u}) \geq \langle f, v - \bar{u}\rangle, \quad \forall v \in K \cap V, \tag{11}$$

where $\langle f, v - \bar{u}\rangle = \int_{\Gamma_1} gv - \int_\Omega f_1 v$.

Let W be a finite dimensional subspace of $L^\infty(\Omega)$ and let $\bar{K} = \{a \in L^\infty : \quad 0 \leq a \leq \lambda\}$ be the set of admissible coefficient where $\lambda >$ is a suitable constant.

Given $(a_0, u_0) \in (K \cap V) \times (\bar{K} \cap W)$, construct a sequence $\{(a_n, u_n)\}$ by solving the following system of finite dimensional variational inequalities: find $(a_{n+1}, u_{n+1}) \in (K \cap V) \times (\bar{K} \cap W)$ such that

$$\int_\Omega \left[h^{-1}(u_{n+1} - u_n)(v - u_{n+1}) + u_{n+1}\nabla(v - u_{n+1})\right.$$
$$\left. - a_{n+1}u_{n+1}(v - u_{n+1})\right] \geq \langle f, v - u_{n+1}\rangle, \ v \in K \cap V \quad \text{(12a)}$$
$$\int_\Omega \left[h^{-1}(a_{n+1} - a_n)(w - a_{n+1}) + u_{n+1}(u_{n+1} - \bar{u})(v - a_{n+1})\right]$$
$$\geq 0, \qquad v \in \bar{K} \cap W. \quad \text{(12b)}$$

It can be shown that for every $h > 0$, the above system of variational inequalities is solvable. Furthermore, there exists $h_0 > 0$ such that the solution is unique for every $0 < h \leq h_0$, the sequence

$$\left\{\int_\Omega \left[(u_n - \bar{u})^2 + (a_n - \bar{a})^2\right]\right\}$$

is decreasing and

$$\sum_{n=0}^{\infty} \int_\Omega |\nabla(u_n - \bar{u})|^2 < \infty.$$

Under suitable assumptions, it holds that $u_n \to \bar{u}$ in $H_1(\Omega)$ and $a_n \to \bar{a}$ in L^∞ where the pair $(\bar{a}, \bar{u}$ satisfies the variational inequality (11). Results concerning the convergence of solutions of the approximate variational inequalities when the finite dimensional subspaces converge are also known to hold.

The proofs of the above result strongly rely on the specific techniques developed for variational inequalities. To explain the basic idea, we define an inner product on the space $H = V \times W$

$$\langle (a, u), (b, v)\rangle = \int_\Omega (uv + ab),$$

and set $\mathcal{H} = (K \cap V) \times (\bar{K} \cap W)$. For $n \geq 0$, define a map $A_, : H \to H^*$ by

$$\langle A_n(a, u), (b, v)\rangle = \int_\Omega \left[\frac{(u - u_n)v + (a - a_n)b}{h} + \nabla u \nabla v - auv + u(u - \bar{u})b\right]$$
$$-\langle f, v\rangle.$$

It turns out that the system (12) is equivalent to finding $(u_{n+1}, a_{n+1}) \in \mathcal{K}$ such that

$$\langle A_n(a_{n+1}, u_{n+1}), (b, v) - (a_{n+1}, u_{n+1})\rangle \geq 0, \quad \forall (v, b) \in \mathcal{H}.$$

The existence results can now obtained by exploring properties of A_n and invoking some known results for variational inequalities.

As an interesting generalization of the above approach Chen, Han, and Schultz[20] proposed a modified asymptotic regularization approach. In their approach, given $(a_0, u_0) \in (K \cap V) \times (\bar{K} \times W)$, a sequence $\{(a_n, u_n)\}$ is constructed by solving the system of variational inequalities of finding $(a_{n+1}, u_{n+1}) \in (K \cap V) \times (\bar{K}_1 \cap W)$ such that

$$\int_\Omega \big[h^{-1}(u_{n+1} - u_n)(v - u_{n+1}) + u_{n+1}\nabla(v - u_{n+1}) \\ - a_{n+1}\bar{u}(v - u_{n+1})\big] \geq \langle f, v - u_{n+1}\rangle, \; v \in K \cap V \qquad (13a)$$

$$\int_\Omega \big[h^{-1}(a_{n+1} - a_n)(w - a_{n+1}) + \bar{u}(u_{n+1} - \bar{u})(v - a_{n+1})\big] \\ \geq 0, \qquad v \in \bar{K}_1 \cap W, \qquad (13b)$$

where $\bar{K}_1 = \{a \in L^2\Omega : \; a \geq 0, \text{a.e. in } \; \Omega\}$ is the set of admissible coefficients.

Notice that the term u_{n+1} in (12) has been replaced by the data $\bar{u}$ in (13). This change brings significant simplification to the analysis and give a form that is easier to implement.

The authors proved results analogous to Anh-Vy.[5] We remark that in[5] the authors proved the convergence in L^∞. This is mainly due to the selection of the set of admissible coefficients $\bar{K}$ which is a priori uniformly bounded, and hence admits a weak* L^∞ convergent subsequence. On the other hand for K_1 no a priori boundedness for the coefficient is assumed.

We conclude this subsection by noticing that although asymptotic regularization seems very promising approach, its numerical applicability is yet to be seen.

5.9. *Method of characteristics*

If u is known in the equation (3a), then this equation can be viewed as a first-order hyperbolic equation in a, which reads:

$$-\nabla a \cdot \nabla u - a\nabla u = f.$$

A natural idea then is to solve the above equation by using the method of characteristics. This approach has been studied in details by Richter.[63] In the following we briefly discuss the idea. Our discussion is on the lines of Richter.[63] The basic assumption in pursuing this approach is

$$\inf_{p\in\Omega}[\max\{|\nabla u(p)|, \Delta u(p)\}] > 0.$$

Richter first studied the case $\inf_\Omega |\nabla u| > 0$. Under this condition, the characteristics are always nondegenerate and integration along the characteristics can be performed. Moreover, each characteristic originates from the inflow. For the condition $\inf \Delta u > 0$ Richter performed a coordinate change along the characteristics

$$t \equiv \int_0^s \frac{\Delta u}{\nabla u} ds'.$$

Thus, along the characteristics, (3a) becomes

$$\frac{da}{dt} + a = \frac{f}{\Delta u}.$$

Then any characteristics originate from the inflow boundary or at a point where $\nabla u = 0$. Combining the above two cases, Richter obtained his main result: Suppose Ω can be divided into two subregions Ω_1 and Ω_2 such that

$$|\nabla u| \geq k_1 \quad x \in \Omega_1 \qquad \Delta u \geq k_2 \quad x \in \Omega_2$$

where k_1 and k_2 are two arbitrary constant. Then (3a) has a unique solution a assuming prescribed values along the inflow boundary Γ_1 and

$$\|a\|_{L^\infty} \leq C(u) \left[\max\{\sup_\Gamma |a|, \frac{\|f\|_{L^\infty}}{k_2}\} + \frac{[u]\|f\|_{L^\infty}}{k_1^2} \right]$$

where

$$[u] = \sup_\Omega u - \inf_\Omega u,$$

$$q_1 = \sup_{\Omega_1} \left\{ -\frac{\Delta u}{k_1} \right\}$$

$$C(u) = \max\{a, \exp\left(\frac{q_1[u]}{k_1}\right)\}.$$

A few results related to this approach are available in[22] and the cited references therein.

5.10. *Vainikko's approach*

A general approach for solving inverse problems was proposed and analyzed by Vainikko.[69] In the following we discuss briefly the main ideas of this approach. We will follow the description by Bruckner et al.[11] For $\Omega \subset \mathbb{R}^d$ $(d \geq 2)$, we consider the following problem

$$-\nabla \cdot (a\nabla u) = f, \quad \text{in } \Omega$$
$$u = 0, \quad \text{on } \Gamma_1,$$
$$a\nabla u \cdot \nu = g, \quad \text{on } \Gamma_2,$$

where $\Omega = \Gamma_1 \cup \Gamma_2$. It is assumed that $u \in H^{1,\infty}(\Omega)$, $f \in L^2(\Omega,)$ $g \in L^2(\Gamma_2)$ and $a \in L^2(\Omega)$ is sought. The above BVP can be represented as the following variational problem

$$\int_\Omega a\nabla u \cdot \nabla w = \int_\Omega fw + \int_{\partial\Omega} gw \quad \forall w \in H^1(\Omega, \partial\Omega), \tag{14}$$

where $H^1(\Omega, \partial\Omega) := \{w \in H^1(\Omega) : \ w(x) = 0 \text{ for } x \in \Gamma_1\} \subset H^1(\Omega)$.

The idea is to introduce an auxiliary problem: Find $\Psi(x)$ such that

$$\begin{aligned} -\Delta\Psi &= f \quad \text{in } \Omega \\ \Psi &= 0 \quad \text{in } \Gamma_1 \\ \nabla\Psi \cdot \nu &= g \quad \text{in } \Gamma_2. \end{aligned}$$

By writing the above BVP into a variational form and comparing it with (14), we get

$$\int_\Omega a\nabla u \cdot \nabla w dx = \int_\Omega \nabla\Psi \cdot \nabla w dx.$$

Let G be the space of gradients of functions $w \in H^1(\Omega, \partial\Omega)$ defined by

$$G = G(\Omega, \partial\Omega) = \{\nabla w : \ w \in H^1(\Omega, \partial\Omega) \subset (L^2(\Omega))^d\}$$

and let $P_G : (L^2(\Omega))^d \to G$ be the orthogonal projection operator. The main problem is: For $T \in \mathcal{L}(L^2(\Omega), G)$, solve the following operator equation

$$T(a) = \nabla\Phi(a)$$

where $\Phi \in H^1(\Omega, \partial\Omega)$ is the solution to the direct problem.

Several features of the above operator equation are given in.[69] The equation is discretized by choosing finite dimensional subspaces and the following minimization problem is solved:

$$\|a_h - a\|_{L^2} = \min_{v_h \in S_h} \|\nabla u \cdot \nabla v_h - a\|.$$

Numerical implementation of this approach is given in[11] .

5.11. *Singular perturbation*

A singular perturbation method was proposed by Alessandrini.[4] The basic idea is to solve the following elliptic BVP for a_ϵ :

$$\begin{aligned} \epsilon\Delta a_\epsilon + \nabla \cdot (a_\epsilon \nabla u) &= 0 \quad \text{in} \quad \Omega \\ a_\epsilon &= a \quad \text{in} \quad \partial\Omega. \end{aligned}$$

Under the assumption that g has a finite number, say N, of relative maxima and minima on $\partial\Omega$, Alessandrini obtained the estimate

$$\left(\int_{\Omega_d} |a - a_\epsilon|^q\right)^{1/q} \leq C\epsilon^{\frac{1}{2qN}}$$

where $q \in [1, \infty)$, $\Omega_d = \{x \in \Omega : \; d(x, \partial\Omega > d)\}$ and C is a constant independent of ϵ. Thus, a_ϵ converges to a on Ω_d in L^q norm as $\epsilon \to 0$.

5.12. *Mollification methods*

In the following we briefly discuss the mollification approach for parameter identification. We follow the presentation of Hinestroza and Murio[38] and concentrate on the following one-dimensional problem:

$$\frac{d}{dx}\left(a\frac{du}{dx}\right) = f, \quad 0 < x < 1 \tag{15}$$

$$a(0)\frac{du}{dx}(0) = 0$$

$$a(1)\frac{du}{dx}(1) = 0. \tag{16}$$

Given functions u and f on the interval $I = [0, 1]$, the coefficient a is identified in some suitable compact set $K \subset I$. If $\delta > 0$ is smaller than the distance from K to the boundary of I, then the functional is introduced

$$j_\delta u = (\rho_\delta * u)(x) = \int_{-\infty}^{\infty} \rho_\delta(x - s)u(s)ds,$$

where

$$\rho_\delta(x) = \frac{1}{\delta\sqrt{\pi}} e^{\frac{-x^2}{\delta^2}}.$$

The value x is chosen in a suitable compact set $K_\delta \subset I$. Under certain assumptions, close to those made by Richter, it is shown that

$$\|a - a_\delta^\epsilon\|_{\infty, K_\delta} \leq C\left(\delta + \frac{\delta}{\lambda(\lambda\delta - 8\epsilon)}\left(\frac{8\epsilon}{\delta} + \lambda\epsilon\right)\right)$$

for some constant $C = C(\lambda, \|u'\|_{\infty, K_\delta}, \|f\|_\infty)$ and an approximate coefficient a_δ^ϵ.

Notice that by picking δ suitably we get $\|a - a_\delta^\epsilon\|_{\infty, K_\delta} \to 0$ as $\epsilon \to 0$.

6. Estimating discontinuous coefficients

Although most of the early approaches mentioned above required some kind of smoothness assumption on the coefficient to be recovered, current research focuses more and more on the recovery of discontinuous coefficients and most recent papers have employed total variation regularization in the context of an optimization problem. Literature on the use of BV regularization has seen a high growth. In the following, we briefly mention a few relevant papers.

The use of BV regularization was promoted by the image processing community. The technical setup is as follows: The classical image problem is to find the original image x in a real Hilbert space H, from the observation of a degraded image

$$y = Lx + \delta$$

where $L : H \to H$ is a bounded linear operator modeling the blurring process and $\delta \in H$ models an additive noise component. This problem is typically formulated as an optimization problem and a regularized version is solved. To capture the sharp edges of the blurred images, Rudin, Osher and Fatemi[64] suggested the use of total variation regularization. Perhaps it was Gutman[35] who initiated the use of BV regularization in connection with parameter identification in a parabolic PDE. An interesting extension of this idea is given by Nashed and Scherzer[59] .

Since then TV regularization has been used extensively in many works. Dobson and Santosa[24] used BV regularization for the reconstruction of "blocky" conductivity profiles. An influential paper by Acar and Vogel[2] studies a general minimization problem in BV setting.

In addition to the papers mentioned above, Chan and Tai[17] presented augmented Lagrangian algorithms for recovering a in (3), with variations depending on the nature of the available data. They used total variation regularization, along with pre-filtering of the (noisy) data, and reported good numerical results even when the data contain a certain degree of noise. In a recent paper, Chan and Tai[18] used the level set methods and BV-regularization for the inverse problems in PDEs.

In an interesting paper, Chen and Zou[21] presented a detailed analysis of a similar algorithm, proving the existence of a solution, convergence of the augmented Lagrangian algorithm, and convergence of the sequence of discretized solutions. The main difference between the Chan-Tai and Chen-Zou results lies in the choice of objective function; Chan and Tai base their work on the L^2 misfit functional, while Chen and Zou use the coefficient-dependent energy norm. Since, as mentioned above, the second functional seems to be much more efficient in the context of output least-squares, it may also be that the Chen-Zou approach has advantages in the constrained optimization setting.

7. Applications to Elasticity Imaging

In recent years elasticity imaging inverse problem emerged as one of the most promising techniques to detect small as well as large cancers. The primary idea behind elasticity imaging is to translate the intuitive notions of palpation into a quantitative form that is amenable to mathematical manipulation. More precisely it can be explained as follows: Using ultrasound it is possible to measure interior displacements in human tissue (for example, breast tissue). Since cancerous tumors are markedly stiffer (around $5-10$ times) in their elastic properties from the surrounding healthy tissue, these tumors can be located by solving an inverse problem of identifying the Lame parameters that describes the elastic properties of the tissue. Due to the need of detecting cancer as early as possible and due to the deadliness of the disease, the elasticity imaging inverse problem has attracted a great deal of attention.

Recall that the system of isotropic elasticity is

$$-\nabla \cdot (2\mu\epsilon_u + \lambda \mathrm{tr}(\epsilon_u)I) = 0 \text{ in } \Omega, \tag{17}$$

where $\epsilon_u = \left(\nabla u + \nabla u^T\right)/2$ is the (linearized) strain tensor and u is the displacement of the elastic body in two or three dimensions.

The technical setup for the elasticity imaging is given through a slightly different formulation which we discuss below.

We also recall that the Lame module are related to the Young's modulus E and Poisson's ration ν through the relationship

$$\mu = \frac{E}{2(1+\nu)}$$

$$\lambda = \frac{\nu E}{(1+\nu)(1-2\nu)}$$

Since human tissue is modeled as an incompressible material, we are interested when ν is close enough to 0.5 and hence when $\lambda \to \infty$. However, in this setting the problem suffers from the so-called locking effect. One remedy for this is to alter the formulation by introducing the pressure function $p = -\lambda(\nabla \cdot u)$. This new equation is added to the system, the resulting system then reads as follows:

$$\begin{aligned} \nabla \cdot (-pI + 0.5\mu\epsilon_u) &= 0 \quad \text{in } \Omega \\ \nabla \cdot u + \frac{p}{\lambda} &= 0 \quad \text{in } \Omega \\ (-pI + 0.5\mu\epsilon_u) \cdot n &= h \quad \text{on } \Gamma_1 \\ u &= g \quad \text{on } \Gamma_2. \end{aligned}$$

Oberai, Gokhale and Feijóo[60] studied the above system and approximated the limit $\lambda \to \infty$ by replacing λ by $\lambda(x) = \beta\mu(x)$ with β chosen to be a large constant ($\approx 10^6$). Therefore, the corresponding inverse problem was only of identifying μ. The authors used L^2-norm based OLS approach and gave numerical examples. Convergence analysis was not given in this work.

In a very interesting paper,[9] Barbone and Bamber discussed many theoretical issues such a uniqueness for the above inverse problem.

Most of the above papers do not deal with theoretical issues such as error estimates and existence results. As far as error estimates are concerned, one of the earliest work on error estimates for any numerical method for the system of isotropic elasticity is by Chen and Gockenbach,[19] where the authors have generalized the method of Kohn and Lowe to the system of isotropic elasticity, proving a similar error bound:

$$\left\{ \|\mu - \mu_h\|_{L^2(\Omega)}, \|\lambda - \lambda_h\|_{L^2(\Omega)} \right\} \leq C \left[h + \frac{\|u - z\|_{H^1(\Omega)}}{h} \right].$$

The fundamental assumption is the following non-degeneracy condition (see Cox and Gockenbach[23]) on the strain $\epsilon = (\nabla u + \nabla u^T)/2$:

$$\min_{\Omega} \min\{|\epsilon_{11} + \epsilon_{22}|, |\epsilon_{12}|\} > 0.$$

8. Other methods and applications

The approaches and the applications mentioned above are only a few main directions which have been pursued by many researchers. There are many other approaches which are not discussed here. In the following we collect a few relevant works.

Essaouini, Nachaoui and Hajji[26] studied a nonlinear inverse problem (cf. (3)) with Cauchy conditions on a part of the boundary and no condition at all on the other part. They propose the use of an iterative boundary element procedure. Baumeister and Kunisch[10] studied a stable method to identify two parameters in a Helmholtz equation. Burger[12] used an iterative regularization scheme for an inverse problem in connection with polymer crystallization. The use of sequential quadratic programming methods for inverse problems is given in papers.[14,15,45,54] Interesting inverse problems in semiconductor devices are studied in.[16] Various optimization techniques were used in[36] to solve nonlinear inverse problems. Hinze[37] proposed a new concept for discretization which seems very promising for the inverse problems. Hsiao and Sprekels[40] studied the elliptic inverse problems from and abstract point of view. Interesting application in car windscreen design is given in.[53] In an interesting paper Manservsi and Gunzburger[58] used a variational inequality formulation to study a inverse problem in elasticity. Yu[73] used a separation argument to obtain necessary optimality conditions for inverse problems. Continuity properties of the regularized solutions are given in.[71] Much recently, Achdou[3] studied an inverse problem for a parabolic variational inequality emerging in connections with American options.

9. Some research directions

In the following we will discuss a few research topics we think deserve special attention.

9.1. *Error estimates*

A critical issue for applications of elliptic inverse problems is the degree of smoothness assumed of the coefficient to be identified. In many cases, the true coefficient may vary sharply, even discontinuously, and the ability to estimate discontinuous coefficients is desirable. On the other hand, error analysis of existing methods tend to assume that the true coefficient has some degree of smoothness. As we evaluate various methods, one concern should be their ability to identify sharply-varying coefficients. A relevant observation is that although a majority of error estimates have been derived for concrete BVPs, there is also plentiful literature devoted to error estimates from an abstract view point,[62] Therefore, it is desirable to obtain error estimates for the methods which are successful in identifying discontinuous coefficients.

9.2. *Optimization methods*

Newton type methods for large, for example in 3D, are too expensive. Therefore, it is natural to compare the performance of other algorithms gaining popularity in optimization communities. A list of these methods will naturally include interior point methods, trust region methods, limited memory BFG etc.

9.3. *Parameter identification for variational and quasi-variational inequality*

Since quasi variational inequality includes variational inequalities and PDEs as a particular cases, it is natural to study identification of coefficients in elliptic and parabolic quasi variational inequalities. It should be mentioned that in context of quasi variational inequalities, the differentiability of the solution operator is more complex and the optimality conditions will require some nontrivial generalizations.

9.4. *Parabolic IBVP*

The inverse problem of identifying the coefficient in parabolic BVP is equally important and hence there is plenty of work available in this direction (see[74]). Kravaris and Seinfeld[52] have studied, in a abstract framework, an output least-squares approach which covered the parameter identification for both parabolic and elliptic IBVPs. Recently, Keung and Zou[46] considered the modified OLS for parabolic problems which included BV-regularization. Theoretical work of Chen and Zou[21] was also extended in[34] to parabolic problems (only for smooth regularization). It will be interesting to extend the convex (energy-norm) OLS, for parabolic problems. Basic ideas given by[52] for OLS seems to be useful.

9.5. *Adaptive methods for inverse problems*

Perhaps one of the most important research directions is a detailed study of various aspects of inverse problems in an adaptive finite element framework. Recently, Bangerth[8] . has shown the usefulness of the adaptive methods in the study of inverse problems.

9.6. *Selection of an optimal regularization parameter*

The overall success of the regularization process depends on the right choice of the regularization parameter. There are many useful results available in the literature (see[55,65]). It will be useful to extend these methods to the selection of the optimal regularization parameter and the optimal smoothing parameter for BV-seminorm regularization.

References

1. R. Acar, Identification of the coefficient in elliptic equations, *SIAM J. Control Optim.*, **31** (1993), 1221–1244.

2. R. Acar and C.R. Vogel, Analysis of bounded variation penalty methods for ill-posed problems, *Inverse Problems*, **10** (1994), 1217–1229.
3. Y. Achdou, An inverse problem for parabolic variational inequality arising in voltality calibration with Americal options, *SIAM J. Control Optim.*, **43**, No. 5, 1583–1615.
4. G. Alessandrini, On the identification of the leading coefficient of an elliptic equation, *Analisi Funzionale e Applicazioni*, **4** (1985), 87–111.
5. D.D. Ang and L.K. Vy, Coefficient identification for an inhomogeneous Helmholtz equation by asymptotic regularization, *Inverse Problems*, **8** (1992), no. 4, 509–523.
6. R.C. Aster, B. Borchers and C.H. Thurber, *Parameter estimation and inverse problems*, Elsevier Press, 2005.
7. H.T. Banks and K. Kunisch, *Estimation Techniques for Distributed Parameter Systems*, Systems & Control: Foundations & Applications, 1. Birkhauser Boston, Inc., Boston, MA, 1989.
8. W. Bangerth, A framework for the adaptive finite element solution of large inverse problems, *SIAM Journal on Scientific Computing*, **30** (2008), 2965-2989.
9. P.E. Barbone and J.C. Bamber, Quantitative elasticity imagin: what can and cannot be inferred from starin images, Phys. Med. Biol., **47**, (2002) 2147-2164.
10. J. Baumeister and K. Kunisch, Identifiability and stability of a two-parameter estimation problem, *Appl. Anal.*, **40** (1991), no. 4, 263–279.
11. G. Bruckner and S. Handrock-Meyer, Langmach, H. An inverse problem from 2D ground-water modeling, *Inverse Problems*, **14**(1998), 835–851.
12. M. Burger, Iterative regularization of a parameter identification problem occurring in polymer crystallization, *SIAM J. Numer. Anal.*, **39** (2001), 1029–1055.
13. M. Burger, Parameter Identification, Lecture Notes, 2005.
14. M. Burger and W. Muhlhuber, Iterative regularization of parameter identification problems by sequential quadratic programming methods, *Inverse Problems*, **18** (2002), no. 4, 943–969.
15. M. Burger and W. Muhlhuber, Numerical approximation of an SQP-type method for parameter identification, *SIAM J. Numer. Anal.*, **40** (2002), no. 5, 1775–1797.
16. M. Burger H.W. Engl, P.A. Markowich and P. Pietra, Identification of doping profiles in semiconductor devices, *Inverse Problems*, **17** (2001), no. 6, 1765–1795.
17. T.F. Chan and X.C. Tai, Identification of discontinuous coefficients in elliptic problems using total variation regularization, *SIAM J. Sci. Comput.*, **25** (2003), 881–904.
18. T.F. Chan and X.C. Tai, Level set and total variation regularization for elliptic inverse problems with discontinuous coefficients, *J. Comput. Phys.*, **193** (2004), 40–66.
19. J. Chen and M.S. Gockenbach, A variational method for recovering planar Lame moduli, *Math. Mech. Solids*, **7** (2002), 191–202.
20. J. Chen, W. Han and F. Schulz, A regularization method for coefficient identification of a non-homogeneous Helmholtz equation, *Inverse Problems*,

10 (1994) 1115–1121.
21. Z. Chen and J. Zou, An augmented Lagrangian method for identifying discontinuous parameters in elliptic systems, *SIAM J. Control Optim.*, **37** (1999), 892–910.
22. C. Chicone and J. Gerlach, A note on the identifiability of distributed parameters in elliptic equations, *SIAM J. Math. Anal.*, **18** (1987), no. 5, 1378–1384.
23. S.J. Cox and M.S. Gockenbach, Recovering planar Lame moduli from a single-traction experiment, *Math. Mech. Solids*, **2** (1997), 297–306.
24. D.C. Dobson and F. Santosa, Recovery of blocky images from noisy and blurred data, *SIAM J. Appl. Math.*, **56** (1996), no. 4, 1181–1198.
25. H.W. Engl, M. Hanke and A. Neubauer, *Regularization of inverse problems*, Mathematics and its Applications, 375. Kluwer Academic Publishers Group, Dordrecht, 1996.
26. M. Essaouini, A. Nachaoui and S. E. Hajji, Numerical method for solving a class of nonlinear elliptic inverse problems, *J. Comput. Appl. Math.*, **162** (2004), no. 1, 165–181.
27. R.S. Falk, Error estimates for the numerical identification of a variable coefficient, *Math. Comp.*, **40** (1983), 537–546.
28. M.S. Gockenbach, The output least-squares approach to estimating Lamé moduli, *Inverse Problems*, (2007) 2437–2455.
29. M.S. Gockenbach and A.A. Khan: *Identification of Lamé parameters in linear elasticity: a fixed point approach,* Journal of Industrial and Management Optimization, Vol. 1, no. 4 (2005).
30. M.S. Gockenbach and A.A. Khan, An abstract framework for elliptic inverse problems. Part 1: an output least-squares approach, Mathematics and Mechanics of Solids, 12, 259-276, 2007.
31. M.S. Gockenbach and A.A. Khan, An abstract framework for elliptic inverse problems. Part 2: an augmented Lagrangian approach,Mathematics and Mechanics of Solids, 14, 2009, 517-539.
32. M.S. Gockenbach, B. Jadamba and A.A. Khan, Equation error approach for elliptic inverse problems with an application to the identification of Lamé parameters, *Inverse Problems in Science and Engineering*, **16** (2008) 349-367.
33. M.S. Gockenbach, B. Jadamba and A. A. Khan: Identification of discontinuous parameters with equation error method, *International Journal of Mathematics and Computer Science*, **1** (2006) 343–359.
34. B. Guo, and J. Zou, An augmented Lagrangian method for parameter identifications in parabolic systems. J. Math. Anal. Appl. 263 (2001), no. 1, 49–68.
35. S. Gutman, Identification of discontinuous parameters in flow equations, *SIAM J. Control Optim.,* **28** (1990), 1049–1060.
36. E. Haber, U. Ascher and D. Oldenburg, On optimization techniques for solving nonlinear inverse problems. Electromagnetic imaging and inversion of the Earth's subsurface, *Inverse Problems,* **16** (2000), 1263–1280.
37. M. Hinze, A variational discretization concept in control constrained optimization: the linear-quadratic case, *Comput. Optim. Appl.*, **30** (2005), no. 1, 45–61.

38. D. Hinestroza and D. A. Murio, Identification of transmissivity coefficients by mollification techniques. I. One-dimensional elliptic and parabolic problems, *Comput. Math. Appl.*, **25** (1993) 59–79.
39. K.-H. Hoffmann and J. Sprekels, On the identification of parameters in general variational inequalities by asymptotic regularization. SIAM J. Math. Anal. 17 (1986), no. 5, 1198–1217.
40. G.C. Hsiao and J.A. Sprekels, A stability result for distributed parameter identification in bilinear systems, *Math. Methods Appl. Sci.* **10** (1988), no. 4, 447–456.
41. K. Ito and K. Kunisch, The augmented Lagrangian method for parameter estimation in elliptic systems, *SIAM J. Control Optim.*, **28** (1990), 113–136.
42. B. Jadamba and A.A. Khan, Error estimates for the inverse problem of identifying variable coefficients by the modified least-squares, *Indian J. Industrial and Applied Mathematics*, **1** (2008), 1–9.
43. B. Jadamba, A.A. Khan and F. Raciti, On the inverse problem of identifying Lamé coefficients in linear elasticity, *Computer and Mathematics with Applications*, **56** (2008) 431-443.
44. T. Kärkkäinen, An equation error method to recover diffusion from the distributed observation, *Inverse Problems*, **13** (1997), 1033–1051.
45. C.T. Kelley and S.J. Wright, Sequential quadratic programming for certain parameter identification problems, *Math. Programming*, **51** (1991), no. 3, (Ser. A), 281–305.
46. Y.L. Keung and J. Zou, Numerical identifications of parameters in parabolic systems, Inverse Problems **14** (1998), no. 1, 83–100.
47. R. Kluge, An inverse problem for *coefficients* in linear equations. uniqueness and iterative solutions, *Abh. Akad. Wiss. DDR, Abt. Math. Naturwiss. Tech., 1981, 2, Akademie-Verlag, Berlin*, (1981) 139–148.
48. I. Knowles, A variational algorithm for electrical impedance tomography, *Inverse Problems,* **14** (1998) 1513–1525.
49. I. Knowles, Uniqueness for an elliptic inverse problem, *SIAM J. Appl. Math.,* **59** (1999), 1356–1370.
50. I. Knowles, Parameter identification for elliptic problems, *J. Comp. Appl. Math.*, **131** (2001) 175–194.
51. R.V. Kohn and B.D. Lowe, A variational method for parameter identification, *RAIRO Model. Math. Anal. Numer.*, **22** (1988), 119–158.
52. C. Kravaris and J.H. Seinfeld, Identification of parameters in distributed parameter systems by regularization, *SIAM J. Control Optim.*, **23** (1985), 217–241.
53. P. Kugler, A parameter identification problem of mixed type related to the manufacture of car windshields, *SIAM J. Appl. Math.,* **64** (2004), 858–877.
54. K. Kunisch and E.W. Sachs, Reduced SQP methods for parameter identification problems, *SIAM J. Numer. Anal.*, **29** (1992), no. 6, 1793–1820.
55. K. Kunisch and J. Zou, Iterative choices of regularization parameters in linear inverse problems, *Inverse Problems*, **14** (1998), no. 5, 1247–1264.
56. T. Lin and E. Ramirez, A numerical method for parameter identification of a boundary value problem, *Appl. Anal.*, **69** (1998), no. 3-4, 349–379.
57. R. Luce and S. Perez, Parameter identification for an elliptic partial differential equation with distributed noisy data, *Inverse Problems,* **15** (1999),

291–307.
58. S. Manservisi and M.A. Gunzburger, A variational inequality formulation of an inverse elasticity problem, *Appl. Numer. Math.,* **34** (2000), 99–126.
59. M.Z. Nashed and O. Scherzer, Stable approximations of a minimal surface problem with variational inequalities, *Abstr. Appl. Anal.* **2** (1997), no. 1-2, 137–161.
60. A.A. Oberai, N.H. Gokhale and G.R. Feijóo, Solution of inverse problems in elasticity imaging using the adjoint method, *Inverse Problems*, **19** (2003), 297–313.
61. G.F. Pinder and E.O. Frind, Application of Galerkin's procedure to acquifer analysis, *Water Resources Research*, **8** (1972), no. 1, 108-120.
62. E. Resmerita, Regularization of ill-posed problems in Banach spaces: convergence rates, *Inverse Problems*, **21** (2005), 1303-1314.
63. G.R. Richter, An inverse problem for the steady state diffusion equation, *SIAM J. Appl. Math.*,**41** (1981), no. 2, 210–221.
64. L.I. Rudin, S. Osher and E. Fatemi, Nonlinear total variation based noise removal algorithm, Phys. D, **60** (1992), 163-191.
65. O. Scherzer, H.W. Engl, H. W. and K. Kunisch, Optimal a posteriori parameter choice for Tikhonov regularization for solving nonlinear ill-posed problems, *SIAM J. Numer. Anal.*, **30** (1993), no. 6, 1796–1838.
66. U. Tautenhahn, A new regularisation method for parameter identification in elliptic problems, *Inverse Problems*, **6** (1990), no. 3, 465–477.
67. U. Tautenhahn and J. Muhs, A new regularization method for nonlinear ill-posed inverse problems, *Wiss. Z. Tech. Univ. Chemnitz*, **34** (1992), no. 1, 111–121.
68. T. Tucciarelli and D.P. Ahlfeld, A new formulation for transmissivity estimation with improved global convergence properties, *Water Resources Research,* **27** (1991) no.2, 243-251.
69. G. Vainikko, On the discretization and regularization of ill-posed problems with noncompact operators, *Numer. Funct. Anal. Optim.,* **13** (1992) 381–396.
70. P.S. Vassilevski, P. S.; Wade, J. G. A comparison of multilevel methods for total variation regularization, *Electron. Trans. Numer. Anal.,* **6** (1997), 255–270.
71. L.W. White and J.Zhou, Continuity and uniqueness of regularized output least squares optimal estimators, *J. Math. Anal. Appl.*, **196** (1994), 53–83.
72. W.W.G. Yeh, Review of Parameter procedures in groundwater hydrology: The inverse problem, *Water Resources Research*, **22** (1986), no. 2, 95-108.
73. W.H. Yu, Necessary conditions for optimality in the identification of elliptic systems with parameter constraints., *J. Optim. Theory Appl.*, **88** (1996), no. 3, 725–742.
74. W.H. Yu and J.H. Seinfeld, Identification of distributed parameter systems with pointwise constraints on the parameters, *J. Math. Anal. Appl.* **136** (1988), no. 2, 497–520.
75. J. Zou, Numerical methods for elliptic inverse problems, *Int. J. Comput. Math.*, **70** (1998), 211–232.

FINITE ELEMENT METHODS FOR HJB EQUATIONS

MESSAOUD BOULBRACHENE

Department of Mathematics and Statistics, Sultan Qaboos University, P.O.Box 36, Muscat 123, OMAN
E-mail: boulbrac@squ.edu.om

The paper surveys recent results on the finite element approximation of Hamilton-Jacobi-Bellman equations. Various methods are analyzed and error estimates in the maximum norm are derived. Also, a finite element monotone iterative scheme for the computation of the approximate solution is given and its geometrical convergence proved.

Keywords: HJB equations, quasivariational inequalities, finite elements, subsolutions, discrete regularity, contraction, error estimates

1. Introduction

This paper is concerned with the finite element approximation of the Hamilton-Jacobi-Bellman equation (HJB)

$$\begin{cases} \max\limits_{1\leq i\leq M} (\mathcal{A}^i u - f^i) = 0 \text{ in } \Omega \\ \dfrac{\partial u}{\partial n} = 0 \qquad\qquad\quad \text{on } \partial\Omega \end{cases} \tag{1}$$

where Ω is a bounded open domain of $\mathbb{R}^N$, $N \geq 1$, with boundary $\partial\Omega$ sufficiently smooth, the f^is are given smooth functions and the $\mathcal{A}^i$s are second order uniformly elliptic operators.

Note that we consider the Neumann problem as the constants belong to $H^1(\Omega)$, the space of trial functions. The Dirichlet case follows the same way with a technical adaptation.

HJB equations are encountered in several applications, for example in stochastic control the solution of (1) characterizes the infimum of the cost function associated to an optimally controlled stochastic switching process without costs for switching. (cf e.g.[1]).

The mathematical analysis of HJB equation has witnessed an intensive activity in the eighties and significant results have been obtained (see [2], [3], [5], [6], [8], [7]).

However, as far as numerical analysis is concerned, very few works are known in the literature.

Indeed, Cortey Dumont [9] studied the finite element approximation, for the first time, for problem (1) but no error estimate was given.

In the last few years, significant results on the finite element approximation have been obtained: Boulbrachene and Haiour [13], derived the first quasi-optimal L^∞ error estimate, by means of an algorithmic approach, then very recently, Boulbrachene and Cortey Dumont established the optimal order making use of the concept of subsolutions and discrete regularity. The noncoercive case has also been investigated in [14], [16] by a contraction fixed point approach.

On the computational side, some numerical algorithms have been developed and analyzed (cf.[4], [17], [13], [18], [19], [20]).

In this article we shall survey recent finite element approximation works carried out for HJB equation (1).

The article is organized as follows: In section 2, we review some fundamental results on elliptic variational inequalities. In section 3, we study the continuous HJB equation and the associated system of quasivariational inequalities. In section 4, as in the continuous case, we carry out the same study for the discrete HJB equation and the associated system of quasi-variational inequalities. In section 5, we introduce a monotone iterative scheme and prove its geometric convergence. Section 6 is devoted to the finite element error analysis. In section 7, we conclude the paper with some interesting open problems.

2. Preliminaries

2.1. *Continuous variational inequalities*

In this section, we shall recall some necessary results related to elliptic variational inequalities (VI) problems. Let $\mathbb{K} = \left\{ v \in H^1(\Omega) : v \leq \psi \right\}$. We denote by $\zeta = \sigma(g, \psi)$ the solution of the VI: Find $\zeta \in \mathbb{K}$ such that

$$a(\zeta, v - \zeta) \geqq (g, v - \zeta) \; \forall v \in \mathbb{K} \tag{2}$$

Theorem 2.1. *(cf.[23], [24]) Let $\mathcal{A}$ be a linear second order elliptic operator associated with the bilinear form $a(.,.)$. Then (2) has one and only one solution. Moreover, $\zeta \in W^{2,p}(\Omega)$, $1 \leq p < \infty$ and*

$$\|\mathcal{A}\zeta\|_{L^\infty(\Omega)} \leq C \quad \textit{(Levy-Stampachia estimate)} \tag{3}$$

Definition 2.1. $z \in K$ is said to be a subsolution for VI (2) if

$$a(z, v) \leq (f, v) \; \forall v \geq 0, v \in H^1(\Omega)$$

Theorem 2.2. *(cf [23], [24]) Let $\mathbb{X}$ be the set of such subsolutions. The solution of VI (2) is the maximum element of $\mathbb{X}$.*

Theorem 2.3. *The mapping σ is monotone, Lipschitz continuous and concave with respect to ψ*

Proof. 1. σ is increasing.

Let ψ and $\widetilde{\psi}$ in $L^\infty(\Omega)$ such that $\psi \leq \widetilde{\psi}$. Then

$$\sigma(\psi) \leq \psi \leq \widetilde{\psi}$$

so $\sigma(\psi)$ is a subsolution for the V.I with obstacle $\widetilde{\psi}$. Then, using Theorem 2.2, we get the desired result.

2. σ is Lipschitz continuous. Let ψ, $\widetilde{\psi}$ in $L^\infty(\Omega)$ and $\zeta = \sigma(\psi)$; $\widetilde{\zeta} = \sigma(\widetilde{\psi})$. Set

$$\Phi = \| \psi - \widetilde{\psi} \|_\infty$$

Then

$$\zeta - \Phi \leq \psi - \Phi \leq \widetilde{\psi}$$

So, $\zeta - \Phi$ is a subsolution for the VI with obstacle $\widetilde{\psi}$. It follows that

$$\zeta - \Phi \leq \widetilde{\zeta} \quad \text{or} \quad \zeta \leq \widetilde{\zeta} + \Phi.$$

Now, interchanging the roles of ψ and $\widetilde{\psi}$, we also get

$$\widetilde{\zeta} \leq \zeta + \Phi.$$

This completes the proof.

3. σ is concave. Let ψ, $\widetilde{\psi}$ in $L^\infty(\Omega)$ and $\theta \in [0; 1]$ and set

$$\sigma_\theta = \sigma(\theta\psi + (1-\theta)\widetilde{\psi})$$

Since $\sigma(\psi) \leq \psi$ and $\sigma(\widetilde{\psi}) \leq \widetilde{\psi}$, it follows that $\theta\sigma(\psi) + (1-\theta)\sigma(\widetilde{\psi})$ is a subsolution for the V.I with obstacle $\theta\psi + (1-\theta)\widetilde{\psi}$. So, using Theorem 2.2, we get the concavity of σ. □

2.2. *Discrete variational inequalities*

Let Ω be decomposed into triangles and let τ_h denote the set of all those elements; $h > 0$ is the mesh size. We assume that the family τ_h is regular and quasi-uniform.

Let $\mathbb{V}_h$ denote the standard piecewise linear finite element space, and $\{\varphi_s\}$, $s = 1, 2, ...m(h)$ the basis of $\mathbb{V}_h$.

Under a discrete maximum principle assumption **d.m.p** (the matrix discretization matrix is an M-Matrix) the above qualitative properties of the continuous VI (2) transfers to the discrete case. Their respective proofs will be omitted as they are identical to their continuous analogous ones.

Let $\mathbb{K}_h = \{v \in \mathbb{V}_h$ such that $v \leq r_h\psi\}$ and $\zeta_h \in \mathbb{K}_h$ be the finite element approximation of ζ defined in (2):

$$a(\zeta_h\,,\, v - \zeta_h) \geqq (f, v - \zeta_h)\ \forall v \in \mathbb{K}_h \tag{4}$$

Definition 2.2. $z_h \in K_h$ is said to be a subsolution for VI (4) if

$$a(z_h, v) \leq (f, v)\forall v \geq 0, v \in H^1(\Omega)$$

Theorem 2.4. *Let $\mathbb{X}_h$ be the set of discrete subsolutions of ζ_h. Then, ζ_h is the maximum element of $\mathbb{X}_h$.*

Theorem 2.5. *Under the d.m.p, the mapping σ_h is increasing, concave and Lipschitz continuous with respect to ψ.*

2.2.1. $\mathbf{L^\infty}$- *Error Estimate*

Theorem 2.6. *(cf [12]) There exists a constant C independent of h such that*

$$\|\zeta - \zeta_h\|_{L^\infty(\Omega)} \;\leq\; Ch^2\,|\log h|^2 \tag{5}$$

3. The Continuous Problem

3.1. *Assumptions, Notations*

We are given functions

$$a^i_{jk}(x),\ b^i_k(x),\ \ a^i_0(x) \in C^2(\bar{\Omega}), x \in \bar{\Omega} \tag{6}$$

such that

$$\sum_{1\leq\ j,\ k\leq\ N} a^i_{jk}(x)\xi_j\xi_k \geqq \alpha \mid \xi \mid^2; \quad (x \in \bar{\Omega},\ \xi \in R^N,\ \ \alpha > 0) \tag{7}$$

$$a_0^i(x) \geqq c_0 \geq 0 \quad (x \in \bar{\Omega};\ \ c_0 > 0) \tag{8}$$

We define the second order uniformly elliptic operators of the form

$$\mathcal{A}^i = \sum_{1\leq\ j,k\leq N} a_{jk}^i(x)\frac{\partial^2}{\partial x_j \partial x_k} + \sum_{k=1}^{N} b_k^i(x)\frac{\partial}{\partial x_k} + a_0^i(x) \tag{9}$$

and the associated bilinear forms: $\forall u, v \in H^1(\Omega)$

$$a^i(u\,,\,v) = \int_\Omega (\textstyle\sum_{1\leq j,k\leq k} a_{jk}^i(x)\frac{\partial u}{\partial x_j}\frac{\partial v}{\partial x_k} + \sum_{k=1}^N b_k^i(x)\frac{\partial u}{\partial x_k}v + a_0^i(x)uv)dx \tag{10}$$

such that:

$$a^i(v\ ,\ v) \geq \delta \parallel v \parallel_{H^1(\Omega)}^2,\ \ \delta > 0,\ \ \forall v \in H^1(\Omega) \tag{11}$$

Let $(.,.)$ denote the scalar product in $L^2(\Omega)$ and $f^{\,i}s$ be nonnegative right-hand sides in $W^{2,\infty}(\Omega)$. Let $W = (w^1, ..., w^M) \in (L^\infty(\Omega))^M$, and $\parallel . \parallel_{L^\infty(\Omega)}$denote the L^∞ norm.We denote by

$$\|W\|_\infty = \max_{1\ \leq\ i\leq\ M} \left\|w^i\right\|_{L^\infty(\Omega)}$$

3.2. *A System of quasivariational inequalities associated with the HJB Equation*

P.L.Lions and J.L.Menaldi [6] proved that the solution of HJB equation (1) can be approximated by the weakly coupled system of quasivariational inequalities (QVIs): Find $U = (u^1, ..., u^M) \in \left(H^1(\Omega)\right)^M$ such that

$$\begin{cases} a^i(u^i, v - u^i) \geqq (f^i, v - u^i)\ \ \forall v \in H^1(\Omega) \\ u^i \leq k + u^{i+1}\ ,\ \ v \leq k + u^{i+1} \\ \text{with } u^{M+1} = u^1 \end{cases} \tag{12}$$

where k is a positive number.

Naturally, the structure of problem (12) is analogous to that of the classical obstacle problem where the obstacle function is replaced by an implicit one, depending upon the solution. The terminology quasivariational inequality [25] being chosen is a result of this remark.

Theorem 3.1. *(cf.[6]) There exists a unique solution $U = (u^1, ..., u^M)$ with $u^i \in W^{2,p}(\Omega), 1 \leq p < \infty$.Moreover, as $k \to 0$, each component of U converges in $C(\bar{\Omega})$to the solution u of HJB equation (1), and $u \in W^{2,\infty}(\Omega)$.*

3.3. *Qualitative Properties*

3.3.1. *Monotonicity*

Lemma 3.1. *(cf.[9].) If* $k \geq \tilde{k}$ *then* $u^i \geq \tilde{u}^i$.

3.3.2. *Lipschitz continuity*

Let k, $\tilde{k}$ be two positive parameters and $u^i = \sigma(f^i, k + u^{i+1})$, $\tilde{u}^i = \sigma(f^i, \tilde{k} + \tilde{u}^{i+1})$ be the corresponding solutions to system (12), respectively.

Theorem 3.2. *Under conditions of lemma 3.1,we have*

$$\max_{1 \leq i \leq M} \| u^i - \tilde{u}^i \|_{L^\infty(\Omega)} \leq \left| k - \tilde{k} \right|$$

Proof.

Set $\Phi = \left| k - \tilde{k} \right|$. For sake of simplicity, we will write $u^i = \sigma(f^i, k)$ instead of $\sigma(f^i, k + u^{i+1})$. Then, thanks to lemma 3.1, we have

$$k \leq \tilde{k} + \Phi$$

implies

$$\sigma(f^i, k) \leq \sigma(f^i, \tilde{k} + \Phi)$$

But

$$\sigma(f^i, \tilde{k}) + \Phi = \tilde{u}^i + \Phi = \sigma(f^i + a_0^i \Phi, \tilde{k} + \Phi)$$

and

$$f^i \leq f^i + a_0^i \Phi \ ; \ (a_0^i(x) > 0)$$

So, applying standard comparison results in VIs, we get

$$\sigma(f^i, \tilde{k} + \Phi) \leq \sigma(f^i + a_0^i \Phi, \tilde{k} + \Phi) = \sigma(f^i, \tilde{k}) + \Phi$$

Hence,

$$\sigma(f^i, k) \leq \sigma(f^i, \tilde{k} + \Phi) \leq \sigma(f^i, \tilde{k}) + \Phi$$

or

$$\sigma(f^i, k) \leq \sigma(f^i, \tilde{k}) + \Phi$$

As the roles of k and $\tilde{k}$ are symmetric, we also have

$$\sigma(f^i, \tilde{k}) \leq \sigma(f^i, k) + \Phi$$

Thus

$$\| u^i - \tilde{u}^i \|_\infty \leq \left| k - \tilde{k} \right| \ \forall i = 1, 2, ..., M \qquad \square$$

As a consequence of the above result, we have the following convergence rate.

Corollary 3.1.

$$\max_{1 \leq i \leq M} \| u^i - u \|_{L^\infty(\Omega)} \leq k$$

3.3.3. ***Subsolutions***

Definition 3.1. $W = (w^1, .., w^M) \in (H^1(\Omega))^M$ is said to be a subsolution for the system of QVIs (12) if

$$\begin{cases} a^i(w^i, v) \leq (f^i, v) \ \forall v \in H^1(\Omega), \ v \geq 0 \\ w^i \leq k + w^{i+1} \\ w^{M+1} = w^1 \end{cases} \tag{13}$$

Theorem 3.3. *(cf. [9]) Let $\mathbb{X}$ denote the set of such subsolutions. The solution of system of QVIs (12) is the maximum element of the set $\mathbb{X}$.*

4. The Discrete Problem

Let $\mathbb{A}^i$ be the matrices with generic coefficients

$$(\mathbb{A}^i)_{ls} = a^i(\varphi_l, \varphi_s), 1 \leq i \leq M; 1 \leq l, s \leq m(h)$$

where, $\{\varphi_l\}$, $l = 1, 2, ...m(h)$ is the basis of $\mathbb{V}_h$. Let F^i be the approximation of f^i:

$$F^i_l = (f^i, \varphi_s), s = 1, ..., m(h)$$

Let r_h be the usual restriction operator defined by

$$\forall v \in C(\Omega) \cap H^1(\Omega), \ r_h v = \sum_{i=1}^{m(h} v_l \varphi_l$$

In the sequel of the paper, we shall make use of the **discrete maximum assumption (d.m.p)**. In other words, we shall assume that the matrices $\mathbb{A}^i$ are M-Matrices.

The d.m.p which has been previously proved in specific cases in [21], has witnessed significant developments since then (see, e.g., [22] and the reference therein).

Under the d.m.p, we shall achieve a similar study to that devoted to the continuous problem. More precisely, we shall show that the qualitative properties and results stated in the previous section are conserved in the discrete case. Their respective proofs will be omitted as they are identical to their continuous analog ones.

The discrete HJB equation consists of finding $u_h \in \mathbb{V}_h$ such that:

$$\max_{1 \leq i \leq M} (\mathbb{A}^i u_h - F^i) = 0 \tag{14}$$

4.1. *A system of quasivariational inequalities associated with the discrete HJB equation*

It is shown in [9] that problem (14) can be approximated by the following system of discrete quasi-variational inequalities (QVIs):

Find $U_h = (u_h^1, ..., u_h^M) \in (\mathbb{V}_h)^M$ such that:

$$\begin{cases} a^i(u_h^i, v - u_h^i) \geqq (f^i, v - u_h^i) \ \forall v \in \mathbb{V}_h \\ u_h^i \leq k + u_h^{i+1}, \ v \leq k + u_h^{i+1} \\ \text{with } u_h^{M+1} = u_h^1 \end{cases} \tag{15}$$

Theorem 4.1. *(cf. [9]) Under the d.m.p, there exists a unique solution to the system of QVIs (15). Moreover, each component of U_h converges in $C(\bar{\Omega})$ to the solution of the discrete HJB equation (14).*

4.1.1. *Monotonicity*

Let k, $\tilde{k}$ be two positive parameters and $u_h^i = \sigma_h(f^i, k)$, $\tilde{u}_h^i = \sigma_h(f^i, \tilde{k})$ be the corresponding solutions to system (15), respectively.

Lemma 4.1. *(cf. [9]) If $k \geq \tilde{k}$ then $u_h^i \geq \tilde{u}_h^i$.*

4.1.2. *Lipschitz continuity*

Theorem 4.2. *Under conditions of lemma 4.1, we have*

$$\max_{1 \leq i \leq M} \left\| u_h^i - \tilde{u}_h^i \right\|_{L^\infty(\Omega)} \leq \left| k - \tilde{k} \right|$$

Proof.

Similar to that of Theorem 3.2. □

As a direct consequence of the above result, we have the following convergence rate.

Corollary 4.1.

$$\max_{1\le i\le M} \left\|u_h^i - u_h\right\|_{L^\infty(\Omega)} \le k$$

4.1.3. *The discrete regularity*

The discrete regularity plays an important role in the regularization of the obstacles appearing in the discrete system of QVIs (15) as it permits to replace the irregular obstacles "$k + u_h^{i+1}$" with $W^{2,p}(\Omega)$ regular ones, and hence preserves the optimal convergence order.

Theorem 4.3. *(cf. [9]) There exists a constant C independent of k and h such that*

$$\left|a(u_h^i, \varphi_s)\right| \le C \left\|\varphi_s\right\|_{L^1(\Omega)} \ \forall i = 1, 2, ..., M \tag{16}$$

Moreover, there exists a family of right-hands side $\left\{g^{1,(h)}, ..., g^{M\ ,(h)}\right\}_{h>0}$ bounded in $L^\infty(\Omega)$ such that each component u_h^i of the solution of the discrete system (15) satisfies the equation

$$a^i(u_h^i, v) = (g^{i,(h)}, v) \ \forall v \in \mathbb{V}_h \tag{17}$$

Remark 4.1. The estimate (16) can be regarded as the discrete counterpart of the Levy-Stampacchia estimate (3) extended to the variational form through the $L^1 - L^\infty$ duality.

Theorem 4.4. *(cf.[9]) Let $u^{i,(h)}$ be the the corresponding continuous counterpart of (17), that is,*

$$a(u^{i,(h)}, v) = (g^{i,(h)}, v) \ \forall v \in H^1(\Omega) \tag{18}$$

Then there exists a constant independent of both k and h

$$\left\|u^{i,(h)}\right\|_{W^{2,p}(\Omega)} \le C \tag{19}$$

and

$$\left\|u^{i,(h)} - u_h^i\right\|_{L^\infty(\Omega)} \le Ch^2 \left|\log h\right| \tag{20}$$

Notation 4.1. From now onward, we will denote by

$$u^{i,(h)} = \sigma(g^{i,(h)}, +\infty)$$

the solution of equation (18) and by

$$u_h^{i,(h)} = \sigma_h(g^{i,(h)}, +\infty)$$

the approximation of $u^{i,(h)}$ as solution of an equation.

4.1.4. ***Discrete subsolution***

Definition 4.1. $W = (w_h^1, .., w_h^M) \in (\mathbb{V}_h)^M$ is said to be a subsolution for the system of QVIs (15) if

$$\begin{cases} a^i(w_h^i, \varphi_s) \leq \ (f^i, \varphi_s) \ \forall \varphi_s; \ s = 1, ..., m(h) \\ w_h^i \leq k + w_h^{i+1} \\ w_h^{M+1} = w_h^1 \end{cases} \tag{21}$$

Theorem 4.5. *(cf. [9]) Let $\mathbb{X}_h$ be the set of discrete subsolutions. Under the* ***d.m.p,*** *the solution of system of QVIs (15) is the maximum element of the set $\mathbb{X}_h$.*

5. Algorithm

Next, we shall construct a monotone iterative scheme and prove its geometrical convergence to the unique solution of this system of QVIs (12).

Let $\mathbb{H}^+ = \left(L_+^\infty(\Omega)\right)^M$ where $L_+^\infty(\Omega)$ denotes the positive cone of $L^\infty(\Omega)$. We define the mapping

$$\begin{aligned} T : \mathbb{H}^+ &\longrightarrow \ \mathbb{H}^+ \\ W &\longrightarrow TW = (\zeta^1, ..., \zeta^M) \end{aligned} \tag{22}$$

such that $\forall i = 1, ..., M$, ζ^i is the solution of the following VI:

$$\begin{cases} a^i(\zeta^i, v - \zeta^i) \geq (f^i, v - \zeta^i) \ \forall v \in H^1(\Omega) \\ \zeta^i \leq k + w^{i+1} \ , \ \ v \leq k + w^{i+1} \\ \text{with} \ \ w^{M+1} = w^1 \end{cases} \tag{23}$$

So, if we denote ζ^i by $\sigma(k + w^{i+1})$, we clearly have

$$TW = [\sigma(k + w^2), \sigma(k + w^3), ..., \sigma(k + w^i), ..., \sigma(k + w^1)]$$

Let $U^0 = (u^{1,0}, ..., u^{M,0})$ be solution to the following equation:

$$a^i(u^{i,0}, v) = (f^i, v) \; \forall \; v \in H^1(\Omega) \, ; \, 1 \leq i \leq M \tag{24}$$

Then, there exists a unique positive solution to problem (24), (^{25}). Moreover, $u^{i,0} \in W^{2,,\infty}(\Omega)$.

5.1. *A continuous monotone iterative scheme*

Now, starting from U^0 solution of (24), we define

$$U^{n+1} = TU^n \;\; ; \; n = 0, 1, .. \tag{25}$$

In view of (22), (23), it is clear that $U^{n+1} = (u^{1,n}, ..., u^{n,M})$ solves the following independent variational inequalities

$$\begin{cases} a^i(u^{i,n+1}, v - u^{i,n+1}) \geqq (f^i, v - u^{i,n+1}) \forall v \in H^1(\Omega) \\ u^{i,n+1} \leq k + u^{i+1,n} , \;\; v \leq k + u^{i+1,n} \\ u^{M+1,n} = u^{1,n}; \, i = 1, 2, ...M \end{cases} \tag{26}$$

which can be solved in parallel.

The convergence of this algorithm stands on some properties of the mapping T.

Proposition 5.1. *The mapping T satisfies*

$$TV \leq TW \quad \forall V \leq W$$

$$TW \geq 0 \quad\;\; \forall \, W \in \mathbb{H}^+$$

$$TW \leq \hat{U}^0 \quad \forall \, W \in \mathbb{H}^+$$

Proof.

1. $TV \; \leq TW \quad \forall V \leq W$. Let $V = (v^1, ..., v^M)$, $W = (w^1, , ..., w^M)$ in $\mathbb{H}^+$ such that $v^i \leq w^i$, $\forall \, i = 1, , ..., M$.

Then since σ is increasing, it follows that $\sigma(k + v^i) \leq \sigma(k + w^i)$.

2. $TW \geq 0$, $\forall W \in \mathbb{H}^+$. This follows directly from the fact that $f^{\,i} \geq 0$ and classical comparison results in elliptic variational inequalities.

3. $TW \leq \hat{U}^0 \quad \forall \, W \in \mathbb{H}^+$.

Let's denote by $\varphi^+ = \max(\varphi\ ,0)$ and $\varphi^- \ = \max(-\varphi\ ,0)$. Then, since ζ^i is solution to (23) and $u^{i,0}$ of (24) belong to $H^1(\Omega)$, we readily have

$$\zeta^i - (\zeta^i - u^{i,0})^+ \in H^1(\Omega)$$

Moreover, as $(\zeta^i - u^{i,0})^+ \geq 0$, it follows that

$$\zeta^i - (\zeta^i - u^{i,0})^+ \leq \zeta^i \leq k + w^{i+1}$$

Therefore, we can take $v = \zeta^i - (\zeta^i - u^{i,0})^+$ as a trial function in (23), which yields

$$a^i(\zeta^i, -(\zeta^i - u^{i,0})^+) \geqq (f^i, -(\zeta^i - u^{i,0})^+)$$

Also, for $v = (\zeta^i - \hat{u}^{i,0})^+$ equation (24) becomes

$$a(u^{i,0}, (\zeta^i - u^{i,0})^+) = (f^i, (\zeta^i - u^{i,0})^+)$$

so, by addition, we obtain

$$-a^i((\zeta^i - u^{i,0})^+, (\zeta^i - u^{i,0})^+\ ) \geqq 0$$

which, by (11), yields

$$(\zeta^i - u^{i,0})^+ = 0$$

Thus

$$\zeta^i \leq u^{i,0} \quad \forall\ i = 1, 2, ..., M$$

i.e.,

$$TW \leq U^0$$

□

Proposition 5.2. *The mapping T is concave on $\mathbb{H}^+$.*

Proof.

Let $\theta \in [0, 1]$. Then we have

$T(\theta V + (1-\theta)W) =$

$$(\sigma(k + \theta v^2 + (1-\theta)w^2), ..., \sigma(k + \theta v^i + (1-\theta)w^i), ..., \sigma(k + \theta v^1 + (1-\theta)w^1))$$

$$(\sigma(\theta(k + v^2) + (1-\theta)(k + w^2)), ..., \sigma(\theta(k + v^i) + (1-\theta)(k + w^i)), ...,$$

$$\sigma(\theta(k + v^1) + (1-\theta)(k + w^1))$$

Then using the concavity of σ, we get

$$T(\theta V + (1-\theta)W)$$

$$\geq \theta.(\sigma(k+v^2), ..., \sigma(k+v^i), ..., \sigma(k+v^1))$$

$$+(1-\theta)(\sigma(k+v^2), ..., \sigma(k+v^i), ..., \sigma(k+v^1))$$

$$\geqq \theta TV + (1-\theta)TW$$

□

Lemma 5.1. *Let* $0 < \lambda < \min\left(\frac{k}{\|U^0\|_\infty}, 1\right)$. *Then we have* $T(0) \geqq \lambda U^0$.

Proof.

Let $\breve{U} = (\breve{u}^1, ..., \breve{u}^M)$ be such that $\breve{u}^i$, $1 \leq i \leq M$ is the solution of the following variational inequality

$$\begin{cases} a^i(\breve{u}^i, v - \breve{u}^i) \geqq (f^i, v - \breve{u}^i) \; \forall v \in H^1(\Omega) \\ \breve{u}^i \leq k, \; v \leq k \end{cases} \tag{27}$$

Then it is clear that

$$v = (\breve{u}^i - \lambda u^{i,0})^- + \breve{u}^i$$

can be taken as a trial function in the VI (27). So taking

$$v = -(\breve{u}^i - \lambda u^{i,0})^-$$

as a trial function in (24) and using the fact that $f^i \geq 0$, we get by addition

$$a^i(\breve{u}^i - \lambda u^{i,0}, (\breve{u}^i - \lambda u^{i,0})^-) \geqq (f^i - \lambda f^i, (\breve{u}^i - \lambda u^{i,0})^-)$$

$$\geqq ((1-\lambda)f^i, (\breve{u}^i - \lambda u^{i,0})^-) \geq 0$$

Thus, by (11)

$$(\breve{u}^i - \lambda u^{i,0})^- = 0$$

i.e.,

$$\breve{u}^i \geqq \lambda u^{i,0} \quad \forall \; i = 1, 2, ..., M$$

which completes the proof. $\square$

Let

$$\mathbb{C} = \{ \ W \in \mathbb{H}^+ \ : \ 0 \leq W \leq U^0 \, \}$$

Proposition 5.3. *Let* $\gamma \in [0\,;1]$ *such that*

$$W - \tilde{W} \leq \gamma W \quad \forall W, \ \tilde{W} \ \in \mathbb{C} \tag{28}$$

Then combining proposition 5.2. and Lemma 5.1, the following holds

$$TW - T\tilde{W} \leq \gamma(1-\lambda)TW \tag{29}$$

Proof.

By (28), we have $(1-\gamma)\,W \ \leq \ \tilde{W}$. Then, using the fact that T is increasing and concave, it follows that

$$(1-\gamma)TW + \gamma T(0)$$

$$\leq T[(1-\gamma)W + \gamma.0]$$

$$\leq T\tilde{W}$$

Finally, using lemma 5.1, we get (29). $\square$

Theorem 5.1. *The sequences* (U^n) *converges decreasingly to the unique solution of system of QVIs (12).*

Proof. The proof will be carried out in four steps.

Step 1. The sequence (U^n) stays in $\mathbb{C}$ and is monotone decreasing.

We know that $U^n = (u^{1,n}, ..., u^{n,M})$ is such that

$$\begin{cases} a^i(u^{i,n}, v - u^{i,n}) \geqq (f^i, v - u^{i,n}) \ \forall v \in H^1(\Omega) \\ u^{i,n} \leq k + u^{i+1,n-1} \ ; v \leq k + u^{i+1,n-1} \\ u^{M+1,n} = u^{1,n} \end{cases} \tag{30}$$

Since $f^i \geq 0$ and $\hat{u}^{i,0} \geq 0$, combining comparison results in variational inequalities with a simple induction, it follows that $\hat{u}^{i,n} \geq 0$ i.e.,

$$U^n \geq 0 \quad \forall n \geq 0 \tag{31}$$

Furthermore, we have

$$U^1 = TU^0 \leq U^0$$

Thus, inductively

$$0 \leq U^{n+1} = TU^n \leq U^n \leq ... \leq U^0 \quad \forall n \geq 0 \tag{32}$$

Step 2. $(\hat{U}^n)$ converges to the solution of the system (12).

From (31), (32), it is clear that

$$\lim_{n\to\infty} u^{i,n}(x) = \bar{u}^i(x) \ , \ x \in \Omega \ \text{ and } \ \bar{u}^i \in \mathbb{C} \tag{33}$$

Moreover, from (31) we have $k + \hat{u}^{i+1,n-1} \geq 0$. Then we can take $v = 0$ as a trial function in (30), which yields

$$\delta \left\|u^{i,n}\right\|^2_{H^1(\Omega)} \leq \ a^i(u^{i,n}, \ u^{i,n}) \leq \ \left\|f^i\right\|_{L^2(\Omega)} \left\|u^{i,n}\right\|_{H^1(\Omega)}$$

or more simply

$$\left\|u^{i,n}\right\|_{H^1(\Omega)} \leq C$$

where C is a constant independent of n.Hence $\hat{u}^{i,n}$ stays bounded in $H^1(\Omega)$ and consequently we can complete (33) by

$$\lim_{n\to\infty} u^{i,n} = u^i \text{ weakly in } H^1(\Omega) \tag{34}$$

Step 3. $\bar{U} = (\bar{u}^1, ..., \bar{u}^M)$ coincides with the solution of system (12). Indeed, since

$$u^{\,i,n}(x) \leq k + u^{\,i,n-1}(x)$$

then (33) implies

$$\bar{u}^i(x) \leq k + \bar{u}^{i+1}(x)$$

Now let $v \leq k + \bar{u}^{i+1}$ then $v \leq k + u^{i,n-1}, \ \forall n \geq 0$. We can therefore take v as a trial function for the VI (30). Consequently, combining (33), (34) with the weak lower semi continuity of $a^i(v, v)$ and passing to the limit in problem (30), we obtain

$$a^i(\bar{u}^i, v - \bar{u}^i) \geqq (f^i, v - \bar{u}^i) \ \forall v \in H^1(\Omega), \ v \leq k + \bar{u}^{i+1}$$

Finally, since $U = (u^1, ..., u^M)$ is the unique solution of (12) , we clearly have

$$U = \bar{U} = (\bar{u}^1, ..., \bar{u}^M) \qquad \square$$

Remark 5.1. From the above proposition, one can observe that the solution of system QVI (12) is a fixed point of T i.e.,

$$U = TU \tag{35}$$

Next, we shall prove that the convergence of the proposed iterative scheme is geometrical.

Theorem 5.2. *There exists a positive constant* $0 < \mu < 1$ *such that*

$$\|U^n - U\|_\infty \leq \mu^n \left\|U^0\right\|_\infty \tag{36}$$

Proof.

We have

$$0 \leq U \leq U^0$$

so

Then, applying (28) with $\gamma = 1$, we get

$$0 \leq TU^0 - TU \leq (1-\lambda)TU^0$$

and by (25), (35)

$$0 \leq U^1 - U \leq (1-\lambda)U^1$$

Now, using (28) again with $\gamma = 1 - \lambda$ it follows that

$$0 \leq TU^1 - TU \leq (1-\lambda)(1-\lambda)TU^1$$

i.e.,

$$0 \leq U^2 - U \leq (1-\lambda)^2 U^2$$

and inductively

$$0 \leq U^n - U \leq (1-\lambda)U^n$$

$$\leq (1-\lambda)^n U^0 \qquad \square$$

As a consequence of the above theorem, we have the following convergence rate for the solution of the HJB equation.

Corollary 5.1.

$$\max_{1\le i\le M} \left\|u-u^{i,n}\right\|_{L^\infty(\Omega)} \le \mu^n \left\|U^0\right\|_\infty + k$$

Proof.

$$\begin{aligned}\left\|u-u^{i,n}\right\|_{L^\infty(\Omega)} &\le \left\|u-u^{i}\right\|_{L^\infty(\Omega)} + \left\|u^i-u^{i,n}\right\|_{L^\infty(\Omega)} \\ &\le k \ + \ \max_{1\le i\le M} \left\|u-u^{i,n}\right\|_{L^\infty(\Omega)} \\ &\le k + \left\|U^n - U\right\|_\infty \\ &\le k + \mu^n \left\|U^0\right\|_\infty\end{aligned}$$

□

5.2. *A discrete monotone iterative scheme*

Next, we shall define a discrete iterative scheme. The proof of its convergence is very similar to that of the continous one (25) and therefore will be omitted. Let's consider the following mapping

$$\begin{aligned} T_h :\mathbb{H}^+ &\longrightarrow (\mathbb{V}_h)^M \\ W &\longrightarrow T_hW = (\zeta_h^1, ..., \zeta_h^M)\end{aligned} \tag{37}$$

where $\zeta_h^i = \sigma_h(k+w^{i+1})$ is the unique solution of the following discrete VI:

$$\begin{cases} a^i(\zeta_h^i, v-\zeta_h^i) \geqq (f^i, v-\zeta_h^i) \ \forall v \in \mathbb{V}_h \\ \zeta_h^i \le r_h(k+w^{i+1})\, ; \ v \le r_h(k+w^{i+1}) \\ \zeta_h^{M+1} = \zeta_h^1 \end{cases} \tag{38}$$

So, we clearly have

$$T_hW = [\sigma_h(k+w^2), \sigma_h(k+w^3), ..., \sigma_h(k+w^i), ...\sigma_h(k+w^1)]$$

Let $U_h^0 = (u_h^{1,0}, ..., u_h^{M,0})$ be the finite element approximation of U^0 defined in (2.20):

$$a^i(u_h^{i,0}\ , v) = (f^i\ , v) \ \forall v \in \mathbb{V}_h \ ; \ \ 1 \le i \le M \tag{39}$$

By the d.m.p and the positivity of f^i, we have $\hat{U}_h^0 \geq 0$. Now, starting from U_h^0, we define

$$U_h^{n+1} = T_h U_h^n \quad , \ n = 0, 1, ... \tag{40}$$

where $U_h^{n+1} = (u_h^{1,n}, ..., u_h^{M,n}) \in \left(H^1(\Omega)\right)^M$ solves the following independent variational inequalities

$$\begin{cases} a^i(u_h^{i,n+1}, v - u_h^{i,n+1}) \geqq (f^i, v - u_h^{i,n+1}) \forall v \in H^1(\Omega) \\ u_h^{i,n+1} \leq k + u_h^{i+1,n}, \ \ v \leq k + u_h^{i+1,n} \\ u_h^{M+1,n} = u_h^{1,n} \end{cases}$$

Proposition 5.4. *T_h is concave on $\mathbb{H}^+$ and possesses the following properties*

$$T_h V \ \leq T_h W \quad \forall V \leq W$$

$$T_h W \geq 0 \ \forall \ W \in \mathbb{H}^+$$

$$T_h W \leq U_h^0 \quad \forall W \in \mathbb{H}^+$$

Let

$$\mathbb{C}_h = \{ \ W \in \ \mathbb{H}^+ \ : 0 \leq \ W \leq U_h^0 \ \}$$

Lemma 5.2. *Let $0 < \lambda < \min(k/ \left\| U_h^0 \right\|_\infty ; 1)$. Then $T_h(0) \geqq \lambda U_h^0$.*

Proposition 5.5. *Let $\gamma \in [0; 1]$ such that*

$$W - \tilde{W} \leq \gamma W \qquad \forall \ W, \ \tilde{W} \ \in \mathbb{C}_h$$

Then we have

$$T_h W - T_h \tilde{W} \leq \ \gamma(1 - \lambda) T_h W$$

Then similarly to Theorem 5.1, we have the following convergence result.

Proposition 5.6. *The sequences (U_h^n) converges decreasingly to the unique solution of system (15).*

Remark 5.2. In view of the above proposition, it is easy to see that the solution of system (15) is a fixed point of T_h i.e.,

$$U_h = T_h U_h \tag{41}$$

Theorem 5.3. *There exists a positive constant* $0 < \mu < 1$ *such that:*

$$\left\|U_h^n - U_h\right\|_\infty \leq \mu^n \left\|U_h^0\right\|_\infty \tag{42}$$

6. The Finite Element Error Analysis

This section is devoted to derive error estimates in the L^∞- norm for both the system of QVIs and HJB equations. For that we shall develop three different methods: The Subsolutiom Method, The Algorithmic Method, and the Fixed Point Method (for the noncoercive problem).

6.1. *The Subsolution Method*

This method consists of constructing a continuous subsolution denoted

$$\beta^{(h)} = (\beta^{1,(h)}, ..., \beta^{M,(h)})$$

such that:

$$\beta^{i,(h)} \leq u^i \quad \text{and} \quad \left\|\beta^{i,(h)} - u_h^i\right\|_{L^\infty(\Omega)} \leq Ch^2 \left|\log h\right|^2, \ \forall i = 1, 2, ..., M$$

and a discrete subsolution $\alpha_h = (\alpha_h^1, ..., \alpha_h^M)$ such that:

$$\alpha_h^i \leq u_h^i \ \text{ and } \ \left\|\alpha_h^i - u^i\right\|_{L^\infty(\Omega)} \leq Ch^2 \left|\log h\right|^2, \ \ \forall i = 1, 2, ..., M$$

The discrete regularity plays a crucial role as it permits to replace the irregular obstacles "$k + u_h^{i+1}$" with $W^{2,p}(\Omega)$ regular ones, and hence preserves the optimal convergence order.

Let us begin with the discrete subsolution, the continuous one will follow the same way but is some how more delicate as it requires the smoothing of the obstacles.

6.1.1. *Construction of a discrete subsolution*

From now on C will denote a constant independent of both h and k. Let us now introduce the following discrete VI:

$$\begin{cases} a^i(\bar{u}_h^i, v - \bar{u}_h^i) \geqq (f^i, v - \bar{u}_h^i) \ \forall v \in \mathbb{V}_h \\ \bar{u}_h^i \leq k + u^{i+1}, \ \ v \leq k + u^{i+1} \\ u^{M+1} = u^1 \end{cases} \tag{43}$$

where u^i is the solution of (12), $i = 1, 2, ..., M$. So, $\bar{u}_h^i = \sigma_h(f^i, k + u^{i+1})$ is nothing but the approximation of $u^i = \sigma(f^i, k + u^{i+1})$ as a solution of an elliptic (VI). Therefore, thanks to Theorem 2.6, we have

$$\left\|u^i - \bar{u}_h^i\right\|_{L^\infty(\Omega)} \leq Ch^2 \left|\log h\right|^2 \tag{44}$$

Theorem 6.1. *There exists a vector function $\alpha_h = (\alpha_h^1, ..., \alpha_h^M)$ such that*

$$\alpha_h^i \leq u_h^i \quad \textit{and} \quad \left\|\alpha_h^i - u^i\right\|_{L^\infty(\Omega)} \leq Ch^2 \left|\log h\right|^2$$

Proof. Indeed, $\bar{u}_h^i$ being solution to the discrete VI (43), it is also a subsolution, that is,

$$\begin{cases} a^i(\bar{u}_h^i, \varphi_s) \leq (f^i, \varphi_s) \ \forall \varphi_s, \ s = 1, 2, ..., m(h) \\ \bar{u}_h^i \leq k + u^{i+1} \\ u_h^{M+1} = u^1 \end{cases}$$

Then

$$\begin{cases} a^i(\bar{u}_h^i, \varphi_s) \leq (f^i, \varphi_s) \ \forall \varphi_s, \ s = 1, 2, ..., m(h) \\ \bar{u}_h^i \leq k + \left\|u^{i+1} - \bar{u}_h^{i+1}\right\|_{L^\infty(\Omega)} + \bar{u}_h^{i+1} \\ \bar{u}_h^{M+1} = \bar{u}_h^1 \end{cases}$$

that is, $(\bar{u}_h^1, ..., \bar{u}_h^M)$ is a subsolution for the system of QVIs (15) with right-hand side $(f^1, ..., f^M)$ and parameter $\tilde{k} = k + \left\|u^{i+1} - \bar{u}_h^{i+1}\right\|_{L^\infty(\Omega)}$.

Let $\bar{U}_h = (\bar{U}_h^1, ..., \bar{U}_h^M)$ be the solution of such a system and $\bar{U}_h^i = \sigma_h(f^i, \tilde{k})$. Then, making use of theorem 4.2, we have

$$\left\|u_h^i - \bar{U}_h^i\right\|_{L^\infty(\Omega)} \leq \left|k - \tilde{k}\right| \leq \left\|u^{i+1} - \bar{u}_h^{i+1}\right\|_{L^\infty(\Omega)}$$

and, due to theorem 4.5,

$$\bar{u}_h^i \leq \bar{U}_h^i \leq u_h^i + \left\|u^{i+1} - \bar{u}_h^{i+1}\right\|_{L^\infty(\Omega)}$$

Now putting

$$\alpha_h^i = \bar{u}_h^i - \left\|u^{i+1} - \bar{u}_h^{i+1}\right\|_{L^\infty(\Omega)}$$

we clearly have

$$\alpha_h^i \leq u_h^i$$

and using (44), we get

$$\left\|\alpha_h^i - u^i\right\|_{L^\infty(\Omega)} \leq \left\|\bar{u}_h^i - u^i\right\|_{L^\infty(\Omega)} + \left\|u^{i+1} - \bar{u}_h^{i+1}\right\|_{L^\infty(\Omega)} \leq Ch^2 \left|\log h\right|^2 \square$$

6.1.2. *Construction of a continuous subsolution*

Let $\bar{u}^i = \sigma(f^i, k + u^{i+1,(h)})$ be the solution of the following continuous VI:

$$\begin{cases} a^i(\bar{u}^i, v - \bar{u}^i) \geqq (f^i, v - \bar{u}^i) \ \ \forall v \in H^1(\Omega) \\ \bar{u}^i \leq k + u^{i+1,(h)}, \ \ v \leq k + u^{i+1(h)} \\ u^{M+1,(h)} = u^{1,(h)} \end{cases} \tag{45}$$

where $u^{i,(h)}$ is the solution of equation (18).Thanks to (19), the obstacles appearing in (45) are in $W^{2,p}(\Omega)$,which will enable us to conserve the optimal convergence order for the above VI as it is shown in the following lemma.

Lemma 6.1.

$$\left\|\bar{u}^i - u_{ih}\right\|_{L^\infty(\Omega)} \leq Ch^2 \left|\log h\right|^2 \tag{46}$$

Proof.

Denote by $\bar{\omega}_{ih} = \sigma_h(f^i, k+u^{i+1,(h)})$ the approximation of $\bar{u}^i = \sigma(f^i, k+ u^{i+1,(h)})$.Then, using (5), we have

$$\left\|\bar{u}^i - \bar{\omega}_{ih}\right\|_{L^\infty(\Omega)} \leq Ch^2 \left|\log h\right|^2$$

On the other hand, combining this with Lipschitz continuity of VI with respect to the obstacle, we get

$$\left\|\bar{\omega}_{ih} - u_h^i\right\|_{L^\infty(\Omega)} \leq \left\|k + \ u^{i+1,(h)} - (k + u_h^{i+1})\right\|_{L^\infty(\Omega)} \leq Ch^2 \left|\log h\right|^2$$

Hence

$$\left\|\bar{u}^i - u_h^i\right\|_{L^\infty(\Omega)} \leq \left\|\bar{u}^i - \bar{\omega}_{ih}\right\|_{L^\infty(\Omega)} + \left\|\bar{\omega}_{ih} - u_h^i\right\|_{L^\infty(\Omega)} \leq Ch^2 \left|\log h\right|^2 \ \square$$

Theorem 6.2. *There exists a vector function* $\beta^{(h)} = (\beta^{1,(h)}, ..., \beta^{M,(h)})$ *such that:*

$$\beta^{i,(h)} \leq u^i \quad \textit{and} \quad \left\|\beta^{i,(h)} - u_h^i\right\|_{L^\infty(\Omega)} \leq Ch^2 \left|\log h\right|^2, \ i = 1, 2, ..., M$$

Proof. Indeed, $\bar{u}^i$ being the solution of the VI (45), it is also a subsolution for the same VI, that is

$$\begin{cases} a^i(\bar{u}^i, v) \leq (f^i, v) \ \ \forall v \in H^1(\Omega), \ v \geq 0 \\ \bar{u}^i \leq k + u^{i+1,(h)} \\ u^{M+1,(h)} = u^{1,(h)} \end{cases}$$

Then

$$\begin{cases} a^i(\bar{u}^i, v) \leq (f^i, v) \ \forall v \in H^1(\Omega), \ v \geq 0 \\ \bar{u}^i \leq k + \left\| u^{i+1,(h)} - \bar{u}^{i+1} \right\|_{L^\infty(\Omega)} + \bar{u}^{i+1} \\ \overline{u}^{M+1} = \overline{u}^1 \end{cases}$$

that is, $(\bar{u}^1, ..., \bar{u}^M)$ is a subsolution for the system of QVIs with right-hand side $(f^1, ..., f^M)$ and parameter $\tilde{k} = k + \left\| u^{i+1,(h)} - \bar{u}^{i+1} \right\|_{L^\infty(\Omega)}$.

Let $\bar{U} = (\bar{U}^1, ..., \bar{U}^M)$ be the solution of such a system. Then, we have

$$\bar{u}^i \leq \bar{U}^i = \sigma(f^i, \tilde{k}) \ \forall i = 1, 2, ..., M$$

On the other hand, due to Theorem 3.2, we have

$$\begin{aligned} \left\| u^i - \bar{U}^i \right\|_\infty &\leq \left| k - \tilde{k} \right| \\ &\leq \left\| u^{i+1,(h)} - \bar{u}^{i+1} \right\|_{L^\infty(\Omega)} \\ &\leq \left\| u^{i+1,(h)} - u_h^{i+1} \right\|_{L^\infty(\Omega)} + \left\| u_h^{i+1} - \bar{u}^{i+1} \right\|_{L^\infty(\Omega)} \end{aligned}$$

So, making use of Theorem 3.3, we get

$$\bar{u}^i \leq \bar{U}^i \leq u^i + \left\| u^{i+1,(h)} - u_h^{i+1} \right\|_{L^\infty(\Omega)} + \left\| u_h^{i+1} - \bar{u}^{i+1} \right\|_{L^\infty(\Omega)}$$

and putting

$$\beta^{i,(h)} = \bar{u}^i - \left\| u^{i+1,(h)} - u_h^{i+1} \right\|_{L^\infty(\Omega)} + \left\| u_h^{i+1} - \bar{u}^{i+1} \right\|_{L^\infty(\Omega)}$$

we clearly have

$$\beta^{i,(h)} \leq u^i$$

Finally, using to (20) and (46), we obtain

$$\begin{aligned} \left\| \beta^{i,(h)} - u_h^i \right\|_{L^\infty(\Omega)} &\leq \left\| \bar{u}^i - u_h^i \right\|_{L^\infty(\Omega)} + \left\| u^{i+1,(h)} - u_h^{i+1} \right\|_{L^\infty(\Omega)} \\ &\quad + \left\| u_h^{i+1} - \bar{u}^{i+1} \right\|_{L^\infty(\Omega)} \\ &\leq C h^2 \left| \log h \right|^2 \end{aligned}$$

□

6.1.3. ***Optimal L^∞ error estimate***

Now, combining Theorems 6.1 and 6.2, we are in position to derive the optimal L^∞ error estimate for both the system (12) and the HJB equation (1).

Theorem 6.3.

$$\|U - U_h\|_\infty \leq Ch^2 |\log h|^2$$

Proof. Indeed, making use of both Theorems 6.1 and 6.2, we have

$$u_h^i \leq \beta^{(h)} + Ch^2 |\log h|^2$$

$$\leq u^i + Ch^2 |\log h|^2$$

$$\leq \alpha_h^i + Ch^2 |\log h|^2$$

Thus

$$\left\|u^i - u_h^i\right\|_{L^\infty(\Omega)} \leq Ch^2 |\log h|^2, \ \forall i = 1, 2, ..$$

which completes the proof. □

Theorem 6.4.

$$\|u - u_h\|_{L^\infty(\Omega)} \leq Ch^2 |\log h|^2$$

Proof. Indeed,

$$\|u - u_h\|_{L^\infty(\Omega)} \leq \left\|u - u^i\right\|_{L^\infty(\Omega)} + \left\|u^i - u_h^i\right\|_{L^\infty(\Omega)} + \left\|u_h^i - u_h\right\|_{L^\infty(\Omega)}$$

$$\leq \left\|u - u^i\right\|_{L^\infty(\Omega)} + \left\|u^i - u_h^i\right\|_{L^\infty(\Omega)} + Ch^2 |\log h|^2$$

so due to Theorems 3.1 and 4.1, we obtain

$$\|u - u_h\|_\infty \leq \lim_{k \to 0} \left\|u - u^i\right\|_{L^\infty(\Omega)} + \lim_{k \to 0} \left\|u^i - u_h^i\right\|_{L^\infty(\Omega)} + Ch^2 |\log h|^2$$

$$\leq Ch^2 |\log h|^2$$

which is the desired result. □

6.2. *The Algorithmic Approach*

The algorithmic approach rests on the geometrical convergence of both the continuous and discrete iterative scheme as well a fundamental lemma which consists of estimating in the L^∞ norm the error between the continuous iterate and the discrete iterate of the iterative scheme.

Let $U^n = (u^{1,n}, ..., u^{M,n})$ be the n-th iterate defined by (25). We introduce a sequence which consists of finding $\tilde{U}_h^n = (\tilde{u}_h^{1,n}, ..., \tilde{u}_h^{M,n})$ such that $\forall n \geq 1\,, \tilde{u}_h^{i,n}$ is the unique solution of the following VI:

$$\begin{cases} a^i(\tilde{u}^{i,n}, v - \tilde{u}^{i,n}) \geqq (f^i, v - \tilde{u}^{i,n}) \; \forall v \in H^1(\Omega) \\ \tilde{u}^{i,n} \leq k + u^{i+1,n-1} \; ; \; v \leq k + u^{i+1,n-1} \\ u^{M+1,n} = u^{1,n} \text{ and } u_h^{i,0} = u_h^{i,0} \end{cases} \tag{47}$$

where $\tilde{U}_h^0 = U_h^0$. So, $\tilde{U}_h^n = T_h U^n$ and, due to standard uniform estimates for linear equation, we have

$$\left\| U^0 - \tilde{U}_h^0 \right\|_\infty = \left\| U^0 - U_h^0 \right\|_\infty \leq Ch^2 \left| \log h \right| \tag{48}$$

Note also that $\forall i = 1, 2, ..., M$, $\tilde{u}_h^{i,n}$ is the finite element approximation of $\hat{u}^{i,n}$. So, making use of estimate (5), the following error estimate holds:

$$\left\| U^n - \tilde{U}_h^n \right\|_\infty \leq Ch^2 \left| \log h \right|^2 \tag{49}$$

Lemma 6.2. *T_h is Lipschitz continuous on $\mathbb{H}^+$.*

Proof.

Let $V = (v^1, ..., v^M)$; $W = (w^1, ..., w^M)$ in $\mathbb{H}^+$. Then

$$T_h V = [\sigma_h(k + v^2), ..., \sigma_h(k + v^i), ..., \sigma_h(k + v^1)]$$

$$T_h W = [\sigma_h(k + w^2), ..., \sigma_h(k + w^i), ..., \sigma_h(k + w^1)]$$

and

$$\left\| T_h V - T_h W \right\|_\infty = \max_{1 \leq i \leq M} \left\| (T_h V)^i - (T_h W)^i \right\|_{L^\infty(\Omega)}$$

$$= \max_{1 \leq i \leq M} \left\| \sigma_h(k + v^i) - \sigma_h(k + w^i) \right\|_{L^\infty(\Omega)}$$

where $(T_h V)^i$ and $(T_h W)^i$ denote the i-th components of the vectors V

and W, respectively. So, since the mapping σ is Lipschitz continuous, it follows that

$$\left\|\sigma_h(k+v^i)-\sigma_h(k+w^i)\right\|_{L^\infty(\Omega)} \leq \left\|(k+v^i)-(k+w^i)\right\|_{L^\infty(\Omega)}$$

$$\leq \left\|v^i - w^i\right\|_{L^\infty(\Omega)}$$

Thus

$$\|T_hV - T_hW\|_\infty \leq \max_{1\leq i\leq M} \left\|v^i - w^i\right\|_{L^\infty(\Omega)} = \|V-W\|_\infty \qquad \square$$

Lemma 6.3

$$\|U^n - U_h^n\|_\infty \leq \sum_{p=0}^{N} \left\|U^p - \tilde{U}_h^p\right\|_\infty$$

Proof.

The proof will be carried out by induction .

Step1. We know that $U_h^1 = T_hU_h^0$; $\tilde{U}_h^1 = T_hU^0$.Then, using Lipschitz continuity of T_h, we get

$$\left\|U^1 - U_h^1\right\|_\infty \leq \left\|U^1 - \tilde{U}_h^1\right\|_\infty + \left\|\tilde{U}^1 - U_h^1\right\|_\infty$$

$$\left\|U^1 - \tilde{U}_h^1\right\|_\infty + \left\|T_hU^0 - T_hU_h^0\right\|_\infty$$

$$\left\|U^1 - \tilde{U}_h^1\right\|_\infty + \left\|U^0 - U_h^0\right\|_\infty$$

$$\leq \sum_{p=0}^{1} \left\|U^p - \tilde{U}_h^p\right\|_\infty$$

step n. Assume that

$$\left\|U^{n-1} - U_h^{n-1}\right\|_\infty \leq \sum_{p=0}^{N-1} \|U^p - U_h^p\|_\infty$$

Then, since

$$\tilde{U}_h^n = T_hU^{n-1} \ ; \ U_h^n = T_hU_h^{n-1}$$

applying Lipschitz continuity of T_h again, we obtain

$$\begin{aligned}
\|U^n - U_h^n\|_\infty \le & \left\|U^n - \tilde{U}_h^n\right\|_\infty + \left\|\tilde{U}_h^n - U_h^n\right\|_\infty \\
\le & \left\|U^n - \tilde{U}_h^n\right\|_\infty + \|T_h U^{n-1} - T_h U_h^{n-1}\|_\infty \\
\le & \left\|U^n - \tilde{U}_h^n\right\|_\infty + \|U^{n-1} - U_h^{n-1}\|_\infty \\
\le & \left\|U^n - \tilde{U}_h^n\right\|_\infty + \textstyle\sum_{p=0}^{N-1} \left\|U^p - \tilde{U}_h^p\right\|_\infty \\
\le & \textstyle\sum_{p=0}^{N} \left\|U^p - \tilde{U}_h^p\right\|_\infty
\end{aligned}$$

This completes the proof. □

6.2.1. ***Quasi-optimal L^∞- Error Estimates***

Theorem 6.5.

$$\|U - U_h\|_\infty \le Ch^2 |\log h|^3$$

Proof.

Making use of Theorems 5.2, 5.3 and lemma 6.3, we get

$$\begin{aligned}
\|U - U_h\|_\infty &\le \|U - U^n\|_\infty + \|U^n - U_h^n\|_\infty + \|U_h^n - U_h\|_\infty \\
&\le \mu^n \|U^0\|_\infty + \mu^n \|U_h^0\|_\infty + \|U^0 - U_h^0\|_\infty + \textstyle\sum_{p=1}^{N} \left\|U^p - \tilde{U}_h^p\right\|_\infty \\
&\le \mu^n \|U^0\|_\infty + \mu^n \|U_h^0\|_\infty + Ch^2 |\log h| + n.Ch^2 |\log h|^2
\end{aligned}$$

Finally taking $\mu^n = h^2$, the desired result follows. □

6.2.2. ***Quasi-optimal L^∞- Error Estimates***

Theorem 6.6.

$$\|u - u_h\|_{L^\infty(\Omega)} \le Ch^2 |\log h|^3$$

Proof. It is exactly the same as that of Theorem 6.4. □

7. The Noncoercive Case

7.1. *The continuous problem*

Let $\mathcal{F}^i(u) = f^i + \lambda u$ and $\mathcal{B}^i = \mathcal{A}^i + \lambda\mathcal{I}$ ($\mathcal{I}$ is the identity operator). One can observe that the noncoercive HJB equation can be solved by considering the following equivalent formulation:

$$\begin{cases} \max\limits_{1\leq i\leq M}(\mathcal{B}^i u - \mathcal{F}^i(u)) = 0 & \text{in} \quad \Omega \\ u = 0 & \text{on} \quad \Gamma \end{cases} \tag{50}$$

where λ is positive number large enough such that the operators $\mathcal{B}^i$ are strongly coercive.

7.1.1. ***Existence and uniqueness***

This can be achieved by characterizing the solution of HJB equation (1) as the unique fixed point of a contraction. Indeed, denoting by $\mathcal{F}^i(w) = f^i + \lambda w$, we introduce the mapping

$$\begin{aligned} \mathbb{S} : & L^\infty(\Omega) \to L^\infty(\Omega) \\ & w \to \mathbb{S}w = \zeta \end{aligned} \tag{51}$$

where ζ is the unique solution of the coercive HJB equations

$$\begin{cases} \max\limits_{1\leq i\leq M}(\mathcal{B}^i \zeta - \mathcal{F}^i(w)) = 0 & \text{in} \quad \Omega \\ \xi = 0 & \text{on} \quad \Gamma \end{cases} \tag{52}$$

Note that the $\mathcal{F}^i(w)'$s play the role of the f^i's in (12). Then, (52) can be approximated by the following system of QVIs: find $(\zeta^1, ..., \zeta^M)$ solution to

$$\begin{cases} b^i(\zeta^i, v - \zeta^i) \geq (\mathcal{F}^i(w), v - \zeta^i) \forall v \in H^1(\Omega) \\ \zeta^i \leq k + \zeta^{i+1}, \;\; v \leq k + \zeta^{i+1}\,, \; i = 1, ..., M \\ \zeta^{M+1} = \xi^1 \end{cases} \tag{53}$$

where

$$b^i(u, v) = a^i(u, v) + \lambda(u, v)$$

So, thanks to Theorem 3.1, (52) has a unique solution and we have

$$\lim_{k\to 0} \left\| \zeta^i - \zeta \right\|_{L^\infty(\Omega)} = 0, \quad \forall i = 1, 2, ...M$$

Lemma 7.1. *Let* w,$\tilde{w}$ *in* $L^\infty(\Omega)$, $(\zeta^1, ..., \zeta^M)$ *and* $(\tilde{\zeta}^1, ..., \tilde{\zeta}^M)$ *be the corresponding solutions to system (53) with right-hand sides* $\mathcal{F}^i(w) = f^i + \lambda w$ *and* $\mathcal{F}^i(\widetilde{w}) = f^i + \lambda \widetilde{w}$, *respectively. Then we have*

$$\max_{1 \leq i \leq M} \left\| \zeta^i - \tilde{\zeta}^i \right\|_{L^\infty(\Omega)} \leq \lambda / (\lambda + \beta) \left\| w - \widetilde{w} \right\|_{L^\infty(\Omega)}$$

Theorem 7.1. *Under conditions of lemma 7.1, the mapping* $\mathbb{S}$ *is a contraction.*

Proof. Indeed, Let $\zeta = \mathbb{S}w$ and $\tilde{\zeta} = \mathbb{S}\widetilde{w}$ be solutions to HJB equation (52) with right-hand sides $\mathcal{F}^i(w) = f^i + \lambda w$ and $\mathcal{F}^i(\widetilde{w}) = f^i + \lambda \widetilde{w}$, respectively. Then

$$\left\| \mathbb{S}w - \mathbb{S}\widetilde{w} \right\|_\infty = \left\| \zeta - \tilde{\zeta} \right\|_\infty$$

$$\leq \left\| \zeta - \zeta^i \right\|_{L^\infty(\Omega)} + \left\| \zeta^i - \tilde{\zeta}^i \right\|_\infty + \left\| \tilde{\zeta}^i - \tilde{\zeta} \right\|_\infty$$

$$\leq \left\| \zeta - \zeta^i \right\|_{L^\infty(\Omega)} + \max_{1 \leq i \leq M} \left\| \zeta^i - \tilde{\zeta}^i \right\|_{L^\infty(\Omega)} + \left\| \tilde{\zeta}^i - \tilde{\zeta} \right\|_{L^\infty(\Omega)}$$

$$\leq \lim_{k\to 0} \left\| \zeta - \zeta^i \right\|_{L^\infty(\Omega)} + \max_{1 \leq i \leq M} \left\| \zeta^i - \tilde{\zeta}^i \right\|_{L^\infty(\Omega)} + \lim_{k\to 0} \left\| \tilde{\zeta}^i - \tilde{\zeta} \right\|_{L^\infty(\Omega)}$$

$$\leq \lambda / (\lambda + \beta) \left\| w - \widetilde{w} \right\|_{L^\infty(\Omega)}$$

Thus, $\mathbb{S}$ is a contraction, and therefore, the solution of HJB equation (1) is its unique fixed point. □

7.2. *The discrete problem*

As in the continuous case, we shall handle the discrete noncoercive problem by transforming (14) into:

$$\max_{1 \leq i \leq M} (\mathbb{B}^i u_h - \mathcal{G}^i(u_h) = 0 \tag{54}$$

where

$$(\mathcal{G}^i(u_h))_l = (f^i + \lambda u_h, \varphi_l), \; l = 1, ..., m(h), \; 1 \leq i \leq M,$$

and $\mathbb{B}^i$ are the matrices by

$$(\mathbb{B}^i)_{ls} = b^i(\varphi_l, \varphi_s), \; l = 1, ..., m(h), 1 \leq i \leq M \tag{55}$$

7.2.1. ***Existence and uniqueness***

As in the continuous case, this can be achieved by characterizing the solution of HJB equation (14) as the unique fixed point of a contraction. Indeed, let us introduce the mapping

$$\begin{aligned} \mathbb{S}_h : \; & L^\infty(\Omega) \to \mathbb{V}_h \\ & w \to \mathbb{S}_h w = \zeta_h \end{aligned} \tag{56}$$

where ζ_h is the unique solution of the coercive HJB equations

$$\max_{1 \leq i \leq M} (\mathbb{B}^i \zeta_h - \mathcal{G}^i(w) = 0$$

which, thanks to Theorem 4.1, can be approximated by the following system of QVIs : find $(\zeta_h^1, ..., \zeta_h^M)$ solution to

$$\begin{cases} b^i(\zeta_h^i, v - \zeta_h^i) \geqq (\mathcal{F}^i(w), v - \zeta_h^i) \; \forall v \in \mathbb{V}_h \\ \zeta_h^i \leq k + \zeta_h^{i+1}, \quad v \leq k + \zeta_h^{i+1}, \; i = 1, ..., M \\ \zeta_h^{iM+1} = \zeta_h^1 \end{cases} \tag{57}$$

and

$$\lim_{k \to 0} \left\| \zeta_h - \zeta_h^i \right\|_{L^\infty(\Omega)} = 0$$

Lemma 7.2. *Let the dmp hold. Then, we have*

$$\max_{1 \leq \; i \; \leq \; M} \left\| \xi_h^i - \tilde{\xi}_h^i \right\|_{L^\infty(\Omega)} \leq \lambda/(\lambda + \beta) \left\| w - \tilde{w} \right\|_{L^\infty(\Omega)} \quad \forall w, \tilde{w} \in L^\infty(\Omega)$$

Proof. Exactly the same as that of lemma 7.1. □

Theorem 7.2. *Under conditions of lemma 7.1, the mapping $\mathbb{S}_h$ is a contraction.*

Proof. Exactly the same as that of Theorem 7.1. □

7.3. *Optimal L^∞- Error estimate*

Next, we will show that the fixed point approach developped above leads to an L^∞ optimal convergence of the approximation.

Let us first introduce the following coercive discrete HJB equation

$$\max_{1\leq i\leq M}(\mathbb{B}^i\bar{\zeta}_h - \mathcal{G}^i(u)) = 0 \tag{58}$$

where u is the solution of (1). So, in view of (56), we clearly have

$$\bar{\zeta}_h = \mathbb{S}_h u \tag{59}$$

Therefore, as problem (58) is the discrete counterpart of problem (50), making use of Theorem 6.4, we have the following error estimate.

$$\| \bar{\zeta}_h - u \|_{L^\infty(\Omega)} \leq Ch^2 |\log h|^2 \tag{60}$$

Theorem 7.3. *Let u and u_h be the solutions of HJB equations (1) and (14), respectively. Then*

$$\| u - u_h \|_{L^\infty(\Omega)} \leq Ch^2 |\log h|^2$$

where C is a constant independent of h.

Proof.

Since $\bar{\zeta}_h = \mathbb{S}_h u$ and $u_h = \mathbb{S}_h u_h$ making use of Theorems 7.1, 7.2, and estimate (60), we obtain

$$\begin{aligned}\| u - u_h \|_{L^\infty(\Omega)} &\leq \| u - \bar{\zeta}_h \|_{L^\infty(\Omega)} + \| \bar{\zeta}_h - u_h \|_{L^\infty(\Omega)} \\ &\leq \| u - \bar{\zeta}_h \|_{L^\infty(\Omega)} + \| \mathbb{S}_h u - \mathbb{S}_h u_h \|_{L^\infty(\Omega)} \\ &\leq Ch^2 |\log h|^2 + \frac{\lambda}{\lambda+\beta} \| u - u_h \|_{L^\infty(\Omega)}\end{aligned}$$

Thus,

$$\| u - u_h \|_{L^\infty(\Omega)} \leq \frac{Ch^2 |\log h|^2}{\lambda/(\lambda+\beta)}$$

□

8. Open Problems

We would like to conclude this article with some interesting open questions on (1):

- The finite element approximation with zero order terms $a_0^i(x)$ equal to zero.
- Error estimates in L^p norms, $2 \leq p < \infty$.
- Parallel algorithms for the parabolic HJB equation.
- The finite element approximation of the Parabolic HJB equation

$$\begin{cases} \dfrac{\partial u}{\partial t} + \max\limits_{1 \leq i \leq M} (\mathcal{A}^i u - f^i) = 0 \text{ a.e. in } \Omega \\ u(0,x) = u_0(x) \qquad \text{in } \Omega \\ u(t,x) = 0 \qquad \text{on } \partial\Omega \ \forall t \in [0,\ T] \end{cases}$$

References

1. W.H. Fleming, R. Rishel, Deterministic and stochastic optimal control. Springer, Berlin (1975)
2. L.C.Evans and A.Friedman, Optimal stochastic switching and the Dirichlet Problem for the Bellman. equations, Transactions of the American Mathematical Society 253, 365-389 (1979)
3. H. Brezis, L.C. Evans, A variational approach to the Bellman-Dirichlet equation for two elliptic operators. Arch. Rational Mech. Anal., 71, 1-14 (1979).
4. P.L. Lions and B. Mercier, Approximation numérique des equations de Hamilton Jacobi Bellman. RAIRO, Anal.Num. 14, 369-393 (1980)
5. P.L. Lions, Resolution Analytique des problemes de Bellman-Dirichlet. Acta Mathematica 146, 151-166 (1981)
6. P.L. Lions and J.L Menaldi, Optimal control of stochastic integrals and Hamilton Jacobi Bellman equations (part I). SIAM control and optimization 20, 58-81(1982).
7. P.L. Lions, N.S. Trudinger, Linear oblique derivative problems for the uniformly elliptic Hamilton-Jacobi-Bellman equation. Math. Z. 191, 1-15 (1986)
8. L.C.Evans, Classical solutions of the Hamilton-Jacobi-Bellman equation for uniformly elliptic operators. Transactions of the American Mathematical Society. 275, 245-255 (1983)
9. P. Cortey Dumont, Sur l' analyse numerique des equations de Hamilton-Jacobi-Bellman. Math. Meth in Appl. Sci 9, 198-209 (1987).
10. P. Cortey Dumont, Analyse numerique de problemes a frontiere libre, Part I, These d' Etat, Universite de Paris VI, (1985).
11. P. Cortey Dumont, Contribution a l' approximation des inequations variationnelles en norme L^∞. C.R.Acad. Sci. Paris Ser. I Math. 296, 17, 753-756. (1983)

12. P.Cortey-Dumont, On the finite element approximation in the L^∞ norm of variational inequalities with nonlinear operators. Num. Math, 47, 45-57(1985).
13. M. Boulbrachene, M. Haiour, The Finite element approximation of Hamilton-Jacobi-Bellman. Computers and Mathematics with Applications, 41, 993-1007 (2001)
14. M. Boulbrachene, B. Chentouf, The Finite element approximation of Hamilton-Jacobi-Bellman: The noncoercive case. Applied Mathematics and Computation 158, 585-592 (2004).
15. M. Boulbrachene, P. Cortey Dumont, Optimal L^∞ error estimate of a finite element approximation of Hamilton-Jacobi-Bellman equations. Numerical Functional Analysis and Optimization. ....(2009).
16. M. Boulbrachene, A contraction approach for noncoercive Hamilton-Jacobi-Bellman equations. Applied Mathematics and Information Science....(2010)
17. R.H.W Hoppe, *Multigrid Methods for Hamilton-Jacobi-Bellman Equations,* Numer. Math. 49, 239-254 (1986).
18. S. Zhou, Z. Zou, A new iterative method for discrete HJB equations. Numer. Math. 111: 159-167 (2008). Numer. Math.111: 159-167 (2008)
19. S. Zhou, Z. Zou, and G-H. C, Domain decomposition method for a system of quasivariational iequalities. Acta Mathematica Applicatae Sinica, English Series, Vol. 25, no. 1, 75-82 (2009).
20. Zhou, Z. Zou, An iterative algorithm for a quasivariational inequality system related to HJB equation. J
21. P.G. Ciarlet, P.A. Raviart, Maximum principle and uniform convergence for the finite element method. Comp. Meth. in Appl. Mech and Eng. 2, 1-20 (1973).
22. J. Karatson, S. Korotov, Discrete maximum principle for finite element solutions of nonlinear elliptic problems with mixed boundary conditions. Numer. Math. 99, 669-698 (2005).
23. D. Kinderlehrer, G.Stampacchia, *An introduction to Variational Inequalities and their applications.* Academic Press (1980).
24. A.Bensoussan, J.L. Lions, Applications des inequations variationnelles en controle stochastique. Dunod, Paris (1978).
25. A.Bensoussan, J.L. Lions, *Impulse control and quasi-variational inequalities,* Gauthier Villars, Paris, (1984).

DYNAMICS AND CONTROL OF UNDERACTUATED SPACE SYSTEMS

K. D. KUMAR* and GODARD

Department of Aerospace Engineering, Ryerson University, 350 Victoria Street, Toronto, Ontario M5B 2K3, Canada

**E-mail: kdkumar@ryerson.ca*

www.ryerson.ca

Feasibility of achieving precise formation or attitude control of space systems in underactuated configurations is explored. Conditions for robustness against unmatched uncertainties and disturbances are derived to establish the regions of asymptotic stabilization. Nonlinear control laws are designed and their performances are validated via numerical simulations to show that precise formation maintenance or attitude stabilization can be achieved in presence of system nonlinearities, variations in initial conditions, and external disturbances, concurrently.

Keywords: Underactuated systems; satellite formation control; satellite attitude control.

1. Introduction

A space system may experience failures of onboard sensors and/or actuators during its operation.[1–4] The present papers examines two underactuated problems. The first problem deals with the satellite formation flying (SFF)[5] while the second problem is on satellite attitude control. The feasibility of achieving precise formation maintenance and efficient formation maneuvering without the need for thrust in the radial or along-track direction is investigated.[6] Next, we examine 3-axis attitude stabilization of a spacecraft orbiting the Earth under arbitrary single-actuation failures. A time-invariant smooth control law is proposed to accommodate single-axis failure cases where there is no control available on either *roll* or *yaw* axis.

The paper is organized as follows: Section 2 introduces the system model and equations of motion for spacecraft formation while Section 3 presents system model and equations of motion for spacecraft attitude. Control laws for spacecraft formation and spacecraft attitude are described in Section 4 and Section 5, respectively. For a detailed assessment of the proposed control strategy, the results of numerical simulations incorporating different mission scenarios are presented in Section 6 and Section 7. Finally, the conclusions of the present study are stated in Section 8.

2. Spacecraft Formation Flying Model and Equations of Motion

The system comprises of a *leader spacecraft* in circular orbit around the Earth and a *follower spacecraft* moving in a relative trajectory about the leader spacecraft. The spacecraft are modeled as point masses. An *Earth centered inertial* (ECI) frame denoted by $\mathfrak{I} - XYZ$, has its origin located at the center of the Earth, with Z_I-axis passing through the celestial North pole, X_I-axis directed towards the vernal equinox, and Y_I-axis completes the right-handed triad (Fig. 1). The orbital motion of the leader spacecraft is defined by $\vec{r_l} \in \mathbb{R}^3$, $\vec{r_l} \triangleq [r_l \; 0 \; 0]^T$, and true anomaly θ. The motion of the follower spacecraft is described relative to the leader spacecraft using a relative local vertical local horizontal (LVLH) frame $\mathfrak{B} - xyz$ fixed at the center of the leader spacecraft with the x-axis pointing along the local vertical, the z-axis taken along normal to the orbital plane, and the y-axis representing the third axis of the right-handed $S - xyz$ frame. $\vec{\rho} \in \mathbb{R}^3$, $\vec{\rho} \triangleq [x \; y \; z]^T$, defines the relative position vector of the follower spacecraft. The motion along x, y, and z will be referred to as *radial*, *along-track*, and *cross-track* motion, respectively.

2.1. *Equations of Motion*

The equations of motion for the leader spacecraft and the relative equations motion of the follower spacecraft with respect to the leader spacecraft can be written as:

$$\ddot{r}_l - r_l\dot{\theta}^2 + \frac{\mu}{r_l^2} = 0, \quad r_l\ddot{\theta} + 2\dot{\theta}\dot{r}_l = 0 \tag{1}$$

$$m_f\ddot{x} - 2m_f\dot{\theta}\dot{y} - m_f\left(\dot{\theta}^2 x + \ddot{\theta} y\right) + m_f\mu\left(\frac{r_l + x}{r_f^3} - \frac{1}{r_l^2}\right) = u_{fx} + F_{dx} \tag{2}$$

$$m_f\ddot{y} + 2m_f\dot{\theta}\dot{x} + m_f\left(\ddot{\theta} x - \dot{\theta}^2 y\right) + m_f\frac{\mu}{r_f^3}y = u_{fy} + F_{dy} \tag{3}$$

$$m_f\ddot{z} + m_f\frac{\mu}{r_f^3}z = u_{fz} + F_{dz} \tag{4}$$

where $r_f = [(r_l + x)^2 + y^2 + z^2]^{1/2}$ is the position of the follower spacecraft, μ is the Earth's gravitational parameter, θ refers to the true anomaly, F_{dj}

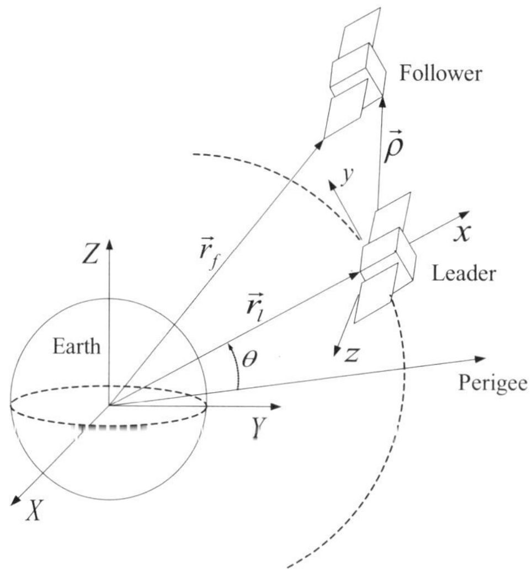

Fig. 1. Geometry of orbit motion of leader and follower spacecraft.

is the net relative perturbations acting on the system, and u_{fj} are the components of the control input vector.

2.2. *Desired Formation Geometry*

Two formation flying designs are considered: (1) circular, and (2) projected circular formations. The desired or commanded states $(x_d, \dot{x}_d, y_d, \dot{y}_d, z_d, \dot{z}_d)$ are taken as the solution of the linearized form of the relative equations of motion when $F_{dj} = 0$, $j = x, y, z$. The following trajectories are considered:

(1) *Circular Formation*: In this formation, the leader and the follower spacecraft maintain a constant separation from each other in three-dimensional space and the formation is mathematically defined as $x^2 + y^2 + z^2 = r_{dc}^2$. The equations of desired trajectory are

$$\begin{Bmatrix} x_d \\ y_d \\ z_d \end{Bmatrix} = \frac{r_{dc}}{2} \begin{bmatrix} \sin(nt + \phi) \\ 2\cos(nt + \phi) \\ \sqrt{3}\sin(nt + \phi) \end{bmatrix} \tag{5}$$

(2) *Projected Circular Formation*: In this formation, the formation is is mathematically defined as $y^2 + z^2 = r_{dpc}^2$. The equations of desired trajectory are

$$\begin{Bmatrix} x_d \\ y_d \\ z_d \end{Bmatrix} = \frac{r_{dpc}}{2} \begin{bmatrix} \sin(nt+\phi) \\ 2\cos(nt+\phi) \\ 2\sin(nt+\phi) \end{bmatrix} \tag{6}$$

where r_{dc} and r_{dpc} are the circular and projected circular formation sizes (radius) respectively, ϕ is the in-plane phase angle between the leader and the follower spacecraft (the initial phase angle is defined, at the time of equator crossing of the leader spacecraft, in the local horizon $y-z$ plane), and n is the mean angular velocity and equals to $\sqrt{\mu/a_c^3}$ (μ_e is the gravitational parameter of the Earth; a_c is the semi-major axis of the leader spacecraft).

2.3. *External Disturbances*

The disturbances in Eqs. (2-4) are time-varying quantities attributed to gravitational field, solar radiation pressure, and third body perturbations. The external disturbance components due to J_2 perturbation are derived as

$$\vec{F}_d = T_{IR}^{-1}[\vec{J}_{2f} - \vec{J}_{2l}] \tag{7}$$

where

$$T_{IR} = \begin{bmatrix} \cos(\Omega_l) & -\sin(\Omega_l) & 0 \\ \sin(\Omega_l) & \cos(\Omega_l) & 0 \\ 0 & 0 & 1 \end{bmatrix} \begin{bmatrix} 1 & 0 & 0 \\ 0 & \cos(i_l) & -\sin(i_l) \\ 0 & \sin(i_l) & \cos(i_l) \end{bmatrix} \begin{bmatrix} \cos(\omega_l+\theta) & -\sin(\omega_l+\theta) & 0 \\ \sin(\omega_l+\theta) & \cos(\omega_l+\theta) & 0 \\ 0 & 0 & 1 \end{bmatrix} \tag{8}$$

$$\vec{J}_{2l} = -\frac{3\mu J_2 R_\oplus{}^2}{2\|\vec{R}_l\|^5} \begin{bmatrix} \{1-\frac{5Z_l{}^2}{\|\vec{R}_L\|^2}\}X_l \\ \{1-\frac{5Z_l{}^2}{\|\vec{R}_L\|^2}\}Y_l \\ \{3-\frac{5Z_l{}^2}{\|\vec{R}_L\|^2}\}Z_l \end{bmatrix} \text{ and } \vec{J}_{2f} = -\frac{3\mu J_2 R_\oplus{}^2}{2\|\vec{R}_f\|^5} \begin{bmatrix} \{1-\frac{5Z_f{}^2}{\|\vec{R}_f\|^2}\}X_f \\ \{1-\frac{5Z_f{}^2}{\|\vec{R}_f\|^2}\}Y_f \\ \{3-\frac{5Z_f{}^2}{\|\vec{R}_f\|^2}\}Z_f \end{bmatrix} \tag{9}$$

where Ω_l, i_l, ω_l, and θ denote the right ascension of the ascending node, orbit inclination, argument of perigee, and true anomaly of the leader spacecraft, respectively. μ is the Earth's gravitational parameter, $R_\oplus$ is the radius of the Earth, and J_2 is second zonal gravitational coefficient, $J_2 = 1.08263 \times 10^{-3}$.

Remark 1: In this study the following assumptions are made, (1) the leader spacecraft remains in an unperturbed elliptical orbit subject to its own controller, (2) all spacecraft in formation have the same ballistic coefficients and area-to-mass ratio and therefore the perturbing accelerations due to aerodynamic drag and solar radiation pressure have negligible effects on the relative motion, and (3) the differential J_2 are perturbing accelerations on the follower spacecraft relative to the leader spacecraft orbit.

3. Spacecraft Attitude Model and Equations of Motion

The system comprises of a *rigid spacecraft* in an elliptical planar trajectory with the Earth's center at one of its foci (Figure 2). We define a local vertical local horizontal (LVLH) orbital reference frame $\mathfrak{L} - x_0\, y_0\, z_0$ with its origin always at the center of mass of the spacecraft (Figure 2). The nodal line represents the reference line in orbit for the measurement of the true anomaly (eccentric orbit) or angle θ (circular orbit). Here the x_0-axis points along the local vertical, the z_0-axis is taken normal to the orbital plane, and the y_0-axis along the orbit direction. The attitude orientation of the body-fixed reference frame, $\mathfrak{B} - x\, y\, z$, relative to the LVLH reference frame, $\mathfrak{L} - x_0\, y_0\, z_0$, is denoted by a set of three successive rotations: α (pitch) about the z-axis, ϕ (roll) about the new y-axis, and finally γ (yaw) about the resulting x-axis.

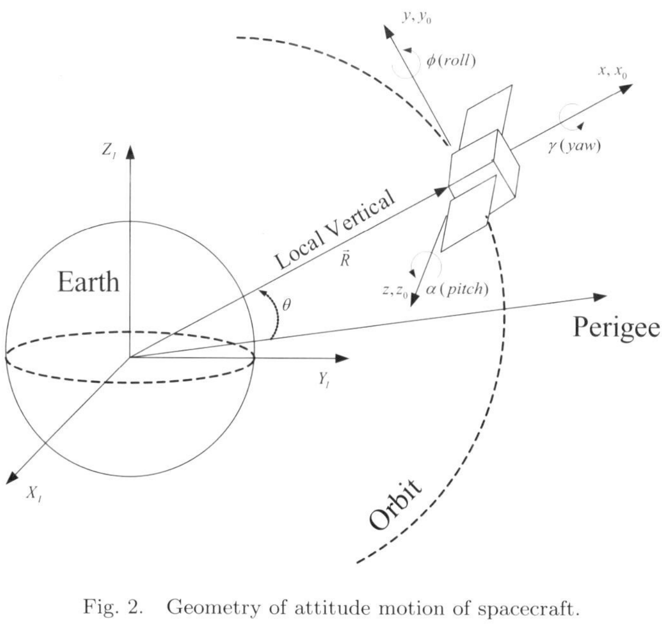

Fig. 2. Geometry of attitude motion of spacecraft.

3.1. *Equations of Motion*

The Lagrangian equations of motion corresponding to the generalized coordinates ($q = [\alpha, \phi, \gamma]^T$) are obtained using the general relation

$$\frac{d}{dt}\left(\frac{\partial T_e}{\partial \dot{q}}\right) - \frac{\partial T_e}{\partial q} + \frac{\partial U_p}{\partial q} = Q_q \tag{10}$$

where Q_q is the generalized force corresponding to the generalized coordinate q while T_e and U_p are the kinetic and potential energies, respectively. The variable of integration t (time) is changed to θ (true anomaly) and the resulting governing nonlinear, coupled ordinary differential equations of motion of the system, after carrying out algebraic operations and nondimensionalization, can be expressed in a general form as follows:

$$q'' = N(q)[F(q, q') + U_{fa} + \tau_e] \tag{11}$$

where $N(q) \in \mathbb{R}^{3\times3}$ and $F(q, q') \in \mathbb{R}^{3\times1}$ are matrices containing nonlinear functions, $q \in \mathbb{R}^3 = [\alpha, \phi, \gamma]^T$, $U_{fa} \in \mathbb{R}^3 = [U_\alpha, U_\phi, U_\gamma]^T$ is the control torque, and $\tau_e \in \mathbb{R}^3 = [\tau_{\alpha,e}, \tau_{\phi,e}, \tau_{\gamma,e}]^T$ represent the external disturbance torques acting on the spacecraft. $(\cdot)'$ and $(\cdot)''$ denote $\mathrm{d}(\cdot)/\mathrm{d}\theta$ and $\mathrm{d}^2(\cdot)/\mathrm{d}\theta^2$, respectively.

We define the following dimensionless parameters:

$$k_1 = \frac{I_z - I_x}{I_y}, \; k_2 = \frac{I_z - I_y}{I_x} \tag{12}$$

$$k_{xz} = \frac{I_x}{I_z} = \frac{1 - k_1}{1 - k_1 k_2}, \; k_{yz} = \frac{I_y}{I_z} = \frac{1 - k_2}{1 - k_1 k_2} \tag{13}$$

The equations of motion derived from the Lagrangian relation are given by

$$\begin{bmatrix} \alpha'' \\ \phi'' \\ \gamma'' \end{bmatrix} = \begin{bmatrix} N_{11} & N_{12} & N_{13} \\ N_{21} & N_{22} & N_{23} \\ N_{31} & N_{32} & N_{33} \end{bmatrix} \left\{ \begin{bmatrix} F_\alpha \\ F_\phi \\ F_\gamma \end{bmatrix} + \begin{bmatrix} U_\alpha \\ U_\phi \\ U_\gamma \end{bmatrix} + \begin{bmatrix} \tau_{\alpha,e} \\ \tau_{\phi,e} \\ \tau_{\gamma,e} \end{bmatrix} \right\} \tag{14}$$

where F_α, F_ϕ, F_γ, and the elements N_{ij} (for $i, j = 1, 2, 3$) are provided in the Appendix for brevity.

3.2. *Underactuated System*

Let the state vector of the system be $X \in \mathbb{R}^{6x1} = [\alpha, \alpha', \phi, \phi', \gamma, \gamma']^T$. The state vector can be split into two parts as $X = [x_1, x_2]^T$ where x_1 and x_2 represents the unactuated and actuated states, respectively. The unactuated states can be further transformed to $x_1 = [x_{10}, x_{11}]^T$, where

$x_{10} \in \mathbb{R}^{3\times 1} = [\alpha, \phi, \gamma]^T$ always. Based on the axis of failure, the nonlinear equation of motion in Eq. (14) can be transformed to

$$\begin{bmatrix} x_{11}' \\ x_2' \end{bmatrix} = \begin{bmatrix} \bar{A}_{11} & \bar{A}_{12} \\ \bar{A}_{21} & \bar{A}_{22} \end{bmatrix} \left\{ \begin{bmatrix} F_1 \\ F_2 \end{bmatrix} + \begin{bmatrix} 0 \\ U_{ua} \end{bmatrix} + \begin{bmatrix} \tau_{1,e} \\ \tau_{2,e} \end{bmatrix} \right\} \tag{15}$$

We now consider the cases of actuation failure to specify state x_{11} as follows:

Case I : $(U_\phi = 0)$ No control authority on *roll*-axis (ϕ) and full control actuation available on *pitch* (α) and *yaw* (γ) axes. For this case $x_{11} = \phi'$, $x_2 = [\alpha', \gamma')]$, and $U_{ua} \in \mathbb{R}^{2\times 1} = [U_\alpha, U_\gamma]^T$. Similarly, $F(q, q') = [F_1, F_2]^T$ where $F_1 = F_\phi$ and $F_2 = [F_\alpha, F_\gamma]^T$. The external disturbance torque components are given by $\tau_{1,e} = \tau_{\phi,e}$ and $\tau_{2,e} = [\tau_{\alpha,e},\ \tau_{\gamma,e}]^T$. The elements of $\bar{A}$ matrix are shown below.

$$\bar{A}_{11} = N_{22};\quad \bar{A}_{12} = \begin{bmatrix} N_{21} & N_{23} \end{bmatrix};\quad \bar{A}_{21} = \bar{A}_{12}^T;\quad \bar{A}_{22} = \begin{bmatrix} N_{11} & N_{13} \\ N_{31} & N_{33} \end{bmatrix} \tag{16}$$

Case II : $(U_\gamma = 0)$ No control authority on *yaw*-axis (γ) and full control actuation available on *pitch* (α) and *roll* (ϕ) axes. For this case $x_{11} = \gamma'$, $x_2 = [\alpha', \phi')]$, and $U_{ua} \in \mathbb{R}^{2\times 1} = [U_\alpha, U_\phi]^T$. Similarly, $F(q, q') = [F_1, F_2]^T$ where $F_1 = F_\gamma$ and $F_2 = [F_\alpha, F_\phi]^T$. The external disturbance torque components are given by $\tau_{1,e} = \tau_{\gamma,e}$ and $\tau_{2,e} = [\tau_{\alpha,e},\ \tau_{\phi,e}]^T$. The elements of $\bar{A}$ matrix are shown below.

$$\bar{A}_{11} = N_{33};\quad \bar{A}_{12} = \begin{bmatrix} N_{31} & N_{32} \end{bmatrix};\quad \bar{A}_{21} = \bar{A}_{12}^T;\quad \bar{A}_{22} = \begin{bmatrix} N_{11} & N_{12} \\ N_{21} & N_{22} \end{bmatrix} \tag{17}$$

4. Design of Control Laws for Spacecraft Formation Flying

A linear system model is derived from the given nonlinear system equations of motion (Eqs. 2-4) assuming the leader spacecraft in a circular orbit as follows:

$$\begin{bmatrix} \dot{x} \\ \dot{y} \\ \dot{z} \\ \ddot{x} \\ \ddot{y} \\ \ddot{z} \end{bmatrix} = \begin{bmatrix} 0 & 0 & 0 & 1 & 0 & 0 \\ 0 & 0 & 0 & 0 & 1 & 0 \\ 0 & 0 & 0 & 0 & 0 & 1 \\ 3n^2 & 0 & 0 & 0 & 2n & 0 \\ 0 & 0 & 0 & -2n & 0 & 0 \\ 0 & 0 & -n^2 & 0 & 0 & 0 \end{bmatrix} \begin{bmatrix} x \\ y \\ z \\ \dot{x} \\ \dot{y} \\ \dot{z} \end{bmatrix} + \frac{1}{m_f} \begin{bmatrix} 0 & 0 & 0 \\ 0 & 0 & 0 \\ 0 & 0 & 0 \\ 1 & 0 & 0 \\ 0 & 1 & 0 \\ 0 & 0 & 1 \end{bmatrix} \begin{bmatrix} u_{fx} \\ u_{fy} \\ u_{fz} \end{bmatrix} \tag{18}$$

In general the above representation is expressed as $\dot{X} = AX + BU_F$, where $X \in \mathbb{R}^{6\times 1}$ is the state vector, $A \in \mathbb{R}^{6\times 6}$, $B \in \mathbb{R}^{6\times 3}$, and $U_F \in \mathbb{R}^3 =$

$[u_{fx}, u_{fy}, u_{fz}]^T$ is the vector of actual control inputs generated by the thrusters. In the case no control force available in the along-track direction ($u_{fy} = 0$), evaluation of the Kalman rank condition for controllability shows that the system is not fully state controllable ($rank = 5 < 6$). Before concluding that the linear approximation is not stabilizable, it is important to determine the eigenvalue associated with the uncontrollable mode. Rearranging terms in Eq. (18) to represent the in-plane dynamics of the SFF system with only radial-axis input gives,

$$\begin{bmatrix} \dot{x} \\ \dot{y} \\ \ddot{y} \\ \ddot{x} \end{bmatrix} = \begin{bmatrix} 0 & 0 & 0 & 1 \\ 0 & 0 & 1 & 0 \\ 0 & 0 & 0 & -2n \\ 3n^2 & 0 & 2n & 0 \end{bmatrix} \begin{bmatrix} x \\ y \\ \dot{y} \\ \dot{x} \end{bmatrix} + \frac{1}{m_f} \begin{bmatrix} 0 \\ 0 \\ 0 \\ 1 \end{bmatrix} u_{fx} \tag{19}$$

Since the pair (A_b, B_b) is not completely state controllable, the nominal system in Eq. (19) can be decomposed into controllable and uncontrollable parts using a transformation matrix, T. The new state vector is given by $Z_b = T\, X_b$ and the open-loop system in the new coordinates has the form,

$$\dot{Z}_b = \bar{A}_b\, Z_b + \bar{B}_b\, u_{fx} \tag{20}$$

where

$$\bar{A}_b = T\, A_b\, T^{-1} = \begin{bmatrix} \bar{A}_{uc} & 0 \\ \bar{A}_{21} & \bar{A}_c \end{bmatrix} \quad \text{and} \quad \bar{B}_b = T\, B_b = \begin{bmatrix} 0 \\ \bar{B}_c \end{bmatrix} \tag{21}$$

The pair $(\bar{A}_c, \bar{B}_c)$ are controllable and all the eigenvalues of $\bar{A}_{uc}$ are uncontrollable. Based on Eq. (21), consider an orthogonal transformation matrix given by

$$T = \begin{bmatrix} 2n & 0 & 1 & 0 \\ 0 & 1 & 0 & 0 \\ 1 & 0 & -2n & 0 \\ 0 & 0 & 0 & -1 \end{bmatrix} \tag{22}$$

Since the rank of $\mathfrak{C}_x = 3 < 4 = n$, the system given by Eq. (19) has 3 controllable modes and 1 uncontrollable mode. Thus Eq. (20) can be written in the form:

$$\begin{bmatrix} \dot{z}_{b1} \\ \cdots \\ \dot{z}_{b2} \\ \dot{z}_{b3} \\ \dot{z}_{b4} \end{bmatrix} = \begin{bmatrix} 0 & \vdots & 0 & 0 & 0 \\ \cdots & \vdots & \cdots & \cdots & \cdots \\ \frac{1}{4n^2+1} & \vdots & 0 & -\frac{2n}{4n^2+1} & 0 \\ 0 & \vdots & 0 & 0 & -4n^2+1 \\ -\frac{2n(1+3n^2)}{4n^2+1} & \vdots & 0 & \frac{n^2}{4n^2+1} & 0 \end{bmatrix} \begin{bmatrix} z_{b1} \\ \cdots \\ z_{b2} \\ z_{b3} \\ z_{b4} \end{bmatrix} + \begin{bmatrix} 0 \\ \cdots \\ 0 \\ 0 \\ -1 \end{bmatrix} u_{fx} \quad (23)$$

where the open-loop eigenvalues of $\bar{A}_b$ are $\{0, 0, \pm j\, n\}$. Based on the definitions provided in Eq. (21), the uncontrollable mode is given by $\dot{z}_{b1} = 0$ and the eigenvalue associated with the uncontrollable mode is 0. Therefore, if the system is formulated with no along-track input, the linearized SFF dynamics possesses one uncontrollable critical mode.

4.1. *Control Objective*

The equations of motion are rewritten in state space form as follows:

$$\begin{bmatrix} \dot{x} \\ \dot{y} \\ \dot{z} \\ \ddot{x} \\ \ddot{y} \\ \ddot{z} \end{bmatrix} = \begin{bmatrix} 0 & 0 & 0 & 1 & 0 & 0 \\ 0 & 0 & 0 & 0 & 1 & 0 \\ 0 & 0 & 0 & 0 & 0 & 1 \\ 3n^2 & 0 & 0 & 0 & 2n & 0 \\ 0 & 0 & 0 & -2n & 0 & 0 \\ 0 & 0 & -n^2 & 0 & 0 & 0 \end{bmatrix} \begin{bmatrix} x \\ y \\ z \\ \dot{x} \\ \dot{y} \\ \dot{z} \end{bmatrix} + \begin{bmatrix} 0 \\ 0 \\ 0 \\ \mu(\frac{1}{r_l^2} - \frac{(r_l+x)}{r_f^3}) - 2n^2 x \\ n^2 y - \frac{\mu y}{r_f^3} \\ n^2 z - \frac{\mu z}{r_f^3} \end{bmatrix} \quad (24)$$

$$+\frac{1}{m_f} \begin{bmatrix} 0 & 0 & 0 \\ 0 & 0 & 0 \\ 0 & 0 & 0 \\ b_1 & 0 & 0 \\ 0 & b_2 & 0 \\ 0 & 0 & b_3 \end{bmatrix} \begin{bmatrix} u_{fx} \\ u_{fy} \\ u_{fz} \end{bmatrix} + \frac{1}{m_f} \begin{bmatrix} 0 & 0 & 0 \\ 0 & 0 & 0 \\ 0 & 0 & 0 \\ 1 & 0 & 0 \\ 0 & 1 & 0 \\ 0 & 0 & 1 \end{bmatrix} \begin{bmatrix} F_{dx} \\ F_{dy} \\ F_{dz} \end{bmatrix}$$

In general form $\dot{X} = AX + E(X) + BU_f + DF_d$, the nonlinear terms in the equation of motion are lumped into $E(X)$, and F_d represent the differential perturbation forces. Now, state vector of the system $X \in \mathbb{R}^{6x1} = [x, y, z, \dot{x}, \dot{y}, \dot{z}]^T$ can be split into two parts as $X = [x_1, x_2]^T$ where x_1 and x_2 represents the unactuated and actuated states, respectively. The unactuated states can be further transformed to $x_1 = [x_{10}, x_{11}]^T$, where $x_{10} \in \mathbb{R}^{3\times 1} = [x, y, z]^T$. Two main cases of actuation failure to determine the state x_{11} is considered as follows:

Case I : ($u_{fx} = 0,\ b_1 = 0,\ b_2, b_3 = 1$) No control force available in the *radial* direction (x) and complete control authority available in the *along-track* (y) and *cross-track* (z) direction. For this case $x_{11} = \dot{x}$, $x_2 = [\dot{y}, \dot{z}]$, and $U \in \mathbb{R}^{2\times 1} = [u_{fy}, u_{fz}]^T$.

Case II : ($u_{fy} = 0,\ b_2 = 0,\ b_1, b_3 = 1$) No control force available in the *along-track* direction (y) and complete control authority available in the *radial* (x) and *cross-track* (z) direction. For this case $x_{11} = \dot{y}$, $x_2 = [\dot{x}, \dot{z}]$, and $U \in \mathbb{R}^{2\times 1} = [u_{fx}, u_{fz}]^T$.

Based on the leader-follower SFF mathematical model developed in the previous section, the relative state vector and the desired relative trajectories are defined as $X(t), X_d(t) \in \mathbb{R}^6$, respectively. The performance measure is defined as the tracking error $e(t) \in \mathbb{R}^6$,

$$e(t) \triangleq X(t) - X_d(t) \tag{25}$$

The controller is derived for circular orbits and therefore, any parameters associated with the elliptic orbit of the leader spacecraft are not considered. For deriving the control laws, the following simplifications to Eqs. (1-4) are made which pertains to the leader in circular orbit, (1) $\ddot{\theta} = 0$, (2) $\dot{\theta} = n = \sqrt{\mu_e/{r_p}^3}$, where r_p is the orbit radius of the leader spacecraft, and (3) r_l is replaced with r_p. The objective is to develop a control algorithm for the SFF mathematical model, Eqs. (2-4), using either radial or along-track thrust combined with cross-track input to drive the relative states of the system to its desired relative trajectories as $t \to \infty$, so that the tracking errors, Eq.(25), converges to zero.

$$\lim_{t\to\infty} X(t) = X_d(t) \tag{26}$$

4.2. *Design of Sliding Manifold*

The sliding surfaces for the SFF are designed for two cases as follows.

4.2.1. *Case I - Complete failure of radial axis thruster*

The linearized HCW model (Eq. 18) in the previous section can be represented in terms of new coordinates $x_1 \in \mathbb{R}^4$ and $x_2 \in \mathbb{R}^2$ as follows

$$\begin{bmatrix} \dot{x}_1 \\ \dot{x}_2 \end{bmatrix} = \begin{bmatrix} A_{11} & A_{12} \\ A_{21} & A_{22} \end{bmatrix} \begin{bmatrix} x_1 \\ x_2 \end{bmatrix} + \begin{bmatrix} 0 \\ B_2 \end{bmatrix} U \tag{27}$$

where x_1 and x_2 are as defined in Case I, $B_2 \in \mathbb{R}^{2\times 2} = I_{2\times 2}$, and the complete forms of A_{ij}, $(i, j = 1, 2)$ are given in the Appendix. Carrying out a similar transformation on the desired trajectory equations (Eqs. 5-6), $X_d = [x_1^d, x_2^d]^T$ gives

$$\begin{bmatrix} \dot{x}_1^d \\ \dot{x}_2^d \end{bmatrix} = \begin{bmatrix} A_{11}^d & A_{12}^d \\ A_{21}^d & A_{22}^d \end{bmatrix} \begin{bmatrix} x_1^d \\ x_2^d \end{bmatrix} \tag{28}$$

Based on Eq. (27) and Eq. (28) the following error dynamics can be formulated:

$$\begin{aligned} \dot{e}_1 &= A_{11}e_1 + A_{12}e_2 + \bar{A}_{11}x_1^d + \bar{A}_{12}x_2^d \\ \dot{e}_2 &= A_{21}e_1 + A_{22}e_2 + B_2U + \bar{A}_{21}x_1^d + \bar{A}_{22}x_2^d \end{aligned} \tag{29}$$

where $\bar{A}_{ij} = A_{ij} + A_{ij}^d$ for i, j = 1, 2, and $e_i = x_i - x_i^d$. By exploiting the coupling between directly actuated and unactuated states, the sliding surface, S, is defined as a function of the tracking errors and desired states.

$$S = \{e_1 \in \mathbb{R}^{4\times1}, e_2 \in \mathbb{R}^{2\times1} : e_2 + K\,e_1 = 0\} \tag{30}$$

where $K \in \mathbb{R}^{2\times4}$ is the weighting matrix. When the system reaches the sliding surface, $S = 0 \;\forall\; t > t_r$, where t_r is the reaching time after which sliding motion starts,

$$e_2 = -K\,e_1 \tag{31}$$

It is important to note that Eq. (31) holds only on the sliding surface and substituting this relation to the reduced order system in Eq. (29) gives

$$\dot{e}_1 = (A_{11} - A_{12}K)e_1 + \bar{A}_{11}x_1^d + \bar{A}_{12}x_2^d \tag{32}$$

Since (A, B) is, by definition a controllable pair it follows directly that the matrix pair (A_{11}, A_{12}) is also controllable. To facilitate the stability analysis, the sliding surface is expressed as:

$$S = \{e \in \mathbb{R}^{6\times1} : \Lambda\,e = 0\} \tag{33}$$

where

$$\Lambda = [K \; I_{2\times2}] = \begin{bmatrix} K_{11} & K_{12} & K_{13} & K_{14} & 1 & 0 \\ K_{21} & K_{22} & K_{23} & K_{24} & 0 & 1 \end{bmatrix} \tag{34}$$

4.2.2. *Case II - Complete failure of along-track thruster*

Based on the error dynamics in Eq. (29) and the complete linear model (Eq. 18), the reduced order system for Case II (excluding the desired trajectory terms - $\bar{A}_{11}x_1^d + \bar{A}_{12}x_2^d$) is given by

$$\begin{bmatrix} \dot{e}_x \\ \dot{e}_y \\ \dot{e}_z \\ \ddot{e}_y \end{bmatrix} = \begin{bmatrix} 0 & 0 & 0 & 0 \\ 0 & 0 & 0 & 1 \\ 0 & 0 & 0 & 0 \\ 0 & 0 & 0 & 0 \end{bmatrix} \begin{bmatrix} e_x \\ e_y \\ e_z \\ \dot{e}_y \end{bmatrix} + \begin{bmatrix} 1 & 0 \\ 0 & 0 \\ 0 & 1 \\ -2n & 0 \end{bmatrix} \begin{bmatrix} \dot{e}_x \\ \dot{e}_z \end{bmatrix} \tag{35}$$

The uncontrollable mode is extracted by representing the reduced order system with a new set of coordinates. In order to facilitate the analysis, linear change of coordinates for Eq. (35) can be obtained using the following transformation matrix

$$T_1 = \begin{bmatrix} 2n & 0 & 0 & 1 \\ 0 & -1 & 0 & 0 \\ 0 & 0 & -1 & 0 \\ -1 & 0 & 0 & 2n \end{bmatrix} \tag{36}$$

By a change of basis using, (1) $z = T_1 e_1$, (2) $\bar{A} = T_1 A_{11} T_1^{-1}$, and (3) $\bar{B} = T_1 A_{12}$, Eq. (35) is transformed into the following lower order system

$$\begin{bmatrix} \dot{z}_1 \\ \dot{z}_2 \end{bmatrix} = \begin{bmatrix} \bar{A}_{uc} & 0 \\ \bar{A}_{21} & \bar{A}_c \end{bmatrix} \begin{bmatrix} z_1 \\ z_2 \end{bmatrix} + \begin{bmatrix} 0 \\ B_c \end{bmatrix} \begin{bmatrix} \dot{e}_x \\ \dot{e}_z \end{bmatrix} \tag{37}$$

where

$$\bar{A}_{uc} = 0; \quad \bar{A}_{21} = \begin{bmatrix} -1 \\ 0 \\ 0 \end{bmatrix}; \quad \bar{A}_c = \begin{bmatrix} 0 & 0 & -2n \\ 0 & 0 & 0 \\ 0 & 0 & 0 \end{bmatrix}; \quad B_c = \begin{bmatrix} 0 & 0 \\ 0 & -1 \\ -1 & 0 \end{bmatrix}$$

with $(\bar{A}_c, B_c)$ a controllable pair. The uncontrollable mode is given by $\dot{z}_1 = 0$ which implies z_1 is a constant. Based on the transformation matrix given by Eq. (36), $z_1 = \dot{e}_y + 2\,n\,e_x$. The objective is to develop a sliding surface that can eliminate the effect of this uncontrollable mode. Hence, using the properties of linear state-space theory, there exists a linear feedback control with the gain matrix $K \in \mathbb{R}^{2\times 3}$ such that $A_c - B_c W$ is Hurwitz.

$$e_2 = -K\,z_2 = -K\,\hat{T}_1 e_1 \tag{38}$$

where $\hat{T}_1 = T_1(2:4,:)$. This result can be utilized to define a sliding surface based on the pair (A_c, B_c) which is stable as opposed to a manifold designed based on (A_{11}, A_{12}). The time-invariant switching function for Case II with a robust component added to alleviate the effect of matched part of the reduced order system and the uncontrollable mode is defined as,

$$S = \{e_1 \in \mathbb{R}^{4\times 1}, e_2 \in \mathbb{R}^{2\times 1} : e_2 + K\,\hat{T}_1 e_1 = 0\} \tag{39}$$

Therefore, asymptotic stabilization of the tracking errors can be guaranteed if the weighting matrix W is appropriately chosen to suppress the influence of the uncontrollable mode. Sufficiently fast error decay when sliding can be ensured by placing the closed-loop eigenvalues of $(A_c - B_c K)$ in the far left-hand half of the complex plane. The error dynamics are represented in the exact same manner as Eq. (29) with $e_1 = [e_x, e_y, e_z, \dot{e}_y]^T$ and $e_2 = [\dot{e}_x, \dot{e}_z]^T$. To facilitate the stability analysis, the sliding surface is expressed as:

$$S = \{e \in \mathbb{R}^{6\times 1} : \Gamma\, e = 0\} \tag{40}$$

where

$$\Gamma = [K\,\hat{T}_1\; I_{2\times 2}] = \begin{bmatrix} -K_{13} & -K_{11} & -K_{12} & 2nK_{13} & 1 & 0 \\ -K_{23} & -K_{21} & -K_{22} & 2nK_{23} & 0 & 1 \end{bmatrix} \tag{41}$$

4.3. *Nonlinear Control Formulation*

The nonlinear equations of motion can be represented in terms of transformed coordinates as follows:

$$\begin{bmatrix} \dot{x}_1 \\ \dot{x}_2 \end{bmatrix} = \begin{bmatrix} A_{11} & A_{12} \\ A_{21} & A_{22} \end{bmatrix} \begin{bmatrix} x_1 \\ x_2 \end{bmatrix} + \begin{bmatrix} E_1(X) \\ E_2(X) \end{bmatrix} + \begin{bmatrix} 0_{4\times 2} \\ I_{2\times 2} \end{bmatrix} U_f + \begin{bmatrix} F_{d1} \\ F_{d2} \end{bmatrix} \tag{42}$$

where $E_1(X)$ and $E_2(X)$ are the nonlinear terms decomposed from $E(X)$ in Eq. 24, $U_f \in \mathbb{R}^{2\times 1} = [u_{fi}, u_{fz}]^T$ is the vector of control inputs, and $F_{d1} \in \mathbb{R}^{4\times 1} = [0, 0, 0, F_{di}]^T$, $F_{d2} \in \mathbb{R}^{2\times 1} = [F_{dj}, F_{dz}]^T$, $\forall\, i, j = x \text{ or } y$ (depending on Case I or Case II), are the differential perturbations.

Next, we assume that the desired reference trajectory, X_d, and the nonlinear component, $E(X)$, in the equations of motion are bounded as

$$\|X_d\| \le \rho_1 \quad \text{and} \quad \|E(X)\| \le \rho_2 \|X\| \tag{43}$$

where $\rho_1 > 0$, and ρ_2 is the Lipschitz constant of the nonlinear vector field associated with $E(X)$. Vaddi[8] showed that the Lipschitz constant can be determined by computing the ratio $\frac{\|E(X)\|}{\|X\|}$ for a particular region of interest (varying formation disc size 1 km – 150 km) and choosing its maximum value. The uncertainties such as relative J_2, magnetic forces, luni-solar perturbations, dynamics of thrusters, perturbations due to thruster misalignment, etc. are assumed to be included in the following chosen uncertainty bound ($\rho_3, \rho_4 > 0$).

$$\|F_d\| \le \rho_3 \|X\| + \rho_4 \tag{44}$$

Next, the control scheme is developed to ensure that the sliding manifold is reached and sliding on the manifold occurs. Based on the sliding manifolds given by Eqs. (33) and (40), the general structure of the surfaces are identical for cases I and II, and therefore the formulation of control scheme will be the same. Then, considering Eq. (33), one can obtain

$$\dot{S} = \Lambda\,\dot{e} = U_f + \Lambda\, A\, X + \Lambda(E(X) + D\, F_d) - \Lambda \dot{X}_d \tag{45}$$

Due to the form of the aforementioned uncertainties given by Eqs. (43) and (44), a second order polynomial function that bounds the lumped term containing nonlinearities and disturbances in the system can be expressed as follows:

$$\begin{gathered}\gamma(t,X,X_d) = \Lambda\left[E(X) + D\,F_d - \dot{X}_d\right] \\ \|\gamma(t,X,X_d)\| \leq \|\Lambda\|\left[(\rho_1+\rho_4) + (\rho_2+\rho_3)\|X\|\right] \leq \rho\|\Lambda\|\Phi = \varphi_3 \\ \Phi = 1 + \|X\|\end{gathered} \tag{46}$$

Carrying out some algebraic manipulations based on Eqs. (45) and (46), the nonlinear control law capable of precise formation-keeping and reconfiguration is given by

$$U_f = -\left[\eta\frac{\varphi_2 S}{\|S\| + \delta} + \Lambda\,A\,X\right] \tag{47}$$

where δ is a small positive scalar specifying the boundary layer thickness that will eliminate chatter if appropriately chosen so that the unmodeled high frequency dynamics are not excited. This choice has no effect on the closed-loop trajectories, except when sliding along the sliding surface S_u, in which case the deadband will strongly influence the high frequency chatter in the control input. The scalar function η depends on the magnitude of the disturbances and uncertainties,

$$\eta = \frac{\varphi_1}{\varphi_2}(\varphi_3 + \varphi_4) \quad \forall \quad [\varphi_1 > 1 \quad \text{and} \quad \varphi_2, \varphi_3, \varphi_4 \geq 0] \tag{48}$$

for some positive constants φ_1, φ_2, φ_3, and φ_4.

4.4. *Stability Analysis*

Theorem 1: *For the underactuated spacecraft formation flying mathematical model in Eq. (42) if, the sliding manifold is chosen as Eq. (33) or Eq. (40), the control law is defined as Eq. (47), and the bounds on the external disturbances and uncertainties on the system is assumed as given by Eq. (46) then the system reaches the sliding surface in finite time for a sufficiently small $\delta > 0$.*

Proof: Consider the Lyapunov function

$$V(S) = \frac{1}{2}S^T S \tag{49}$$

Taking the first derivative of $V(S)$ along the trajectory of the closed-loop system,

$$\dot{V}(S) = S^T\left[\Lambda\,(\dot{X} - \dot{X}_d) + \bar{K}\dot{X}_d\right] \tag{50}$$

Substituting the nonlinear relative equations of motion, and the control law given by Eq. (47), the following relation can be obtained

$$
\begin{aligned}
\dot{V}(S) &= S^T\left[\Lambda\,(A\,X + E(X) + D\,F_d) + U_f + (\bar{K} - \Lambda)\,\dot{X}_d\right] \\
&= S^T\left[-\eta\frac{\varphi_2 S}{\|S\| + \delta} + \gamma(t, X, X_d)\right]
\end{aligned} \tag{51}
$$

Based on Eq. (46), the first derivative of $V(S)$ can be expressed as

$$
\begin{aligned}
\dot{V}(S) &\leq -\eta\frac{\varphi_2\|S\|^2}{\|S\| + \delta} + \|\gamma(t, X, X_d)\|\|S\| \\
&\leq -\eta\frac{\varphi_2\|S\|^2}{\|S\| + \delta} + \varphi_3\|S\|
\end{aligned} \tag{52}
$$

Expressing φ_3 in terms of η, φ_1, φ_2, and φ_4 from Eq. (48) and substituting in Eq. (52),

$$
\begin{aligned}
\dot{V}(S) &\leq -\|S\|\left[\eta\,\varphi_2\,\frac{\|S\|}{\|S\| + \delta} - \frac{\eta\,\varphi_2}{\varphi_1} + \varphi_4\right] \\
&\leq -\varphi_4\|S\| - \eta\,\varphi_2\,\|S\|\left[\frac{\|S\|}{\|S\| + \delta} - \frac{1}{\varphi_1}\right]
\end{aligned} \tag{53}
$$

It is readily obtained from Eq. (53) that, if:

$$
\begin{aligned}
\frac{\|S\|}{\|S\| + \delta} - \frac{1}{\varphi_1} &\geq 0 \\
\|S\| &\geq \frac{\delta}{\varphi_1 - 1}
\end{aligned} \tag{54}
$$

then $\dot{V}(S) < 0$. The condition in Eq. (54) is only satisfied if

$$
V(S) > \frac{1}{2}\left(\frac{\delta}{\varphi_1 - 1}\right)^2 = \epsilon_1 \tag{55}
$$

Based on Eq. (52), where $\frac{\|S_u\|}{\|S_u\| + \delta} \leq 1$ $(\forall\ \delta \geq 0)$, a condition for selecting the gains can be derived as follows:

$$
\eta\,\varphi_2 - \varphi_3 > 0, \quad \varphi_1 > \frac{\varphi_3}{\varphi_3 + \varphi_4} \tag{56}
$$

Using this fact it can be shown that $V'(S_u) \leq -\epsilon_2\sqrt{2V(S_u)}$ for some $\epsilon_2 > 0$. This implies that the sliding boundary layer is reached in finite time. For the case where a small (δ) is chosen, then every solution will eventually enter the set $\Re = \{S_u :\ V(S_u) \leq \epsilon_1\}$ and will be globally uniformly ultimately bounded with respect to the ellipsoid ϵ_1. Thus, it is shown that the continuous control input given by Eq. (47) forces the solutions of the system towards a boundary layer surrounding the sliding surface S in the state space, and the system remains in it thereafter. □

Next we evaluate the properties of the system once the closed-loop error dynamics are constrained to S. To this end, Shyu's stability criterion[9] of the reduced-order system with unmatched uncertainties is introduced in Lemma 1.

Lemma 1: *Consider the reduced order system with uncertainty described by*

$$\dot{x}_1 = (A_{11} - A_{12}K)x_1 + \bar{f}(x_1) \tag{57}$$

where $\bar{f}(x_1)$ *is the unmatched uncertainty. Then, if* $\bar{f}(x_1)$ *satisfies the uniform Lipschitz condition* $\|\bar{f}(x_1^1) - \bar{f}(x_1^2)\| \leq b\|x_1^1 - x_1^2\|$ *where* $0 \leq b \leq 0.5\lambda_{min}(\bar{Q})/\|\bar{P}\|$ *with* $\bar{P}$, $\bar{Q} \in \mathbb{R}^{(n-m)\times(n-m)}$ *which are symmetric, positive-definite matrices satisfying the Lyapunov equation* $(A_{11} - A_{12}K)^T\bar{P} + \bar{P}(A_{11} - A_{12}K) = -\bar{Q}$*, then the uncertain system, Eq. (57), on the sliding surface is asymptotically stable.*

The nonlinear error dynamics of relative motion between a leader spacecraft in circular reference orbit and a follower is rewritten as:

$$\begin{aligned} \dot{e}_1 &= A_{11}e_1 + A_{12}e_2 + \bar{f}_{ru}(X) \\ \dot{e}_2 &= A_{21}e_1 + A_{22}e_2 + B_2U + \bar{f}_{rm}(X) \end{aligned} \tag{58}$$

where $\bar{f}_{ru}(X) \in \mathbb{R}^4$ and $\bar{f}_{rm}(X) \in \mathbb{R}^2$ are the lumped terms containing the unmatched and matched components of uncertainties ($E(X)$, F_d) in the system, respectively. When in sliding mode the system is insensitive to the matched uncertainty,$\bar{f}_{rm}(X)$. The unmatched uncertainties are assumed to be *unknown* but bounded and satisfies, $\|\bar{f}_{ru}(X)\| \leq w_1 + w_2\|X\|$. During the sliding motion $e_2 = -K\,e_1 - \bar{K}\,X_d$ (from Eq. 30), and therefore

$$e = \begin{bmatrix} e_1 \\ -K\,e_1 - \bar{K}\,X_d \end{bmatrix} \Rightarrow X = \begin{bmatrix} 1 \\ -K \end{bmatrix} e_1 + \begin{bmatrix} 0 \\ -\bar{K} \end{bmatrix} X_d + X_d$$
$$\|X\| \leq \sqrt{1 + \|K\|^2}\|e_1\| + (1 + \|\bar{K}\|)\|X_d\| \tag{59}$$

Consequently the bound on the unmatched uncertainty can be written as

$$\|\bar{f}_{ru}(X)\| \leq \varphi_2 + \bar{w}_2\|e_1\| \tag{60}$$

where $\varphi_2 = \rho_1(1 + \|\bar{K}\|)$ (from Eq. 43) and $\bar{w}_2 = w_2\sqrt{1 + \|K\|^2}$. The equation representing the error dynamics confined to the sliding surface is obtained by substituting $S = 0$ in Eq. (58), giving

$$\dot{e}_1 = (A_{11} - A_{12}K)e_1 + \bar{f}_{ru}(X) \tag{61}$$

In the following theorem, based on the sliding surface defined by Eq. (30), a stability criterion for the reduced order system (Eq. 61) is presented. Several cases of asymptotic stabilization of the reduced order system in the presence of unmatched uncertainties, where the vector fields have a special form, have been studied in the literature. The procedure outlined in the book by[10] and Shyu's stability criterion (see Lemma 1) is utilized to present Theorem 2.

Theorem 2: *For the motion constrained to the sliding surface, the trajectory of the reduced-order system (Eq. 61) starting from any initial condition will enter a compact set $\Im$ containing the origin in finite time and the tracking error e_1 will be uniformly ultimately bounded with respect to the ellipsoid*

$$\Im = \left\{ e_1 \in \mathbb{R}^4 \ : \ \|e_1\| \leq \frac{2\varphi_2}{\xi - 2\bar{\varpi}_1} \right\} \tag{62}$$

Then, the reduced-order system is globally asymptotically stable if $\xi > 2\bar{\varpi}_1$, with $\xi \triangleq \lambda_{min}(\bar{Q})/\lambda_{max}(\bar{P})$, where $\bar{P}$, $\bar{Q} \in \mathbb{R}^{4\times 4}$ are positive-definite matrices satisfying the Lyapunov equation

$$(A_{11} - A_{12}K)^T\bar{P} + \bar{P}(A_{11} - A_{12}K) = -\bar{Q} \tag{63}$$

Proof: Consider the Lyapunov function

$$V(e_1) = {e_1}^T\bar{P}e_1 \tag{64}$$

The first derivative of $V(e_1)$ along the motion of Eq. (61) is given by

$$\begin{aligned}
\dot{V}(e_1) &= {e_1}^T\bar{P}\dot{e}_1 + \dot{e}_1^T\bar{P}e_1 \\
&= {e_1}^T\bar{P}\left[(A_{11} - A_{12}K)e_1 + \bar{f}_{ru}(X)\right] \\
&\qquad\qquad + \left[(A_{11} - A_{12}K)e_1 + \bar{f}_{ru}(X)\right]^T\bar{P}e_1 \\
&= {e_1}^T\left[\bar{P}(A_{11} - A_{12}K) + (A_{11} - A_{12}K)^T\bar{P}\right]e_1 + 2e_1^T\bar{P}\bar{f}_{ru}(X) \\
&\leq -{e_1}^T\bar{Q}\,e_1 + 2\|\bar{P}\,e_1\|\,\|\bar{f}_{ru}(X)\|
\end{aligned} \tag{65}$$

Using the Rayleigh principle, the following inequality can be derived:

$$\lambda_{min}(\bar{Q})\|e_1\|^2 \leq {e_1}^T\bar{Q}\,e_1 \leq \lambda_{max}(\bar{Q})\|e_1\|^2 \tag{66}$$

In particular, if $\lambda_{min}(\bar{Q}) \geq 0$ then it follows that $e_1{}^T \bar{Q}\, e_1 \geq 0$ for all e_1. Using Eq. (66) and Eq. (60), Eq. (65) can be expressed as

$$\begin{aligned} \dot{V}(e_1) &\leq -\lambda_{min}(Q)\|e_1\|^2 + 2\lambda_{max}(\bar{P})\|e_1\| \, \|\bar{f}_{ru}(X)\| \\ &\leq -\lambda_{max}(\bar{P}) \left[\xi\|e_1\| - 2\|\bar{f}_{ru}(X)\|\right] \|e_1\| \\ &\leq -\lambda_{max}(\bar{P}) \left[\xi\|e_1\| - 2\,\bar{w}_1\|e_1\| - 2\varphi_2\right] \|e_1\| \end{aligned} \tag{67}$$

Therefore, it is clearly evident from Eq. (67) that $\dot{V}(e_1) < 0$ when e_1 is outside of the set

$$\Im = \left\{ e_1 \in \mathbb{R}^4 \ : \ \|e_1\| \leq \frac{2\varphi_2}{\xi - 2\bar{w}_1} \right\} \tag{68}$$

Analytical estimate of $\lambda_{min}(Q)$ is not needed for numerical simulations because the proposed control law is independent of this parameter. When norm of the unactuated states, $\|e_1\| > \frac{2\varphi_2}{\xi - 2\bar{w}_1}$, then $\dot{V}(e_1)$ decreases; but once the states (e_1) enters the set $\Im$, the states cannot go out of it and hence the unactuated states will be confined to the set $\Im$. For every $e_1(t_0) \in \Im$ then $e_1(t) \in \Im$ for all $t \geq t_0$. Since $\dot{V}(e_1) < 0$, it also follows that if $e_1(t_0) \notin \Im$ then the trajectory will reach $\Im$ in finite time t_r. The system is therefore uniformly ultimately bounded with respect to the ellipsoid $\Im$. □

5. Design of Control Laws for Spacecraft Attitude

We present the procedures for designing the proposed control laws as follows.

5.1. *Design of Sliding Manifold*

Using Eq. (14), we get the equilibrium state vector $X_e = 0$, i.e., $(\alpha_e = \phi_e = \gamma_e = \alpha_e{}' = \phi_e{}' = \gamma_e{}' = 0)$. Considering first order approximation for the system state, we have the linearized equations of motion in state space form as follows:

$$X' = AX + BU \tag{69}$$

where $X \in \mathbb{R}^{6x1} = [\alpha, \alpha', \phi, \phi', \gamma, \gamma']^T$, and the matrices $A \in \mathbb{R}^{6x6}$, $B \in \mathbb{R}^{6x3}$ are described in Eq. (70) with $k_1 = (I_z - I_x)/I_y$ and $k_2 = (I_z - I_y)/I_x$.

$$A = \begin{bmatrix} 0 & 1 & 0 & 0 & 0 & 0 \\ 3\frac{k_2-k_1}{1-k_1k_2} & 0 & 0 & 0 & 0 & 0 \\ 0 & 0 & 0 & 1 & 0 & 0 \\ 0 & 0 & -4k_1 & 0 & 0 & k_1-1 \\ 0 & 0 & 0 & 0 & 0 & 1 \\ 0 & 0 & 0 & 1-k_2 & -k_2 & 0 \end{bmatrix}, \quad B = \begin{bmatrix} 0 & 0 & 0 \\ b_1 & 0 & 0 \\ 0 & 0 & 0 \\ 0 & b_2 & 0 \\ 0 & 0 & 0 \\ 0 & 0 & b_3 \end{bmatrix} \tag{70}$$

When all actuators are healthy, $b_1, b_2, b_3 = 1$, the controllability matrix $C = [B \vdots AB \vdots \cdots \vdots A^5B]$ is of rank 6. The matrix pair (A, B) defining the nominal linear system is also fully state controllable when $b_2 = 0$ ($i.e. b_1 = 1$ and $b_3 = 1$) or $b_3 = 0$ ($i.e. b_1 = 1$ and $b_2 = 1$). Therefore, the linear system is controllable even if the actuation on the *roll*-axis (ϕ) or the *yaw*-axis (γ) fails. If no actuation is available for the decoupled pitch dynamics ($b_1 = 0$), then the system is not fully state controllable. We can now transform the linear system in terms of new coordinates $x_1 \in \mathbb{R}^4$ and $x_2 \in \mathbb{R}^2$, so that Eq. (69) can be transformed to the form given by

$$\begin{bmatrix} x_1' \\ x_2' \end{bmatrix} = \begin{bmatrix} A_{11} & A_{12} \\ A_{21} & A_{22} \end{bmatrix} \begin{bmatrix} x_1 \\ x_2 \end{bmatrix} + \begin{bmatrix} 0 \\ B_2 \end{bmatrix} U \tag{71}$$

This representation separates the actuated and unactuated states based on the failed axis (roll or yaw). By exploiting the coupling between the directly actuated and unactuated states, we define the sliding surface, S_u as a linear combination of the states.

$$S_u = \{x_1 \in \mathbb{R}^{4\times 1}, x_2 \in \mathbb{R}^{2\times 1} : \Lambda_1 x_1 + \Lambda_2 x_2 = 0\} \tag{72}$$

where $\Lambda_1 \in \mathbb{R}^{2\times 4}$ and $\Lambda_2 \in \mathbb{R}^{2\times 2}$ are weights on the states x_1 and x_2 respectively. When the system reaches the sliding surface, $S_u = 0 \ \forall\, t > t_r$, where t_r is the reaching time after which sliding motion starts,

$$x_2 = -{\Lambda_2}^{-1} \Lambda_1 \, x_1 \tag{73}$$

It is important to note that Eq. (73) holds only on the sliding surface and substituting this relation to the reduced order system in Eq. (71) gives

$$x_1' = (A_{11} - A_{12}K)x_1 \tag{74}$$

where $K = -{\Lambda_2}^{-1}\Lambda_1 = [K_1 \ K_2]$.

$$K_1 \in \mathbb{R}^{2\times 3} = \begin{bmatrix} K_{11} & K_{12} & K_{13} \\ K_{21} & K_{22} & K_{23} \end{bmatrix} \quad \text{and} \quad K_2 \in \mathbb{R}^{2\times 1} = \begin{bmatrix} K_{14} \\ K_{24} \end{bmatrix}$$

During an ideal sliding mode behavior, x_2 can be considered as a control signal to stabilize x_1. Therefore, the choice of sliding surface, S_u, clearly affects the dynamics of the reduced order system through the selection of weighting matrix K.

$$S_u = \{x_1 \in \mathbb{R}^{4\times 1}, x_2 \in \mathbb{R}^{2\times 1} : x_2 + Kx_1 = 0\} \tag{75}$$

Remark 1: The weighting matrix K prescribes a desired closed loop behavior of the system [Eq. (74)] and can be determined using any *classical* approaches which provides a full state feedback control scheme for a system represented in state-space form. Since (A, B) is, by definition a controllable pair it follows directly that the matrix pair (A_{11}, A_{12}) is also controllable.

5.2. *Nonlinear Control Formulation*

The lumped term containing the nonlinearities, uncertainties, and disturbances is given by

$$\xi(x_1, x_2) = \left[\bar{A}_{21} + K_2\bar{A}_{11}\right] [F_1 + d_1] + \left[\bar{A}_{22} + K_2\bar{A}_{12}\right] [F_2 + d_2] + K_1 x_{10}' \tag{76}$$

A feasible and practical control scheme should not be designed by including the term $\xi(x_1, x_2)$ directly. One way to account for this in the controller is to assume that the lumped disturbances are bounded and then use the upper bound in the control algorithm design.

$$\|\xi(q, q', q'')\| \le \rho_1 + \rho_2\|x_1\| + \rho_3\|x_2\| = \varphi_3 \tag{77}$$

In order to ensure that the sliding manifold is reached and sliding on the manifold occurs, the continuous nonlinear control law is chosen as

$$U_{ua} = -(\bar{A}_{22} + K_2\bar{A}_{12})^{-1}\left[\eta \frac{\varphi_2 S_u}{\|S_u\| + \delta}\right] \tag{78}$$

where δ is a small positive scalar specifying the boundary layer thickness that will eliminate chatter if appropriately chosen so that the unmodeled high frequency dynamics are not excited.[10] This choice has no effect on the closed-loop trajectories, except when sliding along the sliding surface S_u, in which case details of the dead-band will strongly influence the high frequency chatter in the control input. The nonnegative constant η depends on the magnitude of the disturbances and uncertainties,

$$\eta = \frac{\varphi_1}{\varphi_2}(\varphi_3 + \varphi_4) \tag{79}$$

for some positive constants φ_1, φ_2, φ_3, and φ_4. The steps involved in adequately determining these nonnegative constants are detailed in the next section.

5.3. *Stability Analysis*

Theorem 3: *For the spacecraft mathematical model in Eq. (15) if, the sliding manifold is chosen as Eq. (75), the control law is defined as Eq. (78), and the bounds on the external disturbances, parameter uncertainties, and system nonlinearities are assumed to be governed by Eq. (77), then the closed-loop trajectories of the system will converge in finite time to a neighborhood area of the equilibrium set* $\mathfrak{F}$.

$$\mathfrak{F} \triangleq \left[S_u : \ \|S_u\| \leq \frac{\delta}{\varphi_1 - 1} \right] \tag{80}$$

Proof: Consider the Lyapunov function

$$V(S_u) = \frac{1}{2} {S_u}^T S_u \tag{81}$$

Taking the first derivative of $V(S_u)$ along the trajectory of the closed-loop system,

$$V'(S_u) = {S_u}^T {S_u}' = {S_u}^T \left[{x_2}' + K_1 {x_{10}}' + K_2 {x_{11}}' \right] \tag{82}$$

Substituting the mathematical model, Eq. (15), and the control law, Eq. (78), we get

$$\begin{aligned} V'(S_u) &= {S_u}^T \Big[\left(\bar{A}_{21} + K_2 \bar{A}_{11} \right) \left(F_1 + d_1 \right) \\ &\qquad\qquad + \left(\bar{A}_{22} + K_2 \bar{A}_{12} \right) \left(F_2 + d_2 + U_{ua} \right) + K_1 {x_{10}}' \Big] \\ &= {S_u}^T \left[-\eta \frac{\varphi_2 S_u}{\|S_u\| + \delta} + \xi(x_1, x_2) \right] \end{aligned} \tag{83}$$

Using the property defined based on Eq. (77) and expressing φ_3 in terms of η, φ_1, φ_2, and φ_4 from Eq. (79), we get

$$\begin{aligned} V'(S_u) &\leq \|S_u\| \left[-\eta \frac{\varphi_2 \|S_u\|}{\|S_u\| + \delta} + \|\xi\| \right] \\ &\leq \|S_u\| \left[-\eta \frac{\varphi_2 \|S_u\|}{\|S_u\| + \delta} + \varphi_3 \right] \\ &\leq \|S_u\| \left[-\eta \frac{\varphi_2 \|S_u\|}{\|S_u\| + \delta} - \frac{\eta \varphi_2}{\varphi_1} + \varphi_4 \right] \\ &\leq -\varphi_4 \|S_u\| - \eta \varphi_2 \|S_u\| \left[\frac{\|S_u\|}{\|S_u\| + \delta} - \frac{1}{\varphi_1} \right] \end{aligned} \tag{84}$$

It is readily obtained from Eq. (84) that, if:

$$\frac{\|S_u\|}{\|S_u\| + \delta} - \frac{1}{\varphi_1} \geq 0$$
$$\|S_u\| \geq \frac{\delta}{\varphi_1 - 1} \tag{85}$$

then $V'(S_u) < 0$ when S_u is outside of the set

$$\mathfrak{F} \triangleq \left[S_u : \ \|S_u\| \leq \frac{\delta}{\varphi_1 - 1} \right] \tag{86}$$

The condition in Eq. (85) is only satisfied if

$$V(S_u) > \frac{1}{2}\left(\frac{\delta}{\varphi_1 - 1}\right)^2 = \epsilon_1 \tag{87}$$

Based on the second line Eq. (84), where $\frac{\|S_u\|}{\|S_u\| + \delta} \leq 1 \ (\forall \ \delta \geq 0)$, a condition for selecting the gain φ_3 in relation to φ_1 and φ_4, can be derived such that:

$$\eta\varphi_2 - \varphi_3 > 0, \quad \varphi_1 > \frac{\varphi_3}{\varphi_3 + \varphi_4} \tag{88}$$

Using this fact it can be shown that $V'(S_u) \leq -\epsilon_2\sqrt{2V(S_u)}$ for some $\epsilon_2 > 0$. This implies that the sliding boundary layer is reached in finite time. For the case where a small (δ) is chosen, then every solution will eventually enter the set $\mathfrak{F} = \{S_u : \ V(S_u) \leq \epsilon_1\}$. □

Next we evaluate the properties of the spacecraft motion constrained to S_u. The linear system can be expressed in the general form as

$$\begin{aligned} x_1' &= A_{11}x_1 + A_{12}x_2 + D_1 \\ x_2' &= A_{21}x_1 + A_{22}x_2 + B_2U + d_2 \end{aligned} \tag{89}$$

where $x_1 \in \mathbb{R}^{4\times1}$ and $x_2 \in \mathbb{R}^{2\times1}$ are the actuated and unactuated states, respectively, $A_{11} \in \mathbb{R}^{4\times4}$, $A_{12} \in \mathbb{R}^{4\times2}$, $A_{21} \in \mathbb{R}^{2\times4}$, $A_{22} \in \mathbb{R}^{2\times2}$, $B_2 \in \mathbb{R}^{2\times2}$, $U \in \mathbb{R}^{2\times1}$ is the control input, $D_1 \in \mathbb{R}^{4\times1} = [0, 0, 0, d_1]^T$ and $d_2 \in \mathbb{R}^{2\times1} = [d_{21}, d_{22}]^T$ are the unmatched and matched components of nonlinear uncertainties and disturbances.

We now study the effect of unmatched component (D_1) of the disturbances when the dynamics of the system represents the dynamics of an ideal sliding mode. For convenience we set $\delta = 0$. To determine the spacecraft dynamics on the sliding surface, we can solve $S_u = 0$ for x_2 using Eq. (75) which yields $x_2 = -Kx_1$. We have shown in Theorem 5.1 that a control law exists such that the spacecraft motion can be constrained to S_u. This result can be substituted into Eq. (89) to obtain the following reduced order system

$$x_1' = (A_{11} - A_{12}K)x_1 + D_1 \tag{90}$$

Let $A_c = A_{11} - A_{12}K$. We can ensure that A_c is a stable matrix with eigenvalues containing negative real parts by appropriately choosing K. For asymptotic stability we require that, if P and Q are positive definite matrices, then the solution to the Lyapunov equation [Eq. (91)] will exist because the matrix A_c is stable.

$$PA_c + A_c{}^T P = -Q \tag{91}$$

Theorem 4: *For the motion constrained to the sliding surface, the trajectory of the reduced order system [Eq. (90)] starting from any initial condition will enter a compact set containing the origin in finite time and the states will be uniformly ultimately bounded with respect to the ellipsoid*

$$\Im = \left\{ x_1 : \ \|x_1\| \leq 2\frac{\sup\limits_{D_1 \in \epsilon_3} \|PD_1\|}{\lambda_{min}(Q)} \right\} \tag{92}$$

Proof: Consider the Lyapunov function

$$V(x_1) = x_1{}^T P x_1 \tag{93}$$

The first derivative of $V(x_1)$ along the motion of Eq. (90) is given by

$$V'(x_1) = x_1{}^T(PA_c + A_c{}^T P)x_1 + 2x_1^T P\, D_1 = -x_1{}^T Q\, x_1 + 2x_1^T P\, D_1 \tag{94}$$

Using the Rayleigh principle we know that

$$\lambda_{min}(Q)\|x_1\|^2 \leq x_1{}^T Q\, x_1 \leq \lambda_{max}\|x_1\|^2 \tag{95}$$

In particular, if $\lambda_{min}(Q) \geq 0$ then it follows that $x_1{}^T Q\, x_1 \geq 0$ for all x_1. Based on these conditions, Eq. (94) can be expressed as

$$\begin{aligned} V'(x_1) &\leq -\lambda_{min}(Q)\|x_1\|^2 + 2\|x_1\| \cdot \|PD_1\| \\ &\leq -(\lambda_{min}(Q)\|x_1\| - 2\|PD_1\|)\|x_1\| \end{aligned} \tag{96}$$

It is clearly evident from Eq. (96) that $V'(x_1) < 0$ when x_1 is outside of the set

$$\Im \triangleq \left\{ x_1 : \ \|x_1\| \leq 2\frac{\sup\limits_{D_1 \in \epsilon_3} \|PD_1\|}{\lambda_{min}(Q)} \right\} \tag{97}$$

Analytical estimate of $\lambda_{min}(Q)$ is not required for numerical simulations because the proposed control law is independent of this parameter. For every $x_1(t_0) \in \Im$ then $x_1(t) \in \Im$ for all $t \geq t_0$. Since $V'(x_1) < 0$, it also follows that if $x_1(t_0) \notin \Im$ then the trajectory will reach $\Im$ in finite time t_r. The system is therefore uniformly ultimately bounded with respect to the ellipsoid $\Im$. Explicit consideration of the actuated states (x_2) is not required because it is a well known fact that when in the sliding mode the system is totally insensitive to matched disturbances. This completes the proof. □

Remark 1: The upper bound on the nonlinearities and disturbances, φ_3, for cases I and II were determined from numerical simulations by propagating the nonlinear dynamics along with external disturbances based on the following initial conditions: $[\alpha_0, \phi_0, \gamma_0] = [80°, 80°, 80°]$, and $[\alpha_0', \phi_0', \gamma_0'] = [0.01, 0.01, 0.01]$. The largest value of $1 + \|x_1\| + \|x_2\|$ was chosen. For Case I - $\varphi_3 = 1.5$, and for Case II - $\varphi_3 = 2.0$. The value of universal gain η is determined based on Eq. (79) using this value of φ_3 and the constraint specified in Eq. (88).

6. Numerical Results for Spacecraft Formation Flying

To study the effectiveness and performance of the proposed formation control strategies, the detailed response is numerically simulated using the set of governing equations of motion (Eqs. 2-4) in conjunction with the proposed control law (Eqs. 47). The desired states of the system are given by Eqs. (5) or (6) for circular formation and projected circular formation respectively. The SFF system parameters and the orbital parameters for the leader spacecraft used in the numerical simulations are shown in Table 1.

Table 1. Orbital parameters

Parameters	Values
$\mu\,(\mathrm{km}^3\mathrm{s}^{-2})$	398600
r_p (km)	6878
$\Omega_l,\ i_l,\ \omega_l$ (deg)	0, 45, 0

For all numerical examples presented in this section, the net disturbance force, $F_d(t)$, acting on the system is considered to be differential J_2 based on the formulation presented in Section II(C). The leader spacecraft is assumed to be in an unperturbed circular reference orbit and the differential force on the follower is calculated relative to the leader spacecraft orbit. The control gains ($K\,\varphi_i$) and the boundary layer (δ) used in all simulations for Cases I and II are shown in Table 2.

Table 2. Controller parameters

Control Gains	Case I	Case II
$[\varphi_1, \varphi_2, \varphi_3, \varphi_4]$	[2.0, 4.0, 1.0, 0.5]	[1.5, 1.5, 0.5, 0.2]
$[K_{11}, K_{12}, K_{13}, K_{14}]$	[4, −2, 0, 18]	[−0.0013, 0, 0.0028, −]
$[K_{21}, K_{22}, K_{23}, K_{24}]$	[0, 0, 2, 0]	[0, 0.0013, 0, −]
δ	10^{-5}	10^{-5}

The total control force is assumed to be subjected to saturation limit defined by

$$U_f = \begin{cases} N & \text{if} \quad U_f > N \\ U_f & \text{if} \quad -N < U_f < N \\ -N & \text{if} \quad U_f < -N \end{cases} \tag{98}$$

where $N = 10\,\text{mN}$ for a 10 kg follower spacecraft. The desired relative motion considered for ideal formation keeping is a projected circular formation (PCF) described by Eq. (6), with $r_{dpc} = 1$ km formation radius. The in-plane phase angle (ϕ) between the leader and follower spacecraft is assumed to be zero degree. The initial relative positions for the numerical simulation are computed by substituting $t = 0$ in Eq. (6). The initial velocity components for all states are calculated by taking the time derivative of Eq. (6) and substituting $t = 0$. The initial state vector is given by:

$$X(0) = [0,\ r_{dpc},\ 0,\ 0.5\,n\,r_{dpc},\ 0,\ n\,r_{dpc}]^T \tag{99}$$

6.1. *Formation-keeping and Formation Reconfiguration*

Figure 3 shows relative position errors and thrust demand for formation keeping with no control available in the radial direction in the presence of relative J_2 perturbations. Next, the effectiveness of the proposed control strategies is demonstrated for multiple formation maneuvers. With the same initial conditions as given by Eq. (99), the follower spacecraft moves from a 0.5 km to a 1.5 km (radius) projected circular formation after 5 orbits (Fig. 4). Also, a scenario where the desired geometry is changed from a 0.5 km projected circular formation to a 2 km circular formation is illustrated (Fig. 5). The simulation of extreme cases of initial errors and formation reconfiguration clearly indicate the proposed control scheme is indeed robust to changing operating conditions and ensures precise formation acquisition during reconfiguration maneuvers.

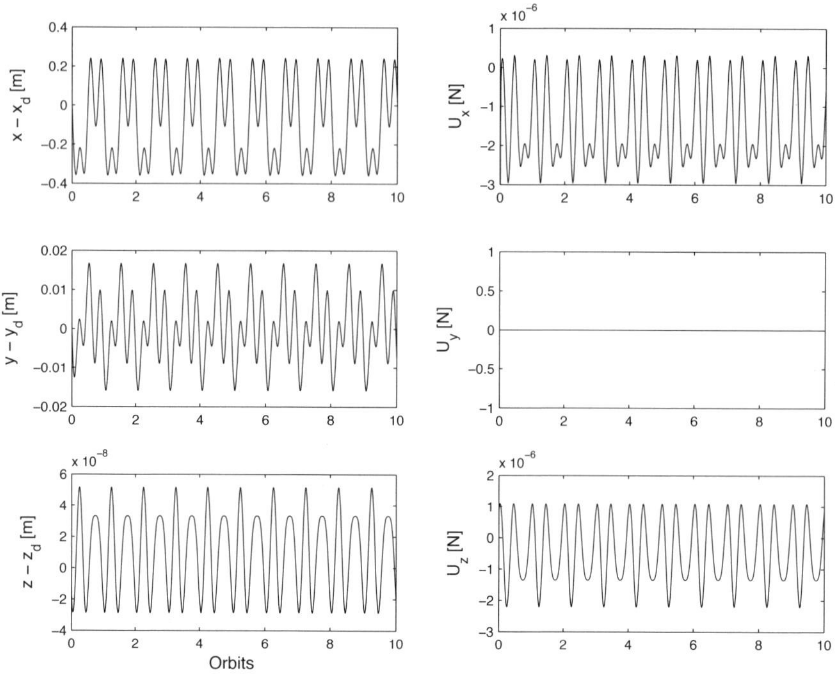

Fig. 3. System response under relative J_2 disturbance.

6.2. *Quantitative Analysis*

Based on the simulation results presented above, the control precision and fuel consumption properties of the proposed control scheme are examined (Table 3 and Table 4). The following scenario is simulated where, (1) the desired reference orbit is a projected circular formation with $r_{dpc} = 1\,\text{km}$, (2) the leader spacecraft is in an unperturbed circular reference orbit, and (3) the follower spacecraft is positioned correctly into the desired orbit (Eq. 99).

Table 3. Formation-keeping steady-state errors

	No Disturbance		Differential J_2	
Errors, m	Case I	Case II	Case I	Case II
$\lvert e_x \rvert_{max}$	1.8×10^{-3}	7.2×10^{-3}	2.2×10^{-2}	2.0×10^{-1}
$\lvert e_y \rvert_{max}$	5.0×10^{-3}	5.0×10^{-4}	5.0×10^{-2}	1.8×10^{-2}
$\lvert e_z \rvert_{max}$	2.0×10^{-10}	1.2×10^{-9}	8.0×10^{-6}	5.0×10^{-8}

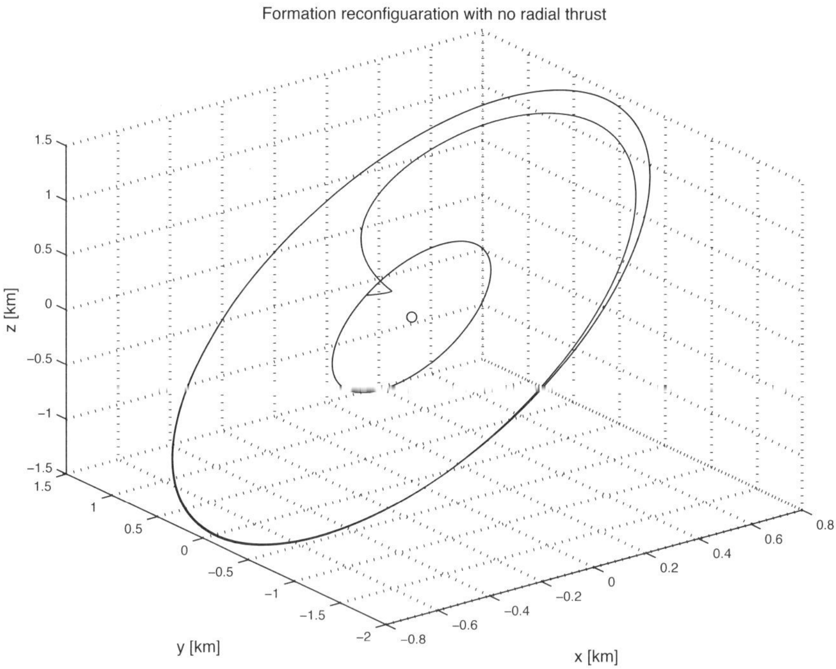

Fig. 4. Case I - Formation reconfiguration from $R_d = 0.5$km to $R_d = 1.5$km.

Table 4. Fuel-consumption per orbit

Fuel Cost	No Disturbance		Differential J_2	
m/s (per orbit)	Case I	Case II	Case I	Case II
ΔV_x	$\cdots$	4.3×10^{-4}	$\cdots$	8.6×10^{-3}
ΔV_y	6.1×10^{-5}	$\cdots$	8.5×10^{-3}	$\cdots$
ΔV_z	1.9×10^{-4}	1.9×10^{-4}	5.5×10^{-3}	5.5×10^{-3}

The proposed control scheme is capable of accomplishing sub-millimeter tracking precision when no external disturbances are considered, while the tracking capability reduces to the order of 10^{-2}m in the presence of differential J_2 perturbations. Note that ΔV is calculated based on the average budget obtained over a period of 10 orbits. For the case of no disturbance, eliminating radial-axis input (Case I) seems to be beneficial in terms of fuel cost when compared to eliminating along-track input (Case II). In the presence of differential J_2 perturbations relative to the leader spacecraft orbit, cases I and II provide similar results as seen in Table 4. The cost required

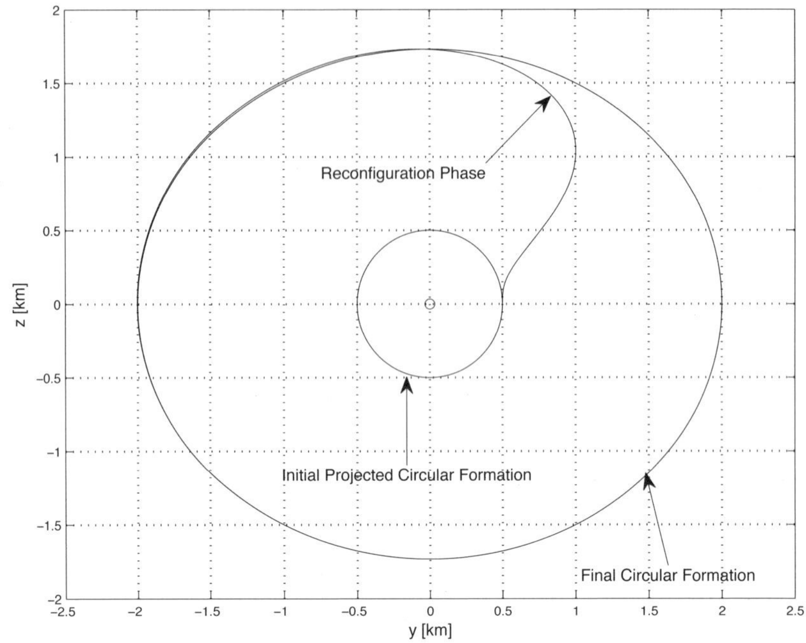

Fig. 5. Case II - Formation reconfiguration from projected circular formation (R_d = 0.5km) to circular formation (R_d = 2.0km).

for formation keeping is calculated as the result of the integral

$$J = \int_0^{\tau} (u_{fx}^2 + u_{fy}^2 + u_{fz}^2)\mathrm{d}\tau \tag{100}$$

where $\tau = 10$ is the number of orbits of the leader spacecraft.

7. Numerical Results for Spacecraft Attitude

To study the effectiveness and performance of the proposed nonlinear control strategies for rigid spacecraft in the event of unexpected actuator failures, the detailed response of the system is numerically simulated using the set of governing equations of motion [Eq. (15] in conjunction with the proposed control law [Eq. (78)]. The system and orbital parameters for the spacecraft along with the initial conditions used in the numerical simulations are shown in Table 5. Unless otherwise stated explicitly, all numerical simulations are based on the parameters stated in Table 5 without considering the product of inertia terms, $I_{xy} = I_{xz} = I_{yz} = 0$, in the inertia matrix.

Table 5. Simulation parameters

Parameters	Values
Orbit	
r_p (km)	6878
μ_e (km^{3}s^{-2})	398600
Spacecraft MOI	
$I_{xx},\ I_{yy},\ I_{zz}$ (kg m^2)	15, 17, 20
Initial Conditions	
$[\alpha_0,\ \phi_0,\ \gamma_0]$	$[80°,\ -40°,\ 40°]$
$[\alpha_0',\ \phi_0',\ \gamma_0']$	[0.001, 0.001, 0.001]

Based on values of the spacecraft moment of inertia we can calculate nondimensional parameters $k_1 = (I_z - I_x)/I_y = 0.3$ and $k_2 = (I_z - I_y)/I_x = 0.2$. These values of moment of inertia are considered for the design of control algorithms. Any change in the moment of inertia of the spacecraft is unknown to the control algorithm. The control gains (φ_i) and the boundary layer (δ) used in all simulations for Cases I and II are shown in Table 6.

Table 6. Controller parameters used for numerical analysis

Control Gains	Case I	Case II
$[\varphi_1,\ \varphi_2,\ \varphi_3,\ \varphi_4]$	[0.40, 0.60, 1.50, 0.40]	[0.15, 0.40, 2.00, 0.15]
$\eta,\ \delta$	1.3, 0.0001	0.8, 0.0001
$[K_{11},\ K_{12},\ K_{13},\ K_{14}]$	[4, 0, 0, 0]	[0.5, 0, 0, 0]
$[K_{21},\ K_{22},\ K_{23},\ K_{24}]$	$[0,\ -2,\ 2,\ -1]$	[0, 2, 2, 1]

It is important to note that a universal gain (η) is calculated based on all φ_i using the formula given by Eq. (79). The sliding plane is given by $S_u = x_2 + K\,x_1$, where $K = [K_1,\ K_2]$ is determined using LQR[10] applied to the reduced order system in Eq. (71). K can be considered as a 'pseudo' feedback matrix that prescribes the required performance of the reduced order system $(A_{11},\ A_{12})$.

Case I: For no actuation available on roll (ϕ) axis, the closed-loop eigenvalues of the reduced order system $[A_{11} - A_{12}K]$ based on values of K in Table 6, are:

$$\lambda_{1,2} = -0.4 \pm 1.0i\,; \qquad \lambda_3 = -2\,; \qquad \lambda_4 = -4 \tag{101}$$

Case II: For no actuation available on yaw (γ) axis, the closed-loop eigenvalues of the reduced order system $[A_{11} - A_{12}K]$ based on values of K in Table 6, are:

$$\lambda_{1,2} = -0.4 \pm 0.2i\,; \qquad \lambda_3 = -2\,; \qquad \lambda_4 = -4 \tag{102}$$

Figure 6 shows the attitude response of the spacecraft in the presence of initial attitude disturbances (Table 5) for Case I when the control law given by (78) are used to stabilize the system. The nonnegative constants in the control law for the case of no actuator failure are chosen as $\Lambda_\alpha = \Lambda_\phi = \Lambda_\gamma = 4$ and $\eta_c = 0.01$. With no control authority available on the ϕ-axis, a control algorithm specifically designed for a spacecraft with healthy actuators fails to stabilize the *roll* motion. The proposed control law (Eq. 78) successfully stabilizes 3-axis attitude of the spacecraft using only two control torques (U_α and U_γ). The driving control torque required for 3-axis stabilization is also presented in Fig. 6. With no external disturbances acting on the spacecraft, motion of the system reaches the sliding surface $S_u = 0$ in finite time which can be analytically determined using the relation

$$t_r \leq \frac{\|S_u(t_0)\|}{2\pi\eta} \text{ orbits} \leq 0.5 \text{ orbit} \tag{103}$$

where $\eta = 1.3$ from Table 6. The angular velocity of the spacecraft is stabilized to $\omega_x = \omega_y = 0$ and $\omega_z = 0.0011\,\text{rad/s}$. According to the coordinate frames selected as shown in Fig. 2 the spacecraft z-axis is normal to the orbit plane and therefore ω_z would be equal to the orbital rate (when $e = 0$).

Next we consider the case where there is no actuation available on the *yaw* axis (Case II). It is clearly evident in Fig. 7 that the conventional control algorithm fails to stabilize the yaw motion motion of the spacecraft with $U_\gamma = 0$. The reason for uncontrollable rotation of the spacecraft about its x-axis can be analytically determined from the zero-dynamics of the *yaw* equation of motion. When $\alpha = \alpha' = \phi = \phi' = 0$ the *yaw* equation of motion is given by

$$\gamma'' + k_2 \sin\gamma\cos\gamma = 0 \tag{104}$$

where $k_2 = (I_z - Iy)/I_x$. From the solution of Eq. (104) and taking γ_0 and $\gamma_0{}'$ as the initial values, the minimum value of γ' is given by

$$\gamma'_{min} = \sqrt{\gamma_0{}'^2 + k_2\sin^2\gamma_0 - k_2} \tag{105}$$

Therefore the initial spin rate to avoid uncontrolled motion of the spacecraft about x-axis can be obtained from Eq. (105) as

$$\gamma_0{}' < |\cos\gamma_0|\sqrt{k_2} \tag{106}$$

With respect to the uncontrollable motion of γ in Fig. 7, determining the spin rate (γ') at the time when *pitch* and *roll* axes stabilize, helps us verify that $\gamma' > |\cos\gamma_0|\sqrt{k_2}$. Therefore, the initial attitude disturbances and the moment of inertias have a significant effect on the uncontrolled response of the system.

The proposed control law explicitly designed to accommodate actuator failure (Eq. 78) successfully stabilizes 3-axis attitude of the spacecraft using only two control torques (U_α and U_ϕ). Although the attitude responses in Fig. 6 and Fig. 7 are simulated with the same initial conditions (Table 6), it is interesting to note the variation in the stabilized responses of the two cases.

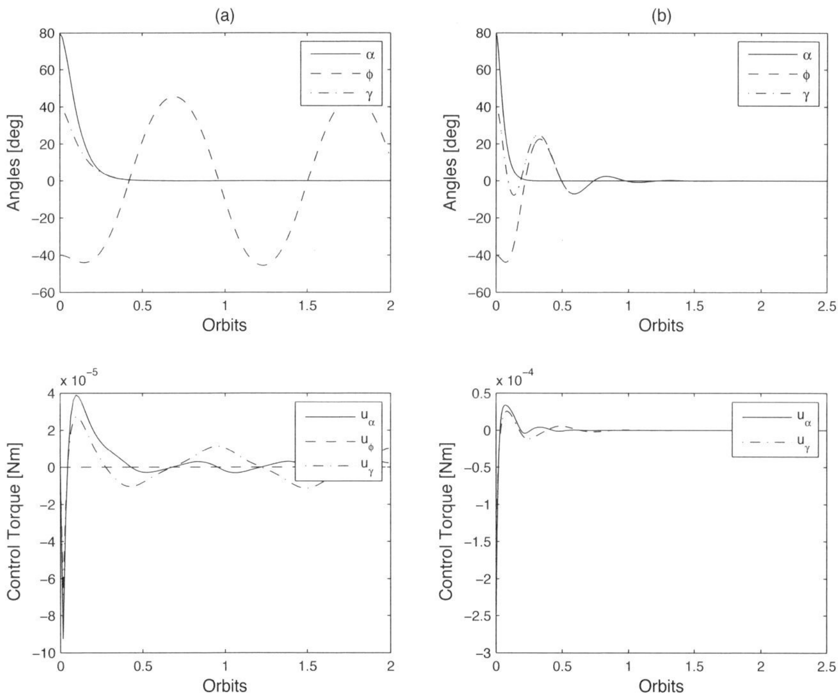

Fig. 6. Comparison between the performance of (a) conventional control algorithm and (b) proposed controller for case I (roll actuator failure).

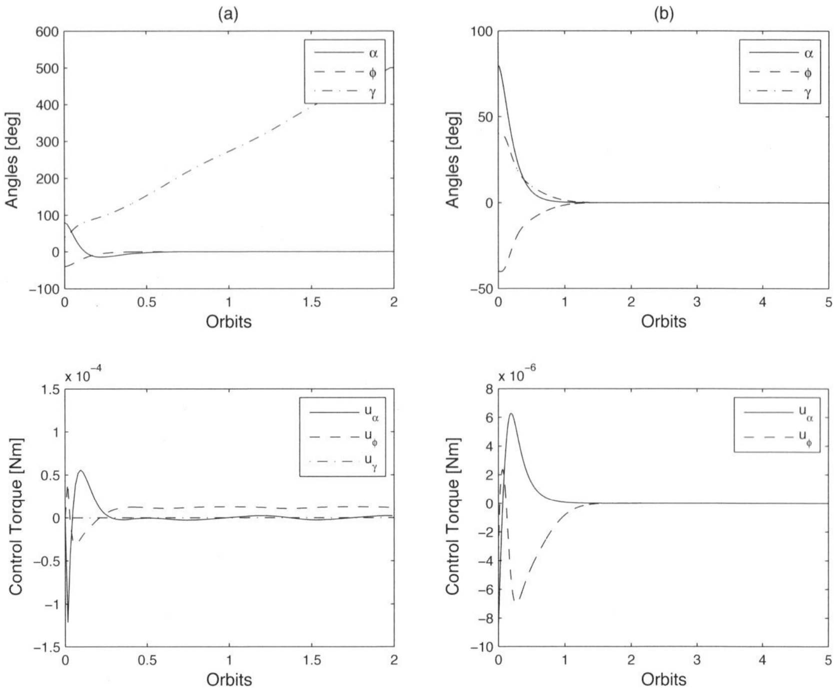

Fig. 7. Comparison between the performance of (a) conventional control algorithm and (b) proposed controller for case II (yaw actuator failure).

8. Conclusions

In this paper, a nonlinear control strategy capable of precision formation control is developed to study two configurations of reduced inputs, where no control force is available in the (1) radial direction, and (2) along-track direction. External disturbances due to differential J_2 is effectively attenuated using the proposed technique. Quantitative analysis of the simulation results show that eliminating the radial axis thrust reduces the fuel cost for formation maintenance. A nonlinear sliding mode control algorithm is developed to stabilize the 3-axis attitude of a spacecraft subject to actuator failures. The numerical simulation results along with the stability analysis establish the robustness of the proposed control scheme in stabilizing the attitude of a spacecraft by dealing with the presence of model uncertainties, time-varying external disturbances, and complete thruster failures simultaneously.

Appendix I

The nonlinear equations of motion represented in terms of transformed coordinates is given by

$$\begin{bmatrix} \dot{x}_1 \\ \dot{x}_2 \end{bmatrix} = \begin{bmatrix} A_{11} & A_{12} \\ A_{21} & A_{22} \end{bmatrix} \begin{bmatrix} x_1 \\ x_2 \end{bmatrix} + \begin{bmatrix} E_1(X) \\ E_2(X) \end{bmatrix} + \begin{bmatrix} 0_{4\times 2} \\ I_{2\times 2} \end{bmatrix} U_f \tag{107}$$

Case I - No control in the radial direction

$$x_1 = [x,\, y,\, z,\, \dot{x}]^T \quad \text{and} \quad x_2 = [\dot{y},\, \dot{z}]^T \tag{108}$$

$$A_{11} = \begin{bmatrix} 0 & 0 & 0 & 1 \\ 0 & 0 & 0 & 0 \\ 0 & 0 & 0 & 0 \\ 3n^2 & 0 & 0 & 0 \end{bmatrix}; \quad A_{12} = \begin{bmatrix} 0 & 0 \\ 1 & 0 \\ 0 & 1 \\ 2n & 0 \end{bmatrix}; \quad A_{21} = \begin{bmatrix} 0 & 0 & 0 & -2n \\ 0 & 0 & -n^2 & 0 \end{bmatrix} \tag{109}$$

$$A_{22} = \begin{bmatrix} 0 & 0 \\ 0 & 0 \end{bmatrix}; \; E_1(X) = \begin{bmatrix} 0 \\ 0 \\ 0 \\ \mu(\frac{1}{r_l^2} - \frac{(r_l + x)}{r_f^3}) - 2n^2 x \end{bmatrix}; \; E_2(X) = \begin{bmatrix} n^2 y - \frac{\mu y}{r_f^3} \\ n^2 z - \frac{\mu z}{r_f^3} \end{bmatrix} \tag{110}$$

Case II - No control in the along-track direction

$$x_1 = [x,\, y,\, z,\, \dot{y}]^T \quad \text{and} \quad x_2 = [\dot{x},\, \dot{z}]^T \tag{111}$$

$$A_{11} = \begin{bmatrix} 0 & 0 & 0 & 0 \\ 0 & 0 & 0 & 1 \\ 0 & 0 & 0 & 0 \\ 0 & 0 & 0 & 0 \end{bmatrix}; \quad A_{12} = \begin{bmatrix} 1 & 0 \\ 0 & 0 \\ 0 & 1 \\ -2n & 0 \end{bmatrix}; \quad A_{21} = \begin{bmatrix} 3n^2 & 0 & 0 & 2n \\ 0 & 0 & -n^2 & 0 \end{bmatrix} \tag{112}$$

$$A_{22} = \begin{bmatrix} 0 & 0 \\ 0 & 0 \end{bmatrix}; \; E_1(X) = \begin{bmatrix} 0 \\ 0 \\ 0 \\ n^2 y - \frac{\mu y}{r_f^3} \end{bmatrix}; \; E_2(X) = \begin{bmatrix} \mu(\frac{1}{r_l^2} - \frac{(r_l + x)}{r_f^3}) - 2n^2 x \\ n^2 z - \frac{\mu z}{r_f^3} \end{bmatrix} \tag{113}$$

Appendix II

In this section, we provide the nonlinear terms in the equation of motion of a rigid body spacecraft orbiting the Earth. The mathematical model of the spacecraft is formulated by considering eccentricity and product of inertia terms in the kinetic energy (T_e) and potential energy (U_p), and subsequently substituting them in the Lagrangian relation (Eq. 10). The proposed control algorithm is designed to accommodate eccentricity and product of inertia terms as uncertainties in the system model. Therefore, the control law is developed based on a simplified spacecraft model with null orbital eccentricity and spacecraft products of inertia (i.e., $e = 0$, $I_{xy} = I_{xz} = I_{yz} = 0$). The nonlinear terms in Eq. (14): F_α, F_ϕ, and F_γ are given by

$$\begin{aligned} F_\alpha &= p_\alpha \cos\phi\cos\gamma + p_\phi \cos\phi \sin\gamma - p_\gamma \sin\gamma \\ F_\phi &= -p_\alpha \sin\gamma + p_\phi \cos\gamma, \quad F_\gamma = p_\gamma \end{aligned} \tag{114}$$

The coefficients p_α, p_ϕ, and p_γ in Eq. (114) are

$$\begin{aligned} p_\alpha = {} & [(1 - k_{xz} + k_{yz})(1+\alpha')\phi' \sin\phi \cos\gamma] - (k_{xz} - k_{yz})(1+\alpha')^2 \sin\phi \\ & \cos\phi \sin\gamma + (1 + k_{xz} - k_{yz})[(1+\alpha')\gamma' \cos\phi \sin\gamma + \phi'\gamma' \cos\gamma] \\ & - 3(k_{xz} - k_{yz})(\cos\alpha \sin\phi \sin\gamma - \sin\alpha\cos\gamma)\cos\alpha\cos\phi \end{aligned}$$

$$\begin{aligned} p_\phi = {} & [(1 - k_{xz} + k_{yz})(1+\alpha')\gamma' \sin\phi \sin\gamma] - (1 - k_{xz})(1+\alpha')^2 \sin\phi \\ & \cos\phi \cos\gamma + (1 - k_{xz} - k_{yz})[(1+\alpha')\gamma' \cos\phi \cos\gamma - \phi'\gamma' \sin\gamma] \\ & + 3(1 - k_{xz})(\cos\alpha \sin\phi \cos\gamma + \sin\alpha \sin\gamma)\cos\alpha\cos\phi \end{aligned}$$

$$\begin{aligned} p_\gamma = {} & [k_{xz} - (1 - k_{yz})\cos 2\gamma](1+\alpha')\phi' \cos\phi - (1 - k_{yz})[(1+\alpha')^2 \cos^2\phi \\ & - \phi'^2]\sin\gamma\cos\gamma + 3(1 - k_{yz})(\cos\alpha\sin\phi\cos\gamma + \sin\alpha\sin\gamma) \\ & (\cos\alpha\sin\phi\sin\gamma - \sin\alpha\cos\gamma) \end{aligned}$$

The elements of the matrix N in Eq. (14) are given by

$$\begin{bmatrix} N_{11} & N_{12} & N_{13} \\ N_{21} & N_{22} & N_{23} \\ N_{31} & N_{32} & N_{33} \end{bmatrix}$$

where $N_{11} = \dfrac{\sin^2\gamma + k_{yz}\cos^2\gamma}{k_{yz}\cos^2\phi}$, $N_{12} = \dfrac{(1 - k_{yz})\sin\gamma\cos\gamma}{k_{yz}\cos\phi}$,

$N_{13} = \dfrac{\sin\phi(\sin^2\gamma + k_{yz}\cos^2\gamma)}{k_{yz}\cos^2\phi}$, $N_{21} = \dfrac{(1 - k_{yz})\sin\gamma\cos\gamma}{k_{yz}\cos\phi}$,

$$N_{22} = \frac{\cos^2\gamma + k_{yz}\,\sin^2\gamma}{k_{yz}},\quad N_{23} = \frac{(1-k_{yz})\sin\gamma\,\cos\gamma\,\sin\phi}{k_{yz}\,\cos\phi},$$

$$N_{31} = \frac{\sin\phi(\sin^2\gamma + k_{yz}\,\cos^2\gamma)}{k_{yz}\,\cos^2\phi},\quad N_{32} = \frac{(1-k_{yz})\sin\gamma\,\cos\gamma\,\sin\phi}{k_{yz}\,\cos\phi},$$

$$N_{33} = \frac{\sin^2\phi(\sin^2\gamma + k_{yz}\,\cos^2\gamma)}{k_{yz}\,\cos^2\phi} + \frac{1}{k_{xz}}.$$

References

1. K. D. Kumar, H. Bang and M. Tahk, *Satellite formation flying using along-track thrust*, in *Acta Astronautica*, Vol. 61 (2007), pp. 553-564.
2. Godard and K. D. Kumar, *Fault tolerant reconfigurable satellite formations using adaptive variable structure techniques*, in *Journal of Guidance, Control, and Dynamics*, Vol. 33, No. 3 (2010), pp. 969-984.
3. P. Tsiotras and V. Doumtchenko, *Control of spacecraft subject to actuator failures: State-of-the-art and open problems*, in *Journal of the Astronautical Sciences*, Vol. 48, No. 2-3 (2003), pp. 337-358.
4. R. Starin, R. Yedavalli, A. Sparks, *Design of a lqr controller of reduced inputs for multiple spacecraft formation flying*, in *Proceedings of the American Control Conference*, (Arlington, VA, USA, 2001).
5. J. Leitner, *Spacecraft formation flying- an overview of missions and technology challenges*, in *Advances in the Astronautical Sciences, Guidance and Control*, AAS 07-031 (Breckenridge, CO, USA, 2007), pp. 125-134.
6. Godard and K. D. Kumar, *Robust attitude stabilization of spacecraft subject to actuator failures*, in *Acta Astronautica*, doi:10.1016/j.actaastro.2010.10.017.
7. R. Brockett, *Asymptotic stability and feedback stabilization, in: Differential Geometric Control Theory*, eds. R. W. Brockett, R. S. Millman, H. H. Sussmann, (Birkhauser, Boston, 1984), pp. 181-208.
8. V. Vaddi, Modelling and control of satellite formations, PhD thesis, Texas A& M University (TX, USA, 2003).
9. K. K. Shyu, Y. W. Tsai, and C. K. Lai, *Sliding mode control for mismatched uncertain systems* in *Electronics Letters*, Vol. 34 (1998), pp. 2359-2360.
10. C. Edwards and S. Spurgeon, *Sliding Mode Control - Theory and Applications* (Taylor and Francis Ltd, London, UK, 1998).

SOME NEW CLASSES OF INVERSE COEFFICIENT PROBLEMS IN ENGINEERING MECHANICS AND COMPUTATIONAL MATERIAL SCIENCE BASED ON BOUNDARY MEASURED DATA *

A. HASANOV

Department of Mathematics and Computer Science
Izmir University, 35350, Izmir - Turkey
E-mail:alemdar.hasanoglu@izmir.edu.tr

Abstract. Three classes of inverse coefficient problems arising in engineering mechanics and computational material science are considered. Mathematical models of all considered problems are proposed within the J_2-deformation theory of plasticity. The first class is related to determination of unknown elastoplastic properties of a beam from a limited number of torsional experiments. The inverse problem here consists of identifying the unknown coefficient $g(\xi^2)$ (plasticity function) in the nonlinear differential equation of torsional creep$-(g(|\nabla u|^2)u_{x_1})_{x_1} - (g(|\nabla u|^2)u_{x_2})_{x_2} = 2\varphi$, $x \in \Omega \subset R^2$, from the torque (or torsional rigidity) $\mathcal{T}(\varphi)$, given experimentally. The second class of inverse problems is related to identification of elastoplastic properties of a $3D$ body from spherical indentation tests. In this case one needs to determine unknown Lame coefficients in the system of PDEs of nonlinear elasticity, from the measured spherical indentation loading curve $\mathcal{P} = \mathcal{P}(\alpha)$, obtained during the quasi-static indentation test. The third model an inverse problem of identifying the unknown coefficient $g(\xi^2(u))$ in the nonlinear bending equation is analyzed. The boundary measured data here is assumed to be the deflections $w_i[\tau_k] := w(\lambda_i;\tau_k)$, measured during the quasi-static bending process, given by the parameter τ_k, $k = \overline{1,K}$, at some points $\lambda_i = (x_1^{(i)}, x_2^{(i)})$, $i = \overline{1,M}$ of a plate. An existence of weak solutions of all direct problems are derived in appropriate Sobolev spaces, by using monotone potential operator theory. Then monotone iteration schemes for all the linearized direct problems are proposed. Strong convergence of solutions of the linearized problems, as well as rates of convergence are proved. Based on obtained continuity property of the direct problem solution with respect to coefficients, and compactness of the

*The results has been announced at the Satellite Conference of International Congress of Mathematicians, 14-17 August, 2010, Delhi - India.

set of admissible coefficients, an existence of quasi-solutions of all considered inverse problems are proved. Some numerical results, useful from the poins of view engineering mechanics and computational material science, are demonstrated.

Key words. inverse coefficient problem, nonlinear monotone potential operator, torsional rigidity, plasticity function, indentation test, bending plate, monotone iteration scheme, convergence, existence of a quasisolution

AMS(MOS) subject classifications. 35R30, 47H50, 74B20

1. Introduction

Determination of unknown materials properties of based on boundary/surface measured data is one of central and actual problems of computational material sciences (see, [4], [10], [18], [25-26] and references therein). Mathematical modeling of these problems leads to inverse coefficient problems for nonlinear PDEs of various types ([9-10], [15], [17-18], [27]]). It is known from inverse problems theory that inverse coefficient problems are most difficult in comparison with all other types inverse problems. Moreover, these problems are severely ill-posed even when the governing equations are linear one ([12], [15]) which means that very close measured output data may correspond to quite different materials (i.e. coefficients).

The present study deals with the following three types of inverse problems governed by nonlinear PDEs, and related to determination of unknown properties of engineering materials based on boundary/surface measured data:

(ICP1): determination of unknown elastoplastic properties of power hardening materials from limited torsional experiment;

(ICP2): identification of unknown elastoplastic properties of engineering materials from indentation tests;

(ICP3): identification of the unknown parameters of an incompressible bending plate from deflections measured at some points.

In view of Materials Theory and Computational Materials Science all these classes of problems can be defined as a *problem of determination of unknown material properties* from boundary/surface measured data. In pactice, these measured data can be given in the forms of additional Diriclet or Neumann types of boundary conditions, or nonlocal additional conditions (i.e. integral operator). From the mathematical theory of inverse problems all the three classes of problems are defined to be as *inverse coefficient problems based on boundary measured output data* [15]. An unknown coefficient is defined to be *an input data.* Introducing the input-output mapping $\Phi : \mathcal{K} \mapsto \Pi$, $\Phi(\mathcal{K}) \subset \Pi$, from the *class of admissible coefficients* K to the *class of admissible measured output data P*, we may formulate all the above problems in the form of the operator equation as follows:

$$\Phi k = \mathcal{P}, \quad k \in \mathcal{K}, \quad \mathcal{P} \in \Pi. \tag{1}$$

The inverse operator $\Phi^{-1} : \Phi(\mathcal{K}) \subset \Pi \mapsto \mathcal{K}$ is not continuous which means ill-posedness of inverse problems, according to [32]. Moreover, it is known that inverse coefficient problems are severely ill-posed [12, 15, 18].

In the first class of problems (defined to be as (ICP1)) one needs to identify elastoplastic properties of a bar (or beam), from limited number of torsional experiments. In elastic behavior case, many analytical, computational and experimental studies has already been presented in scientific literature, especially when small amounts of twisting are reversible, and the beam will return to its original shape after releasing the twisting force. Numerical study of elastoplastic torsion based on finite difference and finite element approaches have been presented in [2-3]. The theory of space curved beams with arbitrary cross-sections with associated finite element formulation is implemented in [8] for three-dimensional beams with elastoplastic material behaviour. A rigid-plastic finite element analysis for torsions of circular, square and rectangular beam sections, related to metal forming, has been presented in [29]. However inverse problems related to determination of elastoplastic properties of a beam during torsional creep is less well known and has received relatively little attention in mathematical as well as engineering literature. A first attempt to study the elastoplastic properties of a bar during quasistatic has been given in [27], and then developed in [16-18].

The second class of inverse problems (defined to be as (ICP2)) is related to determination of unknown elastiplastic properties of materials during spherical indentation testing. Note that the spherical indentation testing is one of extensively used experimental methods to measure the hardness of metal and polymer materials (see, [4], [15], [24], [26], [33] and references therein). The objective here is usually to analyze the indentation curve as dependent on the size of the sample (indent) and indenter relative to the material length parameters, strain hardening, and yield stress to modulus ratio. The idea of relating the mechanical properties of deformable materials to their hardness has first been given in [20]. By using the method of characteristics for hyperbolic equations, the relationship $\sigma_0 = 0.383\tilde{H}_B$ between the Brinell hardness $\tilde{H}_B = \mathcal{P}/(2\pi R\alpha)$, and the yield stress $\sigma_0 > 0$ has been established for a spherically symmetric indentation hardness test. Here and below $\mathcal{P} > 0$, $R > 0$ and $\alpha > 0$ are assumed to be the measured loading force, radius of a spherical indenter and the indentation depth, accordingly. However in the model proposed in [20], the curvature of a contactable surface was ignored, and the problem was considered for a perfectly plastic material. Within the framework of the J_2-deformation theory of plasticity, which is most siutable in the case of small deformations, the inverse problem of identification of unknown elastoplastic properties of engineering materials from an indentation loading curve has been proposed in [10], and then developed in [12, 15]. An experimantal and numerical analysis of ill-conditionedness of this problem has been given in [25].

The third class of inverse problems (defined to be as (ICP3)) is related to determination of unknown elastiplastic properties of a bending plate under the action of normal loads. The mathematical model of this problem leads to the problem of identification of the unknown coefficient in the nonlinear biharmonic (bending) equation from deflections measured at some points on the middle surface of a plate. This inverse problem has first been formulated in [9]. An analysis of the corresponding nonlinear direct problem, based on J_2-deformation theory of plasticity, is given in [14].

The aim of this work is to generalize various studies, related to determination of unknown properties of engineering materials, from mathematical and engineering literatute in order to show that, all the above defined classes of inverse/identification problems can be studied from a common point mathematical and computational poits of view, since they have similar/common distingushed features. Our methodology is based on the variational approach with the monotone potential operator and weak solution theory for PDEs for corresponding nonlinear direct problems. For all inverse problems the quasisolution method is applied to obtain an existence of a quasisolution in appropriate class of admissible coefficients.

The paper is organized as follows. In Section 2 physical and mathematical models for each inverse problems are proposed. In Section 3 we prove that the set of admissible coefficients is compact in the Sobolev space H^1 under the natural conditions of the J_2-deformation theory of plasticity. Mathematical frameworks of the problems (ICP1), (ICP2) and (ICP3) are given Section 4, Section 5 and Section 6, accordingly. In Section 7 parametrization of unknown coefficient and the regularization algorithm are described. Computational results with noise free and noisy data are presented in Section 8. Some concluding remarks are given in the final Section 9.

2. Physical and mathematical models of the inverse problems

2.1. Determination of elastoplastic properties of a beam from limited torsional experiment

Let us consider the torsion of a homogeneous isotropic beam, which cross section occupies the domain $\Omega := (0, l_1) \times (0, l_2)$, $l_i > 0$, under the load (torque) applied. Torsional rigidity is defined as the torque required for per unit angle of twist $\varphi > 0$ per unit length, when the Young's modulus of the material is set equal to one [22]. Specifically, if $u = u(x)$, $x = (x_1, x_2) \in \Omega \subset R^2$, denotes Prandtl stress function, then the torque (or torsional rigidity) is defined to be as the integral of $u = u(x)$ over the domain Ω:

$$T[g](\varphi) = 2 \int_{\Omega} u(x; g; \varphi) dx. \tag{2}$$

The boundary value problem

$$\begin{cases} -\nabla(g(|\nabla u|^2)\nabla u) = 2\varphi, & x \in \Omega \subset R^2, \\ u(x) = 0, & x \in \partial\Omega, \end{cases} \tag{3}$$

represents an elastoplastic torsion of a strain hardening beam, which lower end is fixed, i.e. rigid clamped [24]. The symmetricity axis of the beam is assumed to be parallel to the axix Ox_1 (Figure 1 (left figure)). The torque $\mathcal{T} = \mathcal{T}(\varphi)$, corresponding to the angle of twist $\varphi > 0$ per unit length, is applied to the upper end of the beam.

The function $g = g(\xi^2)$ defined to be the *plasticity function*, describes elastoplastic properties of a homogeneous isotropic material, and $\xi(u) = [(\partial u/\partial x_1)^2 + (\partial u/\partial x_2)^2]^{1/2}$ is the stress intensity,

In view of J_2-deformation theory of plasticity, this function describes elastoplastic properties of a homogeneous isotropic beam, and satisfies the following conditions [18]

$$\begin{cases} \textbf{(i)}\ 0 \le c_0 \le g(\xi^2) \le c_1, \\ \textbf{(ii)}\ g'(\xi^2) \le 0, \\ \textbf{(iii)}\ g(\xi^2) + 2\xi^2 g'(\xi^2) \ge \gamma_0 > 0, \quad \xi \in [\xi_*, \xi^*] \\ \textbf{(iv)}\ g(\xi^2) = g_0, \quad \xi \in [\xi_*, \xi_0], \quad \xi_0 \in (\xi_*, \xi^*). \end{cases} \tag{4}$$

Here $g_0 = 1/G$ is defined to be the shear compliance, $G = E/(2(1+\nu))$ is the elastic shear modulus, $E > 0$ is the Young's modulus and $\nu \in (0, 0.5)$ is the Poisson coefficient. The value $\xi_0^2 = \max_{x \in \Omega} |\nabla u(x)|^2$ is assumed to be the elasticity limit of a material. Here and below the Poisson coefficient $\nu > 0$ is assumed to be known.

In the considered physical model the quasistatic process of torsion is simulated by the monotone increasing values $0 < \varphi_* = \varphi_1 < \varphi_2 < \ldots <$

$\varphi_m = \varphi^*$ of the angle of twist $\varphi \in [\varphi_*, \varphi^*]$, $\varphi_* > 0$. Hence the torque $\mathcal{T} := \mathcal{T}(\varphi)$, which theoretical value is defined by (1), will be considered as a function of the angle $\varphi > 0$. For a given material, i.e. for the given plasticity function $g = g(\xi^2)$, the solution of the nonlinear boundary value problem (3), corresponding to a given value $\varphi \in [\varphi_*, \varphi^*]$ of the angle of twist, will be defined to be as $u(x) := u(x; g; \varphi)$.

The *inverse coefficient problem* (the problem (ICP1)) here consists of determining the unknown coefficient $g = g(\xi^2)$ in the nonlinear elliptic equation (4), from the experimentally given values $\mathcal{T}(\varphi)$ of the torque:

$$\begin{cases} -\nabla(g(|\nabla u|^2)\nabla u) = 2\varphi, \quad x \in \Omega \subset R^2, \\ u(x) = 0, \quad x \in \partial\Omega, \\ 2\int_\Omega u(x; g; \varphi)dx = \mathcal{T}(\varphi). \end{cases} \tag{5}$$

In this contex, for a given angle $\varphi \in [\varphi_*, \varphi^*]$ and $g(\xi^2)$, the nonlinear boundary value problem (4) will be defined to be as the *direct (forward) problem*. The functions $g = g(\xi^2)$ and $\mathcal{T}(\varphi)$ will be defined to be *input data* and *measured output data*, respectively.

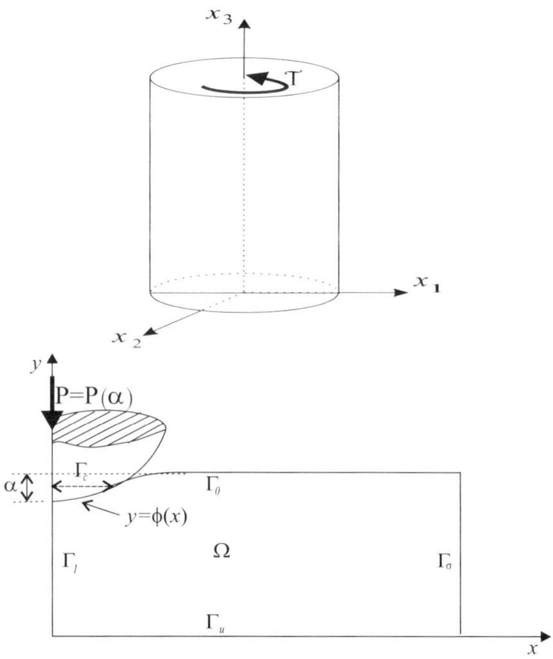

Fig. 1. Geometry of the torsion (left figure) and the spherical indentation (right figure)

2.2. Identification of elastoplastic properties of metallic materials from spherical indentation tests

Let the rigid spherical indenter be loaded with a normal loading force $\mathcal{P}$, into an axially symmetric homogeneous body (sample) occupying the domain $\Omega \times [0, 2\pi]$, in the negative y-axis direction, as shown in (Figure 1 (right figure)). The uniaxial quasi-static indentation testing is simulated by the monotonically increasing value $\alpha > 0$ of the indentation depth. It is assumed that the indentation process is carried out without unloading, moment and friction. For a given value $\alpha \in (0, \alpha^*)$ of the indentation depth the quasi-static axisymmetric indentation process can be modeled by the following contact problem.

The mathematical model of the problem of identification of elastoplastic properties from indentation tests is as follows [10]:

$$\begin{cases} \begin{cases} -\frac{\partial}{\partial x}(x\sigma_{11}(u)) - \frac{\partial}{\partial y}(x\sigma_{12}(u)) + \sigma_{33}(u) = F_1(x,y), \\ -\frac{\partial}{\partial x}(x\sigma_{12}(u)) - \frac{\partial}{\partial y}(x\sigma_{22}(u)) = F_2(x,y), \quad (x,y) \in \Omega \subset R^2; \end{cases} \\ \\ \begin{cases} u_2(x,l_y) \le -\alpha + \varphi(x), \quad \sigma_{22}(u) \le 0, \quad [u_2(x,y) + \alpha - \varphi(x)]\sigma_{22}(u) = 0, \\ \sigma_{12}(u) = 0, \quad (x,y) \in \Gamma_0; \end{cases} \\ \\ \sigma_{11}(u) = 0, \quad \sigma_{12}(u) = 0, \quad (x,y) \in \Gamma_\sigma; \\ u_1(0,y) = 0, \quad \sigma_{12}(u) = 0, \quad (x,y) \in \Gamma_1; \\ \sigma_{12}(u) = 0, \quad u_2(x,0) = 0, \quad (x,y) \in \Gamma_u; \end{cases} \tag{6}$$

$$-2\pi \int_{\Gamma_c(\alpha)} \sigma_{22}(u[\sigma_{\rm i}])x dx = \mathcal{P}(\alpha). \tag{7}$$

Here $u(x,y) = (u_1(x,y), u_1(x,y))$ is the displacement vector, $\Omega = \{(x,y) \in R^2 : 0 < x < l_x, 0 < y < l_y\}$, $l_x, l_y > 0$ is the cross section of the domain occupied by the sample. $\Gamma_\sigma = \{(l_x, y) : 0 < y < l_y\}$, $\Gamma_0 = \{(x, l_y) : 0 \le x \le l_x\}$ is the part of the boundary $\partial\Omega$ of the sample beyond on the contact, where the "free boundary conditions" are given. The symmetry of the sample implies the boundary conditions on $\Gamma_1 = \{(0,y) : 0 < y < l_y\}$. It is assumed that an axisymmetric sample lies on a substrate without friction, as conditions on $\Gamma_u = \{(x,0) : 0 \le x \le l_x\}$ show. Here $\varphi(x) = \sqrt{r_0^2 - x^2}$ is the surface of the spherical indenter, with the radius r_0. The contact conditions in (6), in the form of inequality, means that the contact domain $\Gamma_c(\alpha) = \{(x, l_y) \in \Gamma_0 : u_2(x, l_y) = -\alpha + \varphi(x), \ \ x \in (0, a_c(\alpha))\}$, $\ a_c(\alpha) := \partial\Gamma_c(\alpha)$, depending on the value $\alpha \in [0, \alpha^*]$ of the indentation depth, is also unknown and need to be defined.

The relationship between the components of strain and stress tensors is as follows:

$$\sigma_{ii}(u) = \tilde{\lambda}\theta(u) + 2\tilde{\mu}\varepsilon_{ii}(u),\ i = 1, 2, 3;\quad \sigma_{12}(u) = 2\tilde{\mu}\varepsilon_{12}(u), \tag{8}$$

where $\varepsilon_{11}(u) = \partial u_1/\partial x$, $\varepsilon_{22}(u) = \partial u_2/\partial y$, $\varepsilon_{33}(u) = u_1/x$, $\varepsilon_{12}(u) = 0.5(\partial u_1/\partial y + \partial u_2/\partial x)$, $\theta(u) = (\varepsilon_{11}(u) + \varepsilon_{22}(u) + \varepsilon_{33}(u))/3$ the components of deformation, and

$$\tilde{\lambda} = \lambda + \frac{2\mu g(e_{\mathrm{i}}^2)}{3}, \tilde{\mu} = \mu(1 - g(e_{\mathrm{i}}^2)), \lambda = \frac{E\nu}{(1+\nu)(1-2\nu)},\ \mu = \frac{E}{2(1+\nu)} \tag{9}$$

$\lambda, \mu > 0$ are Lame constants, $E > 0$ is an elasticity modulus, ν is the Poisson coeficient and $G = \mu$ is the modulus of rigidity.

According to the J_2-deformation theory the stress-strain relationship here is assumed to be in the following form:

$$\sigma_{\mathrm{i}}(e_{\mathrm{i}}) = 3G[1 - g(e_{\mathrm{i}}^2)]e_{\mathrm{i}},\quad e_{\mathrm{i}} \in (0, e_{\mathrm{i}}^*), \tag{10}$$

where $e_{\mathrm{i}}(u) = (2/3)\{\sum_{i,j=1,3}^{3}[\varepsilon_{ii}(u) - \varepsilon_{jj}(u)]^2 + 3\varepsilon_{12}^2(u)\}^{1/2}$, is the strain intensity. The function $\sigma_{\mathrm{i}} = \sigma_{\mathrm{i}}(e_{\mathrm{i}})$, given by (10), describes the elastoplastic behaviour of a wide range of engineering materials, and is assumed to be smooth, monotone increasing and concave one (Figure 2 (left figure)):

$$\begin{cases} \sigma_{\mathrm{i}}(e_{\mathrm{i}}) \in C^2(0, e_{\mathrm{i}}^*);\ \frac{d\sigma_{\mathrm{i}}}{de_{\mathrm{i}}} > 0 \text{ (monotonicity)},\quad e_{\mathrm{i}} \in (0, e_{\mathrm{i}}^*); \\ \frac{d\sigma_{\mathrm{i}}}{de_{\mathrm{i}}} - \frac{\sigma_{\mathrm{i}}}{e_{\mathrm{i}}} < 0,\quad e_{\mathrm{i}} \in (0, e_{\mathrm{i}}^*) \text{ (concavity)}; \\ \sigma_{\mathrm{i}} = 3\mu e_{\mathrm{i}},\quad e_{\mathrm{i}} \in (0, e_0) \text{ (pure elastic deformations)}. \end{cases} \tag{11}$$

As a particular case of the stress-strain relationship (10), the power law description

$$g(\xi) = 1 - (\xi/\xi_0)^{0.5(\kappa-1)},\quad \kappa \in [0, 1] \tag{12}$$

is also widely used to approximate the plastic behaviour of metal materials (see, [4], [25-26] and references therein), corresponds to the power hardening materials for which the stress-strain relation is given by the Ramberg-Osgood curve $\sigma_{\mathrm{i}} = \sigma_0(e_{\mathrm{i}}/e_0)^{\kappa}$. Here $\kappa \in [0, 1]$ is a strain hardening exponent. The cases $g(\xi) = 0$ and $g(\xi) = 1 - \sqrt{\xi_0/\xi}$ in (8) correspond to pure elastic ($\kappa = 1$) and perfectly plastic ($\kappa = 0$) materials, respectively.

The *inverse coefficient problem* (the problem (ICP2)) consists of identifying the stress-strain curve $\sigma_{\mathrm{i}} = \sigma_{\mathrm{i}}(e_{\mathrm{i}})$, given by (10), from the measured spherical indentation loading curve $\mathcal{P} = \mathcal{P}(\alpha)$, $\alpha \in (0, \alpha_K)$, obtained during the quasi-static indentation process. Accordingly, for a given function $\sigma_{\mathrm{i}}(e_{\mathrm{i}})$, the unilateral boundary value problem (6) is defined to be the *direct problem.*

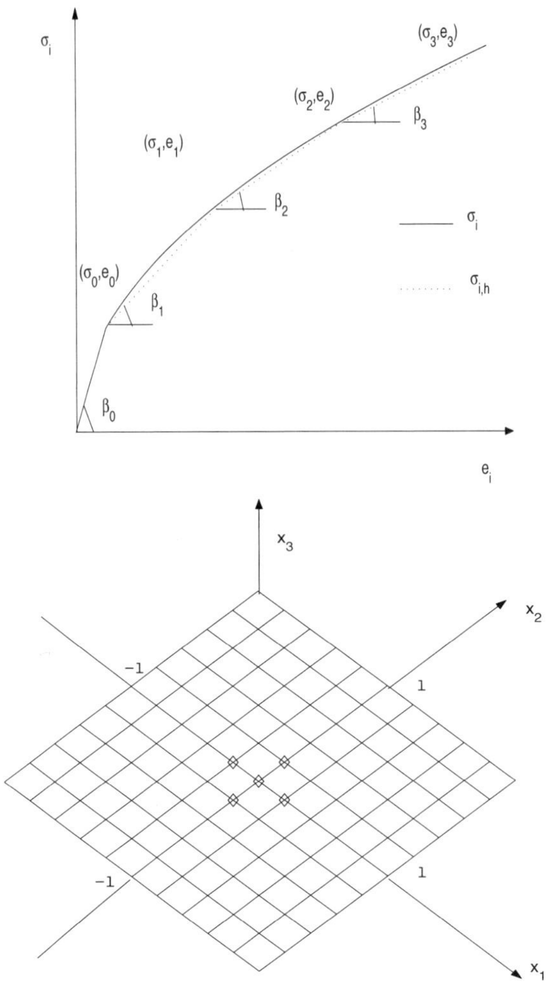

Fig. 2. The function $\sigma_{\mathrm{i}} = \sigma_{\mathrm{i}}(e_{\mathrm{i}})$ (left figure); Geometry of the bending plate under the normal load applied to some points (right figure)

In pure elastic deformations case ($e_{\mathrm{i}} \in (0, e_0)$), only the elasticity modulus $E > 0$ needs to be defined, since the Poisson coefficient $\nu > 0$ is assumed to be known. In plastic deformations case ($e_{\mathrm{i}} > e_0$) the plasticity function $g(e_{\mathrm{i}}^2)$ needs to be reconstructed, as formula (10) shows.

As the problem the problem (ICP1), this inverse problem also is defined to be the *inverse coefficient problem with nonlocal (integral form) measured output data*, according to inverse problems terminology.

Note that indeed the indentation curve $\mathcal{P} = \mathcal{P}(\alpha)$ consists of loading and unloading parts. In the considered model we use only the loading part of this curve, and is assumed that the indentation process is carried out without unloading. Hovewer even in this simplest from the point of view physical model, the inverse coefficient problem (6)-(7) is quite difficult, due to nonlinearity of the differential operator and unilateral boundary conditions (6). Moreover, the problem is severely ill-posed due to the nonlocal additional condition (7).

2.3. Identification of elastoplastic properties of an incompressible bending plate from measured deflections

Let us assume that the coordinate plane Ox_1x_2 is the middle surface of an isotropic homogeneous incompressible plate with thichness $h > 0$. According to occupying the square $\Omega = \{(x_1, x_2) \in R^2 : -l < x_1,\ x_2 < l,\ l > 0\}$. Suppose that the measured values $w_i[\tau_k] := w(\lambda_i; \tau_k)$, $k = \overline{1, K}$, of deflections at the points $\lambda_i = (x_{1i}, x_{2i})$, $i = \overline{1, M}$, correspond to the given values of the external normal load $q(x; \mathcal{T}) > 0$ to the middle surface of the plate (Figure 2 (right figure)). The loading is assumed to be quasistatic, generating by the increasing values $0 < \tau_1 < \tau_2 < ... < \tau_K$ of the loading parameter $\mathcal{T}$. Finally, without loss of generality, assume that an experiment is realized under the rigid clamped boundary conditions $u(x) = \partial u/\partial n = 0$, where n is a unit outward normal to the boundary $\partial\Omega$ of a plate.

Under these conditions the *problem of identification of elastoplastic properties, given by the function $g(\xi^2(u))$, of an incompressible bending plate from measured deflections* (the problem (ICP3)) can be modeled as follows:

$$\begin{cases} Au \equiv \frac{\partial^2}{\partial x_1^2}\left[g(\xi^2(u))\left(\frac{\partial^2 u}{\partial x_1^2} + \frac{1}{2}\frac{\partial^2 u}{\partial x_2^2}\right)\right] + \frac{\partial^2}{\partial x_1 \partial x_2}\left[g(\xi^2(u))\left(\frac{\partial^2 u}{\partial x_1 \partial x_2}\right)\right] \\ +\frac{\partial^2}{\partial x_2^2}\left[g(\xi^2(u))\left(\frac{\partial^2 u}{\partial x_2^2} + \frac{1}{2}\frac{\partial^2 u}{\partial x_1^2}\right)\right] = F(x)\ , \quad x \in \Omega \subset R^2, \\ u(x) = 0, \ \ \frac{\partial u}{\partial n}(x) = 0, \quad x \in \partial\Omega, \end{cases} \tag{13}$$

$$w_i[\tau_k] := w(\lambda_i; \tau_k), \quad i = \overline{1, M}, \quad k = \overline{1, K}. \tag{14}$$

Here $F(x) = 3q(x; \mathcal{T})/h^3$, $q(x; \mathcal{T})$ is the intensity (per unit area) of the load.

The process of bending is assumed to be quasistatic, generating by the increasing values of the normal load $q(x; \mathcal{T})$, corresponding to the values $0 < \tau_1 < \tau_2 < ... < \tau_K$ of the loading parameter $\mathcal{T}$. According to J_2-deformation theory of plasticity the coefficient $g(\xi^2(u))$ of the nonlinear biharmonic equation (11) depends on the effective value of the plate curvature [14]

$$\xi^2(u) = \left(\frac{\partial^2 u}{\partial x_1^2}\right)^2 + \left(\frac{\partial^2 u}{\partial x_2^2}\right)^2 + \left(\frac{\partial^2 u}{\partial x_1 \partial x_2}\right)^2 + \frac{\partial^2 u}{\partial x_1^2}\frac{\partial^2 u}{\partial x_2^2}, \qquad (15)$$

which in turns, depends on the deflection function $u(x)$, i.e. on the solution of problem (11). The coefficient $g(\xi^2(u))$, defined to be as the plasticity function, describes the elastoplasticitic properties of an increasingly hardening plate and also satisfy conditions (4).

Thus the problem (ICP3) is the coefficient identification problem for the nonlinear fourth order (biharmonic) equation.

3. Compactness of the set of admissible coefficients

Let us analyze the compactness of the set of coefficients $\{g(\xi^2)\}$ in the Sobolev space in $H^1[\xi_*, \xi^*]$ [1]. We consider this function in the interval $[\xi_0, \xi^*]$, since in $[\xi_*, \xi_0]$, $\xi_* > 0$, the function $g(\xi^2)$ is a constant. Denote by $\mathcal{G} := \{g(\xi^2) : \xi \in [\xi_*, \xi_0]\}$, the set of admissible coefficients, satisfying the first three conditions of (4).

Lemma 3.1. *Let the set of admissible coefficients $\mathcal{G}$, in addition to conditions **(i)-(iii)** of (4), satisfies also the condition*

$$g'(\xi^2) \text{is a monotone decreasing (or increasing) function.} \qquad (16)$$

Then this set is compact in the Sobolev space in $H^1[\xi_*, \xi^*]$.

Proof. Let us first define the set of coefficients $\mathcal{G}_1 := \{g(\xi^2)\}$ satisfying only the first (boundedness) and second (monotonicity) conditions **(i)** and **(ii)** of (4). In it well known that the class of monotone increasing and uniformly bounded functions is compact in $L_2 \equiv H^0$, according to Tikhonov's lemma [32]. Hence the set $\mathcal{G}_1$ is compact in $H^0[\xi_*, \xi^*]$. Now we use the condition **(iii)** to show the uniform boundedness of the set $\{g'(\xi^2)\}$. We have

$$g(\xi^2) + 2g'(\xi^2)\xi_0^2 \geq g(\xi^2) + 2g'(\xi^2)\xi^2 \geq \gamma_0 > 0, \quad \forall \xi \in [\xi_*, \xi^*],$$

due to $g'(\xi^2) \leq 0$, and since $\xi \geq \xi_0 > 0$. This inequality implies:

$$0 \geq g'(\xi^2) \geq -\frac{c_1 - \gamma_0}{2\xi_0^2}, \quad \xi_0 > 0. \qquad (17)$$

Evidently $c_1 > \gamma_0$, due to $0 < \gamma_0 \leq g(\xi^2) + 2g'(\xi^2)\xi^2 \leq c_1$.

Estimate (18) show that the set $\{g'(\xi^2)\}$ is uniform bounded, in particular in $H^0[\xi_*, \xi^*]$. Now we use condition (17) and apply again Tikhonov's lemma to the set of uniform bounded and monotone decreasing (or increasing) functions $\{g'(\xi^2)\}$. We conclude that this set is compact in $H^0[\xi_*, \xi^*]$. This implies the proof. □

Note that this approach of constructon the compact class of admissible coefficients in the Sobolev space $H^1[\xi_*, \xi^*]$ has been proposed in [12].

The set of coefficients defined by the conditions of Lemma 3.1 will defined as the *compact set of admissible coefficients* $\mathcal{G}_c$.

4. An analysis of the problem (ICP1)

Let $\mathcal{G}_c$ be the above defined compact set of admissible coefficients. Assume that the monotone increasing continuous function $\mathcal{T} = \mathcal{T}(\varphi)$, $\varphi \in [\varphi_*, \varphi^*]$, $\varphi_* > 0$, represents an experimentally given torque. Then the inverse problem (5) (the problem (ICP1)) can be formulated as a solution of the following nonlinear functional equation:

$$T[g](\varphi) := 2\int_\Omega u(x; g; \varphi)dx = \mathcal{T}(\varphi), \quad g \in \mathcal{G}. \tag{18}$$

Here the function $u(x) := u(x; g; \varphi) \in \overset{\circ}{H}{}^1(\Omega)$ is defined to be the weak solution of the direct problem (3), for the given function $g \in \mathcal{G}_c$, and satisfies the following integral identity:

$$\int_\Omega g(|\nabla u|^2)\nabla u \nabla v dx = 2\varphi \int_\Omega v dx, \quad \forall v \in \overset{\circ}{H}{}^1(\Omega). \tag{19}$$

We define (2) (or the left hand side of (18)) as an *input-output map* $T[\cdot] : \mathcal{G} \mapsto \mathbf{T}$ from the class of *admissible coefficients* $\mathcal{G}_c$ to the class $\mathbf{T}$ of output functions $T[g] \in \mathbf{T}$. Then the the problem (ICP1) with the given measured output data $\mathcal{T} = \mathcal{T}(\varphi)$ can be reduced to the solution of the nonlinear operator equation (19) or to inverting the input-output map $T[\cdot] : \mathcal{G} \mapsto \mathbf{T}$.

Using the integral identity (19) we may derive some useful properties of a solution of the problem (ICP1). Specifically, substituting here $v = u$ and then using the nonlocal additional condition (5) we obtain the following characterization of a solution of this inverse problem.

Lemma 4.1. *If the function $g \in \mathcal{G}_c$ is a solution of the problem (ICP1), then it satisfies the following energy identity:*

$$\int_\Omega g(|\nabla u|^2)|\nabla u|^2 dx = \varphi \mathcal{T}(\varphi), \quad u \in \overset{\circ}{H}{}^1(\Omega), \quad \varphi \in [\varphi_*, \varphi^*].$$

Since $g(\xi^2) \geq c_0 \geq 0$, the above identity implies, in particular, that the following boundedness of H^0-norm $\|\nabla u\|_0$ of the solution $u \in \overset{\circ}{H}{}^1(\Omega)$ via the measured output data $\mathcal{T}(\varphi)$:

$$\|\nabla u\|_0^2 \leq c_0^{-1}\varphi^* \mathcal{T}(\varphi^*).$$

Here $\|\nabla u\|_0$ is the norm of the Sobolev space $H^0(\Omega)$ which is equivalent to the norm $\|u\|_1$ of the Sobolev space $H^1(\Omega)$, due to the homogeneous Dirichlet condition (4) [1].

To analyze the input-output map $T[\cdot]:\mathcal{G}\mapsto \mathbf{T}$, first consider the the problem (ICP1) for pure elastic torsion. In this case $g(\xi^2)=1/G$, and we may apply the maximum principle to the linear direct problem

$$\begin{cases} -\Delta u = 2G\varphi, & x\in\Omega, \\ u(s)=0, & s\in\Gamma=\partial\Omega. \end{cases}$$

Since the right hand side $2G\varphi$ is positive, we conclude $u(x;G;\varphi)>0$, $\forall x\in\Omega$. Let us now assume that $G_1>G_2>0$ and denote by $u_i(x):=u(x;G_i;\varphi)$, $i=1,2$, the corresponding solutions of the boundary value problems:

$$\begin{cases} -\Delta u_i = 2G_i\varphi, & x\in\Omega;\quad i=1,2; \\ u_i(s)=0, & s\in\Gamma=\partial\Omega. \end{cases}$$

where $u_i(x)=u(x;G_i;\varphi)$, $i=1,2$. Denoting by $v(x)=u_1(x)-u_2(x)$ we conclude that the function $v(x)$ is the solution of the following problem

$$\begin{cases} -\Delta v = \tilde{F}(x), & x\in\Omega \\ v(s)=0, & s\in\Gamma, \end{cases}$$

where $\tilde{F}(x)=2G_1\varphi-2G_2\varphi>0$. By the maximum principle we conclude that $v(x)>0,\ \ \forall x\in\Omega$ which means $u_1(x)>u_2(x),\ \ \forall x\in\Omega$. Taking into account the definition of the torque we obtain:

$$T[G_1](\varphi):=2\int_\Omega u(x;G_1;\varphi)dx > 2\int_\Omega u(x;G_2;\varphi)dx := T[G_2](\varphi), \quad (20)$$

for $G_1>G_2$.

Thus we get the following lemma.

Lemma 4.2. *For pure elastic torsion the input-output map $T[\cdot]:\mathcal{G}\mapsto\mathbf{T}$ is an isotone one, i.e. monotone increasing and therefore order preserving: $\forall G_1,G_2\in\mathcal{G}$, $G_1>G_2$ implies $T[G_1](\varphi)>T[G_2](\varphi)$, $\forall\varphi\in[\varphi_*,\varphi^*]$.*

This result (the inequality (20)) has a precise physical meaning: for a fixed angle of twist $\varphi\in[\varphi_*,\varphi^*]$, an increase of the shear modulus $G>0$ leads to an increase of the rigidity of a material, and as a result, leads to increase of the torque.

In addition to the above monotonicity property, one can prove that the input-output map $T[\cdot]:\mathcal{G}\mapsto\mathbf{T}$ is also continuous. To show this we need the following estimate:

Lemma 4.3. *Let $u_1(x):=u[x;g_1;\varphi]$, $u_2(x)=u[x;g_2;\varphi]\in\overset{\circ}{H}{}^1(\Omega)$ be solutions of the direct problem (3), corresponding to the given functions $g_1,g_2\in\mathcal{G}$. Then for the input-output map $T[\cdot]:\mathcal{G}\mapsto\mathbf{T}$ the following estimate holds for $\alpha,\ c_\Omega>0$:*

$$\|T[g_1]-T[g_2]\|_{C[\xi_*,\xi^*]}\leq 2\alpha^{-1}c_\Omega\,(meas\ \Omega)^{1/2}\|\nabla u_2\|_0\|g_1-g_2\|_{C[\xi_*,\xi^*]}. \quad (21)$$

Proof. First we estimate the difference $|T(g_1)-T(g_2)|$ by using the Poincare inequality $\|u\|_0 \leq c_\Omega \|\nabla u\|_0$, $c_\Omega \geq 0$:

$$|T[g_1] - T[g_2]| \leq 2 \left| \int_\Omega [u(x; g_1) - u(x; g_2)] dx \right|$$

$$\leq 2 \; (meas \; \Omega)^{1/2} \|u_1 - u_2\|_0 \leq 2 (meas \; \Omega)^{1/2} \; c_\Omega \|\nabla (u_1 - u_2)\|_0. \quad (22)$$

On the other hand, the weak solutions $u_1, u_2 \in \mathring{H}^1(\Omega)$ of the direct problem (3) satisfy the following integral identities:

$$\int_\Omega g_1(|\nabla u_1|^2) \nabla u_1 \nabla v dx = 2\varphi \int_\Omega v dx,$$

$$\int_\Omega g_2(|\nabla u_2|^2) \nabla u_2 \nabla v dx = 2\varphi \int_\Omega v dx,$$

for all $v \in \mathring{H}^1(\Omega)$. Substituting in the first integral identity $v = u_2 - u_1$, and in the second one $v = u_1 - u_2$ we get:

$$\begin{cases} \int_\Omega g_1(|\nabla u_1|^2) \nabla u_1 \nabla (u_2 - u_1) dx = 2\varphi \int_\Omega (u_2 - u_1) dx, \\ \\ \int_\Omega g_2(|\nabla u_2|^2) \nabla u_2 \nabla (u_1 - u_2) dx = 2\varphi \int_\Omega (u_1 - u_2) dx. \end{cases}$$

These identities imply

$$\int_\Omega g_1(|\nabla u_1|^2) \nabla u_1 \nabla (u_1 - u_2) dx = \int_\Omega g_2(|\nabla u_2|^2) \nabla u_2 \nabla (u_1 - u_2) dx.$$

Adding to the both sides the term $- \int_\Omega g_1(|\nabla u_2|^2) \nabla u_2 \nabla (u_1 - u_2)$ we get:

$$\int_\Omega [g_1(|\nabla u_1|^2) \nabla u_1 - g_1(|\nabla u_2|^2) \nabla u_2] \nabla (u_1 - u_2) dx =$$

$$\int_\Omega [g_2(|\nabla u_2|^2) \nabla u_2 - g_1(|\nabla u_2|^2) \nabla u_2] \nabla (u_1 - u_2) dx.$$

On the left hand side of this identity we use the inequality $[g(\xi_1^2)\xi_1 - g(\xi_2^2)\xi_2](\xi_1 - \xi_2) \geq \alpha \|\xi_1 - \xi_2\|^2$ for the function $g_1(\xi^2)$, satisfying conditions (4) (see, [16]). On the right hand side we use the Cauchy inequality $|(p,q)| \leq \|p\|_0 \|q\|_0$. Then we have, for $\alpha > 0$:

$$\alpha \|\nabla (u_1 - u_2)\|_0 \leq \left(\int_\Omega [g_2(|\nabla u_2|^2) - g_1(|\nabla u_2|^2)]^2 |\nabla u_2|^2 dx \right)^{1/2}. \quad (23)$$

Taking into account this inequality on the right hand side of (22) we have the proof. □

Continuity, given by (21), and monotonicity of the input-output map $T[\cdot] : \mathcal{G} \mapsto \mathbf{T}$ implies that this mapping is invertible. Therefore, the problem (ICP1) has a solution, at least for pure-elastic torsion case.

To prove a srtong existence result for the problem (ICP1) let us use a quasisolution approach [21]. For this aim we introduce the cost functional

$$I_1(g) := \left| 2 \int_\Omega u(x; g; \varphi) dx - \mathcal{T}(\varphi) \right|, \quad \varphi \in [\varphi_*, \varphi^*]. \tag{24}$$

and consider the following minimization problem:

$$I_1(g_*) := \min_{g \in \mathcal{G}_c} I_1(g). \tag{25}$$

Theorem 4.1. *Let conditions of Lemma 3.1 hold, and $\mathcal{G}_c$ be the compact set of admissible coefficients. Then minimization problem (25) for the functional (24) has at least one solution.*

Proof. We first use the estimate $\|\nabla u\|_C := \max_{\overline{\Omega}} |\nabla u| \le \eta^*$, $\eta^* > 0$, (see, [16], Lemma 3.1) on the right hand side of (23). Then we have:

$$\alpha \|\nabla(u_1 - u_2)\|_0 \le \eta^* \|g_1 - g_2\|_0, \quad \eta^*, \alpha > 0. \tag{26}$$

Now we estimate the difference $|I_1(g_m) - I_1(g_*)|$, assuming that $\{g_m\} \subset \mathcal{G}_c$ is the sequence of coefficients converging to $g_* \in \mathcal{G}_c$ in H^1-norm. We have:

$$|I_1(g_m) - I_1(g_*)| = \left| \left| 2 \int_\Omega u(x; g_m; \varphi) dx - \mathcal{T}(\varphi) \right| - \left| 2 \int_\Omega u(x; g_*; \varphi) dx - \mathcal{T}(\varphi) \right| \right|$$

$$\le 2 \left| \int_\Omega u(x; g_m; \varphi) dx - \int_\Omega u(x; g_*; \varphi) dx \right| = 2 \left| \int_\Omega [u(x; g_m; \varphi) - u(x; g_*; \varphi)] dx \right|.$$

Applying to the right hand side the Poincare inequality we obtain:

$$|I_1(g_m) - I_1(g_*)| \le 2(meas\ \Omega)^{1/2} \|\nabla(u(\cdot; g_m; \varphi) - u(\cdot; g_*; \varphi))\|_0.$$

This, with (26), implies:

$$|I_1(g_m) - I_1(g_*)| \le 2\alpha^{-1}\eta^*(meas\ \Omega)^{1/2})\|g_m - g_*\|_0, \quad \eta^*, \alpha > 0,$$

$\forall \varphi \in [\varphi_*, \varphi^*]$. Thus, if the sequence of coefficients $\{g_m\} \subset \mathcal{G}_c$ converges to $g_* \in \mathcal{G}_c$ in H^1-norm, then the numerical sequence $\{I_1(g_m)\}$ converges to $I_1(g_*)$. By the Weierstrass's theorem this implies the proof. □

Although the above results assert strict monotonicity and continuity of the input-output map $T[\cdot] : \mathcal{G} \to \mathbf{T}$, this mapping is not continuously invertible. To show this let us assume that $g(\xi^2)$ is given by the formula

$$g(\xi^2) = \begin{cases} 1/G, & \xi^2 \le \xi_0^2, \\ 1/G \ \left(\xi^2/\xi_0^2\right)^{0.5(\kappa-1)}, & \xi_0^2 < \xi^2, \quad \kappa \in (0, 1], \end{cases} \tag{27}$$

which corresponds to the well-known Ramberg-Osgood curve $\sigma_i = \sigma_0(e_i/e_0)^\kappa$ with the strain hardening exponent $\kappa \in [0, 1]$. Evidently, this function satisfies all conditions (4).

Example 4.1. *Ill-conditionedness of the problem (ICP1)*

Two class of (stiff and soft) materials described by (27), with elastic parameters $\langle E = 210(GPa); \xi_0^2 = 0.027\rangle$ and $\langle E = 110(GPa); \xi_0^2 = 0.020\rangle$, are considered. For each class of materials the following two values $\kappa_1 = 0.2$ and $\kappa_2 = 0.7$ of the hardening parameter $\kappa \in [0, 1]$ are taken in the considered example. The plasticity functions $g_r = g_r(\kappa_i)$, $g_s = g_s(\kappa_i)$, $i = 1, 2$, for these materials, with the above data, are shown in the left Figure 3. In order to generate the synthetic data $T[g_r]$, $T[g_s]$, the nonlinear direct problem (3) is numerically solved for the above given plasticity functions. Then approximate values of the corresponding synthetic data $T[g_r]$, $T[g_s]$ are obtained by applying to (3) the numerical integration trapezoidal formula. Results obtained for the two stiff and two soft materials are plotted in the right Figure 3.

These figures show that, for each class of materials the outputs $T[g_r(\kappa_1)]$ and $T[g_r(\kappa_2)]$ (as well as the outputs $T[g_s(\kappa_1)]$ and $T[g_s(\kappa_2)]$) are close enough, although the plasticity functions of these materials, corresponding to each class, are quite different ($\kappa_1 = 0.2$ and $\kappa_2 = 0.7$). For the values $\kappa_1 = 0.2$ and $\kappa_2 = 0.3$ the corresponding the outputs are close enough, especially in the begining plastic deformations. This means ill-posedness of the problem (ICP1). $\square$

The above example show that even for noise free measured output data the problem (ICP1) is ill-conditioned, although, in practice the output data $\mathcal{T} = \mathcal{T}(\varphi)$ can only be given with some measurement error. Thus, the exact fulfillment of the equality in (18) is not possible, and use of the quasisolution approach by introducing the auxiliary functional (24) is necessary.

5. An analysis of the problem (ICP2)

Let us first analyse the direct problem (6). Multiplying the first equation by $u_1(x, y)$ and the second one by $u_2(x, y)$, applying then the Green formula and using the boundary conditions, we define the weak $u \in V$ solution of the nonlinear direct problem (6) as a solution $u \in V$ of the following variational inequality

$$(Au, v - u) \geq (F, v - u), \quad \forall v \in V. \tag{28}$$

Here

$$\begin{aligned}(Au, v) := \int\int_\Omega \{&\lambda\theta(u)\theta(v) + 2\mu\varepsilon_{ij}(u)\varepsilon_{ij}(v) \\ &- 2\mu g(e_{\text{i}}(u))e_{ij}(u)e_{ij}(v)\} \, xdxdy.\end{aligned} \tag{29}$$

is the nonlinear form, (F, u) is the linear form, $e_{ij}(u) = \varepsilon_{ij}(u) - \theta(u)\delta_{ij}/3$

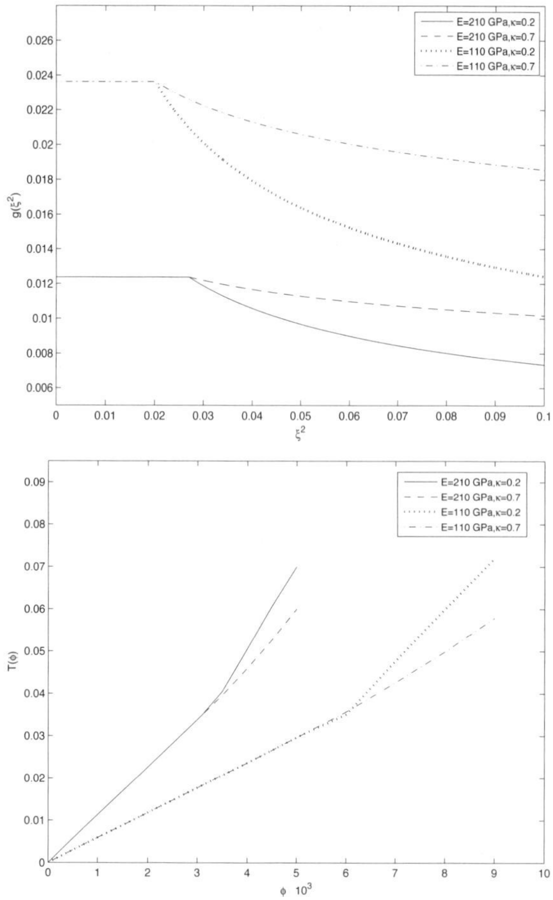

Fig. 3. Plasticity functions for stiff and soft steel materials (left figure) and the corresponding outputs (torques) (right figure)

and $V = \{v \in \mathring{H}^1(\Omega) : u_2(x, l_y) \leq -\alpha + \varphi(x),\ (x,y) \in \Gamma_0\}$ is the closed convex set of admissible displacements, $\mathring{H}^1(\Omega) = \{v \in H^1(\Omega) : v(x,y) = 0,\ (x,y) \in \Gamma_u;\ v_1(x,y) = 0,\ (x,y) \in \Gamma_1\}$ is the subspace of the Sobolev space $H^1(\Omega)$ of vector functions $u(x,y)$ [1], with the norm

$$\|u\|_1 := \left\{ \int \int_\Omega \left[|u|^2 + |\nabla u|^2 \right] x dx dy \right\}^{1/2},$$

taking into account the axially symmetricity of the problem.

Now introduce the nonlinear functional

$$J(u)=\frac{1}{2}\iint_{\Omega}\left\{\lambda\theta(u)^2+2\mu\varepsilon_{ij}(u)\varepsilon_{ij}(u)-3\mu\int_0^{e_{\mathrm{i}}(u)}g(\xi)d\xi\right\}xdxdy, \quad (30)$$

where $u,v\in V$. Calculating the first Gateaux derivative of this functional we get:

$$J'(u;v):=\iint_{\Omega}\{\lambda\theta(u)\theta(v)+2\mu\varepsilon_{ij}(u)\varepsilon_{ij}(v)-2\mu g(e_{\mathrm{i}}^2(u))e_{ij}(u)e_{ij}(v)\}\,xdxdy,$$

where $u,v\in V$.
This shows that the nonlinear operator A, defined by (29), is a potential operator with the potential $J(u)$ defined by (30).

Calculating the second Gateaux derivative and using the relation $e_{ij}(u)=\varepsilon_{ij}(u)-\theta(u)\delta_{ij}/3$ we obtain:

$$\begin{aligned}J''(u;v,h) :=& \int\int_{\Omega}\{\lambda\theta(v)\theta(h)+2\mu\varepsilon_{ij}(v)\varepsilon_{ij}(h)\\ &-2\,\mu[g(e_{\mathrm{i}}^2(u))+2g'(e_{\mathrm{i}}^2(u))e_{\mathrm{i}}^2(u)]e_{ij}(v)e_{ij}(h)\}\,xdxdy\\ =& \int\int_{\Omega}\left\{\left[\lambda+\frac{2}{3}\mu Q(e_{\mathrm{i}}^2(u))\right]\theta(v)\theta(h)\right.\\ &+2\mu\left[1-Q(e_{\mathrm{i}}^2(u))\right]\varepsilon_{ij}(v)\varepsilon_{ij}(h)\}\,xdxdy,\quad u,v,h\in V.\end{aligned}$$

Substituting here $h=v$ and using Korn's inequality we conclude that

$$J''(u;v,v)\geq\gamma_1\|v\|_1^2,\quad\forall h\in V.$$

This means the nonlinear operator A is strong monotone one, i.e.

$$\langle Au-Av,u-v\rangle\gamma_1\geq\|u-v\|_1^2,\quad u,v\in V.$$

This implies an existence and unequeness of the weak solution $u\in V$ of the direct problem.

Note that the direct problem (6) is nonlinear not only due to the presence of the plasticity function $g(e_{\mathrm{i}}^2(u))$. For each value $\alpha>0$ of the penetration depth, the contact zone

$$\Gamma_c(\alpha)=\{(x,l_y)\in\Gamma_0: u_2(x,l_y)=-\alpha+\varphi(x),\quad x\in(0,a_c(\alpha))\},$$
$$a_c(\alpha):=\partial\Gamma_c(\alpha),$$

is also unknown, and needs to be also determined. For this reason two iteration procedures for linearization of the nonlinear direct problem (6) need to be organized: first, with respect to the unknown contact zone, and the second one, with respect to the plasticity function $g=g(e_{\mathrm{i}}^2)$. This processes are described in [15].

Consider now the problem (ICP2). Let $\mathcal{G}_c$ be the compact set of admissible coefficients defined in Section 3. Denote by $u(x;g) \equiv u[\sigma_{\mathrm{i}}]$, $g := g(e_{\mathrm{i}}^2(u)) \in \mathcal{G}_c$, the corresponding solution of the direct problem (28)-(29). To use the quasisolution approach, we introduce the cost functional

$$I_2(g) := \max_{\alpha\in[0,\alpha^*]} \left| 2\pi \int_{\Gamma_c(\alpha)} \sigma_{22}(u(x;g))x dx + \mathcal{P}(\alpha) \right|, \tag{31}$$

taking into account the additional condition (7), and consider the following minimization problem:

$$I_2(g_*) := \min_{g\in\mathcal{G}_c} I_2(g). \tag{32}$$

To obtain an existence of the quasisolution we use the following coefficient continuity result [12].

Theorem 5.1. *Let $\{g_m\} \subset \mathcal{G}_c$, $g_m := g_m(\xi^2)$, $\xi^2 = e_i^2(u)$, be a sequence of coefficients from the compact set of admissible coefficients $\mathcal{G}_c$, and $\{u_m\} \subset V$, $u_m := u(x, y; g_m)$ be the corresponding sequence of solutions of the direct problem (28)-(29). Assume that the sequence of coefficients $\{g_m\}$, $m = 1, 2, 3, ...$, converges to the function $g := g(e_i^2(u)) \subset \mathcal{G}_c$ in the norm of the Sobolev space $H^1[\xi_*, \xi^*]$, as $m \to \infty$. Then the sequence of solutions $\{u_m\} \subset V$ converges weakly in $H^1(\Omega)$ to the solution $u(x;g) \in V$ of the direct problem (28)-(29) which corresponds to the coefficient $g(e_i^2(u)) \subset \mathcal{G}_c$.*

Consider the layer $\Omega_\delta(\Gamma_0) := \{(x, y) \in \Omega : l_y - \delta < y < l_y,\ \delta > 0\}$. The weak convergence of the sequence of solutions $\{u(x; g_m)\} \subset V$ in $H^1(\Omega)$ implies the weak convergence of sequence $\{\sigma_{22}(u(x; g_m))\}$ in $H^0(\Omega)$, in particular in $H^0(\Omega_\delta(\Gamma_0))$. On the other hand, is known that for a function $v(x, y) \in H^0(\Omega)$ satisfying the condition $v_y \in H^0(\Omega)$ the trace $u(x, l_y; g)$ on $\Gamma_0 \subset \partial\Omega$ exists as an element of $H^0(\Gamma_0)$, and continuously depends on $\delta > 0$ (see, [24], Theorem 6.3), i.e.:

$$\int_{\Gamma_0} v^2(x, l_y)dy \le \frac{2}{\delta} \int_{\Omega_\delta(\Gamma_0)} v^2(x, y)dxdy + \delta \int_{\Omega_\delta(\Gamma_0)} v_y^2(x, y)dxdy.$$

Assuming $v(x, y) = \sigma_{22}(u(x;g))$ we can apply the above assertion to the function $\sigma_{22}(u(x;g))$, requiring

$$\frac{\partial}{\partial y}\sigma_{22}(u(x;g)) \in H^0(\Omega_\delta(\Gamma_0)). \tag{33}$$

Theorem 5.2. *Let condition (33) holds, and $\mathcal{G}_c$ be the compact set of admissible coefficients. Then minimization problem (32) for the functional (31) has at least one solution.*

Proof. Consider the difference $|I_2(g_m) - I_2(g)|$:

$$|I_2(g_m) - I_2(g)| = \left| \max_{\alpha\in[0,\alpha^*]} \left| 2\pi \int_{\Gamma_c(\alpha)} \sigma_{22}(u(x;g_m))xdx + \mathcal{P}(\alpha) \right| \right.$$
$$\left. - \max_{\alpha\in[0,\alpha^*]} \left| 2\pi \int_{\Gamma_c(\alpha)} \sigma_{22}(u(x;g))xdx + \mathcal{P}(\alpha) \right| \right|$$
$$\leq 2\pi \max_{\alpha\in[0,\alpha^*]} \int_{\Gamma_c(\alpha)} |\sigma_{22}(u(x;g_m))xdx - \sigma_{22}(u(x;g))|xdx.$$

Due to the above trace theorem the right hand side integral tends to zero, as $m \to \infty$. This means that the functional $I_2(g)$ is continuous. Since $\mathcal{G}_c$ is compact we obtain the proof. □

Thus we obtain the main result related to solvability of the problem (ICP2).

Theorem 5.3. *Let condition (33) holds. Then the problem (ICP2) has at least one solution $g(\xi^2) \in \mathcal{G}_c$ in the set of admissible coefficients $\mathcal{G}_c$.*

6. An analysis of the problem (ICP3)

6.1. Solvability of the direct problem (13)

Let us define the weak solution of the direct problem (13) in the Sobolev space of functions $H^2(\Omega)$ [1]. For this aim we define the subspace

$$\overset{\circ}{H}{}^2(\Omega) = \left\{ v \in H^2(\Omega) : \ u(x) = \frac{\partial u(x)}{\partial n} = 0, \ x \in \partial\Omega \right\},$$

taking into account the homogeneous Dirichlet conditions. Then multiplying the both sides of the biharmonic equation (13) by $v \in \overset{\circ}{H}{}^2(\Omega)$, integrating on Ω and using the Diirichlet conditions we obtain the following integral identity

$$\int_\Omega g(\xi^2(u))H(u,v)dx = \int_\Omega F(x)v(x)dx, \quad \forall v \in \overset{\circ}{H}{}^2(\Omega). \tag{34}$$

Here

$$H(u,v) = \frac{\partial^2 u}{\partial x_1^2}\frac{\partial^2 v}{\partial x_1^2} + \frac{\partial^2 u}{\partial x_2^2}\frac{\partial^2 v}{\partial x_2^2} + \frac{\partial^2 u}{\partial x_1 \partial x_2}\frac{\partial^2 v}{\partial x_1 \partial x_2}$$
$$+\frac{1}{2}\left(\frac{\partial^2 u}{\partial x_1^2}\frac{\partial^2 v}{\partial x_2^2} + \frac{\partial^2 u}{\partial x_2^2}\frac{\partial^2 v}{\partial x_1^2} \right) \tag{35}$$

is a bilinear differential form defined on $H^2(\Omega) \times H^2(\Omega)$. Evidently $H(v,v) = \xi^2(v)$ as formula (15) shows.

The function $u \in \overset{\circ}{H}{}^2(\Omega)$ satifying the integral identity (34) for all $v \in \overset{\circ}{H}{}^2(\Omega)$ is defined to be a weak solution of the direct problem (13).

We introduce the nonlinear and linear functionals

$$\langle Au, v\rangle = \int_\Omega g(\xi^2(u))H(u,v)dx, \quad l(v) = \int_\Omega F(x)v(x)dx \forall v \in \mathring{H}^2(\Omega), \tag{36}$$

to rewrite the *variational problem* (34) in the standard form:

$$\langle Au, v\rangle = l(v), \quad \forall v \in \mathring{H}^2(\Omega). \tag{37}$$

Here and below we assume that the source function satisfies the condition $F \in H^0(\Omega)$.

Lemma 6.1. *The nonlinear functional*

$$J(u) = \frac{1}{2}\int_\Omega \left\{ \int_0^{\xi^2(u)} g(\tau)d\tau \right\} dx, \quad u \in \mathring{H}^2(\Omega). \tag{38}$$

is a potential of nonlinear operator A, defined by (13).

Proof. Calculating the first Gateaux derivative of the functional $J(u)$ we get:

$$\begin{aligned} J'(u;v) &= \frac{1}{2}\int_\Omega \left\{ \frac{d}{dt}\left[\int_0^{\xi^2(u+tv)} g(\tau)d\tau\right]_{t=0} \right\} dx \\ &= \frac{1}{2}\int_\Omega \left\{ g(\xi^2(u+tv))\frac{d}{dt}\xi^2(u+tv)\right\}_{t=0} dx, \end{aligned}$$

$u, v \in \mathring{H}^2(\Omega)$. On the other hand, due to definitions (15) and (35) we conclude

$$\begin{aligned} \frac{d}{dt}\left[\xi^2(u+tv)\right]_{t=0} = 2\Bigg[& \frac{\partial^2 u}{\partial x_1^2}\frac{\partial^2 v}{\partial x_1^2} + \frac{\partial^2 u}{\partial x_2^2}\frac{\partial^2 v}{\partial x_2^2} + \frac{\partial^2 u}{\partial x_1 \partial x_2}\frac{\partial^2 v}{\partial x_1 \partial x_2} \\ & + \frac{1}{2}\left(\frac{\partial^2 u}{\partial x_1^2}\frac{\partial^2 v}{\partial x_2^2} + \frac{\partial^2 u}{\partial x_2^2}\frac{\partial^2 v}{\partial x_1^2}\right)\Bigg] = 2H(u,v). \end{aligned}$$

Hence

$$\langle J'(u), v\rangle = \int_\Omega g(\xi^2(u))H(u,v)dx,$$

and we obtain the proof. □

To derive a solvability result for the variational problem (37), let us analyse monotonicity of the nonlinear biharmonic operator A defined by (13). For this aim we define the energy norm $\|v\|_E$ and the seminorm $|v|_2$ in $H^2(\Omega)$, accordingly:

$$\begin{aligned} \|v\|_E &:= \left\{ \int_\Omega H(v,v)dx \right\}^{1/2}, \\ |v|_2 &:= \left\{ \int_\Omega \left[\left(\frac{\partial^2 v}{\partial x_1^2}\right)^2 + \left(\frac{\partial^2 v}{\partial x_2^2}\right)^2 + \left(\frac{\partial^2 v}{\partial x_1 \partial x_2}\right)^2 \right] dx \right\}^{1/2}. \end{aligned}$$

The following lemma shows that these norms are equivalent.

Lemma 6.2. *If* $v \in \overset{\circ}{H}{}^2(\Omega)$, *then* $|v|_2^2 \le \|v\|_E^2 \le 2|v|_2^2$.

Proof. It is known that for all $v \in \overset{\circ}{H}{}^2(\Omega)$ the following formula holds (see, [28]):

$$\int_\Omega \left[\left(\frac{\partial^2 v}{\partial x_1 \partial x_2} \right)^2 - \frac{\partial^2 v}{\partial x_1^2} \frac{\partial^2 v}{\partial x_2^2} \right] dx = 0,$$

Then taking into account (35) we get

$$\|v\|_E^2 = \int_\Omega H(v,v)dx = \int_\Omega \left[\left(\frac{\partial^2 v}{\partial x_1^2} \right)^2 + \left(\frac{\partial^2 v}{\partial x_2^2} \right)^2 + \left(\frac{\partial^2 v}{\partial x_1 \partial x_2} \right)^2 + \frac{\partial^2 v}{\partial x_1^2} \frac{\partial^2 v}{\partial x_2^2} \right] dx$$

$$= \int_\Omega \left[\left(\frac{\partial^2 v}{\partial x_1^2} \right)^2 + \left(\frac{\partial^2 v}{\partial x_2^2} \right)^2 + 2 \left(\frac{\partial^2 v}{\partial x_1 \partial x_2} \right)^2 \right] dx$$

$$\ge \int_\Omega \left[\left(\frac{\partial^2 v}{\partial x_1^2} \right)^2 + \left(\frac{\partial^2 v}{\partial x_2^2} \right)^2 + \left(\frac{\partial^2 v}{\partial x_1 \partial x_2} \right)^2 \right] dx := |v|_2^2.$$

On the other hand,

$$\int_\Omega H(v,v)dx \le 2 \int_\Omega \left[\left(\frac{\partial^2 v}{\partial x_1^2} \right)^2 + \left(\frac{\partial^2 v}{\partial x_2^2} \right)^2 + \left(\frac{\partial^2 v}{\partial x_1 \partial x_2} \right)^2 \right] dx = |v|_2^2.$$

These two estimates imply the proof. □

By using the equivalence of the norm $\|\cdot\|_2$ and the seminorm $|\cdot|_2$ in $\overset{\circ}{H}{}^2(\Omega)$, we obtain the following

Corollary 6.1. *If* $v \in \overset{\circ}{H}{}^2(\Omega)$, *then*

$$\exists \alpha_1, \alpha_2 > 0, \quad \alpha_1 \|v\|_2 \le \|v\|_E \le \alpha_2 \|v\|_2,$$

i.e. the H^2*-norm and the energy norm are equivalent.*

The lemma permits one to obtain also the following upper estimate

Corollary 6.2. *If* $u, v \in \overset{\circ}{H}{}^2(\Omega)$, *then*

$$\int_\Omega |H(u,v)|dx \le \alpha_2^2 \|u\|_2 \|v\|_2.$$

Proof. We use the Schwartz inequality for the bilinear form $H(u,v)$:

$$(H(u,v))^2 \le H(u,u) \cdot H(v,v).$$

This implies:

$$\int_\Omega |H(u,v)|dx \le \int_\Omega (H(u,u))^{1/2} (H(v,v))^{1/2}$$

$$\le \left(\int_\Omega |H(u,u)|dx \right)^{1/2} \left(\int_\Omega |H(v,v)|dx \right)^{1/2} \le \alpha_2^2 \|u\|_2 \|v\|_2,$$

and we obtain the proof. □

By using these auxiliary results we can prove strong convexity of the potential $J(u)$ of the nonlinear operator A in $\overset{\circ}{H}{}^2(\Omega)$.

Lemma 6.3. *Let the coefficient $g = g(\xi^2)$ of the nonlinear biharmonic equation (13) satisfies the conditions **(i1)-(i3)**. Then the potential of the biharmonic operator A is strongly convex in $\overset{\circ}{H}{}^2(\Omega)$, i.e.*

$$\forall u, v \in \overset{\circ}{H}{}^2(\Omega), \quad \langle J''(u), v, v\rangle \geq \gamma_1 \|v\|_2^2, \quad \gamma_1 > 0. \tag{39}$$

Proof. Calculating the second Gateaux derivative of the functional $J(u)$, defined by (38), we have:

$$\langle J''(u), v, w\rangle := \frac{d}{dt}\langle J'(u+tw), v\rangle|_{t=0} = \frac{d}{dt}\left\{\int_\Omega g(\xi^2(u+tw))H(u+tw, v)dx\right\}_{t=0}$$
$$= \left\{\int_\Omega \left[2g'(\xi^2(u+tw))H(u,w)H(u+tw,v) + g(\xi^2(u+tw))H(w,v)\right] dx\right\}_{t=0}$$
$$= \int_\Omega \left[2g'(\xi^2(u))H(u,w)H(u,v) + g(\xi^2(u))H(w,v)\right] dx.$$

Substituting here $w = v$ and using $H(v, v) = \xi^2(v)$ we find

$$\langle J''(u), v, v\rangle := \int_\Omega \left[g(\xi^2(u))\xi^2(v) + 2g'(\xi^2(u))H^2(u,v)\right] dx, \quad u, v \in \overset{\circ}{H}{}^2(\Omega).$$

Condition **(i2)** and the inequality $H(u, v) \leq H(u, u)H(v, v)$ with the relationship $H(v, v) = \xi^2(v)$ between the differential form $H(v, v)$ and the effective curvature $\xi^2(v)$ imply that $g'(\xi^2(u))H^2(u, v) \geq g'(\xi^2(u))\xi^2(u)\xi^2(v)$. Hence

$$\langle J''(u), v, v\rangle \geq \int_\Omega \left[g(\xi^2(u))\xi^2(v) + 2g'(\xi^2(u))\xi^2(u)\xi^2(v)\right] dx$$
$$= \int_\Omega [g(\xi^2(u)) + 2g'(\xi^2(u))\xi^2(u)]\xi^2(v)dx.$$

By using the condition **(i3)** on the right hand side and applying Corollary 6.1 finally we get

$$\langle J''(u), v, v\rangle \geq \gamma_0 \int_\Omega \xi^2(v)dx = \gamma_0 \int_\Omega H(v, v)dx \geq \gamma_1 \|v\|_2^2, \quad \gamma_1 = \alpha_1^2\gamma_0 > 0.$$

This implies the proof. □

Corollary 6.3. *Let the coefficient $g = g(\xi^2)$ of equation (13) satisfies the conditions **(i1)-(i3)**. Then the nonlinear biharmonic operator A defined by (13) is strongly monotone in $\overset{\circ}{H}{}^2(\Omega)$, i.e.*

$$\forall u, v \in \overset{\circ}{H}{}^2(\Omega), \quad \langle Au - Av, u - v\rangle \geq \gamma_1 \|u - v\|_2^2, \quad \gamma_1 > 0. \tag{40}$$

Corollary 6.4. *Since $A\Theta = \Theta$, where $\Theta \in \overset{\circ}{H}{}^2(\Omega)$ is a zero element, monotonicity (40) of the nonlinear operator A also means its coercivity*

$$\langle Av, v \rangle \geq \gamma_1 \|v\|_2^2, \quad \gamma_1 > 0.$$

Lemma 6.4. *The nonlinear biharmonic operator A is radially continuous (hemicontinuous), i.e. the real valued function $t \to \langle A(u + tv), v \rangle$, for fixed $u, v \in \overset{\circ}{H}{}^2(\Omega)$, is continuous.*

Proof. Since the both $t \mapsto g(\xi^2(u + tw)$, $t \mapsto H((u + tw), v)$ mappings are continuous, proof follows immediately from (36). □

Thus, the potential operator A is radially continuous, strongly monotone and coercive. Applying Browder-Minty theorem [7] we obtain the solvability of the direct problem.

Theorem 6.1. *Let conditions **(i1)**-**(i3)** hold. Then the variational problem (37) has a unique solution in $\overset{\circ}{H}{}^2(\Omega)$.*

6.2. Linearization of the nonlinear direct problem: the monotone iteration scheme and convergence of the approximate solution

To study the inverse coefficient problem, as well as to apply any numerical method for solution of the direct problem, one needs to perform a linearization of the nonlinear problem (36)-(37), and then prove the convergence of the sequence approximate solutions in appropriate Sobolev norm. For this aim we will use so-called convexity argument for nonlinear monotone potential operators, introduced in [13]. For clarity we will explain here the convexity argument in its abstract form.

Let $A : H \mapsto H^*$ be a strongly monotone potential operator defined on the abstract Hilbert space H, and $a(u; \cdot, \cdot)$ be the corresponding bounded, symmetric continuous and coercive the tri-linear form (functional): $a(u; u, v) := \langle Au, v \rangle$, $u, v \in H$, and

$$\begin{cases} \langle Au - Av, u - v \rangle \geq \gamma_1 \|u - v\|_H, & \gamma_1 > 0; \\ |a(u; u, v)| \leq \gamma_2 \|u\|_H \|v\|_H, & \gamma_2 > 0, \quad \forall u, v \in H. \end{cases} \tag{41}$$

Here $\|\cdot\|_H$ is the norm of the space H.

Assume that the functional $J(u)$, $u \in H$ is the potential of the operator A. Then we have $\langle J'(u), v \rangle = a(u; u, v)$.

Definition 6.1. *The monotone potential operator $A : H \to H^*$ defined on the abstract Hilbert space H is said to be satisfy the convexity argument, if the following inequality holds:*

$$\frac{1}{2}a(u; v, v) - \frac{1}{2}a(u; u, u) - J(v) + J(u) \geq 0, \quad \forall u, v \in H. \tag{42}$$

Consider the following variational problem

$$a(u; u, v) = l(v), \quad v \in H, \tag{43}$$

which defines the weak solution $u \in H$ of the abstract operator equation $Au = F$, where $l(v)$ is the linear functional defined by the element $F \in H^*$. Denote by $\Pi(u)$ the potential of the operator equation $Au = F$.

Let us linearize the nonlinear variational problem (43) as follows:

$$a(u^{(n-1)}; u^{(n)}, v) = l(v), \quad \forall v \in H, \quad n = 1, 2, 3, ..., \tag{44}$$

where $u^{(0)} \in H$ is an initial iteration. The function $u^{(n)} \subset H$ is defined to be an approximate solution of the abstract variational problem (43). Evidently, at each nth iteration the variational problem (44) is a linear one, since $u^{(n-1)}$ is known from previous iteration. The iteration scheme (44) is defined to be *the abstract iteration scheme* for the monotone potential operator A.

Theorem 6.2. [13] *Let $A : H \mapsto H^*$ be a strongly monotone potential operator defined on the abstract Hilbert space H, and $a(u; \cdot, \cdot)$ be the corresponding bounded, symmetric continuous and coercive the tri-linear form. If the convexity argument (42) holds, then*
(a1) *the sequence of potentials $\{\Pi(u^{(n)})\} \subset R$, corresponding to the sequence of solutions $\{u^{(n)}\} \subset H$, $n = 1, 2, 3, ..,$ of the linearized problem (44), is a monotone decreasing one;*
(a2) *the sequence of approximate solutions $\{u^{(n)}\} \subset H$ defined by the abstract iteration scheme (44) converges to the solution $u \in H$ of the nonlinear problem (43) in the norm of the space H;*
(a3) *for the rate of convergence the following estimate holds: .*

$$\|u - u^{(n)}\| \leq \frac{\sqrt{2}\gamma_2}{\gamma_1^{3/2}} \left[\Pi(u^{(n-1)}) - \Pi(u^{(n)})\right]^{1/2}, \tag{45}$$

where $\gamma_1, \gamma_2 > 0$ are the constants defined in (41).

To apply the above results to the variational problem (36)-(37) let us introduce the nonlinear functional $v \in \overset{\circ}{H}{}^2(\Omega)$

$$\Pi(u) := J(u) - l(u) = \frac{1}{2} \int_\Omega \left\{ \int_0^{\xi^2(u)} g(\tau) d\tau - 2F(x)u(x) \right\} dx, \tag{46}$$

following Lemma 6.1. The functional $\Pi(u)$ is defined to be as the *potential of the variational problem (36)-(37)*. Consider now the following minimization problem.

Find such a function $u \in \mathring{H}^2(\Omega)$ that

$$\Pi(u) = \min \Pi(v), \quad v \in \mathring{H}^2(\Omega). \tag{47}$$

Evidently this problem is equivalent to the variational problem (37).

Let us linearize now the nonlinear variational problem (36)-(37), according to the linearization scheme (44). The solution $u^{(n)} \in \mathring{H}^2(\Omega)$ of the linearized variational problem

$$a(u^{(n-1)}; u^{(n)}, v) := \int_\Omega g(\xi^2(u^{(n-1)}))H(u^{(n)}, v)dx = \int_\Omega F(x)v(x)dx, \tag{48}$$

where $n = 1, 2, 3, \ldots$ is defined to be an *appoximate solution nonlinear variational problem (36)-(37)*. Here $u^{(0)} \in \mathring{H}^2(\Omega)$ is an initial iteration. Different from the potential $J(u)$, defined by (38), the potential of the linearized operator, defined to be as $J_0(u^{(n)})$, is a quadratic functional, since the left hand side of (48) is a bilinear functional. Thus the potentials of the linearized operator and the linearized problem (48) are defined as follows:

$$\begin{cases} J_0(u^{(n)}) = \frac{1}{2}\int_\Omega g(\xi^2(u^{(n-1)}))H(u^{(n)}, u^{(n)})dx; \\ \Pi_0(u^{(n)}) = J_0(u^{(n)}) - l(u^{(n)}), \quad u^{(n)} \in \mathring{H}^2(\Omega). \end{cases} \tag{49}$$

To apply the above theorem we need, first of all, to the analyze fulfilment of the convexity argument (42) for the nonlinear biharmonic operator defined by (13).

Lemma 6.5. *If the coefficient* $g = g(\xi^2)$ *satisfies the condition* ***(i2)****, then the convexity argument holds for the nonlinear biharmonic operator* A*, defined by (13).*

Proof. Based on definitions of the potentials $J(u)$ and $\Pi(u)$ we calculate the left hand side of inequality (42):

$$\begin{aligned} &\frac{1}{2}a(u; v, v) - \frac{1}{2}a(u; u, u) - J(v) + J(u) \\ &= \frac{1}{2}\int_\Omega g(\xi^2(u))H(v, v)dx - \frac{1}{2}\int_\Omega g(\xi^2(u))H(u, u)dx \\ &-\frac{1}{2}\int_\Omega \left\{ \int_0^{\xi^2(v)} g(\tau)d\tau \right\} dx + \frac{1}{2}\int_\Omega \left\{ \int_0^{\xi^2(u)} g(\tau)d\tau \right\} dx \\ &= \frac{1}{2}\int_\Omega \left\{ g(\xi^2(u))[\xi^2(v) - \xi^2(u)] - \int_0^{\xi^2(v)} g(\tau)d\tau + \int_0^{\xi^2(u)} g(\tau)d\tau \right\} dx. \end{aligned}$$

Now introduce the function

$$Q(t) = \int_0^t g(\tau)d\tau.$$

Due to the condition **(i2)** we conclude $Q''(t) = g'(t) \leq 0$, and hence $Q = Q(t)$ is a concave function:

$$Q'(t_1)(t_2 - t_1) - Q(t_2) + Q(t_1) \geq 0, \quad \forall t_2 > t_1 > 0.$$

Using this inequality on the right hand side of the above integral expression, assuming $\xi^2(u)$ and $\xi^2(v)$ instead of t_1, t_2, respectively, we get

$$g(\xi^2(u))[g(\xi^2(v)) - g(\xi^2(u))] - \int_0^{\xi^2(v)} g(\tau)d\tau + \int_0^{\xi^2(u)} g(\tau)d\tau \geq 0.$$

This implies the proof. □

Evidently the functional

$$a(u; v, w) = \int_\Omega g(\xi^2(u))H(v, w)dx, \quad u, v, w \in \mathring{H}^2(\Omega).$$

satisfies the boundedness condition (41), due to the condition **(i1)**. Then by Corollary 6.2:

$$|a(u; u, v)| \leq c_1 \int_\Omega |H(u, v)|dx \leq c_2\alpha_2^2 \|u\|_2 \|v\|_2 \ .$$

Thus all conditions of Theorem 6.2 hold, and we may apply it to the nonlinear direct problem (36)-(37).

Theorem 6.3. *Let $u \in \mathring{H}^2(\Omega)$ and $u^{(n)} \in \mathring{H}^2(\Omega)$ be the solutions of the nonlinear direct problem (36)-(37), and linearized problem (48), respectively. If the function $g(\xi^2(u))$ satisfies the conditions **(i1)**-(i3) hold, then* ***(b1)*** *the sequence of potentials $\{\Pi_0(u^{(n)})\} \subset R$, defined by (49) and corresponding to the sequence of solutions $\{u^{(n)}\} \subset \mathring{H}^2(\Omega)$, $n = 1, 2, 3, ..$, of the linearized problem (48), is a monotone decreasing one:*

$$\Pi(u^{(n)}) \leq \Pi(u^{(n-1)}), \quad \forall u^{(n-1)}, u^{(n)} \in \mathring{H}^2(\Omega).$$

(b2) *the sequence of approximate solutions $\{u^{(n)}\} \subset \mathring{H}^2(\Omega)$ defined by the iteration scheme (48) converges to the solution $u \in \mathring{H}^2(\Omega)$ of the nonlinear problem (36)-(37) in the norm of the Sobolev space $\mathring{H}^2(\Omega)$;*
(b3) *for the rate of convergence the following estimate holds:*

$$\|u - u^{(n)}\| \leq \frac{\sqrt{2}\gamma_2}{\gamma_1^{3/2}} \left[\Pi_0(u^{(n-1)}) - \Pi_0(u^{(n)})\right]^{1/2}, \tag{50}$$

6.2. Existence of a quasisolution of the problem (ICP3)

Let us denote by $\mathcal{G}_c$ be the set of admissible coefficients $g(\xi^2)$, defined in Section 3. For a given coefficient $g(\xi^2) \in \mathcal{G}_c$ denote by $u(x; g)$ the corresponding solution of nonlinear direct problem (36)-(37). Then for each step of the quasistatic prosess of bending, given by the parameter τ_k, $k = \overline{1, K}$, the inverse coefficient problem can be reformulated as the following nonlinear functional equation:

$$u(x; g) = w_i(\lambda_i; \tau_k), \quad g \in \mathcal{G}. \tag{51}$$

The mapping $\Phi[g] := u(x; g)|_{x=\lambda_i}$, $\Phi : \mathcal{G} \mapsto w_i(\lambda_i; \tau_k)$, is defined to be the *input-output mapping*. In practice an exact equality in (52) is not possible due to measurements errors. For this reason we will introduce the auxiliary functional

$$I(g) = \sum_{k=1}^{K} \sum_{i=1}^{M} [u(x; g) - w_i(\lambda_i; \tau_k)]^2, \quad g \in \mathcal{G}, \tag{52}$$

and consider the following minimization problem:

$$J(g_*) = \inf_{g \in \mathcal{G}} J(g). \tag{53}$$

A solution of this minimization problem will be defined to be as a *quasisolution of the problem (ICP3)*, i.e. the inverse coefficient problem (13)-(14).

To prove an existence of a quasisolution first of all we need analyze continuity of the solution $u(x; g) \in \overset{\circ}{H}{}^2(\Omega)$ of the nonlinear variational problem (36)-(37) with respect to the coefficient $g \in \mathcal{G}_c$.

Lemma 6.6. *Let $\{g_m(\xi^2)\} \subset \mathcal{G}_c$ be a sequence of coefficients from the compact set of admissible coefficients $\mathcal{G}_c$, and $F \in H^0(\Omega)$. Denote by $\{u_m(x; g)\} \subset \overset{\circ}{H}{}^2(\Omega)$ the corresponding sequence of solutions of the variational problem (36)-(37). Assume that the sequence $\{g_m(\xi^2)\}$ converges to the function $g \in \mathcal{G}_0$ in H^1-norm, as $m \to \infty$. Then the sequence of solutions $\{u(x; g_m)\}$ converges to the solution $u(x; g) \in \overset{\circ}{H}{}^2(\Omega)$ of the variational problem (36)-(37) corresponding to the limit function $g \in \mathcal{G}_c$.*

Proof. Let us denote by $u_m^{(n)} \in \overset{\circ}{H}{}^2(\Omega)$, $u_m^{(n)} = u^{(n)}(x; g_m)$, the sequence solutions of the linearized problem (48) corresponding to the sequence of coefficients $\{g_m\} \subset \mathcal{G}_c$:

$$a_m(u^{(n-1)}; u_m^{(n)}, v) := \int_\Omega g_m(\xi^2(u^{(n-1)})) H(u_m^{(n)}, v) dx = \int_\Omega F(x) v(x) dx, \tag{54}$$

$\forall v \in \overset{\circ}{H}{}^2(\Omega)$. Note that the index m in the above bilinear form $a_m(u_m^{(n-1)}; \cdot, \cdot)$ means that in the right hand side integral there is the function $g_m(\xi^2(u^{(n-1)}))$, instead of $g(\xi^2(u^{(n-1)}))$.

Substituting $v = u_m^{(n)}$ in (54) we get:

$$|a_m(u_m^{(n-1)}; u_m^{(n)}, u_m^{(n)})| \le \|F\|_0 \|u_m^{(n)}\|_0 \le \|F\|_0 \|u_m^{(n)}\|_2, \quad u_m^{(n)} \in \overset{\circ}{H}{}^2(\Omega).$$

On the other hand, due to coercitiveness of the bilinear form $a_m(u_m^{(n-1)};\cdot,\cdot)$ we conclude

$$|a_m(u_m^{(n-1)};u_m^{(n)},u_m^{(n)})| \geq \gamma_1 \|u_m^{(n)}\|_2^2, \quad \gamma_1 > 0.$$

These two inequalities imply the uniform boundedness of the sequence $\{u_m^{(n)}\}$:

$$\|u_m^{(n)}\|_2 \leq \|F\|_0/\gamma_1, \quad \gamma_1 > 0$$

in H^2-norm. This implies the weak convergence of the sequence $\{u_m^{(n)}\}$ in $H^2(\Omega)$. Hence there exists such an element $\tilde{u}^{(n)} \in \mathring{H}^2(\Omega)$, that $u_m^{(n)} \rightharpoonup \tilde{u}^{(n)}$ weakly in $H^2(\Omega)$. We need to prove that $\tilde{u}^{(n)} = u^{(n)}(x;g)$, where $g \in \mathcal{G}_c$ is the limit of the sequence $\{g_m\} \subset \mathcal{G}_0$. For this aim let us estimate the difference $|a(u^{(n-1)};\tilde{u}^{(n)},v) - a_m(u^{(n-1)};u_m^{(n)},v)|$:

$$\begin{aligned}
&|a(u^{(n-1)};\tilde{u}^{(n)},v) - a_m(u^{(n-1)};u_m^{(n)},v)| \\
&= \left|\int_\Omega [g(\xi^2(u^{(n-1)}))H(\tilde{u}^{(n)},v) - g_m(\xi^2(u^{(n-1)}))H(u_m^{(n)},v)]dx\right| \\
&\leq \left|\int_\Omega [g(\xi^2(u^{(n-1)}))H(\tilde{u}^{(n)},v) - g_m(\xi^2(u^{(n-1)}))H(\tilde{u}^{(n)},v)]dx\right| \\
&\quad + \left|\int_\Omega [g_m(\xi^2(u^{(n-1)}))H(u^{(n)},v) - g_m(\xi^2(u^{(n-1)}))H(u_m^{(n)},v)]dx\right| \\
&\leq \max_{[\xi_*,\xi^*]} |g(\xi^2(u^{(n-1)})) - g_m(\xi^2(u^{(n-1)}))| \int_\Omega H(\tilde{u}^{(n)},v)dx \\
&\quad + c_1 \int_\Omega H(\tilde{u}^{(n)} - u_m^{(n)},v)dx.
\end{aligned}$$

The first right hand side term tends to zero, $\max_{[\xi_*,\xi^*]} |g(\xi^2(u^{(n-1)})) - g_m(\xi^2(u^{(n-1)}))| \to 0$, since $g_m \to g \in \mathcal{G}_c$ in H^1-norm, as $m \to \infty$. Further, by the weak convergence $u_m^{(n)} \rightharpoonup \tilde{u}^{(n)}$, as $m \to \infty$, in $H^2(\Omega)$, we conclude that the second right hand side also tends to zero. Thus, going to the limit in (54), as $m \to \infty$, we obtain

$$a(u^{(n-1)};\tilde{u}^{(n)},v) := \int_\Omega g_m(\xi^2(u^{(n-1)}))H(\tilde{u}^{(n)},v)dx = \int_\Omega F(x)v(x)dx,$$

$\forall v \in \mathring{H}^2(\Omega)$, i.e. the limit function $\tilde{u}^{(n)}$ is the solution of the linerized variational problem (48). By the uniqueness of the solution of this problem we conclude $\tilde{u}^{(n)} = u^{(n)}(x;g)$.

Thus the convergence $g_m \to g$ of the sequence of coefficients $\{g_m\} \subset \mathcal{G}_0$ in H^1-norm, implies the weak convergence $u_m^{(n)} := u^{(n)}(x;g_m) \rightharpoonup u^{(n)} :=$

$u^{(n)}(x;g)$, $m \to \infty$, in $H^2(\Omega)$ of the approximate solutions $u^{(n)} \in \overset{\circ}{H}{}^2(\Omega)$, defined by (48), of the nonlinear variational problem (36)-(37):

$$|a(u^{(n-1)}; u^{(n)}, v) - a_m(u^{(n-1)}; u_m^{(n)}, v)| \to 0, \quad n \to \infty. \tag{55}$$

The above results permit to conclude that $|a(u;u,v) - a_m(u^{(n-1)}; u_m^{(n)}, v)| \to 0$, as $m, n \to \infty$. Indeed,

$$|a(u;u,v) - a_m(u^{(n-1)}; u_m^{(n)}, v)|$$
$$\le |a(u;u,v) - a(u^{(n-1)}; u^{(n)}, v)| + |a(u^{(n-1)}; u^{(n)}, v) - a_m(u^{(n-1)}; u_m^{(n)}, v)|.$$

The first and second right hand side terms tend to zero, due to Theorem 6.3 and (55), accordingly. This completes the proof. □

Taking into account the compact embedding $H^2(\Omega) \hookrightarrow C^0(\overline{\Omega})$, $\Omega \subset R^2$ [5], we conclude that the sequence solutions $\{u(x; g_m)\} \subset \overset{\circ}{H}{}^2(\Omega)$ converges to the solution $u(x;g) \in \overset{\circ}{H}{}^2(\Omega)$ of the variational problem (36)-(37) in $C^0(\overline{\Omega})$. This means the continuity of the cost functional (52).

Theorem 6.4. *Let conditions of Lemma 6.6 hold. Then the problem (ICP) has at least one solution $g_* \in \mathcal{G}_c$ defined as a solution of the minimization problem (53).*

7. Parametrization of the unknown coefficient $g(\xi^2)$, inversion algorithm and its regularization

The feasibility of the approach given here is that the proposed inversion method is based on a finite number of output measured data, since in engineering practice a limited number of discrete values of the torque $T[g](\varphi)$, indentation curve $\mathcal{P}(\alpha)$ or deflections $w(\lambda_i; \tau_k)$ can only be given. Therefore all these data can only be given as a finite dimensional vectors during a quasistatic process of loading, given by the increasing values of the loading/deflection parameters. For example, in the problem (ICP2), one needs to assume the indentation data $\overline{\mathcal{P}} = \overline{\mathcal{P}}(\overline{\alpha})$, as a finite dimensional vectors $\overline{\alpha} = (\alpha_0, \alpha_1, ..., \alpha_K)$, $\overline{\mathcal{P}} = (\mathcal{P}_0, \mathcal{P}_1, ..., \mathcal{P}_K)$ in $\mathbf{R}^{K+1}$. The monotone increasing values $0 < \alpha_0 < \alpha_1 < ... < \alpha_K$ of the penetration depth generate the quasi-static indentation. For each α_k, the maximal value of the strain intensity $e_{\mathrm{i}}^{(k)}(u)$ for all finite elements or grids will be denoted by e_k. Due to the piecewise linear approximation of the function $u(x)$, in each finite elements or grids the parameter e_k is a constant.

In view of the problem (ICP1), the limited number of discrete values $\mathcal{T}_0, \mathcal{T}_1, ..., \mathcal{T}_M$ of the torque $\mathcal{T} = \mathcal{T}(\varphi)$, corresponding to the monotone increasing values $\varphi_0, \varphi_1, ..., \varphi_M$ of the angle of twist can only be given. Here the values $\mathcal{T}_0$ and $\mathcal{T}_m$, $m = \overline{1,M}$, of the measured torque correspond to pure elastic and mth plastic torsional deformation cases. Note that the elasticity limit (yield stress) $\xi_0^2 = e_0^2$ is assumed to be also unknown, and need to be determined from the above given measured data. For each φ_m, $m = \overline{1,M}$, the torsional state of a material is characterized by the stress intensity $\xi_m := \max_\Omega |\nabla u|$. As noted above, this parameter is a constant in each finite elements or grid, due to the piecewise linear approximation. This situation naturally leads to necessity of piecewise-linear approximation of the unknown function $g(\xi^2)$ (Figure 4, left figure). The piecewise-linear approximation $g_h(\xi^2)$ of the plasticity function has the form:

$$g_h(\xi^2) = \begin{cases} \beta_0 = 1/G, & \xi \in (0, \xi_0]; \\ \beta_0 - \beta_1(\xi^2 - \xi_0^2), & \xi \in (\xi_0, \xi_1]; \\ \beta_0 - \sum_{m=1}^{M-1} \beta_m(\xi_m^2 - \xi_{m-1}^2) - \beta_M(\xi^2 - \xi_{M-1}^2), & \xi \in (\xi_{M-1}, \xi_M]. \end{cases} \tag{56}$$

The unknown parameters (slopes) $\beta_m > 0$ need to be determined step by step, begining from the paramater $\beta_0 = 1/G$. At each mth state, one needs to determine the parameter β_m, by using the measured output data $\mathcal{T}_m$, which corresponds to the value φ_m of the angle of twist. The parameter $\Delta\xi_m = \xi_m^2 - \xi_{m-1}^2$ is defined to be *the state discretization parameter* for the mth state, according to [12, 15].

Thus the pairs (outputs) $\langle \varphi_m, \mathcal{T}_m \rangle$ on the plane $O\varphi, \mathcal{T}$ correspond to the pairs $\langle \xi_m, g_h^{(m)} \rangle$, $g_h^{(m)} = g_h(\xi_m^2)$, on the plane $O\xi g$. Therefore, introducing the $(M+1)$-dimensional vectors $\overline{\mathcal{T}} := (\mathcal{T}_0, \mathcal{T}_1, ..., \mathcal{T}_M) \in R^{M+1}$ and $\overline{\beta} = (\beta_0, \beta_1, ..., \beta_M) \in R^{M+1}$, and the set of unknown parameters

$$\mathcal{B} := \{\overline{\beta} \in R^{M+1} : \ \overline{\beta} = (\beta_0, ..., \beta_M), \beta_0 = 1/G, \beta_1 > ... > \beta_M > 0\}, \tag{57}$$

we can formulate the *disctere (or parametrized) inverse problem* as follows:

$$T_h(\overline{\beta}) = \overline{\mathcal{T}}, \quad \overline{\beta} \in \mathcal{B}. \tag{58}$$

We assume that an appropriate numerical integration formula is applied to the left hand side integral in (18). Hence the operator $T_h : \mathcal{B} \mapsto \{\overline{\mathcal{T}}\}$ is a discrete analogue of the input-output map $T[\cdot] : \mathcal{G} \mapsto \mathbf{T}$. Evidently a solution of the discrete inverse problem (58) may not exists, even in the case of noise free measured output data, due to computational noise factor. We define a weak solution of this problem as a solution of the following minimization problem

$$I_h(\overline{\beta}^*) = \min_{\overline{\beta} \in \mathcal{B}} I_h(\overline{\beta}^*), \tag{59}$$

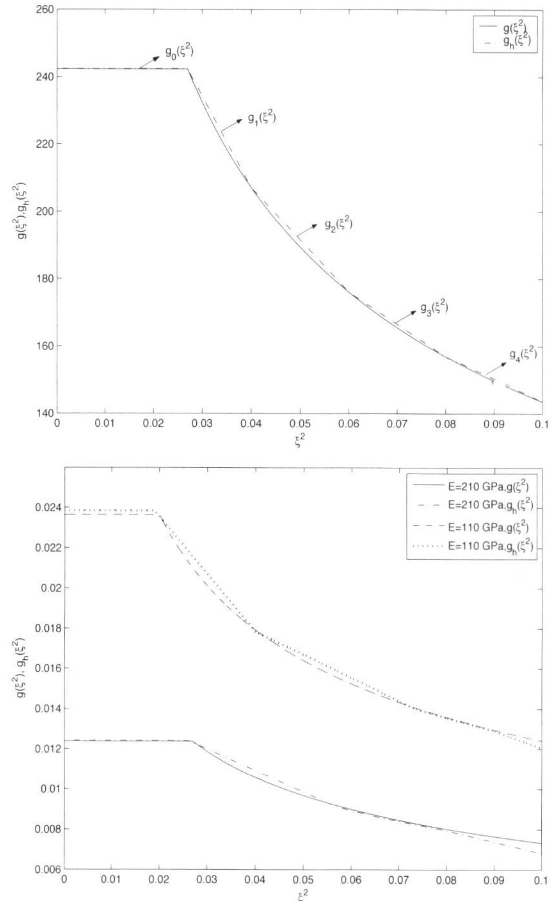

Fig. 4. The plasticity function and its parametrization $g_h(\xi^2)$ (left figure); the plasticity functions $g(\xi^2)$ and the corresponding synthetic noise free data $g_h(\xi^2)$ ($\kappa = 0.2$) for the rigid and sort engineering materials (right figure)

where

$$I_h(\overline{\beta}^*) = \|T_h(\overline{\beta}) - \overline{T}\|_\infty \, . \tag{60}$$

Now let us derive algorithms for determination of the unknown parameters $\beta_0 = 1/G$, ξ_0^2 and β_m, $\overline{1, M}$. In the pure elastic torsion case one needs to determine the shift modulus $G = E/2((1+\nu))$, since the Poisson coefficient is assumed to be known ($\nu = 0.3$).

In the pure elastic case we will use Lemma 4.2. Thus, the algorithm of determining the unknown parameter G is as follows.

Algorithm 1:

(i1) *Choose iterations* $\beta_0^{(1)}$, $\beta_0^{(2)}$, *satisfying the conditions:*

$$T_h[\beta_0^{(1)}] > \mathcal{T}_0 > T_h[\beta_0^{(2)}]; \tag{61}$$

(i2) *Use next iteration* $\beta_0^{(3)} = (\beta_0^{(1)} + \beta_0^{(2)})/2$, *and calculate the theoretical value of the torque* $T_h[\beta_0^{(3)}]$;

(i3) Determine the next iteration by using conditions (25):
 if $T_h[\beta_0^{(3)}] < \mathcal{T}_0$, then $\beta_0^{(4)} = (\beta_0^{(3)} + \beta_0^{(2)})/2$;
 if $T_h[G^{(3)}] > \mathcal{T}_0$, then $\beta_0^{(4)} = (\beta_0^{(3)} + \beta_0^{(1)})/2$;

(i4) *Calculate* $|T_h[G^{(3)}] - \mathcal{T}_0|$;

(i5) *If* $|T_h[\beta_0^{(3)}] - \mathcal{T}_0| < \varepsilon_T$, *then* $\beta_0 = \beta_0^{(3)}$. *Otherwise, continue the steps (i2)-(i4)*;

(i6) *Repeat the process until the fulfilment of the stopping condition*

$$|T_h[\beta_0^{(n)}] - \mathcal{T}_0| < \varepsilon_T. \tag{62}$$

The parameter $\varepsilon_T > 0$ is defined to be a given accuracy for an approximate solution of the inverse discrete problem (59)-(60), in pure elastic torsion.

For each plastic torsion case, we can derive the similar algorithm for determination of the slopes $-\beta_m$, $m = \overline{1, M}$ (see, (56)). In this case the folowing argument is taken into account: an increase of the slope β_m corresponds to an increase of rigidity of a material, and, as a result, to increase of the corresponding theoretical value $T_h[\beta_m]$ of the torque. This means that $\beta_m^{(1)} > \beta_m^{(2)}$ implies $T_h[\beta_m^{(1)}] > T_h[\beta_m^{(2)}]$. Hence for the determination of the unknown parameters β_m similar algorithm can be used (one needs only to write $\beta_m^{(n)}$ instead of $\beta_0^{(n)}$ in the above algorithm). Further, the same stopping condition

$$|T_h[\beta_m^{(n)}] - \mathcal{T}_m| < \varepsilon_T \tag{63}$$

will be used at each mth plastic torsional state.

The found values G_h ($\beta_{0h} = 1/G_h$ and β_{mh} are assumed to be approximate values of the shear modulus $\beta_0 = 1/G$ and the parameters β_m, $m = \overline{0, M}$, respectively.

For determination of the unknown elasticity limit ξ_0 the linearity, in pure elastic case, of the direct problem (3) (as well as the functional equation (18)),corresponding to the parameter $\varphi > 0$ are used. Specifically, let $G_h > 0$ be a given (already determined) parameter, and $\langle \varphi_0, \mathcal{T}_0 \rangle$ is the corresponding output measured data. Denote by $u_0 = u_0(x; G_h; \varphi_0)$ the solution of the corresponding linear direct problem (3). We define the new angle of twist as follows: $\tilde{\varphi}_0 = \delta_0\varphi_0$, where $\delta_0 \in (0,1)$. Then the function $\tilde{u}_0 = \tilde{u}_0(x; G_h; \tilde{\varphi}_0)$ will be solution of the direct problem (3), corresponding to the given parameter $G_h > 0$, and the new angle $\tilde{\varphi}_0$. Due to the linearity of problem (3), $\tilde{u}_0 = \delta_0 u_0$. Further, it follows from definition (2) that the synthetic output measured data $\tilde{\mathcal{T}}_0 = T[G_h](\tilde{\varphi}_0)$ corresponding to the parameter $\tilde{\varphi}_0$ can be defined via the output masured data $\mathcal{T}_0$ as follows: $\tilde{\mathcal{T}}_0 = \delta_0\mathcal{T}_0$. Thus, in pure elastic torsional case, the synthetic output data $\tilde{\mathcal{T}}_0 = T[G_h](\tilde{\varphi}_0)$ can be explicitely obtained from the output measured data $\mathcal{T}_0$ by the formula $\tilde{\mathcal{T}}_0 = \delta_0\mathcal{T}_0$.

This property will be used in the following algorithm of determining of the unknown elasticity limit ξ_0^2.

Algorithm 2:

(i1) *Choose the iteration* $\varphi_0^{(1)}$, *satisfying the condition* $\varphi_0^{(1)} > \varphi_0$;

(i2) *Use the synthetic output data* $T_0^{(1)} = \delta_1\mathcal{T}_0$, $\delta_1 = \varphi_0/\varphi_0^{(1)}$, *and apply Algorithm 1 to find* $\beta_0^{(1)} = 3G^{(1)}$;

(i3) *Calculate the relative error*

$$\delta G^{(1)} = |G_h - G^{(1)}|/G_h; \tag{64}$$

(i4) *(a) If* $\delta G^{(1)} < \varepsilon_G$, *use the next iteration* $\varphi_0^{(2)} > \varphi_0^{(1)}$;

(b) If $\delta G^{(1)} > \varepsilon_G$, *use the next iteration* $\varphi_0^{(2)} = 0.5[\varphi_0^{(1)} + \varphi_0]$;

(i5) *Use the synthetic output data* $T_0^{(2)} = \delta_2\mathcal{T}_0$, $\delta_2 = \varphi_0/\varphi_0^{(2)}$, *and repeat the items (i2)-(i3);*

(i6) *Repeat the process until the fulfilment of the following conditions*

$$\delta G^{(n)} < \varepsilon_G, \quad \delta G^{(n+1)} > \varepsilon_G, \quad \Delta\xi_0^{(n)} := \left|\frac{\xi_0^{(n)} - \xi_0^{(n+1)}}{0.5[\xi_0^{(n)} - \xi_0^{(n+1)}]}\right| < \varepsilon_\xi. \tag{65}$$

Here $\xi_0^{(n)} = \max_\Omega |\nabla u^{(n)}(x)|$, $\xi_0^{(n+1)} = \max_\Omega |\nabla u^{(n+1)}(x)|$, and the functions

$$u^{(n)}(x) = u(x; G^{(n)}; \varphi_0^{(n)}), \quad u^{(n+1)}(x) = u(x; G^{(n+1)}; \varphi_0^{(n+1)})$$

are the solutions of corresponding direct problems.

The found value $\xi_{0h} = 0.5[\xi_0^{(n)} + \xi_0^{(n+1)}]$ is assumed to be an approximate value of the the elasticity limit ξ_0.

The above algorithms hold true for the problems (ICP2) and (ICP3) as well (one needs to use the output measured data, $\mathcal{P}(\alpha)$ and $w(\lambda_i; \tau_k)$, accordingly, instead of $T[g](\varphi)$).

6.2. Generating noise free synthetic output data and reconstruction of the parameters β_0 and β_1: the problem (ICP1)

In all the numerical examples below two types of engineering materials defined to be as rigid ($E = 210(GPa)$) and soft ($E = 110(GPa)$), are used. The plasticity function for these materials and its piecewise-linear approximation are given by formulas (56) and (27), correspondingly. To generate the noise free synthetic output data $\langle \varphi_m, \mathcal{T}_m \rangle$, the direct problem (3) was solved by the variational finite-difference scheme using the given function $g_h(\xi^2)$, given by (56), and the input data φ_m, given in Table 1. Here $m = 0$ correspond to pure elastic state, and the numbers $m = 1, 2, 3, 4$ correspond to plastic ones. Then the values $T[g](\varphi_m)$ are calculated by applying numerical integration formula to the integral (2). These the values are then assumed to be the noise free synthetic data in subsequent computational experiments.

The first series of numerical examples are related to implementation of Algorithm 1 for finding the shift modulus $G = E/2((1+\nu))$ (Here and in all computational experiments below the Poisson coefficient is taken to be $\nu = 0.3$). For different values of the stopping parameter $\varepsilon_T = 10^{-2}, 10^{-3}, 10^{-4}$, the corresponding relative errors in the found values of the shift modulus G_h was between $\delta G_h = 10^{-2}$ and $\delta G_h = 10^{-3}$. The number of iterations was $n = 4 \div 7$. Numerical resuts show that the relative error δG_h decreases proportionally by decreasing the stopping parameter ε_T, due to linearity of the problem in pure elastic case. Thus for $\varepsilon_T = 10^{-4}$ the relative errors δG_h for the both materials are of same order 10^{-4}.

In the second numerical example, the above found values G_h separately have been used as an input data in Algorithm 2, for determination of the unknown elasticity limit ξ_0^2 for the rigid material. With the first input data $G_{h,1} = 76.50$ the found elasticity limit is $\xi_{0h,1}^2 = 0.022$, while computed from the direct problem solution, with the same input data, elasticity limit is $\xi_{0,1}^2 = 0.024$. Hence the relative error is $\delta\xi_{0h,1}^2 = 1.9 \times 10^{-1}$. Further, in the case of the second input data $G_{h,2} = 78.75$, which corresponds to the stopping parameter $\varepsilon_T = 10^{-3}$, the found elasticity limit is $\xi_{0h,2}^2 = 0.023$, while computed from the direct problem solution elasticity limit is $\xi_{0,2}^2 = 0.025$. The relative error is $\delta\xi_{0h,2}^2 = 1.4 \times 10^{-1}$. Although further increase of the stopping parameter improve the reconstructed value of the elasticity limit ξ_0^2, the order of the relative error $\delta\xi_{0h,3}^2$ remains the same. Specifically, in the third case, which corresponds to the stopping parameter $\varepsilon_T = 10^{-4}$, the found by Algorithm 2 value of the elasticity limit and the relative error are $\xi_{0,3}^2 = 0.026$ and $\delta\xi_{0h,3}^2 = 3.7 \times 10^{-2}$, respectively. Similar results are obtained for the soft material.

The above numerical results permit to conclude that the relative errors $\delta G_h = 3.3 \times 10^{-3}$ and $\delta \xi^2_{0h} = 3.7 \times 10^{-2}$ in the reconstructed values of the shear modulus G and elasticity limit ξ^2_0, correspond to the value $\varepsilon_T = 10^{-4}$ of the *stopping parameter*. This value will be used in subsequent computational experiments. From the point of view natural experiments, it is interesting to define also the level of accuracy in the measured values φ_i, corresponding to the value $\varepsilon_T = 10^{-4}$ of the stopping parameter. For this aim the linear direct problem ($g(\xi^2) = 1/G$, $G = 80.77$) was solved for a given two values φ_1, φ_2, with $|\varphi_1 - \varphi_2| = 3.5 \times 10^{-3}$. The absolute error was found $\varepsilon_T := |T[g](\varphi_1) - T[g](\varphi_2)| = 6.5 \times 10^{-3}$. Therefore in the case of pure elastic deformations, the accepted below value $\varepsilon_T = 10^{-4}$ of the stopping parameter corresponds to the absolute error $\delta\varphi = 10^{-3}$ in the measured values of the angle of twist, which can be defined as an accuracy of the torsional experiment for pure elastic case. Almost same result were obtained for the soft material $G = 42.30$. Specifically, the above value $\varepsilon_T = 6.5 \times 10^{-3}$ correspond to the absolute error $|\varphi_1 - \varphi_2| = 5.2 \times 10^{-3}$ in the measured values of the angle of twist.

Table 1. Synthetic noise free data for rigid and soft materials ($\kappa = 0.2$)
($E = 210(GPa)$, $\xi^2_0 = 0.027$) ($E = 110(GPa)$, $\xi^2_0 = 0.02$)

$m =$	0	1	2	3	4		0	1	2	3	4
$\varphi_m \times 10^3$	2.0	3.5	3.6	3.7	3.8		4.0	6.0	6.3	6.5	6.6
$T[g](\varphi_m) \times 10^2$	2.26	4.05	4.22	4.41	4.61		2.36	3.71	4.03	4.27	4.39
$\xi_{mh} \times 10^2$	1.0	5.7	6.6	8.0	9.3		1.22	5.40	7.33	9.01	9.80
$g(\xi^2_{mh}) \times 10^2$	1.24	9.2	8.7	8.0	7.5		2.36	1.59	1.41	1.29	1.25

Consider now the problem identification of the unknown parameter β_1 in (56), by using the synthetic noise free data given in Table 1. In the case of rigid material, the first plastic state ($m = 1$) is generated by the value $\varphi_1 = 3.5 \times 10^{-3}$ of the angle of twist. Algorithm 1 is implemented for determination of the unknown parameter (slope) β_1. The reconstructed values, corresponding to the stopping parameters $\varepsilon_T = 10^{-2};\ 10^{-3};\ 10^{-4}$, are found to be $\beta_{1h} = 0.131; 0.120; 0.104$, respectively. The relative errors ($\delta\beta_{1h} = |\beta_1 - \beta_{1h}|/\beta_1$) are $\delta\beta_{1h} = 1.8 \times 10^{-1}; 1.6 \times 10^{-1}; 2.8 \times 10^{-2}$. Evidently, these errors decrease by decreasing the stopping parameter ε_T. In the case of the soft material, with the same $\kappa = 0.2$, the reconstructed parameters and the relative errors, corresponding to the above values of the stopping parameter, are $\beta_{1h} = 0.240; 0.233; 0.228$, and $\delta\beta_{1h} = 2.6 \times 10^{-2}$; 4.4×10^{-2}; 1.3×10^{-2}, respectively. The similar decrease of the relative error is observed in this case. Further, in the case of first plastic state, one also needs to define the level of accuracy $\delta\varphi$ in the measured values of the angle of twist, corresponding to the given value of the stopping parameter ε_T. As the computational experiments show, the absolute error $\varepsilon_T := |T[g](\varphi_1) - T[g](\varphi_2)| = 6.5 \times 10^{-3}$ corresponds to the absolute errors $|\varphi_1 - \varphi_2| = 4.5 \times 10^{-3}$ and $|\varphi_1 - \varphi_2| = 2.4 \times 10^{-3}$, for the rigid and soft materials, respectively.

6.3. Ill-posedness of the problems: regularization of the inversion algorithm

Consider first the numerical examples related to reconstruction of the unknown coeficient $g_h(\xi^2)$ for rigid and soft materials in *the problem (ICP1)*. The reconstructed coeficients plotted in Figure 4 (right figure), correspond to the synthetic noise free data $\langle\varphi_m, \mathcal{T}_m\rangle$, $m = \overline{0,4}$, given in Table 1 for the rigid and soft materials, with $\kappa = 0.2$. Note that the values of the angles of twist φ_m in Table 1 are choosen so that to guarantee the uniformness of the state discretization parameters $\Delta\xi_{mh} = \xi^2_{mh} - \xi^2_{m-1h}$ for $m = 2, 3, 4$. As the results plotted in the figure show, the reconstructions for the both materials are satisfactory, due to noise free data. Specifically, the relative errors, defined by $\delta g_h = \|(g - g_h)/g\|_{\infty,h}$, are $\delta g_h = 7.0 \times 10^{-2}$, $\delta g_h = 4.2 \times 10^{-2}$, for the rigid and soft materials, correspondingly. *However, the relative error $\delta\beta_m = |\beta_m - \beta_{mh}|/\beta_m$ increases by increasing the number of states, i.e. by increasing the values φ_m of the angle of twist.* This can precisely be observed from Table 2. In the case of small values of the state discretization parameter $\Delta\xi_{mh}$ these errors can lead to the divergence of the iteration algorithm. Thus, for $\Delta\xi_{mh} = 0.4 \times 10^{-2}$ divergence arises immediately at the first plastic state. This situation is illustrated in Figure 5 (left figure, the first $\star$- point). Two times increase of the state discretization step ($\Delta\xi^{(1)}_{h,m} = 2\Delta\xi_{mh}$) also leads to divergence of the iteration algorithm. Only the three times increase ($\Delta\xi^{(3)}_{mh} = 3\Delta\xi_{mh}$) of the state discretization step leads to the convergence of the iteration algorithm (the first $\circ$-point in the left Figure 5).

Similar situation arises in the case of *the problem (ICP2)*. Reconstruction of the stress-strain curve $\sigma_i = \sigma_i(e_i)$, given by (10) and (56) is illustrated in the right Figure 5, which again shows that, the inversion algorithm may not converge for arbitrary values of the state discretization step $\Delta\xi_{mh} = \xi^2_{mh} - \xi^2_{m-1h}$, if the stopping parameter $\varepsilon_T > 0$, defined in (63), is choosen to be small enough and given in advance. To guarantee the convergence of the algorithm one needs to increase either the parameter ε_T or the step $\Delta\xi_{mh}$. The both situations evidently will lead to loss of accuracy.

Table 2. Relative errors $\delta\beta_m$ corresponding to the increasing values φ_m from Table 1

(a) Rigid material ($E = 210(GPa)$; $\kappa = 0.2$) (b) Soft material ($E = 110(GPa)$ $\kappa = 0.2$)

m	β_m	β_{mh}	$\delta\beta_m$		β_m	β_{mh}	$\delta\beta_m$
1	0.107	0.104	2.8×10^{-2}		0.228	0.225	1.3×10^{-2}
2	0.058	0.052	1.0×10^{-1}		0.095	0.090	5.2×10^{-2}
3	0.046	0.040	1.3×10^{-1}		0.066	0.058	1.2×10^{-1}
4	0.036	0.028	2.2×10^{-1}		0.054	0.044	1.8×10^{-1}

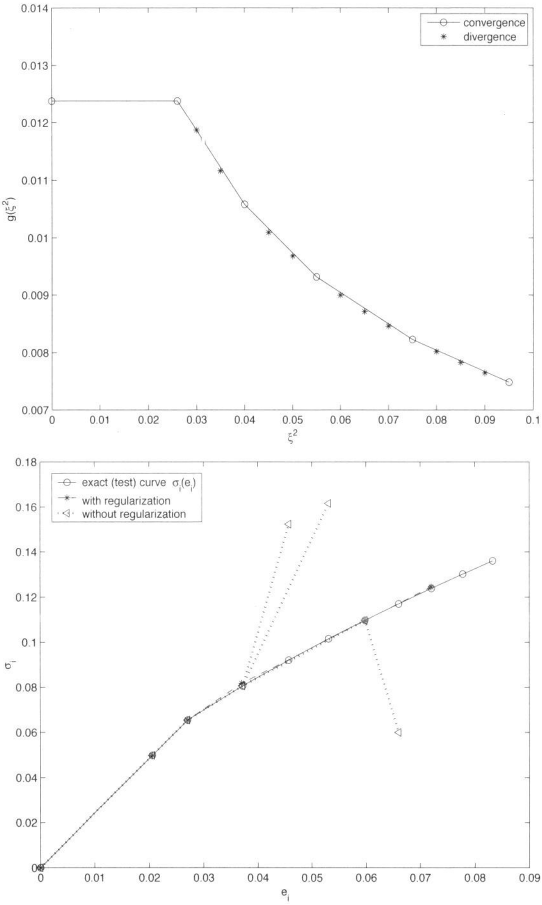

Fig. 5. Divergence of the inversion algorithm in small values of the state discretization parameter $\Delta\xi_{mh}$: the problem (ICP1) (left figure) and the problem (ICP2) (right figure)

To analyze this situation, first of all one needs to emphasize two distinguished features of all the three inverse problem. First, solvability of the inverse problem is a result of monotonicity of the plasticity function $g(\xi^2)$, which for disrete model (58) means fufilment of the monotonicity condition (57) for slopes β_k, during the iteration process. Due to closeness of neighbouring slopes for the developed plasticity states, and due to computational errors in β_k, in pactice this condition may not be fulfilled. Second, all previously determined parameters ξ^2_{mh} are included, as formulas (56) shows, in the next state. Subsequently computational errors are compounded, and as a result, the second noise factor - *computational noise factor* arises. By increasing the number of states the influence of this factor will be increased. Hence for small values of state discretization parameters the inverse problem may become unstable after some m-th state. The left and right figures 5 is an illustration of this phenomenon, when the above algorithm is directly applied to the inverse problems. These results show that the inversion algorithm may not converge for arbitrary values of the state discretization parameters and within the strong fulfment of the monotonicity condition (57).

In order to guarantee stability and convergence of the inversion method two regularization schemes - *relaxation of the monotonicity condition for slopes* β_k, and *optimization of the state discretization parameters* $\Delta\xi_{mh} = \xi^2_{mh} - \xi^2_{m-1h}$ - have been added to proposed inversion algorithm. Introducing the *relaxation parameter* $\delta_\beta > 0$ we will require that the unknown parameters β_k satisfy the following relaxed monotonicity condition

$$\beta_k + \delta_\beta \geq \beta_{k-1} \geq \beta_k - \delta_\beta, \quad k = \overline{1, K}, \quad \delta_\beta > 0. \tag{66}$$

Hence we define the new set of admissible unknown parameters will be defined as follows:

$$\mathcal{B}_\delta := \{\overline{\beta}_\delta \in \mathbf{R}^{K+1} : \ \beta_0 > 0, \ \ \beta_k + \delta_\beta \geq \beta_{k-1} \geq \beta_k - \delta_\beta, \quad k = \overline{1, K}, \quad \delta_\beta > 0\}.$$

the instead of $\mathcal{B}$, defined by (57).

An optimal choice of the state discretization parameters has been realized as follows. Let us assume that the inversion algorithm converges at $(k-1)$-th plastic state, but diverges at k-th plastic state, i.e. neither condition (57) nor the condition (63) hold. In this case the parameter $\Delta\xi_{mh}$ is eliminated from the consideration and taking the new state discretization parameter $\tilde{\Delta}\xi_{mh} = \Delta\xi_{m+1,h} - \Delta\xi_{m-1,h}$ instead of the parameter $\Delta\xi_{mh}$. Then the iteration process is repeated anew from the $(m-1)$-th state. This process is repeated until the fulfilment of conditions (62) and (66) and the found step $\tilde{\Delta}\xi_{k+1,h} = \Delta\xi_{m+m_s,h} - \Delta\xi_{m-1,h}$ is assumed to be a new state discretization parameter for the m-th state. This modification leads to the natural selection of the state discretization parameters, also allows us to minimize the number of measurements.

To illustrate an effectiveness of the inversion algoritm with the above regularization consider the problem (ICP1).

Let $\{\Delta\xi_{mh}\}$, $m = \overline{1, M}$, be an initial set of state discretization parameters corresponding to an experimentally given set $\{\varphi_m\}$ of angles of twist. Assume that the iteration algirithm converges at the $(m-1)$th torsional state, but diverges at the mth state $(1 < m < M)$, which means condition (63) does not hold. In this case the parameter $\Delta\xi_{mh}$ is eliminated from the consideration, and the new state discretization parameter $\Delta\xi^{(1)}_{mh} = \Delta\xi_{mh} + \Delta\xi_{m+1h}$ is defined. Then the inversion Algoritm 2 is applied with the new parameter $\Delta\xi^{(1)}_{mh}$. In the case of divergence, again the new state discretization parameter $\Delta\xi^{(2)}_{mh} = \Delta\xi_{mh} + \Delta\xi_{m+1h} + \Delta\xi_{m+2h}$ is defined, and the process is repearted. In the case of convergence, the value $\tilde{\Delta}\xi_{mh} = \Delta\xi^{(1)}_{mh}$ is defined to be an optimal state discretization parameter for the mth state. At the next $(m+1)$th state the same procedure is applied. This natural selection of the state discretization parameters is illustrated in Table 3. The input data here are taken from Table 1 and corresponds to the scheme given in Figure 5. As Table 3 shows, the found new values of the state discretization parameters are as follows:

$$\begin{aligned}
\tilde{\Delta}\xi_{1h} &= \Delta\xi_{1h} + \Delta\xi_{2h} + \Delta\xi_{3h},\\
\tilde{\Delta}\xi_{2h} &= \Delta\xi_{4h} + \Delta\xi_{5h} + \Delta\xi_{6h},\\
\tilde{\Delta}\xi_{3h} &= \Delta\xi_{7h} + \Delta\xi_{8h} + \Delta\xi_{9h} + \Delta\xi_{10h},\\
\tilde{\Delta}\xi_{4h} &= \Delta\xi_{11h} + \Delta\xi_{12h} + \Delta\xi_{13h} + \Delta\xi_{14h}.
\end{aligned}$$

The above natural selection of the state discretization parameters guarantees convergence of the inversion algorithm for each initially given set of output measured data $\{\langle\varphi_m, \mathcal{T}_m\rangle\}$.

Table 3. (a) Regularization of the inversion algorithm applied to the example illustrated in the left Figure 5

m	1*	2*	3	4*	5*	6	7*
$\xi_i \times 10^2$	3.0	3.5	4.0	4.5	5.0	5.5	6.0
$\Delta\xi_i \times 10^2$	0.5	0.5	0.5	0.5	0.5	0.5	0.5
Divergence	⋆	⋆		⋆	⋆		⋆
Convergence			○			○	

Table 3. (b) Regularization of the inversion algorithm applied to the example illustrated in the left Figure 5

m	8*	9*	10	11*	12*	13*	14
$\xi_i \times 10^2$	6.5	7.0	7.5	8.0	8.5	9.0	9.5
$\Delta\xi_i \times 10^2$	0.5	0.5	0.5	0.5	0.5	0.5	0.5
Divergence	⋆	⋆		⋆	⋆	⋆	
Convergence			∘				∘

8. An implementation of the inversion algorithm to the problems (ICP1) and (ICP2) with noisy output data

Let us consider the problem (ICP1). Since in practice a finite number of measured values of the torque $\mathcal{T}$ can only be given with some measurement error: $\mathcal{T}_\gamma = \mathcal{T} + \gamma\, \mathcal{T}$, where $\gamma \in R$ is the noise level. In this computational experiment the following noisy levels are used: $\gamma = 0.03, -0.05, \pm 0.07$.

For the exact and noisy (synthetic) output data shown in the left Figure 6, for the rigid and soft materials, the reconstructed coefficients are plotted in the right Figure 6. In all cases the relative errors, defined to be as $\delta g_h = \| (g - g_h)/g\|_{\infty,h}$, were obtained as $\delta g_h = 6.5 \div 8.0 \times 10^{-3}$. This show that the proposed algorithm allows to identify the unknown coefficient with enough high accuracy, not only for the noisy free, but also for the noisy data.

Consider now the problem (ICP2). In experiments the indentation curve can only be given with some measurement error $\gamma = 1 \div 5\%$ [4,6, 26], as the finite number of measured data $\mathcal{P}_m^\gamma = \mathcal{P}_m \pm \gamma \mathcal{P}_m$, $m = \overline{1, M}$. Here $\mathcal{P}_m = \mathcal{P}(\alpha_m)$ is an exact data corresponding to m-th indentation step. In this case, in addition to the parameters $\varepsilon_P > 0$ and $\delta_\beta > 0$, the stability of the inversion algorithm with respect to the noise factor $\gamma > 0$ also arises. Two types of power hardening materials (with elasticity parameters $E = 110GPa$, $e_0 = 0.020$ and $E = 210GPa$, $e_0 = 0.027$) are selected to examine the proposed inversion method. The indentation curves for these materials, obtained by from the numerical solution of the direct problem (6) with $r_0 = 0.5$, $\kappa = 0.5$, are shown in Fig. 7. The noisy data $\mathcal{P}_m^\gamma$ have been generated by taking $\gamma = 0.05$ and $\gamma = -0.1$. All these data have been then used as a synthetic data for identification of the curve $\sigma_{\mathrm{i},h}^{(m)} = \sigma_{\mathrm{i},h}^{(k)}(e_\mathrm{i})$, by using the piecewice approximation (56) in formula (10). The results are shown in the Figures 8.

For noise free data $\mathcal{P}_m$ the reconstructions of the stress-strain curve have enough accuracy, in the both cases. Thus, the relative error $\varepsilon_\sigma = (|(\sigma^{(m)} - \sigma_h^{(m)})/\sigma^{(m)}|) \times 100\%$ in the reconstructed curve, for the soft material, is about $\varepsilon_\sigma = 3\%$, except the fifth plastic state (the last $\star$-point in the left Figure 8). At this state the relative error was $\varepsilon_\sigma = 10\%$. This is due to the

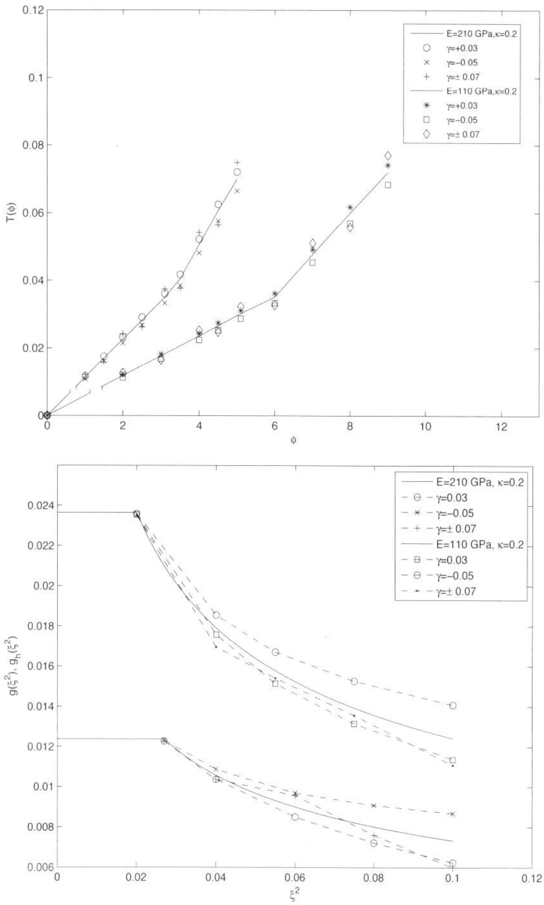

Fig. 6. (The exact and noisy output data (left figure) and the reconstructed coefficients (right figure): the problem (ICP2)

compounded errors in (56) from previous states, as was suggested above.

For the rigid power hardening material the reconstructed of the stress-strain curves are plotted the right in Figure 8. In the case of the noisy data, an accuracy of the reconstructed curve for the first to fourth plastic states is about $\varepsilon_\sigma = 1\%$, and $\varepsilon_\sigma = 6\%$, for the fifth plastic state. For the noisy data $\mathcal{P}_m^\gamma$ accuracy of the reconstruction decreases, in particular in the begining plastic states. For the noise factors $\gamma = 0.05; -0,1$ the relative errors rise up to $\varepsilon_\sigma = 12\%$ and $\varepsilon_\sigma = 7\%$, for the both types of materials, respectively. Note that, the noisy data also can change the optimal selection of the state discretization parameters, as shows the examples.

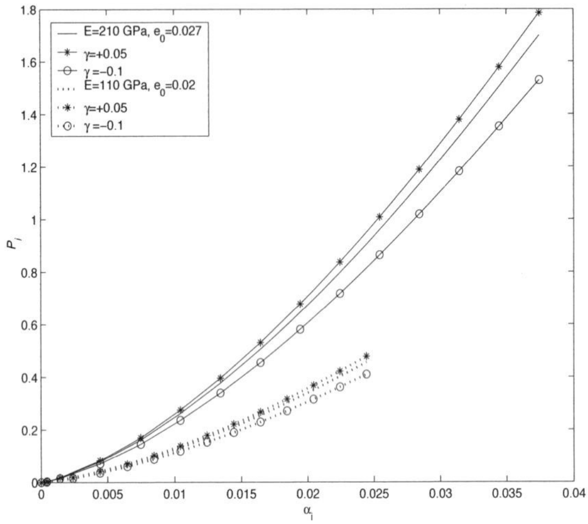

Fig. 7. (Noise free and noisy indentation curves for power hardening materials

The obtained results show that the presented inversion algorithm applied to the problem (ICP2) is also feasible in the presence of a noise factor $\gamma = 5 \div 10\%$.

9. Conclusions

The new inversion method for identifying the elastoplastic properties of a strain hardening materials within the range of J_2-deformation theory is proposed. Three types of inverse problems, with different governing equations and different measured output data, arising in engineering mechanics and computational material science are analyzed. Mathematical models of these inverse problems lead to inverse coefficient problem for nonlinear steady-state equations. An analysis of the corresponding direct problems, as well as all the three the inverse problems are proposed. Ill-conditionedness of the considered inverse problem has been carried out theoretically, as well as computationally. The compact set of admissible coefficients constructed here consistent with the assumptions of the physical model.

The presented inversion algorithm using the optimal selection of state discretization parameters is a new and natural regularization algorithm for such a class of inverse (or computational material diagnostics) prob-

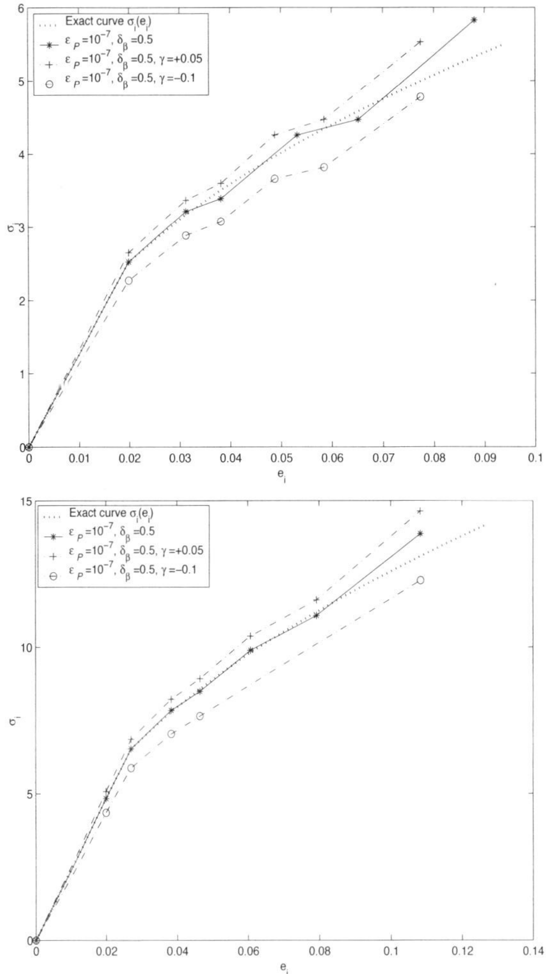

Fig. 8. The the reconstructed stress-strain curves for soft ($E = 110GPa$, $e_0 = 0.020$) (left figure) and rigid ($E = 210GPa$, $e_0 = 0.027$) (right figure) materials from the noise free and noisy data given in Figure 7

lems. The inversion method with this regularization algorithm permit one not only to determine effectively the elastoplastic properties from the the measured loading curve, but also suggests how one needs to simulate and select the limited number of experimental data (torque, loading curve, etc.). The demonstrated numerical results for different engineering materials show that the presented inversion method allows to determine the plasticity function with enough high accuracy from the noise free as well as noisy measured output data.

Finally, it should be emphasized that this study mainly focuses on the theoretical and computational aspects of the inverson method for considered here indentation problems. Further applications of the method, by taking into account such a physical and engineering aspects as geometrical non-linearity, elastic deformations of the indenter, use of the unloading part of the indentation curve, will be especially useful and will be essential part of the future research.

Acknowledgements

The author is grateful to Dr. Zahir Muradoglu and Salih Tatar for the assistance in providing the computational experiments. The work was supported by the International Center for Theoretical Physics (Trieste, Italy).

References

1. R. A. Adams, *Sobolev Spaces*, Academic Press, New York (1975).
2. J.A. Bland, Implementation of an algorithm for elastoplastic torsion, Adv. Engnr.Software, 17 (1993), 61-68.
3. P. Boonlualohr, S. Valliappan, Elastoplastic Torsion of Anisotropic Bars, J. Engineering Mechanics Division, 102 (1976), 995-1008.
4. Y.P. Cao and J. Lu, A new method to extract the plastic properties of metal materials from an instrumented spherical indentation loading curve, Acta Mater., 52 (2004), 4023-4032.
5. P. Ciarlet, *Finite-Element Method for Elliptic Problems*, Amsterdam, North-Holland (1978).
6. N.A. Fleck, J.W. Hutchinson, A reformulation of strain gradient plasticity, J. Mech. Phys. Solids 49 (2001), 2245-2271.
7. H. Gajewski, K. Greger, K. Zacharias, *Nichtlineare Operator Gleichungen und Operator Differential Gleichungen*, Akademic-Verlag, Berlin (1974).
8. F. Gruttmann, R. Sauer, W. Wagner, Theory and numerics of three-dimensional beams with elastoplastic material behaviour, Int. J. Numer. Meth. Engnr. 48 (2000) 1675 - 1702.
9. A. Hasanov, A. Mamedov, An inverse problem related to the determination of elastoplastic properties of a plate, *Inverse Problems,* **10**, 601-615 (1994).
10. A. Hasanov, An Inverse coefficient problem for an elasto-plastic medium, *SIAM J. Appl. Math.*, **55**(1995), 1736-1752.
11. A. Hasanov, Inverse coefficient problems for monotone potential operators, *Inverse Problems*, **13**(1997), pp. 1265-1278.
12. A. Hasanov, Inverse coefficient problems for elliptic variational inequalities with a nonlinear monotone operator, *Inverse Problems*, **14**(1998), pp. 1151-

1169.
13. A. Hasanov, Convexity argument for monotone potential operators, *Nonlinear Analysis: TMA,* **47**, 906-918 (2000).
14. A. Hasanov, Variational approach to non-linear boundary value problems for elasto-plastic incompressible bending plate, *Int. J. Non-Linear Mech.,* **42** 711-721 (2007).
15. A. Hasanov, An inversion method for identification of elastoplastic properties for engineering materials from limited spherical indentation measurements, *Inverse Problems in Science and Engineering*, **15(6)**(2007), pp. 601-627.
16. A. Hasanov, A. Erdem, Determination of unknown coefficient in a nonlinear elliptic problem related to the elasto-plastic torsion of a bar, *IMA J. Appl. Math.*, **73(4)**(2008), pp. 579-591.
17. A. Hasanov, S. Tatar, Semi-analytic inversion method for determination of elastoplastic properties of power hardening materials from limited torsional experiment, *Inverse Problems in Science and Engineering*, **18**(2010), pp. 265-278.
18. A. Hasanov, S. Tatar, An inversion method for identification of elastoplastic properties of a beam from torsional experiment, *Int. J. Non-Linear Mech.*, **45** (2010), pp. 562-571.
19. A. Hasanov, An introduction to inverse source and coefficient problems for PDEs based on boundary measured data, in *Mathematics in Science and Technology,* Abstracts of the Satellite Conference of International Congress of Mathematicians, (Eds.: A.H. Siddigi, R.C. Singh, P. Manchanda), Delhi - India, 14-17 August, 2010.
20. A.J. Ishlinski (1944). Axisymmetrical problem of plasticity and the Brinell test. *Prikl. Mat. Mekh.*, **8**, 204-224.
21. V.K. Ivanov, V.V. Vasin, V.P. Tanana, *Theory of Linear Ill-Posed Problems and Its Applications*, Nauka, Moscow, 1978.
22. L.M. Kachanov, The Theory of Greep, National Lending Library for Sciences and Technology, Boston Spa, Yarkshire, England, 1967.
23. A. Kufner, S. Fučik, *Nonlinear Differential Equations Studies in Applied Mathematics 2* Amsterdam: Elsevier Scientific Publ. Comp. (1980).
24. L. Liu, N. Ogasawara, N. Chiba, X. Chen, Can indentation test measure unique elastoplastic properties, *J. material Research,* **24** (2009), pp. 784-800.
25. O.A. Ladyzhenskaya, *Boundary Value Problems in Mathematical Physics*, Springer, New York, (1965).
26. D. Ma, C.W. Ong, J. Lu and J. He, Methodology for the evaluation of yield stress and hardening behaviour of metallic materials by indentation with spherical tip. *J. Appl. Phys.*, **94(1)** (2003), pp. 288-294.
27. A. Mamedov, An Inverse problem related to the determination of elastoplastic properties of a cylindirical bar, Int. J. Non-Linear Mech. 30(1995) 23-32.
28. S. G. Mikhlin, *Variational Methods in Mathematical Physics*, Pergamon, New York, (1964).
29. Y.-B. Park, D.-Y. Yang, Analysis of torsional deformation by rigid-plastic finite element method using recurrent boundary conditions, J. Mater. Processing Tech. 182 (2007), 303-311.

30. A. A. Samarskii, V. B. Andreev, *Difference Methods for Elliptic Problems*, Nauka, Moscow (1976)(in Russian).
31. A. E. Showalter, *Monotone Operators in Banach Space and Nonlinear Partial Differential Equations*, Mathematical Surveys and Monographs, vol. 49, Amer. Math. Soc., Providence (1996).
32. A. Tikhonov and V. Arsenin (1977). *Solution of Ill-Posed Problems.* John Wiley, New York.
33. Y. Wei and J.W. Hutchinson, Hardness trends in micron scale indentation, J. Mech. Phys. Solids, 51(2003) 2037-2056.

SOME RECENT DEVELOPMENTS ON MATHEMATICAL ASPECT OF WAVELETS

P. MANCHANDA

Department of Mathematics
Guru Nanak Dev University, Amritsar, India
E-mail: pmanch2k1@yahoo.co.in

MEENAKSHI

Dev Samaj College for Women
Ferozepur, India
E-mail: meenakshi_wavelets@yahoo.com

Wavelet theory is a refinement of Fourier analysis. Replacement of Fourier analysis by this theory very often yield much better results. The basic theory of wavelets has been invented between 1983 and 1995. Applications of wavelets have been studied in diverse fields such as numerical simulation of partial differential equations, modeling real world problems, signal and image processing specially more accurate understanding of medical signals such as EEG and ECG. This theory has been also used in oil exploration, analyzing meteorological data and prediction of financial time series. In the recent past, three mathematical aspects of this theory have been investigated namely: nonuniform multiresolution analysis, where $\mathbb{Z}$ the set of translation parameters in scaling function is replaced by its subset which is not a group [6]; effect of the replacement of whole real line $\mathbb{R}$ by half positive line $\mathbb{R}_+$ in the definition of multiresolution analysis [5] and vector valued wavelet and multiresolution analysis, where scalar scaling function in the classical definition is replaced by vector valued function [1,15].

In the review article we present a resume of the results obtained by the above authors and our own results obtained recently and presented in the conferences: sattelite conference of ICM 2010 held in Delhi and ICM 2010[12,14].

Keywords: multiresolution analysis, non-uniform multiresolution analysis, Walsh-Fourier transform

1. Introduction

The multiresolution analysis is known as the heart of the wavelet theory. The concept of multiresolution analysis provides a very elegant tool for the construction of wavelets i.e. the functions $\psi \in L^2(\mathbb{R})$ having the property that the collection of functions $\{2^{j/2}\psi(2^j x - n)\}_{j,n\in\mathbb{Z}}$ forms a complete orthonormal system for $L^2(\mathbb{R})$ where $\mathbb{Z}$ denote the set of all integers [3]. A multiresolution analysis on the real line $\mathbb{R}$, introduced by Mallat [10] is an increasing sequence of closed subspaces $\{V_j\}_{j\in\mathbb{Z}}$ of $L^2(\mathbb{R})$ such that $\cap_{j\in\mathbb{Z}}V_j = \{0\}$, $\cup_{j\in\mathbb{Z}}V_j$ is dense in $L^2(\mathbb{R})$, and which satisfies $f(x) \in V_j$ if and only if $f(2x) \in V_{j+1}$. Furthermore, there should exist an element $\phi \in V_0$ such that the collection of integer translates of ϕ, $\{\phi(x-n)\}_{n\in\mathbb{Z}}$ is a complete orthonormal system for V_0.

In the definition of multiresolution analysis the dilation factor of 2 can be replaced by an integer $N \geq 2$ and one can construct $N-1$ wavelets to generate the whole space $L^2(\mathbb{R})$. A similar generalization of multiresolution analysis can be made in higher dimensions by considering matrix dilations [9]. Gabardo and Nashed [6] considered a generalization of the notion of multiresolution analysis, which is called nonuniform multiresolution analysis(NUMRA) and is based on the theory of spectral pairs. Farkov [5] has extended the notion of multiresolution analysis on locally compact abelian groups and has constructed orthogonal wavelets with compact support on locally compact abelian groups by the scaling function associated with this multiresolution analysis. The approach adopted by Farkov is connected with Walsh-Fourier transform and the elements of M-band wavelet theory. Chen and Cheng [1] introduced vector-valued multiresolution analysis and orthogonal vector-valued wavelets. They find the necessary and sufficient condition for the existence of orthogonal vector-valued wavelets.

A basic technique in the standard theory of wavelets is to construct a multiresolution analysis based on a scaling function ϕ whose Fourier transform is defined by

$$\hat{\phi}(\xi) = \prod_{k=1}^{\infty} m_0(\frac{\xi}{2^k}) \tag{1}$$

where $m_0(\xi) = \sum_{|k|\leq M} a_k e^{-2\pi i \xi k}$ is a trigonometric polynomial satisfying the conditions $m_0(0) = 1$ and [3,4,17]

$$|m_0(\xi + \frac{1}{2})|^2 + |m_0(\xi)|^2 = 1.$$

The scaling relation $\hat{\phi}(2\xi) = m_0(\xi)\hat{\phi}(\xi)$ is then automatically satisfied. If the orthonormality of the system of functions $\{\phi(x-n)\}_{n\in\mathbb{Z}}$ in $L^2(\mathbb{R})$ is satisfied then V_0 can be defined as a closed linear span of the collection $\{\phi(x-n)\}_{n\in\mathbb{Z}}$. The subspaces V_j for $j \in \mathbb{Z}$ can be defined as

$$f(x) \in V_j \;\Leftrightarrow\; f(2x) \in V_{j+1}.$$

It can be shown that the collection $\{V_j\}_{j\in\mathbb{Z}}$ satisfies all the properties of a multiresolution analysis and the scaling function and wavelet function have compact support.
Cohen's condition provide the necessary and sufficient condition for the orthonormality of the collection $\{\phi(x-n)\}_{n\in\mathbb{Z}}$.

Theorem 1.1. [2] *Let m_0 be a 1-periodic trigonometric polynomial satisfying $m_0(0) = 1$ as well as $|m_0(\xi+\frac{1}{2})|^2 + |m_0(\xi)|^2 = 1$ and ϕ is defined by (1). Then the collection $\{\phi(x-n)\}_{n\in\mathbb{Z}}$ is orthonormal in $L^2(\mathbb{R})$ if and only if there exists a compact set $K \subset \mathbb{R}$ containing a neighborhood of 0 and a constant $c > 0$ such that*

$$\sum_{k\in\mathbb{Z}} \delta_k * \psi_K = 1 \quad \textit{and}$$

$$|m_0(\frac{\xi}{2^k})| \geq c, \;\; \forall \xi \in K \; \forall k \geq 1.$$

In the above theorem δ_k and $*$ denote the Dirac mass at k and $*$ the usual convolution product respectively.

This paper is organized as follows: In section 2 we review the main results obtained by Gabardo and Nashed in [6,7], the definition of NUMRA, the necessary and sufficient condition for the existence of associated wavelets and the analogue of Cohen's condition. In section 3 we explain certain results of Walsh-Fourier analysis. We present a brief review of generalized Walsh functions, Walsh-Fourier transforms and its various properties, multiresolution p-analysis in $L^2(\mathbb{R}_+)$ introduced by Farkov [5]. In section 4 we present a review of results obtained in [1,15] on vector-valued multiresolution analysis and present an algorithm for constructing a class of compactly supported orthogonal vector-valued wavelets. Nonuniform multiresolution analysis on positive half line is defined in section 5 and a necessary and sufficient condition for the existence of associated wavelet is given. In section 6 we construct nonuniform multiresolution analysis on positive half line starting from a Walsh polynomial m_0 satisfying appropriate conditions and showing that the scaling function φ defined, via the Fourier transform, by the corresponding infinite product

$$\tilde{\varphi}(\xi) = \prod_{k=1}^{\infty} m_0(\frac{\xi}{N^k})$$

belong to $L^2(\mathbb{R}_+)$. We construct a nonuniform multiresolution analysis on positive half line with a compactly supported scaling function φ. We find the analogue of Cohen's condition for nonuniform multiresolution on positive half line which gives necessary and sufficient condition for the orthonormality of the system $\{\varphi(x \ominus \lambda)\}_{\lambda\in\Lambda_+}$ where $\Lambda_+ = \{0, r/N\} + \mathbb{Z}_+$, $N > 1$ is an integer and r is an odd integer with $1 \le r \le 2N-1$ such that r and N are relatively prime.

2. Nonuniform Multiresolution analysis

Gabardo and Nashed [6] considered a generalization of the notion of multiresolution analysis, which is called nonuniform multiresolution analysis (NUMRA) and is based on the theory of spectral pairs. In this set up, the associated subspace V_0 of $L^2(\mathbb{R})$ has, as an orthonormal basis, a collection of translates of the scaling function ϕ of the form $\{\phi(x-\lambda)\}_{\lambda\in\Lambda}$ where $\Lambda = \{0, r/N\} + 2\mathbb{Z}$, $N \ge 1$ is an integer and r is an odd integer with $1 \le r \le 2N-1$ such that r and N are relatively prime and $\mathbb{Z}$ is the set of all integers. The main results of Gabardo and Nashed deal with necessary and sufficient condition for the existence of associated wavelets [6] and extension of Cohen's theorem [7].

Let us recall the definitions of NUMRA and associated set of wavelets:

Definition 2.1. Given an integer $N \ge 1$ and an odd integer r with $1 \le r \le 2N-1$ such that r and N are relatively prime, an associated nonuniform multiresolution analysis (abbreviated NUMRA) is a collection $\{V_j\}_{j\in\mathbb{Z}}$ of closed subspaces of $L^2(\mathbb{R})$ satisfying the following properties:

(a) $V_j \subset V_{j+1}\ \forall\ j \in \mathbb{Z}$,
(b) $\cup_{j\in\mathbb{Z}} V_j$ is dense in $L^2(\mathbb{R})$,
(c) $\cap_{j\in\mathbb{Z}} V_j = \{0\}$,
(d) $f(x) \in V_j$ if and only if $f(2Nx) \in V_{j+1}$,
(e) There exists a function $\phi \in V_0$, called a scaling function such that the collection $\{\phi(x-\lambda)\}_{\lambda\in\Lambda}$ where $\Lambda = \{0, r/N\} + 2\mathbb{Z}$ is a complete orthonormal system for V_0.

It is worth noticing that when $N = 1$, one recovers from definition above the standard definition of a one-dimensional MRA with dyadic dilation. When $N > 1$, the dilation factor of $2N$ ensures that $2N\Lambda \subset 2\mathbb{Z} \subset \Lambda$. However, the existence of associated wavelets with the dilation $2N$ and translation set Λ is no longer guaranteed as is the case in the standard setting.

Given a NUMRA, we denote by W_m the orthogonal complement of V_m in V_{m+1}, for any integer m. It is clear from (a), (b) and (c) of Definition

2.1 that

$$L^2(\mathbb{R}) = \oplus_{m\in\mathbb{Z}} W_m.$$

As is the case in the standard situation, the main purpose of NUMRA is to construct orthonormal basis of $L^2(\mathbb{R})$ given by appropriate translates and dilates of a finite collection of functions, called the associated wavelets.

Definition 2.2. A collection $\{\psi_k\}_{k=1,2,\ldots,2N-1}$ of functions in V_1 will be called a set of wavelets associated with a given NUMRA if the family of functions $\{\psi_k(x-\lambda)\}_{k=1,\ldots,2N-1,\lambda\in\Lambda}$ is an orthonormal system for W_0.

It is shown in [6] that given a set of wavelets $\{\psi_k\}_{k=1,2,\ldots,2N-1}$ associated with a given NUMRA, the collection $\{(2N)^{m/2}\psi_k((2N)^m x - \lambda)\}_{\lambda\in\Lambda,m\in\mathbb{Z},k=1,2,\ldots,2N-1}$ forms a complete orthonormal system for $L^2(\mathbb{R})$.

The following result proved in [6] provides the simple necessary and sufficient conditions for the existence of the associated set of wavelets:

Theorem 2.1. *Consider a NUMRA with associated parameters N and r, as in Definition 1.1, such that the corresponding space V_0 has an orthonormal system of the form $\{\phi(x-\lambda)\}_{\lambda\in\Lambda}$, where $\Lambda = \{0, r/N\} + 2\mathbb{Z}$, $\hat{\phi}$ satisfies the scaling relation*

$$\hat{\phi}(2N\xi) = m_0(\xi)\hat{\phi}(\xi),$$

where $\hat{\phi}$ denotes the Fourier transform of a function ϕ and m_0 has the form

$$m_0(\xi) = m_0^1(\xi) + e^{-2\pi i\xi r/N} m_0^2(\xi),$$

for some locally L^2, $1/2$-periodic functions m_0^1 and m_0^2. Define M_0 by

$$M_0(\xi) = |m_0^1(\xi)|^2 + |m_0^2(\xi)|^2.$$

Then, each of the following conditions is necessary and sufficient for the existence of associated wavelets $\psi_1, \ldots, \psi_{2N-1}$:

(a) M_0 is $1/4$-periodic.
(b) $\sum_{k=0}^{N-1} \delta_{k/2} * \sum_{j\in\mathbb{Z}} \delta_{jN} * |\hat{\phi}|^2 = 1$.
(c) For any odd integer m, we have

$$\int_{\mathbb{R}} \phi(x)\overline{\phi(x-m/N)}\, dx = 0.$$

Gabardo and Nashed [7] provides an extension to the setting of NUMRAs of the standard construction of wavelet analysis which consists in constructing a multiresolution analysis starting from a trigonometric polynomial m_0 of the form

$$m_0(\xi) = m_0^1(\xi) + e^{-2\pi i\xi r/N} m_0^2(\xi), \tag{2}$$

where the integers r and N satisfy $N \geq 1$, $1 \leq r \leq 2N - 1$, r is odd, r and N are relatively prime and $m_0^1(\xi)$ and $m_0^2(\xi)$ are $\frac{1}{2}$-periodic trigonometric polynomials. The polynomial m_0 satisfies $m_0(0) = 1$ and the following conditions:

$$\sum_{p=0}^{2N-1} M_0(\xi + p/4N) = 1, \tag{3}$$

$$\sum_{p=0}^{2N-1} \alpha^p M_0(\xi + p/4N) = 0, \tag{4}$$

where $\alpha = e^{-\pi ir/N}$ and $M_0(\xi) = |m_0^1(\xi)|^2 + |m_0^2(\xi)|^2$.

Then a compactly supported function $\phi \in L^2(\mathbb{R})$ is obtained which satisfies the scaling relation

$$\hat{\phi}(2N\xi) = m_0(\xi)\hat{\phi}(\xi). \tag{5}$$

Now, it is necessary to determine the orthonormality of the system of functions $\{\phi(x - \lambda)\}_{\lambda \in \Lambda}$ where $\lambda = \{0, r/N\} + 2\mathbb{Z}$. If the orthonormality requirement is satisfied, one can define V_0 as the closed linear span of the collection $\{\phi(x - \lambda)\}_{\lambda \in \Lambda}$ and V_j for $j \in \mathbb{Z}$ is defined as

$$f(x) \in V_j \Leftrightarrow f(x/(2N)^j) \in V_0 \tag{6}$$

so that the conditions (d) and (e) of Definition 2.1 hold. Also the equation 5 implies that (a) holds. The remaining two conditions (b) and (c) follow from the Proposition 3.3 and 3.4 proved in [7].

The remaining issue is to prove the orthonormality of the family $\{\phi(x - \lambda)\}_{\lambda \in \Lambda}$. Gabardo and Nashed proved the analogue of Cohen's result for NUMRAs, which gives a sufficient condition for the orthonormality of the collection $\{\phi(x - \lambda)\}_{\lambda \in \Lambda}$.

The following theorem by Gabardo and Nashed [7] generalizes Cohen's result:

Theorem 2.2. *Let m_0 be a trigonometric polynomial of the form* (2) *which satisfies $m_0(0) = 1$ together with the conditions* (3) *and* (4). *Let ϕ be defined by the formula* (5) *and let $\Lambda = 0, r/N + 2\mathbb{Z}$. Then a sufficient condition for* *the collection $\{\phi(x-\lambda)\}_{\lambda\in\Lambda}$ to be orthonormal in $L^2(\mathbb{R})$ is the existence of a constant $c>0$ and of a compact set $K\subset\mathbb{R}$ that contains a neighborhood of the origin and satisfies*

$$\sum_{k=0}^{N-1}\delta_{k/2}*\sum_{j\in\mathbb{Z}}\delta_{Nj}*\chi_K=1,$$

such that

$$|m_0(\xi/(2N)^k)|\geq c\ \ \forall\xi\in K,\ \forall k\geq 1.$$

Furthermore, if the function M_0 defined by $M_0(\xi)=|m_0^1(\xi)|^2+|m_0^2(\xi)|^2$ is 1/4- periodic, the condition is also necessary.

3. Multiresolution p-analysis on $\mathbb{R}_+$

Let p be a fixed natural number greater than 1. As usual, let $\mathbb{R}_+=[0,+\infty)$ and $\mathbb{Z}_+=\{0,1,\ldots\}$. Denote by $[x]$ the integer part of x. For $x\in\mathbb{R}_+$ and for any integer j

$$x_j=[p^jx](\text{mod }p),\quad x_{-j}=[p^{1-j}x](\text{mod }p), \tag{7}$$

where $x_j,x_{-j}\in\{0,1,\ldots,p-1\}$. It is clear that for each $x\in\mathbb{R}_+$, $\exists k=k(x)$ in $\mathbb{N}$ such that $x_{-j}=0\ \forall\ j>k$.

Consider on $\mathbb{R}_+$ the addition defined as follows:

$$x\oplus y=\sum_{j<0}\xi_jp^{-j-1}+\sum_{j>0}\xi_jp^{-j}$$

with

$$\xi_j=x_j+y_j(\text{mod}\,p),\ \ j\in\mathbb{Z}\backslash\{0\},$$

where $\xi_j\in\{0,1,2,\ldots,p-1\}$ and x_j, y_j are calculated by (7).
For $p=2$, $\oplus$ was introduced by N.J.Fine, see for example [13]. For $x\in\mathbb{R}_+$ and $j\in\mathbb{N}$ we define the numbers $x_j,x_{-j}\in\{0,1\}$ as follows:

$$x_j=[2^jx](\text{mod}\,2),\ \ x_{-j}=[2^{1-j}x](\text{mod}\,2), \tag{8}$$

where [.] denotes the integral part of $x\in\mathbb{R}_+$.
x_j and x_{-j} are the digits of the binary expansion

$$x=\sum_{j<0}x_j2^{-j-1}+\sum_{j=0}x_j2^{-j}. \tag{9}$$

(For dyadic x, we obtain an expansion with finitely many non-zero terms)
For fixed $x, y \in \mathbb{R}_+$, we set

$$x \oplus y = \sum_{j<0} |x_j - y_j| 2^{-j-1} + \sum_{j>0} |x_j - y_j| 2^{-j},$$

where x_j, y_j are defined in (8). By definition
$x \ominus y = x \oplus y$ (because $x \oplus x = 0$).

The binary operation $\oplus$ identifies $\mathbb{R}_+$ with the group G_2 (dyadic group with addition modulo 2) and is useful in the study of dyadic Hardy classes and image processing, see for example [8,13].
For $x \in [0, 1)$, let $r_0(x)$ is given by

$$r_0(x) = \begin{cases} 1 & x \in [0, 1/p) \\ \varepsilon_p^j & x \in [jp^{-1}, (j+1)p^{-1}), j = 1, 2, \dots, p-1 \end{cases} \tag{10}$$

where $\varepsilon_p = \exp\left(\frac{2\pi i}{p}\right)$.
The extension of the function r_0 to $\mathbb{R}_+$ is defined by the equality $r_0(x + 1) = r_0(x)$, $x \in \mathbb{R}_+$. Then the **generalized Walsh functions** $\{w_m(x)\}$ $(m \in \mathbb{Z}_+)$ are defined by

$$w_0(x) \equiv 1 \text{ and } w_m(x) = \prod_{j=0}^{k} (r_0(p^j x))^{\mu_j},$$

where

$$m = \sum_{j=0}^{k} \mu_j p^j, \quad \mu_j \in \{0, 1, \dots, p-1\}, \ \mu_k \neq 0.$$

(The classical Walsh system corresponds for the case $p = 2$)
For $x, w \in \mathbb{R}_+$, let

$$\chi(x, w) = \exp\left(\frac{2\pi i}{p} \sum_{j=1}^{\infty} (x_j w_{-j} + x_{-j} w_j)\right), \tag{11}$$

where x_j and w_j are given by (7).
We observe that

$$\chi(x, \frac{m}{p^{n-1}}) = \chi(\frac{x}{p^{n-1}}, m) = w_m(\frac{x}{p^{n-1}}) \ \forall \ x \in [0, p^{n-1}), \ m \in \mathbb{Z}.$$

The **Walsh-Fourier transform** of a function $f \in L^1(\mathbb{R}_+)$ is defined by

$$\tilde{f}(w) = \int_{\mathbb{R}_+} f(x)\overline{\chi(x,w)}\,dx, \tag{12}$$

where $\chi(x,w)$ is given by (11).
If $f \in L^2(\mathbb{R}_+)$ and

$$J_a f(w) = \int_0^a f(x)\overline{\chi(x,w)}\,dx \quad (a > 0), \tag{13}$$

then $\tilde{f}$ is defined as limit of $J_a f$ in $L^2(\mathbb{R}_+)$ as $a \to \infty$.

Properties of Walsh-Fourier transform
If $f \in L^2(\mathbb{R}_+)$, then $\tilde{f} \in L^2(\mathbb{R}_+)$ and

$$\|\tilde{f}\|_{L^2(\mathbb{R}_+)} = \|f\|_{L^2(\mathbb{R}_+)}.$$

If $x, y, w \in \mathbb{R}_+$ and $x \oplus y$ is p-adic irrational, then

$$\chi(x \oplus y, w) = \chi(x,w)\chi(y,w), \tag{14}$$

see [8]. Thus for fixed x and w, the equality (14) holds for all $y \in \mathbb{R}_+$ except for countably many. It is well known that systems $\{\chi(\alpha,.)\}_{\alpha=0}^{\infty}$ and $\{\chi(.,\alpha)\}_{\alpha=0}^{\infty}$ are orthonormal bases in $L^2[0,1]$.
Let $\{w\}$ denotes the fractional part of w. For any $\varphi \in L^2(\mathbb{R}_+)$ and $k \in \mathbb{Z}_+$, we have

$$\begin{aligned} \int_{\mathbb{R}_+} \varphi(x)\overline{\varphi(x \ominus k)}\,dx &= \sum_{l=0}^{\infty} \int_l^{l+1} |\tilde{\varphi}(w)|^2 \overline{\chi(k,\{w\})}\,dw \\ &= \int_0^1 \left(\sum_{l\in\mathbb{Z}_+} |\tilde{\varphi}(w+l)|^2 \right) \overline{\chi(k,w)}\,dw. \end{aligned} \tag{15}$$

Therefore, a necessary and sufficient condition for a system $\{\varphi(. \ominus k)/k \in \mathbb{Z}_+\}$ to be orthonormal in $L^2(\mathbb{R}_+)$ is

$$\sum_{l\in\mathbb{Z}_+} |\tilde{\varphi}(w+l)|^2 = 1 \;\; \text{a.e.} \tag{16}$$

Multiresolution p-analysis in $L^2(\mathbb{R}_+)$ defined by Farkov [5] is as follows:

Definition 3.1. A multiresolution p-analysis in $L^2(\mathbb{R}_+)$ is a sequence of closed subspaces $V_j \subset L^2(\mathbb{R}_+)$ $(j \in \mathbb{Z})$ such that the following hold:

(i) $V_j \subset V_{j+1}$ for all $j \in \mathbb{Z}$.
(ii) $\cup V_j$ is dense in $L^2(\mathbb{R}_+)$ and $\cap V_j = \{0\}$.
(iii) $f(.) \in V_j \quad \Leftrightarrow \quad f(p.) \in V_{j+1}$ for all $j \in \mathbb{Z}$.
(iv) $f(.) \in V_0 \quad \Leftrightarrow \quad f(. \oplus k) \in V_0$ for all $k \in \mathbb{Z}_+$.
(v) There is a function $\varphi \in L^2(\mathbb{R}_+)$ such that $\{\varphi(. \ominus k)/k \in \mathbb{Z}_+\}$ is an orthonormal basis of V_0.

The function φ is called a scaling function in $L^2(\mathbb{R}_+)$.

Farkov has given a general construction of compactly supported orthogonal p-wavelets in $L^2(\mathbb{R}_+)$ arising from scaling filters with p^n many terms. For all integer $p \geq 2$ these wavelets are identified with certain lacunary Walsh series on $\mathbb{R}_+$.

Let $\varphi \in L^2(\mathbb{R}_+)$ satisfies the refinement equation

$$\varphi(x) = p \sum_{\alpha=0}^{p^n-1} a_\alpha \varphi(px \ominus \alpha). \tag{17}$$

We get

$$\tilde{\varphi}(w) = m_0(p^{-1}w)\tilde{\varphi}(p^{-1}w), \tag{18}$$

where

$$m_0(w) = \sum_{\alpha=0}^{p^n-1} a_\alpha \overline{\chi(\alpha, w)}. \tag{19}$$

Suppose that

$$m_0(0) = 1.$$

Put

$$\Delta_s^{(n)} = [sp^{-n}, (s+1)p^{-n}) \text{ for } s \in \mathbb{Z}_+.$$

Then $m_0(w)$ is constant on $\Delta_s^{(n)}$ for each s and $m_0(w) = 1$ on $\Delta_0^{(n)}$. It follows from equation (18) that

$$\tilde{\varphi}(w) = \prod_{j=1}^{\infty} m_0(p^{-j}w), \quad w \in \mathbb{R}_+. \tag{20}$$

We note that $m_0(p^{-j}w) = 1$ as $p^{-j}w \in \Delta_0^{(n)}$.

We say that a function $f : \mathbb{R}_+ \to \mathbb{C}$ is W- continuous at a point $x \in \mathbb{R}_+$,

if for each $\epsilon > 0$ there exists $\delta > 0$ such that $|f(x \oplus y) - f(x)| < \epsilon$ for $0 < y < \delta$.
Suppose that E is a W- compact set in $\mathbb{R}_+$. The notion $E \equiv [0,1)(\mathrm{mod}\,\mathbb{Z}_+)$ means that for each $x \in [0,1)$ there exists $k \in \mathbb{Z}_+$ such that $x \oplus k \in E$. Let μ denotes the Lebesgue measure on $\mathbb{R}_+$.

The following theorem by Farkov generalizes A.Cohen's result:

Theorem 3.1. *Let*

$$m_0(w) = \sum_{\alpha=0}^{p^n-1} a_\alpha \overline{\chi(\alpha, w)}$$

be a polynomial satisfying the following conditions:

(a) $m_0(0) = 1$.
(b) $\sum_{j=0}^{p^n-1} |m_0(sp^{-n} \oplus jp^{-1})|^2 = 1$ *for* $s = 0, 1, \ldots, p^{n-1} - 1$.
(c) There exists a W-compact set E such that $0 \in$ int(E), $\mu(E) = 1$, $E \equiv [0,1)(mod\,\mathbb{Z}_+)$ and

$$\inf_{j \in \mathbb{N}} \inf_{w \in E} |m_0(p^{-j}w)| > 0.$$

If the Walsh-Fourier transform of $\varphi \in L^2(\mathbb{R}_+)$ can be written as

$$\tilde{\varphi}(w) = \prod_{j=1}^{\infty} m_0(p^{-j}w),$$

then φ is a scaling function in $L^2(\mathbb{R}_+)$.

4. Vector-valued multiresolution analysis

Chen and Cheng [1] introduced the concept of vector-valued multiresolution analysis and orthogonal vector-valued wavelets. They derive a necessary and sufficient condition on the existence of orthogonal vector-valued wavelets and also presented the construction of a compactly supported orthogonal vector-valued wavelets.

Let s be a constant such that $2 \leq s \in \mathbb{Z}$. $\mathbf{I}_s$ and $\mathbf{O}$ stand for the $s \times s$ identity matrix and zero matrix respectively. $L^2(\mathbb{R}, \mathbb{C}^s)$ denote the set of all vector-valued functions $\mathbf{h}(t)$ defined as:

$$L^2(\mathbb{R}, \mathbb{C}^s) = \left\{\mathbf{h}(t) = (h_1(t), h_2(t), \ldots, h_s(t))^T : h_\nu(t) \in L^2(\mathbb{R}), \nu = 1, 2, \ldots, s\right\}$$

where T denotes the transpose.

For $\mathbf{h} \in L^2(\mathbb{R}, \mathbb{C}^s)$, $\|\mathbf{h}\|$ is the norm of the operator $\mathbf{h}$, i.e. $\|\mathbf{h}\| = \left(\sum_{\nu=1}^{s} \int_{\mathbb{R}} |h_\nu(t)|^2 \, dt \right)^{1/2}$, and the integration of $\mathbf{h}(t)$ is defined by $\hat{\mathbf{h}}(\eta) = \int_{\mathbb{R}} \mathbf{h}(t).e^{-it\eta} \, dt$.

For two vector-valued functions $\mathbf{h}, \boldsymbol{\eta} \in L^2(\mathbb{R}, \mathbb{C}^s)$, their symbolic inner product is defined by

$$< \mathbf{h}(.), \boldsymbol{\eta}(.) >= \int_{\mathbb{R}} h(t)\eta(t)^* \, dt, \tag{21}$$

where $*$ means the complex conjugate and the transpose.

A sequence $\{\mathbf{h}_k(t)\} \subset \mathbf{X} \subset L^2(\mathbb{R}, \mathbb{C}^s)$ is called an orthogonal set of $\mathbf{X}$, if it satisfies

$$< \mathbf{h}_k(.), \mathbf{h}_n(.) >= \delta_{k,n} \mathbf{I}_s, \quad k, n \in \mathbb{Z}, \tag{22}$$

where $\delta_{k,n}$ is the Kronecker symbol such that $\delta_{k,n} = 1$ when $k = n$ and $\delta_{K,n} = 0$ when $k \neq n$.

Definition 4.1. A function $\mathbf{h}(t) \in \mathbf{X} \subset L^2(\mathbb{R}, \mathbb{C}^s)$ is an orthogonal vector-valued function in $\mathbf{X}$ if its translations $\{\mathbf{h}(t-k)\}_{k \in \mathbb{Z}}$ is an orthonormal set in $\mathbf{X}$, i.e.

$$< \mathbf{h}(. - k), \mathbf{h}(. - n) >= \delta_{k,n} \mathbf{I}_s, \quad k, n \in \mathbb{Z}. \tag{23}$$

Definition 4.2. A sequence $\{\mathbf{h}_k(t)\}_{k \in \mathbb{Z}} \subset \mathbf{X} \subset L^2(\mathbb{R}, \mathbb{C}^s)$ is called an orthonormal basis of $\mathbf{X}$ if it satisfies (22), and for any $\boldsymbol{\Lambda}(t) \in \mathbf{X}$, there exists a unique sequence of $s \times s$ constant matrices $\{A_k\}_{k \in \mathbb{Z}}$ such that

$$\boldsymbol{\Lambda}(t) = \sum_{k \in \mathbb{Z}} A_k \mathbf{h}_k(t). \tag{24}$$

The expansion (24) is also called the Fourier series expansion of $\boldsymbol{\Lambda}(t)$.

Let $\boldsymbol{\phi}(t) = (\phi_1(t), \phi_2(t), \ldots, \phi_s(t))^T \in L^2(\mathbb{R}, \mathbb{C}^s)$ satisfy the following refinement equation:

$$\boldsymbol{\phi}(t) = \sum_{k \in \mathbb{Z}} P_k \boldsymbol{\phi}(2t - k), \tag{25}$$

where $\{P_k\}_{k \in \mathbb{Z}}$ is a $s \times s$ constant matrix sequence that has only finite terms.

Define a closed subspace $V_j \subset L^2(\mathbb{R}, \mathbb{C}^s)$ by

$$V_j = \text{clos}_{L^2(\mathbb{R},\mathbb{C}^s)}(\text{span}\{\boldsymbol{\phi}(2^j t - k) : k \in \mathbb{Z}\}), \quad j \in \mathbb{Z}. \tag{26}$$

Vector-valued multiresolution analysis defined by Chen and Cheng is as follows:

Definition 4.3. $\boldsymbol{\phi}(t)$ defined by (25) generates a vector-valued multiresolution analysis $\{V_j\}_{j\in\mathbb{Z}}$ of $L^2(\mathbb{R}, \mathbb{C}^s)$, if the sequence $\{V_j\}_{j\in\mathbb{Z}}$ defined in (26) satisfies:

(a) $\ldots \subset V_{-1} \subset V_0 \subset V_1 \subset \ldots$,
(b) $\cap_{j\in\mathbb{Z}} V_j = \{\mathbf{0}\}$, $\cup_{j\in\mathbb{Z}} V_j$ is dense in $L^2(\mathbb{R}, \mathbb{C}^s)$, where $\mathbf{0}$ is the zero vector of $L^2(\mathbb{R}, \mathbb{C}^s)$,
(c) $\mathbf{h}(t) \in V_0$ if and only if $\mathbf{h}(2^j t) \in V_j \ \forall j \in \mathbb{Z}$,
(d) there exists $\boldsymbol{\phi}(t) \in V_0$ such that the sequence $\{\boldsymbol{\phi}(t-k), k \in \mathbb{Z}\}$ is an orthonormal basis of V_0. The vector-valued function $\boldsymbol{\phi}(t)$ is called a scaling function of the vector-valued multiresolution analysis.

On taking the Fourier transform on both sides of (25), and assuming that $\hat{\boldsymbol{\phi}}(\eta)$ is continuous at zero, we have

$$\hat{\boldsymbol{\phi}}(\eta) = \alpha(\eta/2)\hat{\boldsymbol{\phi}}(\eta/2), \quad \eta \in \mathbb{R}, \tag{27}$$

where

$$\alpha(\eta) = \frac{1}{2} \cdot \sum_{k\in\mathbb{Z}} P_k. \exp\{-ik\eta\}. \tag{28}$$

Let W_j, $j \in \mathbb{Z}$ denote the orthogonal complement of V_j in V_{j+1} and there exists a vector-valued function $\boldsymbol{\psi}(t) \in L^2(\mathbb{R}, \mathbb{C}^s)$ such that the translations and dilations of $\boldsymbol{\psi}(t)$ form a Riesz basis of W_j i.e.

$$W_j = \text{clos}_{L^2(\mathbb{R},\mathbb{C}^s)}(\text{span}\{\boldsymbol{\psi}(2^j t - k) : k \in \mathbb{Z}\}), \quad j \in \mathbb{Z}. \tag{29}$$

Since $\boldsymbol{\psi}(t) \in W_0 \subset V_1$, there exists a unique finitely supported sequence $\{B_k\}_{k\in\mathbb{Z}}$ of $s \times s$ constant matrices such that

$$\boldsymbol{\psi}(t) = \sum_{k\in\mathbb{Z}} B_k \boldsymbol{\phi}(2t - k). \tag{30}$$

Let

$$\beta(\eta) = \frac{1}{2} \sum_{k\in\mathbb{Z}} B_k.\exp\{-ik\eta\}. \tag{31}$$

Then the equation (30) becomes

$$\hat{\boldsymbol{\psi}}(\eta) = \beta(\eta/2)\boldsymbol{\phi}(\hat{\eta}/2), \quad \eta \in \mathbb{R}. \tag{32}$$

The following theorem proves the existence of orthogonal vector-valued wavelets:

Theorem 4.1. *Let $\boldsymbol{\phi}(t)$ defined in (25) is an orthogonal vector-valued scaling function. Assume that $\boldsymbol{\psi}(t) \in L^2(\mathbb{R}, \mathbb{C}^s)$ and $\alpha(\eta)$ and $\beta(\eta)$ are defined respectively by (28) and (31). Then $\psi(\mathbf{t})$ is an orthogonal vector-valued wavelet function associated with $\boldsymbol{\phi}(t)$ if and only if*

$$\alpha(\eta)\beta(\eta)^* + \alpha(\eta+\pi)\beta(\eta+\pi)^* = \boldsymbol{O}, \quad \eta \in \mathbb{R}.$$

$$\beta(\eta)\beta(\eta)^* + \beta(\eta+\pi)\beta(\eta+\pi)^* = \boldsymbol{I_s}, \quad \eta \in \mathbb{R}.$$

The following theorem proved in[1] present an algorithm for the construction of compactly supported orthogonal vector-valued wavelets:

Theorem 4.2. *Let $\boldsymbol{\phi}(t) \in L^2(\mathbb{R}, \mathbb{C}^s)$ be a 3-coefficient compactly supported orthogonal vector-valued scaling functions satisfying the following refinement equation:*

$$\boldsymbol{\phi}(t) = P_0\boldsymbol{\phi}(2t) + P_1\boldsymbol{\phi}(2t-1) + P_2\boldsymbol{\phi}(2t-2).$$

Assume that there exists an integer n, $0 \le n \le 2$, such that the matrix A defined in the following equation , is not only an invertible matrix but also an Hermitian matrix:

$$A^2 = (2\mathbf{I}_s - P_nP_n^*)^{-1}P_nP_n^*.$$

Define

$$\begin{cases} B_j = AP_j, & j \ne n, \\ B_j = -A^{-1}P_j, & j = n, \ \ j, n \in \{0, 1, 2\} \end{cases}$$

and

$$\boldsymbol{\psi}(t) = B_0\boldsymbol{\phi}(2t) + B_1\boldsymbol{\phi}(2t-1) + B_2\boldsymbol{\phi}(2t-2).$$

Then $\boldsymbol{\psi}(t)$ is an orthogonal vector-valued function associated with $\boldsymbol{\phi}(t)$.

5. Nonuniform multiresolution analysis on positive half line

Definition 5.1. For an integer $N > 1$ and an odd integer r with $1 \le r \le 2N-1$ such that r and N are relatively prime, an associated nonuniform multiresolution analysis on positive half line is a sequence of closed subspaces $V_j \subset L^2(\mathbb{R}_+)$, $j \in \mathbb{Z}$ such that the following properties hold:

(i) $V_j \subset V_{j+1} \ \ \forall j \in \mathbb{Z}$,

(ii) $\cup_{j\in\mathbb{Z}}$ is dense in $L^2(\mathbb{R}_+)$,
(iii) $\cap_{j\in\mathbb{Z}} V_j = \{0\}$,
(iv) $f(.) \in V_j \;\Leftrightarrow\; f(N.) \in V_{j+1}\ \forall j \in \mathbb{Z}$,
(v) There exists a function $\varphi \in V_0$ such that $\{\varphi(x \ominus \lambda), \lambda \in \Lambda_+\}$ where $\Lambda_+ = \{0, r/N\} + \mathbb{Z}_+$, is a complete orthonormal system for V_0.

The function φ is called a scaling function in $L^2(\mathbb{R}_+)$.

When $N > 1$, the dilation factor of N ensures that

$$N\Lambda_+ \subset \mathbb{Z}_+ \subset \Lambda_+.$$

By the conditions (iv) and (v) of the Definition 5.1

$$\varphi_{1,\lambda}(x) = N^{1/2}\varphi(Nx \ominus \lambda),\ \ \lambda \in \Lambda_+$$

constitute an orthonormal basis in V_0.
Since $V_0 \subset V_1$, the function $\varphi \in V_1$ and has the Fourier expansion

$$\varphi(x) = \sum_{\lambda\in\Lambda_+} h_\lambda (N)^{1/2}\varphi(Nx \ominus \lambda)$$

where $h_\lambda = \displaystyle\int_{\mathbb{R}_+} \varphi(x)\overline{\varphi_{1,\lambda}(x)}\,dx$.
This implies that

$$\varphi(x) = N\sum_{\lambda\in\Lambda_+} a_\lambda \varphi(Nx \ominus \lambda),\ \ \sum_{\lambda\in\Lambda_+} |a_\lambda|^2 < \infty$$

where $a_\lambda = (N)^{-1/2}h_\lambda$.

Lemma 5.1. *Consider a nonuniform multiresolution analysis on positive half line as in Definition 5.1. Let $\psi_0 = \varphi$ and suppose $\exists\ N-1$ functions ψ_k, $k = 1, 2, \ldots, N-1$ in V_1 such that the family of functions $\{\psi_k(x \ominus \lambda)\}_{\lambda\in\Lambda_+, k=0,1,\ldots,N-1}$ forms an orthonormal system in V_1. Then the system is complete in V_1.*

Proof. Since $\psi_k \in V_1$, $k = 0, 1, \ldots, N-1$ there exists the sequences $\{h^k_\lambda\}_{\lambda\in\Lambda_+}$ satisfying $\sum_{\lambda\in\Lambda_+} |h^k_\lambda|^2 < \infty$ such that

$$\psi_k(x) = \sum_{\lambda\in\Lambda_+} h^k_\lambda N^{1/2}\varphi(Nx \ominus \lambda).$$

This implies that

$$\psi_k(x) = N\sum_{\lambda\in\Lambda_+} a^k_\lambda \varphi(Nx \ominus \lambda),\ \ \sum_{\lambda\in\Lambda_+} |a^k_\lambda|^2 < \infty$$

where $a_\lambda^k = (N)^{-1/2} h_\lambda^k$.
On taking the Walsh-Fourier transform, we have

$$\tilde{\psi_k}(\xi) = m_k(\frac{\xi}{N})\tilde{\varphi}(\frac{\xi}{N}), \tag{33}$$

where $m_k(\xi) = \sum_{\lambda \in \lambda_+} a_\lambda^k \overline{\chi(\lambda, \xi)}$.
Since $\Lambda_+ = \{0, r/N\} + \mathbb{Z}_+$, we can write

$$m_k(\xi) = m_k^1(\xi) + \overline{\chi(\frac{r}{N}, \xi)} m_k^2(\xi), \ k = 0, 1, \ldots, N-1 \tag{34}$$

where m_k^1 and m_k^2 are locally L^2 functions.
According to [8] for $\lambda \in \Lambda_+$, where $\Lambda_+ = \{0, r/N\} + \mathbb{Z}_+$ and by assumption we have

$$\begin{aligned}
\int_{\mathbb{R}_+} \psi_k(x)\overline{\psi_l(x \ominus \lambda)}\, dx &= \int_{\mathbb{R}_+} \tilde{\psi_k}(\xi)\overline{\tilde{\psi_l}(\xi)}\,\overline{\chi(\lambda, \xi)}\, d\xi \\
&= \delta_{kl}\delta_{0\lambda},
\end{aligned}$$

where δ_{kl} denotes the Kronecker delta.
Define

$$h_{kl}(\xi) = \sum_{j \in \mathbb{Z}_+} \tilde{\psi_k}(\xi + Nj)\overline{\tilde{\psi_l}(\xi + Nj)}, \quad 0 \le k, l \le N-1.$$

If $\lambda \in \mathbb{Z}_+$, we have

$$\begin{aligned}
\int_{\mathbb{R}_+} \psi_k(x)\overline{\psi_l(x \ominus \lambda)}\, dx &= \int_{[0,N]} h_{kl}(\xi)\overline{\chi(\lambda, \xi)}\, d\xi \\
&= \int_{[0,1]} \Big[\sum_{p=0}^{N-1} h_{kl}(\xi + p)\Big]\overline{\chi(\lambda, \xi)}\, d\xi.
\end{aligned}$$

On taking $\lambda = \frac{r}{N} + n$ where $n \in \mathbb{Z}_+$, we obtain

$$\begin{aligned}
\int_{\mathbb{R}_+} \psi_k(x)\overline{\psi_l(x \ominus \lambda)}\, dx &= \int_{[0,N]} \overline{\chi(n + \frac{r}{N}, \xi)}\, h_{kl}(\xi)\, d\xi \\
&= \int_{[0,N]} \overline{\chi(n, \xi)}\,\overline{\chi(r/N, \xi)} h_{kl}(\xi)\, d\xi \\
&= \int_{[0,N]} \overline{\chi(n, \xi)}\,\overline{\chi(r/N, \xi)}\Big[\sum_{p=0}^{N-1} \overline{\chi(r/N, p)} h_{kl}(\xi + p)\Big]\, d\xi.
\end{aligned}$$

By the orthonormality of the system $\{\psi_k(x \ominus \lambda)\}_{\lambda\in\Lambda_+,k=0,1,\dots,N-1}$ we conclude that

$$\sum_{p=0}^{N-1} h_{kl}(\xi+p) = \delta_{kl}, \tag{35}$$

$$\text{and} \quad \sum_{p=0}^{N-1} \overline{\chi(r/N,p)} h_{kl}(\xi+p) = 0$$

$$\text{i.e.} \quad \sum_{p=0}^{N-1} \alpha^p h_{kl}(\xi+p) = 0, \tag{36}$$

where $\alpha = \overline{\chi(r/N,1)}$, since $\overline{\chi(r/N,p)} = \left[\overline{\chi(r/N,1)}\right]^p$ for $p = 0,1,\dots,N-1$.

Now we will express the conditions (35) and (36) in terms of m_k as follows

$$\begin{aligned}
h_{kl}(N\xi) &= \sum_{j\in\mathbb{Z}_+} \tilde{\psi}_k(N\xi+Nj)\overline{\tilde{\psi}_l(N\xi+Nj)} \\
&= \sum_{j\in\mathbb{Z}_+} \tilde{\psi}_k[N(\xi+j)]\overline{\tilde{\psi}_l[N(\xi+j)]} \\
&= \sum_{j\in\mathbb{Z}_+} m_k(\xi+j)\overline{m_l(\xi+j)}|\tilde{\varphi}(\xi+j)|^2 \\
&= [m_k^1(\xi)\overline{m_l^1(\xi)} + m_k^2(\xi)\overline{m_l^2(\xi)}] \sum_{j\in\mathbb{Z}_+} |\tilde{\varphi}(\xi+j)|^2 \\
&\quad + \Big[m_k^1(\xi)\overline{m_l^2(\xi)} \sum_{j\in\mathbb{Z}_+} \chi(r/N,\xi+j)|\tilde{\varphi}(\xi+j)|^2\Big] \\
&\quad + \Big[m_k^2(\xi)\overline{m_l^1(\xi)} \sum_{j\in\mathbb{Z}_+} \overline{\chi(r/N,\xi+j)}|\tilde{\varphi}(\xi+j)|^2\Big].
\end{aligned}$$

Therefore,

$$\begin{aligned}
h_{kl}(N\xi) &= [m_k^1(\xi)\overline{m_l^1(\xi)} + m_k^2(\xi)\overline{m_l^2(\xi)}] \sum_{j=0}^{N-1} h_{00}(\xi+j) \\
&\quad + \Big[m_k^1(\xi)\overline{m_l^2(\xi)}\chi(r/N,\xi) \sum_{j=0}^{N-1} \chi(r/N,j)h_{00}(\xi+j)\Big] \\
&\quad + \Big[m_k^2(\xi)\overline{m_l^1(\xi)\chi(r/N,\xi)} \sum_{j=0}^{N-1} \overline{\chi(r/N,j)}h_{00}(\xi+j)\Big] \\
&= m_k^1(\xi)\overline{m_l^1(\xi)} + m_k^2(\xi)\overline{m_l^2(\xi)}.
\end{aligned}$$

By using the last identity and the equations (35) and (36), we obtain

$$\sum_{p=0}^{N-1}\left[m_k^1(\frac{\xi+p}{N})\overline{m_l^1(\frac{\xi+p}{N})}+m_k^2(\frac{\xi+p}{N})\overline{m_l^2(\frac{\xi+p}{N})}\right]=\delta_{kl}, \tag{37}$$

and

$$\sum_{p=0}^{N-1}\alpha^p\big[m_k^1(\frac{\xi+p}{N})\overline{m_l^1(\frac{\xi+p}{N})}+m_k^2(\frac{\xi+p}{N})\overline{m_l^2(\frac{\xi+p}{N})}\big]=0, \tag{38}$$

$0\le k,l\le N-1$.
Both of these conditions together are equivalent to the orthonormality of the system $\{\psi_k(x\ominus\lambda)\}_{\lambda\in\Lambda_+,k=0,1,\ldots,N-1}$.

The completeness of the system $\{\psi_k(x\ominus\lambda)\}_{\lambda\in\Lambda_+,k=0,1,\ldots,N-1}$ in V_1 is equivalent to the completeness of the system $\left\{\frac{1}{N}\psi_k(\frac{x}{N}\ominus\lambda)\}_{\lambda\in\Lambda_+,k=0,1,\ldots,N-1}\right\}$ in V_0. For a given arbitrary function $f\in V_0$, by assumption $\exists$ a unique function $m(\xi)$ of the form $\sum_{\lambda\in\Lambda_+} b_\lambda\overline{\chi(\lambda,\xi)}$ where $\sum_{\lambda\in\Lambda_+}|b_\lambda|^2<\infty$ such that $\tilde{f}(\xi)=m(\xi)\tilde{\varphi}(\xi)$.
Hence, in order to prove the claim, it is enough to show that the system of functions

$$S=\{\overline{\chi(N\lambda,\xi)}m_k(\xi)\}_{\lambda\in\Lambda_+,k=0,1,\ldots,N-1}$$

is complete in $L^2[0,1]$.
Let $g\in L^2[0,1]$, therefore $\exists$ locally L^2 functions g_1 and g_2 such that

$$g(\xi)=g_1(\xi)+\overline{\chi(r/N,\xi)}g_2(\xi).$$

Assuming that g is orthogonal to all functions in S, we then have for any $\lambda\in\Lambda_+$ and $k\in\{0,1,\ldots,N-1\}$ that

$$\begin{aligned}0&=\int_{[0,1]}\overline{\chi(\xi,N\lambda)}m_k(\xi)\overline{g(\xi)}\,d\xi\\&=\int_{[0,1]}\overline{\chi(\xi,N\lambda)}[m_k^1(\xi)\overline{g_1(\xi)}+m_k^2(\xi)\overline{g_2(\xi)}]\,d\xi\end{aligned} \tag{39}$$

Taking $\lambda=m$ where $m\in\mathbb{Z}_+$ and defining

$$w_k(\xi)=m_k^1(\xi)\overline{g_1(\xi)}+m_k^2(\xi)\overline{g_2(\xi)},\ \ k=0,1,\ldots,N-1,$$

we obtain

$$0 = \int_{[0,1]} \overline{\chi(\xi, Nm)} w_k(\xi)\, d\xi$$

$$= \int_{[0,\frac{1}{N}]} \overline{\chi(\xi, Nm)} \sum_{j=0}^{N-1} w_k(\xi + j/N)\, d\xi.$$

Since this equality holds for all $m \in \mathbb{Z}_+$, therefore

$$\sum_{j=0}^{N-1} w_k(\xi + j/N) = 0 \text{ for a.e. } \xi. \tag{40}$$

Similarly, on taking $\lambda = m + \frac{r}{N}$ where $m \in \mathbb{Z}_+$, we obtain

$$0 = \int_{[0,1]} \overline{\chi(\xi, Nm)}\, \overline{\chi(\xi, r)} w_k(\xi)\, d\xi$$

$$= \int_{[0,\frac{1}{N}]} \overline{\chi(\xi, Nm)}\, \overline{\chi(\xi, r)} \sum_{j=0}^{N-1} \alpha^j w_k(\xi + j/N)\, d\xi.$$

Hence we deduce that

$$\sum_{j=0}^{N-1} \alpha^j w_k(\xi + j/N) = 0 \text{ for a.e. } \xi,$$

which proves our claim. □

If $\psi_0, \psi_1, \ldots, \psi_{N-1} \in V_1$ are as in Lemma 5.1, one can obtain from them an orthonormal basis for $L^2(\mathbb{R}_+)$ by following the standard procedure for construction of wavelets from a given MRA [4,9,16,17]. It can be easily checked that for every $m \in \mathbb{Z}$, the collection $\{N^{m/2}\psi_k(N^m x \ominus \lambda)\}_{\lambda\in\Lambda_+, k=0,1,\ldots,N-1}$ is a complete orthonormal system for V_{m+1}.

Given a NUMRA on positive half line, we denote by W_m the orthogonal complement of V_m in V_{m+1}, $m \in \mathbb{Z}$. It is clear from (i), (ii) and (iii) of Definition 5.1 that

$$L^2(\mathbb{R}_+) = \oplus_{m\in\mathbb{Z}} W_m$$

where $\oplus$ denotes the orthogonal direct sum with the inner product of $L^2(\mathbb{R}_+)$.

From this it follows immediately that the collection $\{N^{m/2}\psi_k(N^m x \ominus \lambda)\}_{\lambda\in\Lambda_+, m\in\mathbb{Z}, k=1,2,\ldots,N-1}$ forms a complete orthonormal system for $L^2(\mathbb{R}_+)$.

Definition 5.2. A collection $\{\psi_k\}_{k=1,2,\dots,N-1}$ of functions in V_1 will be called a set of wavelets associated with a given nonuniform multiresolution analysis on positive half line if the family of functions $\{\psi_k(x \ominus \lambda)\}_{k=1,2,\dots,N-1,\lambda\in\Lambda_+}$ is an orthonormal system for W_0.

The following theorem proves the necessary and sufficient condition for the existence of associated set of wavelets to nonuniform multiresolution analysis on positive half line.

Theorem 5.1. *Consider a nonuniform multiresolution analysis on a positive half line with associated parameters N and r, as in Definition 5.1, such that the corresponding space V_0 has an orthonormal system of the form $\{\varphi(x \ominus \lambda)\}_{\lambda\in\Lambda_+}$ where $\Lambda_+ = \{0, r/N\} + \mathbb{Z}_+$, $\tilde{\varphi}$ satisfies*

$$\tilde{\varphi}(\xi) = m_0(\xi/N)\tilde{\varphi}(\xi/N), \tag{41}$$

and m_0 has the form

$$m_0(\xi) = m_0^1(\xi) + \overline{\chi(r/N, \xi)} m_0^2(\xi), \tag{42}$$

for some locally L^2 functions m_0^1 and m_0^2. M_0 is defined as

$$M_0(\xi) = |m_0^1(\xi)|^2 + |m_0^2(\xi)|^2. \tag{43}$$

Then a necessary and sufficient condition for the existence of associated wavelets $\psi_1, \psi_2, \dots, \psi_{N-1}$ is that M_0 satisfies the identity

$$M_0(\xi + 1) = M_0(\xi). \tag{44}$$

We refer to[11] for the proof of this theorem. The main purpose in the next section is to construct Nonuniform multiresolution analysis on positive half line starting from a Walsh polynomial m_0 satisfying appropriate conditions and finding suitable analogue of Cohen's conditions.

6. Construction of Nonuniform multiresolution analysis on positive half line

Our goal in this section is to construct nonuniform multiresolution analysis on a positive half line starting from a polynomial m_0 of the form

$$m_0(\xi) = m_0^1(\xi) + \overline{\chi(r/N, \xi)} m_0^2(\xi), \tag{45}$$

where $N > 1$ is an integer and r is an odd integer with $1 \leq r \leq 2N - 1$ such that r and N are relatively prime and $m_0^1(\xi)$ and $m_0^2(\xi)$ are locally L^2

Walsh polynomials. The scaling function φ associated with given nonuniform multiresolution analysis on positive half line should satisfy the scaling relation

$$\tilde{\varphi}(\xi) = m_0(\xi/N)\tilde{\varphi}(\xi/N). \tag{46}$$

Define

$$M_0(\xi) = |m_0^1(\xi)|^2 + |m_0^2(\xi)|^2,$$

and suppose

$$\sum_{p=0}^{N-1} M_0(\xi + p/N) = 1 \tag{47}$$

$$\text{and } \sum_{p=0}^{N-1} \alpha^p M_0(\xi + p/N) = 0 \tag{48}$$

where $\alpha = \overline{\chi(r/N, 1)}$. It follows from (46) that

$$\tilde{\varphi}(\xi) = \prod_{k=1}^{\infty} m_0\left(\frac{\xi}{N^k}\right). \tag{49}$$

Also assume that $m_0(0) = 1$ in order for the infinite product $\prod_{k=1}^{\infty} m_0\left(\frac{\xi}{N^k}\right)$ to converge pointwise. For an arbitrary function m_0 of the form (45) the conditions (47) and (48) imply that $|m_0| \leq 1$ a.e. Since if $|m_0(\xi)| > 1$ for a fixed ξ, then $|m_0^1(\xi)| + |m_0^2(\xi)| > 1$ and thus $|M_0(\xi)| > \frac{1}{2}$.
We obtain the inequalities

$$\sum_{p=1}^{N-1} M_0(\xi + p/N) < \frac{1}{2}$$

and

$$|\sum_{p=1}^{N-1} M_0(\xi + p/N)| = |M_0(\xi)| > \frac{1}{2}.$$

which yields to a contradiction.

Theorem 6.1. *Let m_0 be a polynomial of the form* (45) *where m_0^1 and m_0^2 are locally square integrable functions and M_0 satisfy* (47) *and* (48). *Let φ be defined by* (49) *and assume that the infinite product defining $\tilde{\varphi}$ converges a.e. on $\mathbb{R}_+$. Then the function $\varphi \in L^2(\mathbb{R}_+)$.*

Proof. Consider the integrals

$$A_1 = \int_0^N M_0(\xi/N)\, d\xi,$$

and

$$A_k = \int_0^{N^k} M_0\big(\frac{\xi}{N^k}\big) \prod_{j=1}^{k-1} \big|m_0\big(\frac{\xi}{N^j}\big)\big|^2\, d\xi \ \text{ for } k \geq 2.$$

$$A_1 = \int_0^N M_0(\xi/N)\, d\xi = \int_0^1 \sum_{p=0}^{N-1} M_0\big(\frac{\xi}{N} + \frac{p}{N}\big)\, d\xi = 1,$$

and

$$\begin{aligned} A_k &= \int_0^{N^k} M_0\big(\frac{\xi}{N^k}\big) \prod_{j=1}^{k-1} \big|m_0\big(\frac{\xi}{N^j}\big)\big|^2\, d\xi \\ &= \int_0^{N^{k-1}} \sum_{p=0}^{N-1} \big|m_0\big(\frac{\xi}{N^{k-1}} + p\big)\big|^2 M_0\big(\frac{\xi}{N^k} + \frac{p}{N}\big) \prod_{j=1}^{k-2} \big|m_0\big(\frac{\xi}{N^j}\big)\big|^2\, d\xi. \end{aligned}$$

We find that

$$\begin{aligned} &\textstyle\sum_{p=0}^{N-1} \big|m_0\big(\frac{\xi}{N^k} + p\big)\big|^2 M_0\big(\frac{\xi}{N^k} + \frac{p}{N}\big) \\ &= \Big[\big|m_0^1\big(\frac{\xi}{N^{k-1}}\big)\big|^2 + \big|m_0^2\big(\frac{\xi}{N^{k-1}}\big)\big|^2\Big] \sum_{p=0}^{N-1} M_0\big(\frac{\xi}{N^k} + \frac{p}{N}\big) \\ &\quad \overline{+m_0^1\big(\frac{\xi}{N^{k-1}}\big)} m_0^2\big(\frac{\xi}{N^{k-1}}\big)\, \chi\big(\frac{r}{N}, \frac{\xi}{N^{k-1}}\big) \sum_{p=0}^{N-1} \alpha^p M_0\big(\frac{\xi}{N^k} + \frac{p}{N}\big) \\ &\quad +m_0^1\big(\frac{\xi}{N^{k-1}}\big) \overline{m_0^2\big(\frac{\xi}{N^{k-1}}\big)}\, \overline{\chi\big(\frac{r}{N}, \frac{\xi}{N^{k-1}}\big)} \sum_{p=0}^{N-1} \alpha^{-p} M_0\big(\frac{\xi}{N^k} + \frac{p}{N}\big) \\ &= \big|m_0^1\big(\frac{\xi}{N^{k-1}}\big)\big|^2 + \big|m_0^2\big(\frac{\xi}{N^{k-1}}\big)\big|^2 \\ &= M_0\big(\frac{\xi}{N^{k-1}}\big). \end{aligned}$$

This shows that

$$A_k = \int_0^{N^{k-1}} M_0\big(\frac{\xi}{N^{k-1}}\big) \prod_{j=1}^{k-2} \big|m_0\big(\frac{\xi}{N^j}\big)\big|^2\, d\xi = A_{k-1}.$$

It follows that, for all k

$$A_k = A_{k-1} = A_{k-2} = \ldots = A_1 = 1.$$

Hence

$$\begin{aligned}\int_0^{N^k} |\tilde{\varphi}(\xi)|^2 \, d\xi &\le \int_0^{N^k} \prod_{j=1}^{k-1} \left|m_0\left(\frac{\xi}{N^j}\right)\right|^2 d\xi \\ &\le \int_0^{N^k} 2M_0\left(\frac{\xi}{N^k}\right) \prod_{j=1}^{k-1} \left|m_0\left(\frac{\xi}{N^j}\right)\right|^2 d\xi \\ &= 2A_k = 2.\end{aligned}$$

Since k is arbitrary, it follows that $\varphi \in L^2(\mathbb{R}_+)$. $\sqcup$

We will construct nonuniform multiresolution analysis on a positive half line from a polynomial m_0 of the form (45) which satisfies (47) and (48) and also the condition $m_0(0) = 1$.
By Theorem 6.1 we obtain a compactly supported function $\varphi \in L^2(\mathbb{R}_+)$ [4] which satisfies

$$\tilde{\varphi}(N\xi) = m_0(\xi)\tilde{\varphi}(\xi). \tag{50}$$

Now, it is necessary to determine the orthonormality of the system of functions $\{\varphi(x \ominus \lambda)\}_{\lambda \in \Lambda_+}$ in $L^2(\mathbb{R}_+)$ where $\Lambda_+ = \{0, r/N\} + \mathbb{Z}_+$.
If the orthonormality condition is satisfied, we can define

$$V_0 = \overline{span}\{\varphi(x \ominus \lambda)\}_{\lambda \in \Lambda_+}$$

and V_j for $j \in \mathbb{Z}$ is defined as

$$f(x) \in V_j \quad \Leftrightarrow \quad f\left(\frac{x}{N^j}\right) \in V_0 \tag{51}$$

so that (iv) and (v) of Definition 5.1 hold.
Also equation (50) implies that (i) also holds. The remaining two conditions (ii) and (iii) follow from the results of the Theorems 6.2 and 6.3 which are analogues of results in standard theory [4,9].

For an integer m, let $\varepsilon_m(\mathbb{R}_+)$ denotes the collection of all functions f on $\mathbb{R}_+$ which are constant on $[sN^{-m}, (s+1)N^{-m})$ for each $s \in \mathbb{Z}_+$. Further we set,

$$\tilde{\varepsilon_m}(\mathbb{R}_+) = \{f : f \text{ is W-continuous and } \tilde{f} \in \varepsilon_m(\mathbb{R}_+)\}$$

and

$$\varepsilon(\mathbb{R}_+) = \cup_{m=1}^{\infty}\varepsilon_m(\mathbb{R}_+), \quad \tilde{\varepsilon}(\mathbb{R}_+) = \cup_{m=1}^{\infty}\tilde{\varepsilon_m}(\mathbb{R}_+).$$

The following properties are true:

(1) $\varepsilon_m(\mathbb{R}_+)$ and $\tilde{\varepsilon_m}(\mathbb{R}_+)$ are dense in $L^q(\mathbb{R}_+)$ for $1 \le q \le \infty$.
(2) If $f \in L^1(\mathbb{R}_+) \cup \varepsilon_m(\mathbb{R}_+)$, then supp $\tilde{f} \subset [0, N^m]$.
(3) If $f \in L^1(\mathbb{R}_+) \cup \tilde{\varepsilon_m}(\mathbb{R}_+)$, then supp $f \subset [0, N^m]$.

For $\varphi \in L^2(\mathbb{R}_+)$, we put

$$\varphi_{j,\lambda}(x) = N^{j/2}\varphi(N^j x \ominus \lambda), \ j \in \mathbb{Z}, \ \lambda \in \Lambda_+.$$

Let P_j be the orthogonal projection of $L^2(\mathbb{R}_+)$ to V_j.

Theorem 6.2. *Let $\Lambda_+ = \{0, r/N\} + \mathbb{Z}_+$. Suppose that $\varphi \in L^2(\mathbb{R}_+)$ is such that the collection $\{\varphi(x \ominus \lambda)\}_{\lambda\in\Lambda_+}$ is an orthonormal system in $L^2(\mathbb{R}_+)$ with closed linear span V_0 and V_j is defined by (51) then $\cap_{j\in\mathbb{Z}}V_j = \{0\}$.*

Proof. Let $f \in \cap_{j\in\mathbb{Z}}V_j$. Given an $\epsilon > 0$ and a continuous function u which is compactly supported in some interval $[0, R]$, $R > 0$ and satisfies $\|f - u\|_2 < \epsilon$. Then we have

$$\|f - P_j u\|_2 \le \|P_j(f - u)\|_2 \le \|f - u\|_2 < \epsilon,$$

so that

$$\|f\|_2 < \|P_j u\|_2 + \epsilon.$$

Using the fact that the collection $\{N^{j/2}\varphi(N^j x \ominus \lambda)\}_{\lambda\in\Lambda_+}$ is an orthonormal bases for V_j.

$$\begin{aligned}\|P_j u\|_2^2 &= \sum_{\lambda\in\Lambda_+} |< P_j u, \varphi_{j,\lambda} >|^2 \\ &= (N)^j \sum_{\lambda\in\Lambda_+} \left|\int_0^R u(x)\overline{\varphi(N^j x \ominus \lambda)}\, dx\right|^2 \\ &\le (N)^j \|u\|_\infty^2 R \sum_{\lambda\in\Lambda_+} \int_0^R |\varphi(N^j x \ominus \lambda)|^2\, dx,\end{aligned}$$

where $\|u\|_\infty$ denotes the supremum norm of u. If j is chosen small enough so that $RN^j \le 1$, then

$$\begin{aligned}\|P_j u\|_2^2 &\le \|u\|_\infty^2 \int_{S_{R,j}} |\varphi(x)|^2\, dx \\ &= \|u\|_\infty^2 \int_{\mathbb{R}_+} I_{S_{r,j}}(x)|\varphi(x)|^2\, dx, \qquad (52)\end{aligned}$$

where $S_{R,j} = \cup_{\lambda\in\Lambda_+}\{y\ominus\lambda / y\in[0,RN^j]\}$ and $I_{S_{r,j}}$ denotes the characteristic function of $S_{r,j}$.
It can be easily checked that

$$\lim_{j\to-\infty} I_{S_{R,j}}(x) = 0 \ \forall x \notin \Lambda_+.$$

Thus from equation (52) by using the dominated convergence theorem, we get

$$\lim_{j\to-\infty} \|P_j u\|_2 = 0.$$

Therefore, we conclude that $\|f\|_2 < \epsilon$ and since ϵ is arbitrary $f = 0$ and thus $\cap_{j\in\mathbb{Z}} V_j = \{0\}$. □

Theorem 6.3. *Let $\varphi \in L^2(\mathbb{R}_+)$ is such that the collection $\{\varphi(x\ominus\lambda)\}_{\lambda\in\Lambda_+}$ is an orthonormal system in $L^2(\mathbb{R}_+)$ with closed linear span V_0 and V_j is defined by* (51) *and assume that $\tilde{\varphi}(\xi)$ is bounded for all ξ and continuous near $\xi = 0$ with $|\tilde{\varphi}(0)| = 1$, then $\overline{\cup_{j\in\mathbb{Z}} V_j} = L^2(\mathbb{R}_+)$.*

Proof. Let $f \in \big(\cup_{j\in\mathbb{Z}} V_j\big)^{\perp}$.
Given an $\epsilon > 0$ we choose $u \in L^1(\mathbb{R}_+) \cap \varepsilon(\mathbb{R}_+)$ such that $\|f - u\|_2 < \epsilon$.
Then for any $j \in \mathbb{Z}_+$, we have

$$\|P_j f\|_2^2 = \langle P_j f, P_j f\rangle = \langle f, P_j f\rangle = 0,$$

and so

$$\|P_j u\|_2 = \|P_j(f-u)\|_2 \le \|f-u\|_2 < \epsilon.$$

Then we put $g(\xi) = \tilde{u}(\xi)\overline{\tilde{\varphi}(N^{-j}\xi)}$ for some function g of the form

$$g(\xi) = g_1(\xi) + \overline{\chi(\xi, \frac{r}{N})} g_2(\xi),$$

where g_1 and g_2 are locally square integrable, 1/2-periodic functions.

If $g(\xi)$ has the expansion of the form $\sum_{\lambda\in\Lambda_+} c_\lambda \overline{\chi(\xi,\lambda)}$ on the set $[0,1]$, then

$$\begin{aligned} c_\lambda &= \int_0^1 g(\xi)\chi(\xi,\lambda)\, d\xi \\ &= \int_{\mathbb{R}_+} \tilde{u}(\xi)\overline{\tilde{\varphi}(N^{-j}\xi)}\chi(\xi,\lambda)\, d\xi, \quad \lambda\in\Lambda_+. \end{aligned}$$

If $\lambda \in \mathbb{Z}_+$, we have

$$\int_0^1 2g_1(\xi)\chi(\xi,\lambda)\, d\xi = \int_0^1 \sum_{k\in\mathbb{Z}_+} \tilde{u}(\xi+k)\overline{\tilde{\varphi}(\frac{\xi}{N^j} + \frac{k}{N^j})}\chi(\xi,\lambda)\, d\xi.$$

Therefore

$$g_1(\xi) = \frac{1}{2} \sum_{k \in \mathbb{Z}_+} \tilde{u}(\xi + k) \overline{\tilde{\varphi}(\frac{\xi}{N^j} + \frac{k}{N^j})}.$$

On taking $\lambda = \frac{r}{N} + m$ where $m \in \mathbb{Z}_+$, we obtain

$$g_2(\xi) = \frac{1}{2} \sum_{k \in \mathbb{Z}_+} \tilde{u}(\xi + k) \overline{\tilde{\varphi}(\frac{\xi}{N^j} + \frac{k}{N^j})}\, \overline{\chi(\xi + k, \frac{r}{N})}.$$

Therefore

$$g(\xi) = \frac{1}{2} \sum_{k \in \mathbb{Z}_+} \tilde{u}(\xi + k) \overline{\tilde{\varphi}(\frac{\xi}{N^j} + \frac{k}{N^j})} (1 + \alpha^k),$$

where $\alpha = \overline{\chi(r/N, 1)}$.
Since the collection $\{N^{j/2} \varphi(N^j x \ominus \lambda)\}_{\lambda \in \Lambda_+}$ is an orthonormal basis for V_j, if in addition $\tilde{u}$ has compact support, for large values of j

$$\|P_j u\|_2^2 = \sum_{\lambda \in \Lambda_+} \left|\langle u, \varphi_{j,\lambda} \rangle\right|^2 = \int_{\mathbb{R}_+} |u(\xi)|^2 |\tilde{\varphi}(N^{-j}\xi)|^2 \, d\xi.$$

By Lebesgue dominated convergence theorem as $j \to \infty$, the expression on R.H.S converges to $|\tilde{\varphi}(0)|^2 \|\tilde{u}\|_2^2$. Therefore

$$\epsilon > \|P_j u\|_2 = \|\tilde{u}\|_2 = \|u\|.$$

Consequently

$$\|f\|_2 < \epsilon + \|u\|_2 < 2\epsilon.$$

Since ϵ is arbitrary, therefore $f = 0$. □

The analogue of Cohen's condition

On the basis of above construction it is necessary to determine the orthonormality of the system of functions $\{\varphi(x \ominus \lambda)\}_{\lambda \in \Lambda_+}$. We are going to consider the analogue of Cohen's condition for nonuniform multiresolution analysis on positive half line.

Theorem 6.4. *Let m_0 be a polynomial of the form* (45) *which satisfies $m_0(0) = 1$ together with the conditions* (47) *and* (48). *Let φ be defined by the formula* (49) *and $\Lambda_+ = \{0, r/N\} + \mathbb{Z}_+$. Then the following are equivalent:*

(1) $\exists$ a W-compact set E such that $0 \in int(E)$, $\mu(E) = 1$, $E \equiv [0,1](mod \mathbb{Z}_+)$ and

$$inf_{j \in \mathbb{N}}\; inf_{w \in E}\; |m_0(N^{-j} w)| > 0.$$

(2) The system $\{\varphi(x \ominus \lambda)\}_{\lambda\in\Lambda_+}$ *is orthonormal in* $L^2(\mathbb{R}_+)$.

We refer to [11] for the proof of this theorem.

7. Nonuniform vector-valued multiresolution analysis

The concept of nonuniform vector-valued multiresolution analysis introduced by A.H.Siddiqi and P.Manchanda[14] is defined as follows:

Given integers $N \geq 1$ and r odd with $1 \leq r \leq 2N-1$ such that r and N are relatively prime $\Lambda = \{0, r/N\} + 2\mathbb{Z}$ where $\mathbb{Z}$ denotes the set of integers. Let $\boldsymbol{\phi}(t) = (\phi_1(t), \phi_2(t), \ldots, \phi_s(t))^T \in L^2(\mathbb{R}, \mathbb{C}^s)$ where $L^2(\mathbb{R}, \mathbb{C}^s)$ is as in section 4 satisfy the following refinement equation :

$$\boldsymbol{\phi}(t) = \sum_{\lambda\in\Lambda} A_\lambda \boldsymbol{\phi}(2t - \lambda), \tag{53}$$

where $\{A_\lambda\}_{\lambda\in\Lambda}$ is a $s \times s$ constant matrix sequence that has only finite terms.

Define a closed subspace $V_j \subset L^2(\mathbb{R}, \mathbb{C}^s)$ by

$$V_j = \operatorname{clos}_{L^2(\mathbb{R},\mathbb{C}^s)}(\operatorname{span}\{\boldsymbol{\phi}(2^j t - \lambda) : \lambda \in \Lambda\}), \quad j \in \mathbb{Z}. \tag{54}$$

Definition 7.1. We say that $\boldsymbol{\phi}(t)$ defined by (53) generates a nonuniform vector-valued multiresolution analysis $\{V_j\}_{j\in\mathbb{Z}}$ of $L^2(\mathbb{R}, \mathbb{C}^s)$, if the sequence $\{V_j\}_{j\in\mathbb{Z}}$ defined in (54) satisfies:

(a) $\ldots \subset V_{-1} \subset V_0 \subset V_1 \subset \ldots$,
(b) $\cap_{j\in\mathbb{Z}} V_j = \{\mathbf{0}\}$, $\cup_{j\in\mathbb{Z}} V_j$ is dense in $L^2(\mathbb{R}, \mathbb{C}^s)$, where $\mathbf{0}$ is the zero vector of $L^2(\mathbb{R}, \mathbb{C}^s)$,
(c) $\mathbf{h}(t) \in V_j$ if and only if $\mathbf{h}(2Nt) \in V_{j+1}$ $\forall j \in \mathbb{Z}$,
(d) there exists $\boldsymbol{\phi}(t) \in V_0$ such that the sequence $\{\boldsymbol{\phi}(t - \lambda), \lambda \in \Lambda\}$ is an orthonormal basis of V_0. The vector-valued function $\boldsymbol{\phi}(t)$ is called a scaling function of the vector-valued multiresolution analysis.

Note that when $N = 1$, one recovers from the above definition the definition of vector-valued multiresolution analysis with dilation factor equals to 2.

Given a nonuniform vector-valued multiresolution analysis let W_m denotes the orthogonal complement of V_m in V_{m+1}, for any integer m. It is clear from (a) and (b) of Definition 7.1 that

$$L^2(\mathbb{R}, \mathbb{C}^s) = \oplus_{m\in\mathbb{Z}} W_m.$$

The main purpose of nonuniform vector-valued multiresolution analysis is to construct Riesz basis of $L^2(\mathbb{R}, \mathbb{C}^s)$ given by appropriate translates and dilates of a finite collection of functions, called the associated wavelets.

Definition 7.2. A collection $\{\psi_k\}_{k=1,2,\ldots,2N-1}$ of functions in V_1 will be called a set of wavelets associated with a given nonuniform vector-valued multiresolution analysis if the family of functions $\{\psi_k(x - \lambda)\}_{k=1,\ldots,2N-1,\lambda\in\Lambda}$ is Riesz system for W_0.

References

1. Q.Chen and Z.Cheng, *A study on compactly supported orthogonal vector-valued wavelets and wavelet packets*, Chaos, Solitons and Fractals 31:1024-1034, 2007.
2. A.Cohen, *Ondelettes, analysis multirésolutions et filters miroir en quadrature*, Ann. Inst. H. Poinc., Anal. non lineaire 7:439-459, 1990.
3. A.Cohen and R.D.Ryan, *Wavelets and Multiscale Signal Processing (R.J.Knops and K.W.Morton eds.)*, Appl. Math. Math. Comp., Vol. 11, Chapman & Hall, London/ New York, 1995.
4. I.Daubechies, *Ten Lectures on Wavelets*, CBMS 61, SIAM, Philadelphia, 1992.
5. Y.A.Farkov, *Orthogonal p-wavelets on* $\mathbb{R}_+$, Proceedings of International Conference Wavelets and Splines, pp $4-26$, St. Peterberg State University, St. Peterberg, 2005.
6. J.P.Gabardo and M.Z.Nashed, *Nonuniform multiresolution analysis and spectral pairs*, Journal of Functional Analysis $158: 209-241$, 1998.
7. J.P.Gabardo and M.Z.Nashed, *An anlogue of Cohen's condition for nonuniform multiresolution analyses*, Contemporary Mathematics $216 : 41-61$, 1998.
8. B.I.Golubov, A.V.Efimov and V.A.Skvortsov, *Walsh series and transforms*, Kluwer, Dordrecht, 1991.
9. W.R.Madych, *Some elementary properties of multiresolution analysis in* $L^2(\mathbb{R}^n)$, in Wavelets: a tutorial in theory and applications (C.K.Chui eds.), Academic Press, Boston, pp $259-294$, 1992.
10. S.Mallat, *Multiresolution approximation and wavelet orthonormal bases of* $L^2(\mathbb{R})$, Trans. Amer. Math. Soc. $315 : 69-87$, 1989.
11. P.Manchanda, Meenakshi and A.H.Siddiqi, *Wavelets associated with Nonuniform Multiresolution analysis on positive half line*, preprint, 2010.
12. P.Manchanda and A.H.Siddiqi, *Vector-valued wave packets*, abstract-short communication, International Congress of Mathematicians 2010, Hyderabad, 19-27 August 2010, pp 543.
13. F.Schipp, W.R.Wade and P.Simon (with assistance by J.Pal), *Walsh series: An introduction to Dyadic Harmonic Analysis*, Adam Hilger Ltd., Bristol and New York, 1990.
14. A.H.Siddiqi and P.Manchanda, *Wavelets associated with non-uniform vector*

valued multiresolution analysis, abstract-short communication, International Congress of Mathematicians 2010, Hyderabad, 19-27 August 2010, pp 728.
15. L.Sun and Z.Cheng, *Construction of a class of compactly supported orthogonal vector-valued wavelets*, Chaos, Solitons and Fractals 34:253-261, 2007.
16. D.Walnut, *An Introduction to Wavelet Analysis*, Birkhäuser, Boston, 2001.
17. P.Wojtaszczyk, *A Mathematical Introduction to Wavelets*, London Mathematical Society Student Texts 37, Cambridge University Press, Cambridge, UK, 1997.

RELEVANCE OF WAVELETS AND INVERSE PROBLEMS TO BRAIN *

A.H.SIDDIQI[a], H.K.SEVINDIR[b†], Z.ASLAN[c], C.YAZICI[d]

[a] *Department of Mathematics, Sharda University,*
32,34, Knowledge Park-III, Greater Noida-201306, Delhi NCR, India
E-mail: siddiqi.abulhasan@gmail.com

[b] *Department of Mathematics, University of Kocaeli,*
Umuttepe Yerleskesi Fen Edebiyat Fakultesi, Izmit, Kocaeli 41380, Turkey
E-mail: hkodal@kocaeli.edu.tr

[c] *Department of Computer Engineering, Istanbul Aydin University,*
Besyol Mah.Inn Cad.No:38 Sefaky-Kkekmece, Istanbul 34295, Turkey
E-mail: zaferaslan@aydin.edu.tr

[d] *Department of Mathematics Education, University of Kocaeli,*
Umuttepe Yerleskesi Egitim Fakultesi, Izmit, Kocaeli 41380, Turkey
E-mail: cuneyt.yazici@kocaeli.edu.tr

A human brain is the most important organ which controls the functioning of the body including heartbeat and respiration. It is an extremely complex system. Electroencephalography (EEG) is the recording of electrical activity along the scalp produced by the firing of neurons within the brain. In clinical context EEG refers to the recording of the brain's spontaneous electrical activity over a short period of time, say 20-40 minutes, as recorded from multiple electrodes placed on the scalp. The main application of EEG is in the case of epilepsy, as epileptic activity can create clear abnormalities on a standard EEG study. A secondary clinical use of EEG is in coma, Alzheimer's disease, encephalopathies, and brain death. However in the recent years EEG is also being used to design the brain of a robot. Mathematical concepts specially methods for numerical solution of partial differential equations with boundary conditions, inverse problem methods and wavelet analysis have found prominent position in the study of EEG. The present paper is devoted to this theme and will highlight the role of wavelet methods. It will also include the results

*This work is supported by Kocaeli University Scientific Research Foundation (Project number: 2010/003).

†Corresponding Author Phone: (+90)2623032115 Fax: (+90)2623032003

obtained in our research project.

Keywords: EEG; Epilepsy; Wavelet; Inverse Problems.

1. Introduction

The outline of the paper is as follows. In the next section an introduction to inverse problems is given. In the third section wavelet methods will be introduced. Under this section brief information on multiresolution analysis, wavelet spectrum and wavelet cross-correlation will be given. Wavelet methods for inverse problems in signal and image processing will be studied in the fourth section. Wavelet-vaguelette decomposition approach will be introduced in the same section. In the fifth section wavelets for EEG (Electroencephalograhy - recording of brain activity) related inverse problems will be investigated. In the same section after giving a brief information on brain and EEG, direct problem and inverse problem for EEG source localization will be explained. Also usage of wavelet methods in medicine with references for further reading and some open problems are included. In Appendix some case studies regarding wavelet analysis of data obtained through Kocaeli University's Medical School via the project will be presented.

2. Introduction to Inverse Problems

Let X and Y be spaces having appropriate structure, say a Banach Space or a Hilbert space. Given $x \in X$ and $T : X \to Y$, direct problem consists of finding Tx such that

$$y = Tx \tag{1}$$

Given an observed output y, finding an input x that produces it is called an inverse problem, i.e.,

$$x \in X \to y = Tx \in Y \tag{2}$$

or equivalently, given a desired output z, finding an input x that produces an output y that is as "close" to z as possible:

$$\min_{x_i \in X} \| Tx - z \|_{L_2} .$$

Remark here that an inverse problem is to find out "x" such that Eq. (2) holds (at least approximately), where T is the operator describing explicitly relationship between the data "y" and the model parameters x and is representation of the physical system from which the inverse problem is generated. The operator T is called "forward operator" or "observable operator" or "observation function".

If T is a linear operator then the inverse problem is called linear inverse problem, otherwise non-linear inverse problem:

$$g(x) = \int_a^b K(x,t)f(t)dt \Leftrightarrow Tf = g \tag{3}$$

is an example of a linear inverse problem where $K(x,t)$ is the kernel of the integral equation, $g(x)$ is the data and $f(x)$ is the model parameter.

The common thread among inverse problems, signal analysis, and moment problems is a canonical problem: recovering an object (function, signal, picture) from partial or indirect information about the object. In seismology studying the behaviour of elastic waves propagating through the earth which are produced by earthquakes and tsunamis, etc. These sources produce different types of seismic waves which travel through rock, and provide an effective way to image both sources and structures deep within the earth. This is an inverse problem.

In describing the heat conduction in a material occupying a three dimensional domain Ω whose temperature is kept zero at the boundary, the temperature distribution "u" after a sufficiently long time is modeled by

$$\begin{aligned} -\nabla(q(u)\nabla(u)) &= f(u), \quad x \in \Omega \\ u &= 0, \quad x \in \partial\Omega \end{aligned} \tag{4}$$

where f denotes internal heat sources and q is the spatially varying heat conductivity. If one can not measure q directly, one can try to determine q from internal measurements of the temperature u or from boundary measurements of the heat flux. Eq. (4) with unknown q and given u is non-linear although direct problem is linear. The inverse problem in the context of the BVP Eq. (4) is to estimate the coefficient q from a measurement z of the solution u.

In recent years the field of inverse problems has emerged as one of the most active branches of applied mathematics. Certainly the main reason behind this is the ever-growing number of real-world situations that are being modeled and studied in a unified framework of inverse problems. However, the theory of mathematical aspects of inverse problems are also challenging and require a fine blending of various branches of mathematics. We cite the following references for updated information [15, 20, 25, 26, 27, 28, 36, 37].

A number of approaches to the aforementioned inverse problem have been proposed in the literature; most of them involve either regarding Eq.

(4) as a hyperbolic PDE in q or posing an optimization problem whose solution is an estimate of q. Furthermore, the approach of reformulating Eq. (4) as an optimization problem is divided into two possibilities, namely either formulating the problem as an unconstrained optimization problem or treating it as a constrained optimization problem, in which the BVP itself is the constraint.

3. Wavelet Methods

The continuous wavelet transform (CWT) of a function with respect to some local base function (wavelet) is defined as

$$W(a,b) = W_W f(b,a) = \frac{1}{\sqrt{a}} \int_{-\infty}^{\infty} f(t)\Psi^*(\frac{t-b}{a})dt, \quad a > 0 \tag{5}$$

where y^* is the complex conjugate of y. The parameter b and a are called as translation (shifting) and dilation parameters respectively. The wavelet behaves like a window function. At any scale a, the wavelet coefficients $W(a,b)$ can be obtained by convolving $f(t)$ and a dilated version of the wavelet. To be a window and to recover from its inverse wavelet transform (IWT), $y(t)$ must satisfy

$$\Psi(0) = \int_{-\infty}^{\infty} \Psi(t)dt = 0. \tag{6}$$

Although $W(b,a)$ provides space-scale analysis rather than space-frequency analysis, proper scale-to-frequency transformation allows analysis that is very close to space-frequency analysis. Reducing the scale parameter 'a' reduces the support of the wavelet in space and hence covers higher frequencies and vice-versa therefore '$1/a$' is a measure of frequency. The parameter 'b' indicated the location of the wavelet window along the space axis thus changing (b,a) enables computation of the wavelet coefficients on the entire frequency plane.

Scalograms are the graphical representation of the square of the wavelet coefficients for the different scales. They are isometric view of sequence of the wavelet coefficients versus wavelength. A scalogram clearly shows more details, identifies the exact location at a particular depth, and detects low frequency cyclicity of the signal. The scalogram surface highlights the location (depth) and scale (wavelength) of dominant energetic features within the signal, say of gamma rays, bulk density and neutron porosity of a well log.

The combinations of the various vectors of coefficients at different scales (wavelengths) form the scalogram. The depth with the strongest coefficient indicates the position that the particular wavelength change is taking place. The scalogram provides a good space-frequency representation of the signal.

3.1. *Multiresolution Analysis*

The continuous wavelet transform (CWT) is useful for cyclicity determination. However, for other applications, it was found that the CWT was not easy to apply. Meyer discovered that by using certain discrete values of the two parameters a and b, an orthonormal basis can be obtained. The basis is of the form

$$\{2^{s/2}\Psi(2^s t - k)\}_{s,k\in Z}$$

Thus, a will be of the form 2^{-s} and b of the form $k2^{-s}$. With these values of a and b the discrete wavelet transform (DWT) becomes

$$W_\Psi f(k2^{-s}, 2^{-s}) = 2^{s/2}\int_{-\infty}^{\infty} f(t)\Psi(2^s t - k)dt$$

By discretizing the function $f(t)$, and assuming the sampling rate to be 1, the integral above can be approximated by

$$W_\Psi f(k2^{-s}, 2^{-s}) \approx 2^{s/2}\sum f(n)\Psi(2^s t - k).$$

The discrete wavelet transform (DWT) plays an important role in dividing a complicated signal into several simpler ones and analyze them separately. This concept is known as multiresolution analysis (MRA). Here, the function is decomposed at various levels of approximations and resolutions. As a result, a time series signal can be decomposed into a low frequency approximation and several medium-to-high frequency details. Each individual approximation or detail can be analyzed separately, depending on the application.

The discrete wavelet transform belongs to the multiresolution analysis. It is a linear transformation with a special property of time and frequency localization at the same time. It decomposes the given signal series onto a set of basis functions of different frequencies, shifted each other and called wavelets. Unlike the discrete Fourier transform the discrete wavelet transform is not a single object. In reality, it hides a whole family of transformations. The individual members of the family are determined by the choice of so-called mother wavelet function. The goal of discrete wavelet transform is to decompose arbitrary signal $f(t)$ into a finite summation of wavelets at different scales (levels) according to the expansion

$$f(t) = \sum_{j} \sum_{k} c_{ij} \Psi(2^j t - k) \tag{7}$$

where $c_{j,k}$ is a new set of coefficients and $\Psi(2^j t - k)$ is the wavelet of j th level (scale) shifted by k samples. The set of wavelets of different scales and shifts can be generated from the single prototype wavelet, called mother wavelet, by dilations and shifts. What makes the wavelet bases interesting is their self-similarity: Every function in wavelet basis is a dilated and shifted version of one (or possibly few) mother functions. In practice the most often used are the orthogonal or bio-orthogonal wavelets, for which the set of wavelets forms an orthogonal or bi orthogonal base.

Let us denote the discrete form of the original signal vector by f and by $A_j f$ the operator that computes the approximation of f at resolution 2^j. Let $D_j f$ denote the detailed signal,$D_j f = A_{j+1} f - A_j f$ at the resolution 2^j. It was shown by Mallat in [12] that both operations $A_j f$ and $D_j f$ can be interpreted as the convolution of the signal of previous resolution and the finite impulse response of the quadrature mirror filters: the high pass $(\tilde{G})$ of coefficients $(\tilde{g})$ and the low pass $(\tilde{H})$ of coefficients $(\tilde{h})$

$$A_j f = \sum_{k=-\infty}^{\infty} \tilde{h}(2n - k) A_{j+1} f(2n),$$

$$D_j f = \sum_{k=-\infty}^{\infty} \tilde{g}(2n - k) A_{j+1} f(2n).$$

These operations performed for values of j, from 1 to J, deliver the coefficients of the decomposition at different levels (scales) and different resolutions of the original vector f and form the analysis of the signal. The most often used discrete wavelet analysis scheme uses Mallat pyramid algorithm.

As a result of such transformation we get the set of coefficients representing the detailed signals D_j at different levels j, $(j = 1, 2, \ldots, J)$ and the residue signal $A_j f$ at the level J. All of them are of different resolutions, appropriate to the level. The coefficients of $D_j f$ can be interpreted as the high frequency details, that distinguish the approximation of f at two subsequent levels of resolution. On the other hand, the signal $A_j f$ represents the coarse approximation of the vector f.

The next step is the transformation of the detailed signals $D_j f$ $(j = 1, 2, \ldots, J)$ and the coarse approximation signals $A_j f$ into the original resolution. It is done by using special filters G and H associated with the analysis filters $(\tilde{G})$ and $(\tilde{H})$ by the quadrature and reflection relationships. This is so-called reverse Mallat pyramid algorithm, forming the reconstruction of the original signal. As a result we get the decomposed signals of each level in the original resolution. The recovery of the original signal $f(n)$ in each time instant n is then performed by simply adding the appropriate wavelet coefficients and the coarse approximation. At J-level decomposition we have

$$f(n) = D_1(n) + D_2(n) + \ldots + D_j(n) + A_j(n) \tag{8}$$

3.2. *Wavelet Spectrum*

The total energy contained in a signal is defined as

$$E = \int_{-\infty}^{\infty} |f(t)|^2 dt = \|f\|^2.$$

Two dimensional wavelet energy density function is defined as $E(a,b) = W(b,a)$. It signifies the relative contribution of the energy contained at a specific scale "a" and location "b". The wavelet energy density function $E(a,b)$ can be integrated across "a" and "b" to recover the total energy in the signal using admissibility constant c_g as follows

$$E = \frac{1}{c_g} \int_{-\infty}^{\infty} \int_{0}^{\infty} |W(a,b)|^2 \frac{da}{a^2} db.$$

Wavelet spectrum denoted by $E(a)$ is defined as

$$E(a) = \frac{1}{c_g} \int_{-\infty}^{\infty} |W(a,b)|^2 db.$$

The wavelet spectrum $E(a)$ has a power law behavior $E(a) \approx a^{\lambda}$.

Wavelet Spectrum $E(a)$ defines the energy of the wavelet coefficient (wavelet transform) for scale 'a'. Peaks in $E(a)$ highlights the dominant energetic scales within the signal.

The total energy contained in a 2D signal is defined as

$$E = \int_{c_1}^{d_1} \int_{c_2}^{d_2} |f(x,y)|^2 dx dy = \|f\|^2.$$

For discrete f, the total energy E is given as

$$E = \sum_{m} \sum_{n} |f(m,n)|^2 = \|f\|^2.$$

It may be noted that the wavelet transform of a given signal can be reconstructed. Furthermore, the total energy of the given signal and its wavelet transform are identical. If $E(\text{Total})$ is considered to be the total energy of the signal then relative energy is given by

$$E_{\text{Relative}} = \frac{E(\text{Energy of the level to be considered})}{E(\text{Total})}.$$

Scalogram is the graphical representation of the square of the wavelet coefficient versus wavelength. It clearly shows more details and direct low frequency cyclicity of the signal. It may be noted that scalogram is nothing but a 2-dimension wavelet energy density function.

3.3. *Wavelet Cross-Correlation*

Two signals are said to be correlated if they are linearly associated, in other words if their wavelength spectrum a certain scale or wavelength are linearly associated. Broadly speaking graphs of a versus $E(a)$ for two signals are similar (increase or decrease together).

4. Wavelet Methods for Inverse Problems in Signal and Image Processing

As we have seen in Section 2, in an inverse problem we wish to estimate an unknown function (signal) $f(t)$ with the help of an observed data (information) $(Tf)(t)$, where T is some linear operator. Assume that the data are observed at discrete points t_i and are corrupted by noise, so observed data $y(t)$ are

$$y(t) = (Tf)(t) + \varepsilon(t) \tag{9}$$

where $\varepsilon(t)$ represents unwanted quantity (noise). Thus the inverse problem under consideration is the problem of estimating f from noisy data y in the model in Eq. (9).

For most such inverse problems one cannot recover f simply as $\tilde{f} = T^{-1}y$ in practice; either because the inverse operator T^{-1} does not exist at all, or because it is an unbounded operator, which means small changes in y would cause large changes in $T^{-1}y$ (ill posed problem). Classical methods are discussed to solve such problems in references mentioned in Section 2.

Donoho proposed wavelet methods to solve such problems where T is linear or nonlinear [38]. Essentially one uses the expansion f in a wavelet series, constructing a corresponding vaguelette series (to be defined below) for Tf , then estimating the coefficients using appropriate methods (one of such is thresholding). For usage of either kinds of expansion/decomposition such as wavelet-vaguelette, wavelet frames, shearlets we cite references given above.

4.1. *Outline of Wavelet-vaguelette Decomposition Approach*

In wavelet vaguelette decomposition method wavelet expansion of the unknown function f is written as

$$f = \sum_j \sum_k \langle f, \Psi_{j,k} \rangle \Psi_{j,k} \tag{10}$$

Let $\Phi_{j,k} = T\Psi_{j,k}$ for some operators T there exists constants $\tilde{beta}_{j,k}$ such that the set of scaled functions

$$v_{j,k} = \Psi_{j,k}/\tilde{\beta}_{j,k}$$

forms a Riesz basis in L_2 norm, i.e., there exist two constants $0 < A \leq B < \infty$ such that

$$A \sum_j \sum_k c_{j,k}^2 \leq \| \sum c_{j,k} v_{j,k} \|^2 \leq B \sum_j \sum_k c_{j,k}^2 \tag{11}$$

for all square summable sequences $c_{j,k}$. The functions $v_{j,k}$ are called vaguelettes.

If wavelet basis $\Psi_{j,k}$ is chosen appropriately, any function g in the range of T can be expanded in a vaguelette series as

$$g = \sum_j \sum_k \langle g, u_{j,k} \rangle v_{j,k} \tag{12}$$

where $u_{j,k}$ is dual vaguelette basis satisfying

$$T^* u_{j,k} = \tilde{\beta}_{j,k} \Psi_{j,k}.$$

$u_{j,k}$ and $u_{j,k}$ are biorthogonal, that is, $\langle v_{j,k}, u_{l,m} \rangle = \delta_{j,l}\delta_{k,m}$.

If we observe the signal Tf without noise, we could expand it in a vaguelette series:

$$Kf = \sum_j \sum_k \langle Kf, u_{j,k} \rangle v_{j,k} \tag{13}$$

and recover the original function f as

$$f = \sum_j \sum_k \langle Kf, u_{j,k} \rangle \tilde{\beta}_{j,k}^{-1} \Psi_{j,k} = \sum_j \sum_k \langle Kf, \tilde{\Psi}_{j,k} \rangle \Psi_{j,k} \tag{14}$$

where $\tilde{\Psi}_{j,k} = u_{j,k}/\tilde{\beta}_{j,k}$ and, hence $T^* \tilde{\psi}_{j,k} = \psi_{j,k}$. (14) is the key formula for estimating f from Kf by the wavelet-vaguelette method.

In the case of noisy data, we expand the observed signal y in terms of vaguelettes, with coefficients $\hat{b}_{j,k} = \langle y, \tilde{\Psi}_{j,k} \rangle$ which satisfy

$$\hat{b}_{j,k} = b_{j,k} + w_{j,k} \tag{15}$$

from (14) and

$$w_{j,k} = \langle \varepsilon, \tilde{\Psi}_{j,k} \rangle$$

are the vaguelette decomposition of a white noise.

5. Wavelets for EEG (Electroencephalograhy) Related to Inverse Problems

Since the 1930s electrical activity of the brain has been measured by surface electrodes connected to the scalp. Potential differences between these electrodes were then ploted as a function of time named EEG (Electroencephalogram). Mathematicial tools and techniques are used to find the underlying sources which generate the EEG. This activity is known as EEG source localization. It consists of solving a direct and inverse problem.

Solving direct problem starts from a given electrical source configuration representing active neurons in the head., then the potentials at the electrodes are calculated for this configuration. The inverse problem endeavors to find the electrical source which generates a measured EEG. By solving the inverse problem, repeated solutions of the forward problem for different source configurations are needed.

Mathematical technology which has been used to solve direct and inverse problems related to EEG include Poissons equation, the finite element method (FEM), the boundary element method (BEM), the finite difference method (FDM), fast solver for the matrix equation, multigrid methods for direct problem and fairly good number of methods. We cite two important reviews [7] and [6]. We also refer to a recent paper where wavelet methods have been used to analyze EEG. Wavelet methods have not been used in two reviews cited above. There is a research project of Kocaeli University, Turkey, taken by Dr.Hulya K. Sevindir on applications of wavelet methods to EEG data taken at the Hospital of Kocaeli University. Besides a medical doctor from the medical schools hospital, the authors of this paper are involved in this project.

5.1. *Introduction to Brain and Electroencephalogram (EEG)*

Brain is the portion of the vertebrate central nervous system that is enclosed within the cranium, continuous with the spinal cord, and composed of gray matter and white matter. It is the primary center for the regulation and control of bodily activities, receiving and interpreting sensory impulses, and transmitting information to the muscles and body organs. It is also the seat of consciousness, thought, memory, and emotion. In short, brain is the most important organ which controls the functioning of the human body including heart beat and respiration.

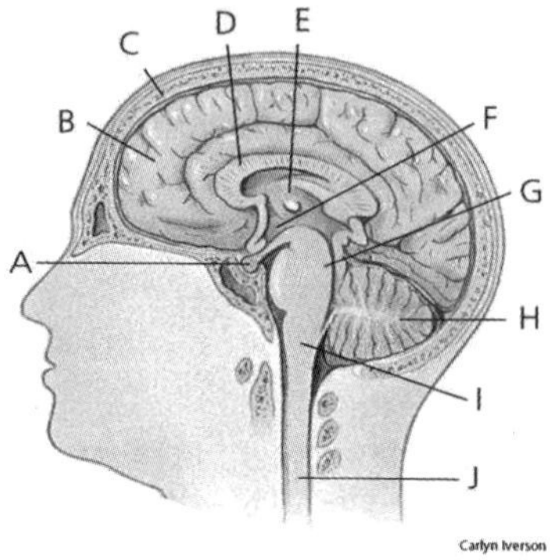

Fig. 1. Brain A. pituitary gland B. cerebrum C. skull D. corpus callosum E. thalamus F. hypothalamus G. pons H. cerebellum I. medulla J. spinal cord

Brain is an extremely complex system. The cerebral cortex of the human brain contains roughly 15-33 billion neurons, perhaps more, depending on gender and age, linked with up to 10,000 synaptic connections each. Each cubic millimeter of cerebral cortex contains roughly one billion synapses. These neurons communicate with one another by means of long protoplasmic fibers called axons, which carry trains of signal pulses called action potentials to distant parts of the brain or body and target them to specific recipient cells. Methods of observation such as EEG recording and functional brain imaging tell us that brain operations are highly organized, while single unit recording can resolve the activity of single neurons, but how individual cells give rise to complex operations is unknown.

Electroencephalography (EEG) is the recording of electrical activity along the scalp produced by the firing of neurons within the brain. In clinical contexts, EEG refers to the recording of the brain's spontaneous electrical activity over a short period of time, usually 20-40 minutes, as recorded from multiple electrodes placed on the scalp. In neurology, the main diagnostic application of EEG is in the case of epilepsy, as epileptic activity can create clear abnormalities on a standard EEG study. A secondary clinical use of EEG is in the diagnosis of coma, encephalopathies, and brain death. EEG used to be a first-line method for the diagnosis of tumors, stroke and other focal brain disorders, but this use has decreased with the advent of anatomical imaging techniques such as MRI and CT.

A routine clinical EEG recording typically lasts 20-30 minutes (plus preparation time) and usually involves recording from scalp electrodes. Routine EEG is typically used in the following clinical circumstances:

- to distinguish epileptic seizures from other types of spells, such as psychogenic non-epileptic seizures, syncope (fainting), sub-cortical movement disorders and migraine variants,
- to differentiate "organic" encephalopathy or delirium from primary psychiatric syndromes such as catatonia,
- to serve as an adjunct test of brain death,
- to prognosticate, in certain instances, in patients with coma,
- to determine whether to wean anti-epileptic medications. Both effects are independent and additive.

At times, a routine EEG is not sufficient, particularly when it is necessary to record a patient while he/she is having a seizure. In this case, the patient may be admitted to the hospital for days or even weeks, while EEG is constantly being recorded (along with time-synchronized video and audio recording). A recording of an actual seizure (i.e., an ictal recording, rather than an inter-ictal recording of a possibly epileptic patient at some period between seizures) can give significantly better information about whether or not a spell is an epileptic seizure and the focus in the brain from which the seizure activity emanates. Epilepsy monitoring is typically done:

- to distinguish epileptic seizures from other types of spells,
- to characterize seizures for the purposes of treatment,
- to localize the region of brain from which a seizure originates for work-up of possible seizure surgery.

If a patient with epilepsy is being considered for resective surgery, it is often necessary to localize the focus (source) of the epileptic brain activity with a resolution greater than what is provided by scalp EEG. This is because the cerebrospinal fluid, skull and scalp smear the electrical potentials recorded by scalp EEG. In these cases, neurosurgeons typically implant strips and grids of electrodes (or penetrating depth electrodes) under the dura mater, through either a craniotomy or a burr hole. The recording of these signals is referred to as electrocorticography (ECoG), subdural EEG (sdEEG) or intracranial EEG (icEEG)–all terms for the same thing. The signal recorded from ECoG is on a different scale of activity than the brain activity recorded from scalp EEG. Low voltage, high frequency components that cannot be seen easily (or at all) in scalp EEG can be seen clearly in ECoG. Further, smaller electrodes (which cover a smaller parcel of brain surface) allow even lower voltage, faster components of brain activity to be seen. Some clinical sites record from penetrating microelectrodes.

Besides above usage of EEG, it can also be used to develope information extraction technology to remotely control a robot. Honda Motor Co. reports that they have good results doing so [24].

A different method to study brain function is functional magnetic resonance imaging (fMRI). There are some benefits of EEG compared to fMRI. For interested reader see [34] and [35].

EEG data is almost always contaminated with biological and enviromental artifacts. Electrical signals detected along the scalp by an EEG, but that originate from non-cerebral origin are called artifacts. The amplitude of artifacts can be quite large relative to the size of amplitude of the cortical signals of interest. This is one of the reasons why it takes considerable experience to correctly interpret EEGs clinically. In addition to artifacts generated by the body, many artifacts originate from outside the body. Movement by the patient may cause electrode pops, spikes originating from a momentary change in the impedance of a given electrode.

Recently, independent component analysis techniques have been used to correct or remove EEG contaminates. These techniques attempt to "unmix" the EEG signals into some number of underlying components. There are many source separation algorithms, often assuming various behaviors or natures of EEG. Regardless, the principle behind any particular method usually allow "remixing" only those components that would result in "clean" EEG by nullifying (zeroing) the weight of unwanted components.

5.2. *Direct Problem for EEG Source Localization*

Hallez *et al* [7] have presented a detailed review of the direct problem. In this review they have elaborated EEG source localization and have shown that this phenomenon is modeled by Poissons equation with its boundary conditions. Different methods are discussed for solving this model with appropriate boundary conditions. The role of well known numerical methods such as Finite difference methods, Finite element methods, Boundary element methods is highlighted in solving these boundary value problems modelling the phenomena. The number of unknowns in the FEM and FDM can easily exceed the million and thus lead to large but sparse linear systems. As the number of unknowns is too large to solve the system in a direct manner, iterative solvers need to be used. Some popular iterative solvers are discussed such as successive over-relaxation (SOR), conjugate gradient method (CGM) and algebraic multigrid methods (AMG). In this review paper physics of EEG, neurophysiology and the generators of EEG are also discussed.

We present here formulation of the main direct problem. In symbolic terms, the EEG forward problem is that of finding, in a reasonable time, the scalp potential $g(r, r_{dip})$ at an electrode positioned on the scalp at r due to a single dipole with dipole moment $d = de_d$ (with magnitude d and orientation e_d),positioned at r_{dip}. This amounts to solving Poisson's equation to find the potentials $V(r)$ on the scalp for different configurations of r_{dip} and d. For multiple dipole sources, the electrode potential would be

$$V(r) = \sum_i g(r, r_{dip_i}, d_i) = \sum_i g(r, r_{dip_i}, e_{d_i}) d_i.$$

In practice, one calculates a potential between an electrode and a reference (which can be another electrode or an average reference). For N electrodes and p dipoles:

$$\mathbf{V} = \begin{bmatrix} V(r_1) \\ \vdots \\ V(r_N) \end{bmatrix} = \begin{bmatrix} g(r_1, r_{dip_1}, e_{d_1}) & \cdots & g(r_1, r_{dip_p}, e_{d_p}) \\ \vdots & \ddots & \vdots \\ g(r_N, r_{dip_1}, e_{d_1}) & \cdots & g(r_N, r_{dip_p}, e_{d_p}) \end{bmatrix} \begin{bmatrix} d_1 \\ \vdots \\ d_p \end{bmatrix}$$

$$= G(\{r_j, r_{dip_i}, e_{d_i}\}) \begin{bmatrix} d_1 \\ \vdots \\ d_p \end{bmatrix}$$

where $i = 1, \ldots, p$ and $j = 1, \ldots, N$. Here V is a column vector. For N electrodes, p dipoles and T discrete time samples:

$$\mathbf{V} = \begin{bmatrix} V(r_1, 1) & \cdots & V(r_1, T) \\ \vdots & \ddots & \vdots \\ V(r_N, 1) & \cdots & V(r_N, T) \end{bmatrix} = G(\{r_j, r_{dip_i}, e_{d_i}\}) \begin{bmatrix} d_{1,1} & \cdots & d_{1,T} \\ \vdots & \ddots & \vdots \\ d_{p,1} & \cdots & d_{p,T} \end{bmatrix}$$

$$= G(\{r_j, r_{dip_i}, e_{d_i}\})D$$

where V is now the matrix of data measurements, G is the gain matrix and D is the matrix of dipole magnitudes at different time instants.

More generally, a noise or perturbation matrix n is added $V = GD + n$. In general for simulations and to measure noise sensitivity, noise distribution is a gaussian distribution with zero mean and variable standard deviation. Further details can be found in [7].

We would like to remark here that wavelet based numerical methods have not been applied to above mentioned boundary value problems. The wavelet methods which have been discussed by various speakers in this workshop may be applied to the present situation and performance can be compared with classical methods.

5.3. *Inverse Problem for EEG Source Localization*

In a recent paper Grech *et al* [6] have reviewed the inverse problem in EEG source analysis. The main modal relevant to this workshop is presented here. In symbolic terms, the EEG direct problem is that of finding, in a reasonable time, the potential $g(r, r_{dip})$ at an electrode positioned on the scalp at a point having position vector r due to a single dipole with dipole moment $d = de_d$ (with magnitude d and orientation e_d) positioned at r_{dip} (see Figure 2). This amounts to solving Poisson's equation to find the potentials V on the scalp for different configurations of r_{dip} and d. For multiple dipole sources, the electrode potential would be

$$m(r) = \sum_i g(r, r_{dip}, d_i).$$

Assuming the principle of superposition, this can be rewritten as

$$\sum_i g(r, r_{dip_i})(d_{ix}, d_{iy}, d_{iz})^T = \sum_i g(r, r_{dip_i})d_i e_i$$

where $g(r, r_{dip})$ now has three components corresponding to the Cartesian x, y, z directions, $d_i = (d_{ix}, d_{iy}, d_{iz})$ is a vector consisting of the three dipole magnitude components, ^{T} denotes the transpose of a vector, $d_i = \|d_i\|$ is the dipole magnitude and $e_i = d_i/\|d_i\|$ is the dipole orientation. In practice, one calculates a potential between an electrode and a reference (which can be another electrode or an average reference).

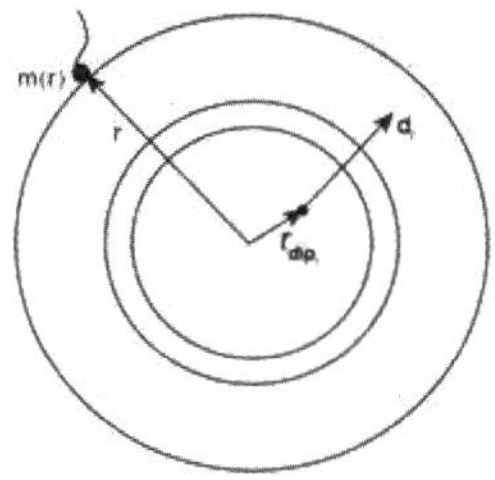

Fig. 2. A three layer head model.

For N electrodes and p dipoles:

$$\mathbf{m} = \begin{bmatrix} m(r_1) \\ \vdots \\ m(r_N) \end{bmatrix} = \begin{bmatrix} g(r_1, r_{dip_1}) & \cdots & g(r_1, r_{dip_p}) \\ \vdots & \ddots & \vdots \\ g(r_N, r_{dip_1}) & \cdots & g(r_N, r_{dip_p}) \end{bmatrix} \begin{bmatrix} d_1 e_1 \\ \vdots \\ d_p e_p \end{bmatrix}$$

where $i = 1, \ldots, p$ and $j = 1, \ldots, N$. Each row of the gain matrix G is often referred to as the lead-field and it describes the current flow for a given electrode through each dipole position. For N electrodes, p dipoles and T discrete time samples:

$$\mathbf{M} = \begin{bmatrix} m(r_1, 1) & \cdots & m(r_1, T) \\ \vdots & \ddots & \vdots \\ m(r_N, 1) & \cdots & m(r_N, T) \end{bmatrix} = G(\{r_j, r_{dip_i}\}) \begin{bmatrix} d_{1,1}e_1 & \cdots & d_{1,T}e_1 \\ \vdots & \ddots & \vdots \\ d_{p,1}e_p & \cdots & d_{p,T}e_p \end{bmatrix}$$

$$= G(\{r_j, r_{dip_i}\})D$$

where M is the matrix of data measurements at different times $m(r, t)$ and D is the matrix of dipole moments at different time instants.

In the formulation above it was assumed that both the magnitude and orientation of the dipoles are unknown. However, based on the fact that apical dendrites producing the measured field are oriented normal to the surface, dipoles are often constrained to have such an orientation. In this case only the magnitude of the dipoles will vary and the formulation above can therefore be re-written as:

$$\mathbf{M} = \begin{bmatrix} g(r_1, r_{dip_1})e_i & \cdots & g(r_1, r_{dip_p})e_p \\ \vdots & \ddots & \vdots \\ g(r_N, r_{dip_1})e_1 & \cdots & g(r_N, r_{dip_p})e_p \end{bmatrix} \begin{bmatrix} d_1 \\ \vdots \\ d_p \end{bmatrix}$$

$$= G(\{r_j, r_{dip_i}, e_i\}) \begin{bmatrix} d_{1,1} & \cdots & d_{1,T} \\ \vdots & \ddots & \vdots \\ d_{p,1} & \cdots & d_{p,T} \end{bmatrix} = G(\{r_j, r_{dip_i}, e_i\})D$$

where D is now a matrix of dipole magnitudes at different time instants. This formulation is less underdetermined than that in the previous structure. Generally a noise or perturbation matrix n is added to the system such that the recorded data matrix M is composed of: $M = GD + n$.

Under this notation, the inverse problem then consists of finding an estimate of the dipole magnitude matrix given the electrode positions and scalp readings M and using the gain matrix G calculated in the forward problem. In what follows, unless otherwise stated, $T = 1$ without loss of generality.

The EEG inverse problem is an ill-posed problem because for all admissible output voltages, the solution is non-unique (since $p \gg N$ and unstable (the solution is highly sensitive to small changes in the noisy data). There are various methods to remedy the situation. As regards the EEG inverse problem, there are six parameters that specify a dipole: three spatial coordinates (x, y, z) and three dipole moment components (orientation angles (θ, φ) and strength d), but these may be reduced if some constraints are placed on the source.

Several methods discussed by earlier speakers could be applied to the inverse problem of EEG, especially the wavelet methods. In the review paper cited above all methods are discussed except the wavelet method. Thus it appears that wavelet method for the inverse problem of EEG is an open question and one should try to study this method. The following references provide updated account of this study [11, 3, 1, 17, 14, 20, 10, 23, 18, 5, 19, 9, 8].

5.4. *Wavelets in Medicine*

Seizures are sudden surge of electrical activity in the brain that usually affects how a person feels or acts for a short duration of time. Epilepsy is a common neurological disorder characterized by recurrent seizures. There are more than fourty types of epilepsy which can be characterized by their different energy distribution in different levels of decomposition using wavelet transform (See Section 2). Electroencephalogram (EEG) is a record of electrical potential generated by cerebral cortex nerve cells. Recorded EEG provides graphical exhibition of the spatial distribution of the changing voltage field. Epileptic seizures are characterized by various events of electrical activity which may rapidly change with time and exhibit different frequency content.

Drake *et al* [4] reported that seizure patients have a decreased power at high frequencies (8.25-30 Hz) relative to lower frequency (0.25-8 Hz), low amplitude pattern at high frequency named electrodermal events appear as a key phenomenon at focal seizure onset. These patterns are characterized by a decrease of signal voltage and increase of signal frequency.

D'Attellis *et al* in [3] essentially initiated the study of identification of epileptic events in EEG using wavelet analysis (See Section 2) time localization and characterization of epileptic form events and computational efficiency of the method. The spline wavelet transform has been used. A concept of energy is introduced and different types of epileptic events have been characterized in terms of this energy. The detection is made when the energy is bigger than a threshold value defined for each level. In this paper the performance of the algorithm proposed with the help of spline biorthogonal wavelets and algorithms based especially on orthogonal wavelets are compared. In the same paper it was suggested that computational techniques based on wavelet theory may be incorporated in the automatic analysis of EEG signals to deal with the problem of extraction features containing relevant information.

It is well known by now that about 1% of the world population is suffering from epilepsy and 30% of epileptic patients are not cured by medication and may need surgery. For surgery careful analysis of EEG is essential. As seen earlier EEG records can provide valuable insight and improved understanding of the mechanisms causing epileptic disorders. It is also well established by now that wavelet transform is particularly effective for representing various aspects of non-stationary signals such as trends, discontinuities, and repeated patterns where other signal processing approaches fail or are not as effective.

In [1] Adeli analyzed epileptic EEG records available at http://kdd.ics. uci.edu/ databases /eeg/eeg.html using Daubechies 4 and harmonic wavelets. The capability of multiresolution analysis known as mathematical microscope is tested especially with Daubechies 4 and harmonic wavelets which have been found experimentally very appropriate for wavelet analysis of spike and wave EEG signals. This analysis depicts physiological processes undergoing in the brain in epilepsy onset. Better understanding of the dynamics of the human brain through EEG analysis is expected according to this investigation.

In [17] Rosso *et al* have reviewed wavelet based informational tools for quantitative EEG record analysis. Relative wavelet energies, wavelet entropies, and wavelet statistical complexities are introduced and are used in the characterization of EEG (scalp) records corresponding to a class of epileptic seizures. In this study it has been shown that the epileptic recruitment observed during seizure development is well described in terms of the relative energies.

In a recent long paper [11] a wavelet based approach in the analysis of biomedical signals is studied which exploits the capability of wavelet transform to separate the signal energy among different frequency bands (different scales), realizing a good compromise between temporal and frequency resolution. The two important aspects of this paper are to present a mathematical formalization of energy calculation (different kinds of energy have been introduced) from wavelet coefficients in order to obtain uniformly time distributed atoms of energy across all the scales and then to study different classes of biomedical signals including EEG. This study helps us to know epileptic brain electrical activity, with aim of identifying typical patterns of energy distribution during the seizure. In this study EEG records from epileptic patients acquired at Bellaria Hospital in Bologna in Italy were analyzed by using Daubechies 4. It may noted here that an important aspect in the analysis of EEGs during epilepsy is the energy distribution among the different details; this redistribution may be indicative of changes in the characteristics of EEG signals which, in turn, may represent specific events in the course of seizure. They have confined their analysis to the resolution levels 4-7 which corresponds to frequency range 0.75 and 12Hz. approximately due to the reasons given below. It is well known that analysis of EEG is represented by artifacts due to muscular activity. Muscular activity induces artifacts in the EEG, which are especially localized high frequencies, and which often blunt information of neural origin and consequently low and medium frequencies EEGs are preferred.

One of the most dreadful features of epilepsy is the random nature of seizures. In patients seizure occur suddenly, without external, previously detected, precipitant. A system which can predict seizures would allow some preventing measures to keep the risk of seizures to a minimum. These measures could improve the quality of life of patients. In view of this, authors of [5] initiated work on prediction of epileptic seizures using accumulative energy in a multiresolution framework. In the references [10, 5, 9, 19, 8] attempts have been made to develop wavelet based methods in conjunction with other methods to predict seizures. In our group we are studying wavelet based energetic approach of Kocaeli University data and trying to develop reliable prediction methods for seizures based on the data available with us.

5.5. *Selected References*

Besides the references given at the end, we cite the following references for updated information. For direct problem case (modelling by Poisson equation) [7] can be cited. For inverse problem related to Poisson equation [6] and [15] can be cited. Under analysis and miscellaneous the papers [1, 2, 3, 10, 11, 14, 17, 21, 23] and for prediction with wavelet, wavelet-cum-ANFIS and wavelet-cum-Neuro-Fuzzy [4, 8, 9, 10, 16, 19] can be listed.

5.6. *Open Problems*

(1) What has been done till today on the data of other countries using wavelet will be tried for the data of Kocaeli University provided by the Kocaeli University's Medical School. Research papers by [11, 3, 1, 17] will be basis for our studies. This is a vast field which requires intensive study.

(2) In Section 2 we have seen inverse problem of EEG and various methods for solving it. It is clear that wavelet methods have not been used to solve inverse problem related to EEG. The work of Prof. M.Z.Nashed [15] provides us guidelines for solving inverse problem related to EEG using wavelet methods.

(3) We know that wavelet methods have been used to solve partial differential equations with boundary value conditions, see e.g. book of A.H.Siddiqi [20] and Dr.Mani Mehra [13]. Direct problem of EEG can be solved by using wavelet techniques discussed in these references providing better results at specified scale.

(4) As we have seen, problems related to brain are studied using PDE

with appropriate boundary conditions or with the help of time series. Reliability of these two methods could be investigated and compared.

Appendix Case Study

In this paper we have analyzed thirty signals of three categories. First group (s1-s10) corresponds to the signals having no symptoms of epilepsy in their electroencephalogram (EEG), second group (s11-s20) corresponds to the signal having single epileptic seizure in their EEG and third group (s21-s30) corresponds to the signal having sequence of epileptic seizures in their EEG.

Analysis of signal s25 by wavelet transforms

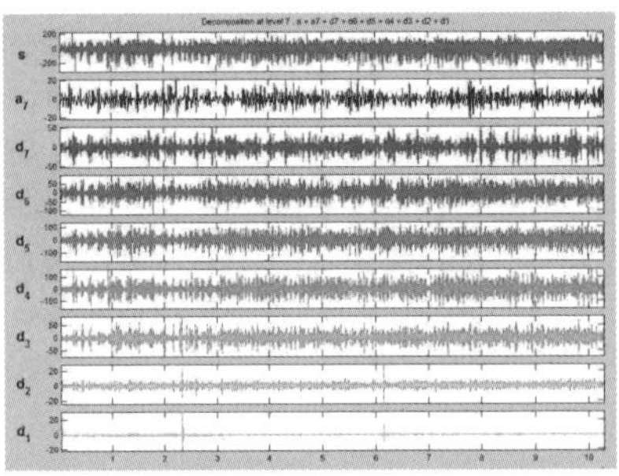

Fig. 3. s25 analysis using 'db10'

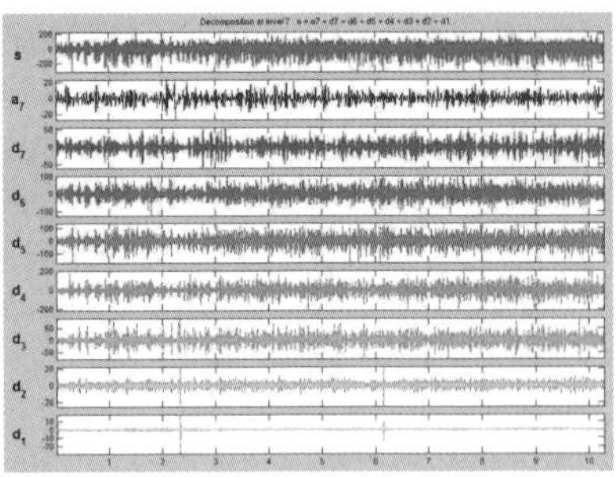

Fig. 4. s25 analysis using 'coif4'

From above decomposition of the signal s25 using db10 and coif4, it can be easily observed that the signal decomposition is quite similar except for the detailed level 1 which contains very small amount of energy as compared to the other detailed levels. Hence changing the mother wavelet has very small effect on the decomposition of the signal.

3-Dimensional visualization of EEG signal

With the help of wavelet transform, it is possible to analyze time and frequency at a time along with the amplitude plot. This makes analysis of signal very easy as compared to spectrum provided by Fourier transform e.g. amplitude of any peak can be determined along with its frequency at a time from same plot.

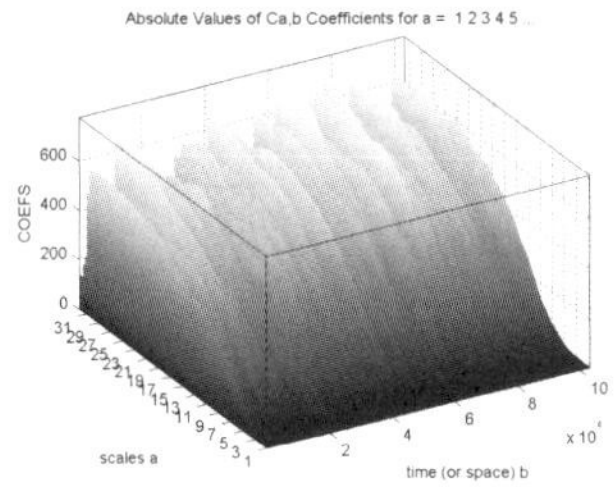

Fig. 5. 3-Dimensional Visualization of s25

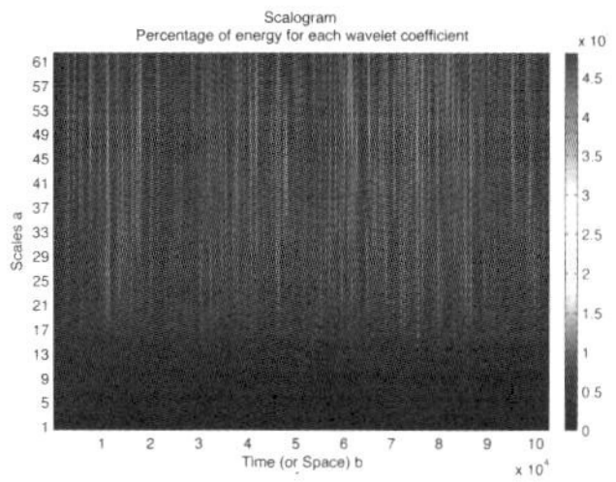

Fig. 6. Frequency-Time Plot of s25

Energy Computation and EEG Analysis

In this research work, relative wavelet energy corresponding to different band of frequencies of epileptic patient is calculated and is compared to the relative energy distribution of a person having no symptoms of epilepsy using MATLAB. For calculating relative energy db10 wavelet is used.

In Table 1, row represents the signals (s1 to s10) and column represents relative energy in approximate (a7) and detailed levels (d1-d7).Entries in the matrix are in percentage (relative energy) showing their total contribution to the signal energy. These signals are of normal human being in which majority of energy content is concentrated in lower frequency band i.e. approximate level of wavelet decomposition and very less amount of energy is shared by high frequency signal.

Now analyzing the energy levels of the signal in which epileptic seizure occurred only once. Table 2 below shows that there is an energy transfer from approximate level to detail level i.e. now some major portion of the energy is contained by high frequency component of the EEG signals which is a characteristic of epileptic seizure i.e. there is an energy transfer from low frequency component to high frequency component.

Table 1. Table 1. Relative energy distribution in non-epileptic EEG signals

	d1	d2	d3	d4	d5	d6	d7	a7
S1	0.0334	0.1419	1.3319	7.3304	14.4934	3.8816	1.5219	71.2654
S2	0.0004	0.0277	0.0751	0.6409	3.1432	5.1494	1.6651	89.2980
S3	0.0423	0.2453	2.9036	17.7078	5.1810	2.7608	2.9175	68.2417
S4	0.0183	0.1789	1.5836	4.3190	4.2371	8.8468	1.4465	79.3699
S5	0.0510	0.0237	0.5060	3.1641	3.2330	1.4525	8.5090	83.0607
S6	0.0291	0.5157	1.2825	2.1287	1.9477	4.4790	4.2514	85.3659
S7	0.0279	0.1895	2.7740	13.3421	8.0650	2.0215	2.8864	70.6936
S8	0.0004	0.0169	0.0580	0.5336	7.2668	5.7495	0.6891	85.6857
S9	0.0471	0.0855	1.0215	4.6689	5.6763	5.3544	4.2884	78.8580
S10	0.0392	0.3217	1.3989	3.3492	4.1335	4.9630	3.8433	81.9512

Table 2. Table 2. Relative energy distribution in single seizure epileptic EEG signals

	d1	d2	d3	d4	d5	d6	d7	a7
S11	0.0021	0.0947	2.3308	26.7369	25.5774	20.3462	10.1937	14.7182
S12	0.0586	0.4162	2.3742	8.5767	25.2817	43.4494	4.8862	14.9569
S13	0.0083	0.2447	5.0308	8.6989	27.7278	14.2981	7.3948	36.5965
S14	0.0143	0.0960	1.4914	14.9424	27.8668	24.0007	13.8599	17.7286
s15	0.0013	0.0275	0.1897	1.0162	7.6810	23.5638	16.3855	51.1349
S16	0.0581	0.1668	4.5553	9.4586	22.7161	11.3294	8.4004	43.3154
S17	0.0352	0.1900	3.6503	20.4685	26.1663	12.8692	4.9087	31.7118
S18	0.0126	0.1323	1.6804	15.6393	28.3191	10.7967	4.2711	39.1485
S19	0.0155	0.2618	2.7697	21.2343	32.1086	18.0549	7.8356	17.7197
S20	1.3330	0.3934	3.5659	6.1014	12.6269	14.3218	4.1937	57.4639

Now analyzing the EEG signal having sequence of spike i.e. now the patient is suffering from epilepsy seizure. Now most of the energy content is shifted to detailed levels and energy is considerably reduced in approximate level.

On analyzing the above tables for all three groups of signals, energy transfer from approximate level to detailed levels (d4, d6) can be observed. This proves that during the seizure, in EEG signal, high frequency component are dominating as compared to the low frequency signal.

Cross-correlation

Cross-correlation gives the similarity between two signals. Applying the concept of cross-correlation to s9 and s10, we get results showing that there

Table 3. Table 3. Relative energy distribution in EEG signals having sequence

	d1	d2	d3	d4	d5	d6	d7	a7
S21	0.0142	0.1037	5.6530	33.7127	30.6331	21.6553	5.6002	2.6277
S22	0.0016	0.0357	2.3568	24.8919	34.1427	26.4745	7.8432	4.2535
S23	0.0016	0.0559	1.5332	15.8848	29.5901	39.4139	9.7807	3.7398
S24	0.0068	0.0592	1.9956	22.9827	33.0698	31.4255	8.1007	2.3599
S25	0.0024	0.0492	2.7363	24.1476	34.3231	29.3186	8.0204	1.4023
S26	0.0041	0.2629	8.3963	38.9510	33.2099	14.4360	3.6358	1.1039
S27	0.0081	0.0462	1.3358	21.0487	33.5500	31.9619	8.3568	3.6926
S28	0.0016	0.0574	1.4378	15.9367	32.9950	37.5294	9.5269	2.5151
S29	0.0010	0.0508	2.8008	24.2717	34.6544	30.1317	6.5542	1.5354
S30	0.0052	0.2724	6.8588	33.2147	40.2686	15.8056	2.7417	0.8329

is a great similarity in the energy content in various decomposition levels of these two signals.

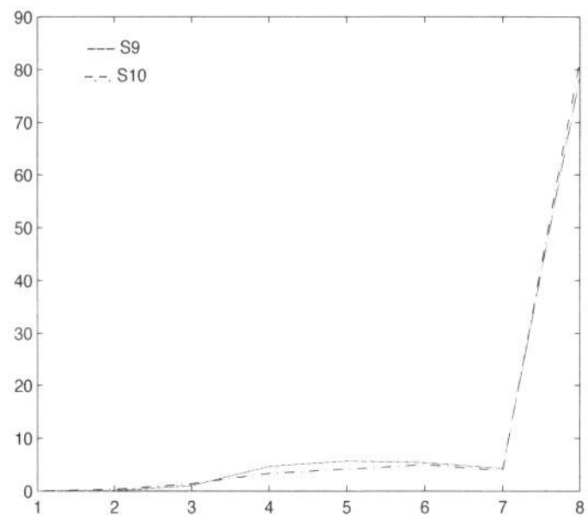

Fig. 7. Plot Showing energy distribution between two signals s9 and s10

Plotting cross-correlation for s13 and s18 will tell us about energy distribution similarity in these two signals.

Here it can be observed that on the basis of energy distribution, there is a similarity between these two signals. Plotting for signal s25 and s26 which contains sequence of spikes, we get the same similarity in these two signals s25 and s26.

This means that same type of EEG signal will have same type of cross-correlation on the basis of energy distribution.

Now plotting cross-correlation for all three types of signal, i.e., s9, s18 and s25.

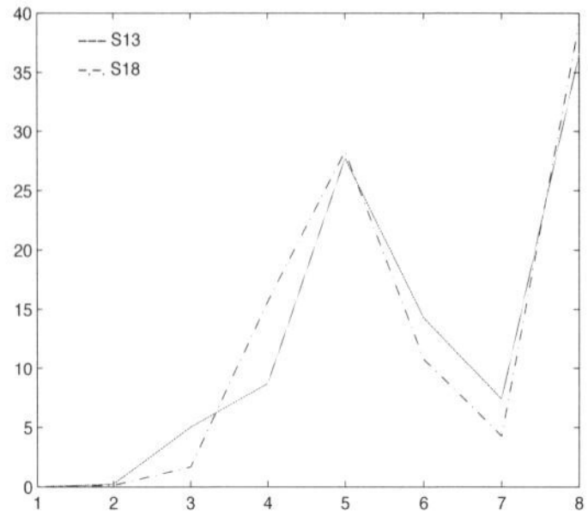

Fig. 8. Plot Showing relative energy distribution between two signals s13 and s18

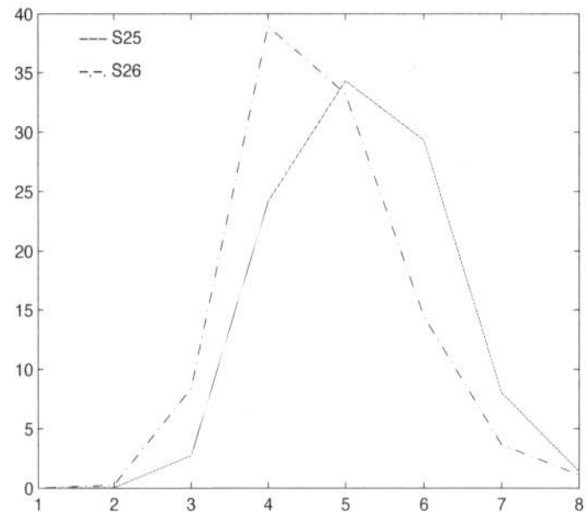

Fig. 9. Plot Showing relative energy distribution between two signals s25 and s26

It can be observed that the relative energy cross correlation of a sampled data s9, s18 and s25 representing normal, single spike and multiple spike, respectively, have different relative energy at approximate signal shown at instant 8 in the above figure. On the basis of approximate energy distribution we can predict the epileptic signal.

Conclusion

From above tables and cross correlation plots, it can be concluded that characteristics of signal having no symbol of epilepsy, single seizure and multiple seizure have different energy content. During epilepsy, EEG contains majority of high frequency signals which can be observed from the relative energy distribution which shows that they are concentrated in high

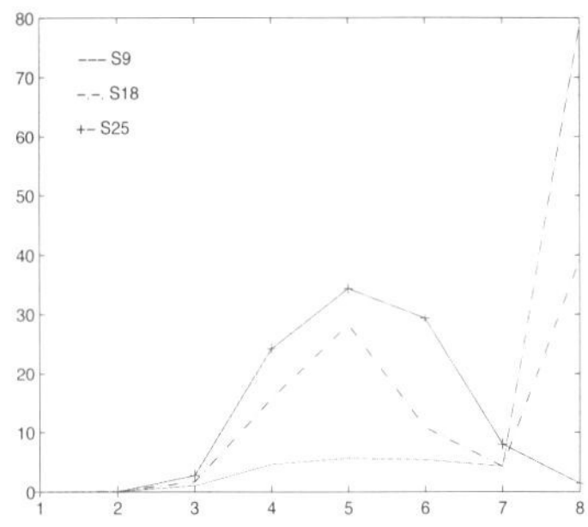

Fig. 10. Plot Showing energy distribution between three signals s9, s18 and s25

frequency levels. There is also very less similarity between different groups of EEG signals because energy distribution is varying in these groups.

Hence epileptic signal can be predicted as signal in which there is a decrease of energy in approximate signals (a7) or increased value of signal energy in detailed signals (d4, d5, d6, d7).

References

1. H. Adeli, Z. Zhou, N. Dadmehr, *Analysis of EEG records in an epileptic patient using wavelet transform*, (Journal of Neuroscience Methods, 123, pp. 69–87, 2003).
2. M. Akin, *Comparison of wavelet transform and FFT methods in the analysis of EEG signals*, (Journal of Medical Systems, 26, No.3., 2003).
3. C. E. DÁttellis, S. I. Isaacson, and R. O. Sirne, *Detection of epileptic events in electroencephalograms using wavelet analysis*, (Annals of Biomedical Engineering,pp. 286–293, 1997).
4. M.E.Drake, H. Padamadan, S.A. Newell,*Interictal quantitative EEG in epilepsy*, (Seizure, 39–42, 1998).
5. S. Gigola, F. Ortiz, C.E. DAttellis, W. Silva, S. Kochen,*Prediction of epileptic seizures using accumulated energy in a multiresolution framework*, (Journal of Neuroscience Methods, 138,pp.107–111, 2004).
6. R. Grech, T. Cassar, J. Muscat, K. P. Camilleri,S. G. Fabri, M. Zervakis, P. Xanthopoulos, V. Sakkalis and B. Vanrumste,*Review on solving the inverse problem in EEG source analysis*,(Journal of NeuroEngineering and Rehabilitation, 5:25,doi:10.1186/1743–0003-5-25, 2008).
7. H Hallez, B Vanrumste, R. Grech, J Muscat, W. De Clercq, A. Vergult, Y. D'Asseler, K. P Camilleri, S. G Fabri, S. Van Huffel and I. Lemahieu,*Review on solving the forward problem in EEG source analysis*, (Journal of NeuroEngineering and Rehabilitation, 4:46, pp. 107–111, doi: 10.1186/1743-0003-4-46 2007).

8. W. Hsu,*EEG-based motor imagery classification using neuro-fuzzy prediction and wavelet fractal features*, (Journal of Neuroscience Methods, 189, pp. 295–302, 2010).
9. S.P. Kumar, N. Sriraam, P. G. Benakop, B. C. Jinaga, *Entropies based detection of epileptic seizures with artificial neural network classifiers*, (Expert Systems with Applications, 37,3284–3291,2010).
10. C.A.M. Lima, Andr L.V. Coelho, Sandro Chagas,*Automatic EEG signal classification for epilepsy diagnosis with relevance vector machines*, (Expert Systems with Applications, 36, 10054-10059, 2009).
11. E Magosso, M Ursino, A Zaniboni and E Gardella, *A wavelet-based energetic approach for the analysis of biomedical signals: Application to the electroencephalogram and electro-oculogram*, (Applied Mathematics and Computation Volume, 207, Issue 1, pp. 42–62, 2009).
12. S.G. *Mallat,A theory for multiresolution signal decomposition: The wavelet representation*, (IEEE Trans. On Pattern Analysis and Machine Intell., Vol. 2, No. 7, pp. 674–693, 1989).
13. M.Mehra, *Wavelets and differential equations - A short review*, (AIP Conference Proceedings,Vol. 1146, pp. 241-252, 2009).
14. H.A. Al-Nashash, J.S. Paul, W.C. Ziai, D.F. Hanley, N.V. Thakor, *Wavelet entropy method for EEG analysis: Application to global brain injury*, (Proceedings of the 1st international IEEE, 2009).
15. Z.Nashed, *Applications of wavelets and kernel methods in inverse problems*, (In Costanda,Nashed and Rolins (eds), Integral Methods in Science and Engineering Theoretical and Practical Aspects,Birkhausser, pp. 189–197, 2006).
16. S. Osowski, K. Garanty, *Forecasting of the daily meteorological pollution using wavelets and support vector machine*, (Engineering Applications of Artificial Intelligence, 20, pp. 745–755, 2007).
17. O.A. Rosso, M.T. Martin, A. Figliola, K. Keller, A. Plastino, *EEG analysis using wavelet-based information tools*, (Journal of Neuroscience Methods 153, pp. 163-182, 2006).
18. A.Sengur, *Wavelet transform and adaptive neuro-fuzzy inference system for color texture classification*, (Expert Systems with Applications 34, pp. 2120-2128, 2008).
19. A.Subasi, *Application of adaptive neuro-fuzzy inference system for epileptic seizure detection using wavelet feature extraction*, (Computers in Biology and Medicine 37, pp. 227–244, 2007)
20. A.H.Siddiqi, *Applied functional analysis*, (Chapter 10, Anamaya, 2010).
21. A.H.Siddiqi, A. Chandiok, V.Singh Bhadouria, *Analysis and prediction of energy distribution in electroencephalogram (EEG) using wavelet transform*, (IVth. International Workshop on Applications of Wavelets to Real World Problems: IWW2009, ISBN No: 978-605-4158-12-6, Anamaya Publishing Company, India, France, 2010 (in press)).
22. M.E. Tagluk, M. Akin, N. Sezgin, *Classification of sleep apnea by using wavelet transform and artificial neural networks*, (Expert Systems with Applications, V. 37, pp. 1600–1607, 2010).
23. E.D. Ubeyli, *Combined neural network model employing wavelet coefficients*

for EEG signals classification, (Digital Signal Processing 19, 297–308, 2009).
24. http://search.japantimes.co.jp/cgi-bin/nb20090401a2.html
25. I. Victor, *Inverse problems for partial differential equations*, (Springer Science + Business Media, Inc, New York, 2006).
26. W. Freden, M.Z. Nashed, T. Sonar (Eds.) *Handbook of geomathematics*, (Springer Verlag, Berlin, Heidelberg, 2010).
27. I. Daubechies, M. Defrise, and Christine De Mol, *An iterative thresholding algorithm for linear inverse problems with sparsity constraint*, (Comm. of Pure and Appl. Math. 57, 1413–1541, 2004).
28. I. Daubechies, M. Farnasier and I. Loris, *Accelerated projected gradient method for linear inverse problems with sparsity constraints*, (Journal of Fourier Analysis and Applications, 14 (5-6), 764–792, 2004).
29. Z. Shen, *Wavelet frames and image restorations*, (Proc. of ICM 2010, 2835-2863, 19-27 August 2010).
30. M.Z. Nashed, *Inverse problems, moment problems and signal processing*, (Invited talk, Satellite Conference ICM 2010 on Mathematics in Science and Technology, 14-19 August 2010).
31. P.Maas, *Wavelets and inverse problems*, (Invited talk, Satellite Conference ICM 2010 on Mathematics in Science and Technology, 14-19 August 2010).
32. A. Cohen, M. Hoffmann and M. Reib, *Adaptive Wavelet-Galerkin methods for inverse problems*, (SIAM J. Num. Anal. 42, 1479-1501, 2004).
33. F. Abramovich and B.W. Silverman, *Wavelet decomposition approaches to statistical inverse problems*, (Biometrika 85, pp. 115–129, 1998).
34. M. Bottema, B. Moran, and S. Suvorova, *An application of wavelets in tomography*, (Digital Signal Processing 8, 244–254, 1998).
35. G.R. Easky, F. Colonna, D. Labate, *Improved random based imaging using the Shearlet transform*, (Preprint, 2010).
36. J.L. Starck, F. Murtagh, J.M. Fadili, *Sparse image and signal processing*, (Cambridge (Chapter 1), 2010).
37. A.H. Siddiqi *et al*, Proc.Int. Conf. August 2010, Mathematics in science and technology, (World Scientific Publisher, 2010/2011).
38. D.L.Donoho, *Nonlinear solution of linear inverse problems by wavelet-vaguelette decomposition*, (Appl. Comput. Harmonic Analysis, 2,101–126, 1995).

WAVELETS AND INVERSE PROBLEMS

KAVITA GOYAL and MANI MEHRA*

Indian Institute of Technology Delhi, Hauz Khas,
New Delhi–110116, India
E-mail:kavita-ma@student.iitd.ac.in
** E-mail:mmehra@maths.iitd.ac.in*

Inverse problem is to deduce cause from effect. A wide variety of scientific problems ends with the situation where if m is desirable quantity, instead of m we have $G(m)$ accessible for s ome operator G. Inverse problem is to find out m given $G(m)$. There are large number of traditional methods available for solving inverse problems and some methods for solving inverse problems have been formulated using wavelets which are better than traditional methods in many senses because of special properties of wavelets (e.g. localizing property, compact support, adaptivity etc.). In this review article initially we are discussing general approach to solve an inverse problem and then we will move on to the wavelet methods to solve inverse problem.

Keywords: Inverse problem, Singular values of a matrix, Wavelets, Vaguelettes, Discrete wavelet transform (DWT), Least square problem.

1. Introduction

An inverse problem is a task that often occurs in many branches of Mathematics and Science. Physically, inverse problem is the problem that consist of finding an unknown property of an object or a medium, from the observation of response of this object to a probing signal. Mathematically an inverse problem is

$$d = G(m), \tag{1}$$

where for given data 'd' we have to find out the model parameter 'm'. There are many approaches to solve the inverse problems with very appropriate results but still with many limitations. Recently, wavelets have shown their usefulness in many fields of Mathematics, Science and Engineering. The field of inverse problems is also within the impact area of wavelets. Many people have used different kinds of wavelet to solve inverse problems[1] and

PDEs[2] arising in the physical world and have produced better results than the traditional methods. Nowadays many new wavelets have been discovered (e.g. Daubechies wavelets,[3] interpolating wavelets,[4] second generation wavelets,[5] diffusion wavelets[6]) and we believe that with these wavelets we can generate wavelet methods to solve inverse problems which will be proved better than the existing methods in many respects. This review article has been written in such a way that even a novice to this field can have an idea about the field after reading it. The format goes like this, first of all we will introduce the concept and examples of inverse problems and a general idea of how to solve an inverse problem. Next we will throw light on how wavelet methods are proved to be better in solving inverse problems as compared to traditional existing methods. Then we will conclude our discussion by looking at some future challenges.

2. Inverse Problems

Mathematically, inverse problems can be formulated in more understandable way as follows:

$$\text{Data} \Rightarrow \text{Model Parameters}$$

i.e. an inverse problem is to find out 'm' such that Eq. (1) holds (at least approximately), where 'G' is the operator describing explicitly relationship between the data 'd' and the model parameters 'm' and is a representation of the physical system from which the inverse problem is generated. This operator G is called **'Forward Operator'** or **'Observation Operator'** or sometimes **'Observation Function'**.

Depending on the type of the operator G, the inverse problems can be categorized in the following two categories:

- **Linear inverse problem:** If the forward operator G is a linear operator then Eq. (1) becomes linear inverse problem.
  Example: Fredholm's first kind integral equation

$$d(x) = \int_a^b g(x, y)m(y)dy, \tag{2}$$

  where $g(x, y)$ is the kernel of the integral equation, $d(x)$ is the data and $m(y)$ is the model parameter.

- **Non linear inverse problem:** If the forward operator G is a non linear operator then Eq. (1) becomes non linear inverse problem.
  Example: In describing the heat conduction in a material occupying a three dimensional domain Ω whose temperature is kept zero at the boundary, the temperature distribution 'u' after a sufficiently long time is modeled by

$$-\nabla.(q(x)\nabla(u)) = f(x), \;\; x \in \Omega \qquad (3)$$
$$u = 0 \text{ on } \partial\Omega,$$

  where f denotes internal heat sources and q is the spatially varying heat conductivity. If one cannot measure q directly, one can try to determine q from internal measurements of the temperature u or from boundary measurements of the heat flux $q * \partial u/\partial n$. This kind of problems occurs in geophysical applications and non-destructive material testing. Note that Eq. (3) with unknown q and given u is nonlinear although the direct problem of computing u given q is linear.

Physically, inverse problem is to deduce cause from an effect. Consider a physical system, there will be input and output parameters related to this system. If all the parameters were known perfectly, then for a given input we can predict the output very easily and this is what we do most of the times. It may happen, however, that some of the parameters characterizing the system are not known, being inaccessible to direct measurements. If it is important to know what these parameters are, in order to understand the system, then we have to infer them by observing the outputs from the system corresponding to special inputs. Thus in an inverse problem we seek cause (the system parameters), given the effects (the outputs of the system for given special inputs).

2.0.1. *Examples*

- If an acoustic plane wave is scattered by an obstacle, and one observes the scattered field far away from the obstacle or in some exterior region, then the inverse problem is to find out the shape and the material properties of the obstacle from the observed scattered field. Such problems are important in identification of flying objects (airplanes, missiles etc.), objects immersed in water (submarines, paces of fishes etc.) and in many other situations.
- One of the central example of inverse problem is tomography. Tomography is imaging by sections through the use of waves of energy. To-

mography is used in geophysics, medicine, metallurgy, radiology, biology, astrophysics, seismology and in many other branches of science.

- In geophysics one sends an acoustic wave from the surface of the earth and collects the scattered field on the surface for various positions of the source of the field, for a fixed frequency or for several frequencies. The inverse problem, in this case, is to find out the inhomogeneities from the study of scattered field. This inhomogeneity can be an oil deposit, a cave or a mine in case of geophysics, in medicine it can be a tumor or some abnormality in human body, and in the field of metallurgy it can be a hole in the metal.
- In seismology we study the behavior of elastic waves propagating through the earth which are produced by earthquakes and tsunamis etc. to study earth's interior structure. Earthquakes, and other sources (e.g. tsunamis), produce different types of seismic waves which travel through rock, and provide an effective way to image both sources and structures deep within the earth.

  For details of above examples one can refer.[7] If one is able to find the inhomogeneities in the medium by processing the scattered field on the surface as explained in above examples, then one does not have to drill a hole in the medium. This in turns avoid the expensive and destructive evaluations. The practical advantages of remote sensing makes inverse problems more important. It is interesting to note that when astronomer Urbain Le Verrier worked out the math to successfully predict where the planet Neptune would be discovered in the night sky back in 1846, he was really solving an inverse problem. By that, he used the observations that had been recorded about Uranus' position in its orbit to infer how that orbit might have been affected by what was then a hypothetical eighth planet in the solar system, and where it would have to be to create the observed effects.

3. General procedure for solving an inverse problem

Whenever we formulate inverse problem mathematically, we typically find that the problem amounts to that of determining one or more coefficients in the differential equation, or system of differential equations, given partial knowledge of of certain special solutions of equations. For example in seismology, the propagation of wave in the earth is governed by equations of elasticity, a system of partial differential equations in which material properties of earth manifest themselves as coefficient functions in the equation. The measurements we make amounts to the knowledge of special solutions of the equation at special points (e.g. those points which are on the surface of earth). In a very simplified case inverse problem of seismology reduces to find the coefficient in following **Eikonal equation**

$$a^2(x_3)((\partial_1\tau)^2 + (\partial_3\tau)^2) - 1 = 0,$$

where ∂_j is partial differentiation w.r.t. to x_j, τ is the travel time of seismic waves and the coefficient $a(x_3)$ is density of earth. Any numerical technique applied to solve above problem will ultimately ends up with the problem of solving Eq. (1) for m. Hence our ultimate aim is to solve Eq. (1) for m.

Understanding ill posed problems is an integral part of the subject of inverse problems, because inverse problems are typically ill posed problems. The term Well posed problem was introduced by 'Jacques Hadamard'. He believed that mathematical models of physical phenomenon should have the following properties:

(1) Existence: Solution of the mathematical model of the problem should exist.

(2) Uniqueness: Solution of the mathematical model should be unique.

(3) Stability: The Solution of the mathematical model should depend continuously on the data in some reasonable space.

Examples of well posed problems are: Dirichlet's problem for Laplace equation, heat equation with specified initial conditions etc. Problems that are not well posed in the sense of Hadamard are termed as ill posed problems. For example inverse heat equation and deducing a previous distribution of temperature from the final data are ill posed problems in the sense that solution is highly sensitive to the changes in the final data.

It should be noted that among the three conditions of well posed problems suggested by Hadamard (i.e. existence, uniqueness and stability), the condition of stability is most often violated by inverse problems. Now to solve an inverse problem we need to know how ill posed problems are solved. Our problem is of the form of Eq. (1), where $G : X \to Y$ is a compact linear or non linear operator. In the worst case an ill posed problem will violate all three conditions due to Hadamard.

- Non Uniqueness can be treated by taking into account some information source related to our system. For example if we are solving inverse problem (arising in Tomography) to find out the properties of earth's interior, then we can use travel time of earthquake as extra constraint to the problem to overcome the problem of non-uniqueness.

- Instability and the possible non-existence can be treated by regularization techniques.

Regularization?

Practically, we can never have an exact data for a physical problem available with us. The data available is perturbed with noise because of the errors in the measurements and the limitations of the measuring instruments. Even if their deviation from the exact data is small, algorithms developed for well-posed problems then fail in case of a violation of the third Hadamard condition, since data as well as round-off errors may then be amplified by an arbitrarily large factor. In order to overcome these instabilities one has to use regularization methods, which in general terms replace an ill-posed problem by a family of neighboring well-posed problems. Typically regularization involves including additional assumptions such as smoothness of the solution. Mathematically, regularization can be explained as follows:

We take a family of bounded (linear or non-linear) regularization operators

$$R_\alpha : Y \to X, \quad \alpha > 0, \tag{4}$$

with the property

$$\lim_{\alpha \to 0} R_\alpha(G(x)) = x, \quad for \quad all \quad x \in X, \tag{5}$$

i.e. the operators $R_\alpha G$ converges point wise to the identity as $\alpha \to 0$. Here α is called the regularization parameter. If R_α satisfies Eq. (5), then the family of operators R_α is called regularization strategy. In the presence of the data error of size δ we calculate solution of Eq. (1) with d replaced by d^δ i.e. we calculate

$$m^\delta = R_{\alpha(\delta)} d^\delta,$$

with regularization parameter $\alpha(\delta)$ depending on $\delta > 0$. Of course, we would like to choose the regularization parameter in such a way that our approximate solution tend toward the true solution if the data error tend to zero. This motivates the following definition:

A strategy for the choice of the parameter α depending on the error level δ is called regular, if for all $d \in G(X)$ and for all $d^\delta \in Y$ with $\| d^\delta - d \| \leq 0$, the following holds:

$$R_\alpha d^\delta \to G^{-1} d \text{ as } \delta \to 0.$$

We will call a set of regularization operators R_α with regular strategy for the inversion of Eq. (1) as convergent regularization. Tikhonov regularization is most commonly used regularization technique for ill posed problems and is defined as follows:

Tikhonov regularization: Tikhonov defines a regularization scheme with $\| R_\alpha \| \leq 1/2\alpha$. It is regular provided $\alpha(\delta)$ is chosen such that

$$\alpha(\delta) \to 0 \text{ and } \delta^2/\alpha(\delta) \to 0 \text{ for } \delta \to 0.$$

The very important problem of option pricing in financial mathematics can be solved using Tikhonov regularization.[8] Another technique by which we can solve inverse problems is minimum norm solutions, which is briefly discussed below:

Minimum norm solution: Consider the inverse problem $d = G(m)$, where $G : X \to Y$ is a bounded linear operator between the two normed spaces X and Y. For $\delta > 0$ and $d \in Y$ an element $m_0 \in X$ is called a minimum norm solution of the problem with discrepancy δ, if $\| Gm_0 - d \| \leq \delta$ and

$$\| m_0 \| = \inf\{\| m \| : \| Gm - d \| \leq \delta\}.$$

The theorem which motivates the above technique to solve inverse problems is the following:

Theorem 3.1.

Minimum Norm Theorem:_Let X and Y be two Hilbert spaces. If $G : X \to Y$ has dense range in Y, then for each $d \in Y$ there is a unique minimum norm solution m_0 of $d = G(m)$ with discrepancy δ. The minimum norm solution m_0 can be calculated by_

$$m_0 = (\alpha I + G^*G)^{-1}G^*d,$$

where α is a zero of the function

$$H(\alpha) := \| (\alpha I + G^*G)^*d - d \|^2 - \delta^2.$$

Note: For an injective linear operator G, minimum norm solutions defines a regular strategy for the choice of the regularization parameter α for Tikhonov regularization.

4. Different approaches to solve inverse problems

4.1. *Truncated singular value decomposition*

TSVD (Truncated singular value decomposition) is the most frequently used method for the solution of linear ill-posed least square problems. In this method, we seek solution $\tilde{m}$ of linear inverse problem in Eq. (1) which minimize the value of $\| d - Gm \|^2$ in the least square sense. Suppose we compute the singular value decomposition of G, i.e. we find the left singular vectors $\{u_k\}_{k=1}^{m}$, right singular vectors $\{v_k\}_{k=1}^{n}$ and the singular values λ_k such that

$$G = \sum_{k=1}^{r} \lambda_k u_k v_k^T .$$

Since $\{u_k\}_{k=1}^{m}$ form a basis of data space, we may write the data d as a linear combination:

$$d = \sum_{k=1}^{m} u_k (u_k^T d).$$

Then, given any m, we have

$$\begin{aligned} \| d - Gm \|^2 &= \| \sum_{k=1}^{m} u_k (u_k^T d) - \sum_{k=1}^{r} \lambda_k u_k (v_k^T) \|^2 \\ &= \| \sum_{k=1}^{r} u_k \{u_k^T d - \lambda_k (v_k^T m)\} + \sum_{k=r+1}^{m} u_k (u_k^T d) \|^2 . \end{aligned}$$

Using the theorem of Pythagoras (since the vectors $\{u_k\}$ are orthogonal),

$$\| d - Gm \|^2 = \sum_{k=1}^{r} | u^T d - \lambda_k (v_k^T m) |^2 + \sum_{k=r+1}^{m} | u_k^T d |^2 .$$

Choosing $\tilde{m}$ so as to minimize $\| d - Gm \|^2$ is now straight forward. The second term on the right hand side is the square of the perpendicular distance of d to the image of G, and is completely unaffected by the choice of m. The first term on the right hand side can be reduced to zero (its minimum possible value) by choosing $\tilde{m}$ such that

$$v_k^T \tilde{m} = u_k^T d/\lambda_k \quad for \quad k = 1, 2, ..., r.$$

Whether or not this completely determines $\tilde{m}$ depends on if $r = n$ or $r < n$. For $r = n$, we will get the unique solution.

In the theory discussed so far, we have drawn a sharp line between the eigenvalues of $G^T G$ which are non- zero and those which are zero. In practice when the eigenvalues of $G^T G$ are sorted in decreasing order, there is a smooth transition from the large eigenvalues through the small eigenvalues to tiny eigenvalues. It is important to note that the small singular values always create a problem. Let us suppose that the measured data d may be written as sum of the transformed image $G(m)$ and a noise vector n, therefore,

$$d = G(m) + n. \tag{6}$$

The vector m represent the true image and $G(m)$ is the data that would have been obtained in the absence of noise. Neither of these quantities is known in practice, but the aim of reconstruction is to find a vector $\tilde{m}$ approximating m. Now using the SVD (singular value decomposition) of the matrix G and the least-square solution which we obtained above, we will get

$$\tilde{m} = m + \sum_{k=1}^{n} (u_k^T) v_k / \lambda_k.$$

We have seen that the reconstruction $\tilde{m}$ is the sum of true image i.e. m and the terms due to noise. The error term along the direction of v_k in the image space arises from the component of noise in the direction of u_k in the data space divided by the singular value λ_k. If we now suppose that some of the singular values λ_k are small, this division will give a very large component, often completely swamping the component of m in that direction. Thus when there are small singular values, the simple SVD regularization technique can give bad reconstructions. It is better to consider small singular values as being effectively zero, and to regard the components along such directions as being free parameters which are not determined by the data. This is what we do in the method TSVD.[9,10]

Although the TVSD explained above is one of the oldest and most frequently used technique for solving inverse problems, it has still some serious drawbacks. The essential drawback of TSVD are the following (note that all the drawbacks are connected with the polynomial character of the basis functions that are used in TSVD):

(1) The continuous dependence of the approximate solution on G, is trivially given due to the finite dimension (bandlimitedness) of the restricted solution space.
(2) Whole data set available has to be used completely for the calculation of each orthogonal (Fourier) coefficient of investigated function. This means TVSD is always connected with global smoothing of available data. Moreover, in the case of large data sets numerical effort will become immense.
(3) A variation of only a few data in a small region requires the new computation of all coefficients.
(4) Due to global nature of the basis functions used in TVSD space-dependent measuring accuracies cannot be sufficiently taken into account.
(5) We can note that the basis used above is entirely defined by the operator G and ignores the specific physical nature of the problem under study. For example, for stationary operators the corresponding eigenfunctions generate a Fourier sine and cosine basis. Fourier series are appropriate for smooth spatially homogeneous functions, but do not provide a good information of the inhomogeneous signals which are smooth in one region and have rapid variations in others. Hence using TSVD, while dealing with problems where G is inhomogeneous will not give us accurate results and in most of the real world problems the operator is usually inhomogeneous.

Now if we use wavelet methods for solving inverse problems then we can theoretically obtain non-bandlimited solutions and also the localizing properties of the scaling functions and the wavelets used as basis functions resolve first four problems discussed above.

4.2. *Generalized wavelet-Galerkin or projection method*

Generalized wavelet-Galerkin or projection method to solve inverse problem (described in Eq. (6) where G is an operator from X to Y (which are assumed to bo Hilbert spaces)) is defined by family of subspaces $\{X_h \subset X_{h'} \subset X\}$ for $h' < h$ and $\{Y_h \subset Y\}$ growing in size with step width h. We determine the approximate solution m_h in the space X_h by solving

$$\langle Gm_h|v\rangle_Y = \langle d|v\rangle_Y \quad \forall \quad v \in Y_h. \tag{7}$$

The basic question is to estimate the quality of the approximation $\| m - m_h \|_X$ and to determine the optimal step width h such that this quantity becomes minimum.

Now let us fix basis for X_h and Y_h

$$X_h = \text{span}\{u_j | j \in I_h\},$$

$$Y_h = \text{span}\{v_j | j \in I_h\}.$$

If $m_h = \sum_{j \in I_h} x_j u_j$, then Eq. (7) is transformed to the following system:

$$G_h x = y, \tag{8}$$

where $x = \{x_j | j \in I_h\}$, $(G_h)_{j,k} = \langle G u_k | v_j \rangle_Y$ and $y = \{y_j | j \in I_h\}$ with $y_j = \langle d | v_j \rangle_Y$. We then obtain the corresponding regularization operator R_h

$$R_h : Y \to X,$$

$$d \mapsto R_h d = m_h.$$

Now if we suppose that space X_h has wavelet basis then by the results in[1] we get that optimal step width h_{opt} can easily calculated and the corresponding error $\| m - m_{h_{opt}} \|$ is asymptotically bounded.

It is to be noted that the Galerkin method suffers from the drawback of being unstable in many cases (e.g. in case of inhomogeneous data. consider[11] for details). In response to above limitation of Galerkin method and fifth limitation of TSVD wavelet-domain linear inversion and wavelet-vaguelette decomposition methods has been proposed which are discussed below:

4.3. *Wavelet domain linear inversion*

Usually while solving an inverse problem, the data set encountered is inhomogeneous. Traditional approaches in linear inversion (such as TSVD, least square deconvolution or interpolation) are often based on certain homogeneity and may face difficulties while dealing with non-homogeneous data. Most of the inhomogeneous data sets can be shown to lie in the Besov function spaces and are characterized by their smoothness (differentiability). Contrary to fourier transforms, wavelets form an unconditional basis for Besov spaces, allowing for a new generation of linear inversion schemes which incorporate smoothness information of the data sets.

Besov space: A Besov space $B^s_{p,q}(R)$ is a complete quasi-normed space which is a Banach space for $1 \leq p, q \leq \infty$.

let $n = 0, 1, 2, ...$ and $s = n + \alpha$ with $0 < \alpha \leq 1$, the Besov space $B^s_{p,q}(R)$

contains all the functions f such that

$$f \in W_p^n \quad \text{and} \quad \int_0^\infty \mid W_p^2(f^{(n)}, t)/t^\alpha \mid^q dt/t < \infty,$$

where $W_p^2(f,t) = \sup_{|h| \leq t} \parallel \Delta_{h^2 f} \parallel_p$ and $\Delta_h = f(x-h) - f(x)$.
The Besov space $B_{p,q}^s(R)$ is equipped with the norm

$$\parallel f \parallel_{W_p^n(R)} + (\int_0^\infty \mid W_p^2(f^{(n)}, t)/t^\alpha \mid^q dt/t)^{1/q}.$$

Now we consider the linear inversion problem in Eq. (6) where the data set d is in some Besov space $B_{p,q}^\beta$, n is the possible noise in the experiment. The solution of the inverse problem involves minimizing:

$$\parallel n \parallel_{B_{p,q}^\beta} = \parallel d - Gm \parallel_{B_{p,q}^\beta}, \tag{9}$$

over all the possible values of m. For unique solution we provide an extra constraint to the problem, which usually comes in the form of smoothness constraint in which we minimize the norm of m (i.e. $\parallel m \parallel_{p\prime,q\prime}^{\beta'}$) in some function space. In order to obtain the solution of inverse problem we must find a vector model $\hat{m}$, which simultaneously minimizes a norm on the noise vector and the unknown model:

$$\hat{m} = min[\parallel (d - Gm) \parallel_{B_{p,q}^\beta} + \parallel m \parallel_{B_{p\prime,q\prime}^{\beta'}}] \tag{10}$$

In transforming the inverse problem to wavelet domain we change our problem (i.e. $d = G(m) + n$) to

$$d = GW^{-1}Wm + n \text{ or } d = \tilde{G}\tilde{m} + n,$$

where $\tilde{G} = GW^{-1}$ is the wavelet transform of each row of the matrix G and $\tilde{m} = Wm$, where W is discrete wavelet transform and W^{-1} is inverse discrete wavelet transform. This expresses the inverse problem in terms of wavelet coefficients. Then, we need to redefine the minimization problem in Eq. (10) in terms of the wavelet coefficients. The problem now is that Besov norm in the wavelet domain is equivalent to the norm of the function in state domain, not equal to it. Thus $A \parallel m \parallel_{B_{p,q}^\beta} \parallel \leq \parallel \tilde{m} \parallel_{B_{p,q}^\beta} \leq B \parallel m \parallel_{B_{p,q}^\beta}$ for some constants A and B. Only in case of an orthogonal transform is the energy in both the domains equal. Therefore, orthogonal wavelets (e.g. orthogonal fractional spline wavelets) are chosen .
Then,

$$\hat{m} = min[\parallel d - \tilde{G}\tilde{m} \parallel_{B_{p,q}^\beta} + \parallel \tilde{m} \parallel_{B_{p\prime,q\prime}^{\beta'}}]. \tag{11}$$

The simplest and easiest inverse methods assumes that $p = q = 2$. In this special case the Besov spaces reduce to simpler Sobolov spaces and it is much easier to perform inversion in Sobolov spaces because we can use standard least square methods. Important problem of well logging in geology can be solved using above explained method as in.[12]

4.4. *Wavelet-Vaguelette decomposition*

Term vaguelette was introduced by Meyer in[13] to describe a collection of functions which are wavelet-like. Wavelet Vaguelette decomposition method was proposed by Donoho in.[14] It is based on the expansion of the unknown function m as wavelet series. Wavelet series is generated by translations and dilations of a single function ψ called the mother wavelet. In wavelet vaguelette decomposition method we write the wavelet expansion of the unknown function m as:

$$m = \sum_j \sum_k \langle m, \psi_k^j \rangle \psi_k^j,$$

where $\psi_k^j = 2^{j/2}\psi(2^j x - k)$. Let $\Psi_k^j = G\psi_k^j$, for some operators G there exist constants $\tilde{\beta}_k^j$ such that the functions $v_k^j = \Psi_k^j / \tilde{\beta}_k^j$ forms a Riesz basis in L_2 norm, that is there exist two constants $0 < A \leq B < \infty$ such that

$$A \sum_j \sum_k (c_k^j)^2 \leq \| \sum c_k^j v_k^j \|^2 \leq B \sum_j \sum_k (c_k^j)^2, \tag{12}$$

for all square summable sequences $\{c_k^j\}$. The functions v_k^j are called vaguelettes. Operators satisfying Eq. (12) include integration, fractional integration and Radon transformation. If we choose the basis ψ_k^j properly, then any function g in the range of G can be written as

$$g = \sum_j \sum_k \langle g, u_k^j \rangle v_k^j,$$

where $\{u_k^j\}$ is dual vaguelette basis satisfying $G^* u_k^j = \tilde{\beta}_k^j \psi_k^j$. The dual basis $\{u_k^j\}$ and $\{v_k^j\}$ are orthogonal i.e. $\langle v_k^j, u_m^l \rangle = \delta_{jl}\delta_{km}$. Thus, the signal $G(m)$ is expanded in vaguelette series as:

$$G(m) = \sum_j \sum_k \langle Gm, u_k^j \rangle v_k^j,$$

and then recover the original function m as

$$\begin{aligned} m &= \sum_j \sum_k \langle Gm, u_k^j \rangle (\tilde{\beta}_k^j)^{-1} \psi_k^j \\ &= \sum_j \sum_k \langle Gm, \tilde{\Psi}_k^j \rangle \psi_k^j, \end{aligned} \tag{13}$$

where $\tilde{\Psi}_k^j = u_k^j / \beta_k^j$ and hence $G^* \tilde{\Psi}_k^j = \psi_k^j$. The formula (13) is main formula for wavelet-vaguelette decomposition method. In the case of noisy data, the observed signal $y = G(m) + n$ is expanded in terms of vaguelettes, with coefficients $\hat{b}_k^j = \langle y, \Psi_k^j \rangle$ which satisfy

$$\hat{b}_k^j = b_k^j + w_k^j, \tag{14}$$

where $b_k^j = \langle Gm, \tilde{\Psi}_k^j \rangle$ are the noiseless vaguelette coefficients from Eq. (12) and $w_k^j = \langle \epsilon, \tilde{\Psi}_k^j \rangle$ are the vaguelette decomposition of white noise.

Using the central limit theorem of probability theory, from Eq. (14), we have $\tilde{b}_k^j \sim N(b_k^j, \sigma_0^2 \parallel \tilde{\Psi}_k^j \parallel^2)$ for some σ_0^2. Construct rescaled coefficients $(\hat{b}_k^j)^0 = \hat{b}_k^j / \parallel \tilde{\Psi}_k^j \parallel$, which all have same variance σ_0^2. Now we will apply threshold on $(\hat{b}_k^j)^0$, either using the soft threshold function

$$\delta_\lambda(x) = sign(x)(\mid x \mid - \lambda)_+,$$

or using the hard threshold function

$$\delta_\lambda(x) = \begin{cases} x & : \quad \mid x \mid > \lambda \\ 0 & : \quad otherwise \end{cases},$$

for some threshold value $\lambda \geq 0$. Note that the thresholding used above is non-linear as compared to thresholding in TSVD (linear weighting of eigenvalues). Also note that there are thresholding techniques which can exploit some features of representation of the function in a particular space (e.g. sparsity) as in.[15] Mapping the threshold coefficients back into the wavelet expansion in original space yields the resulting wavelet-vaguelette decomposition estimator $\hat{m}_\lambda^{WVD}$:

$$\hat{m}_\lambda^{WVD} = \sum_j \sum_k \parallel \tilde{\Psi}_k^j \parallel \delta_\lambda (\hat{b}_k^j)^0 \psi_k^j. \tag{15}$$

In a matrix formulation the method proceeds as follows: an orthogonal wavelet transform matrix W is constructed, where each row is a discrete wavelet. G then operates on each individual wavelet to produce what is called a vaguelette:

$$GW^T = V^T \Gamma. \tag{16}$$

Each column in the matrix V^T is a discrete vaguelette and is normalized to unit energy. Each normalization factor has been put on the diagonal matrix Γ. Moving W to the other side of Eq. (16), the wavelet-vaguelette decomposition (WVD) obtained is:

$$G = V^T \Gamma W. \tag{17}$$

Entries of Γ are called quasi-singular values. W and Γ are always invertible, but V^T is invertible only if G is invertible. If V^T is invertible, its inverse is called U.

Now WVD is applied to the solution of a linear inverse problem. The problem is Eq. (6), where m is a function which is to be estimated, n is the noise and d is the noise contaminated data. A solution to this problem is obtained by minimizing

$$\| d - Gm \|_{L_2} . \tag{18}$$

If G is rectangular and $G^T G$ is invertible,this leads to

$$m_{est} = (G^T G)^{-1} d. \tag{19}$$

If G is square and invertible we have

$$m_{est} = G^{-1} d. \tag{20}$$

Substituting Eq. (6) in Eq. (20) we obtain

$$m_{est} = m + G^{-1} n. \tag{21}$$

From the Eq. (21), it is observed that the solution is contaminated with colored noise ($i.e. G^{-1} n$), which can be very large while dealing with ill posed problems. The traditional way to solve such problems is regularization where we redefine the minimization problem in Eq. (18) by adding one of the following constraints

$$\min_m \| m \|_{L_2} \quad \text{or} \quad \min_m \| Lm \|_{L_2} \tag{22}$$

where L is usually a differential operator. Assuming that G is invertible and square, its inverse via WVD is represented as

$$G^{-1} = W^T \Gamma^{-1} U. \tag{23}$$

Substituting Eq. (23) into Eq. (20)

$$m_{est} = W^T \Gamma^{-1} U d. \tag{24}$$

Noting that $U = V^{-1} = \Gamma W K^{-1}$, substituting in Eq. (24)

$$m_{est} = W^T \Gamma^{-1} \Gamma W K^{-1} d. \tag{25}$$

Equation (25) is actually same as Eq. (20); the first four matrices cancel themselves. A non-linear thresholding operator, Θ_T is defined as

$$\Theta_T[\cdot] = \begin{cases} m_j & : \ | m_j | > T \\ 0 & : \ | m_j | \leq T. \end{cases} \tag{26}$$

Then, Θ_T is substituted into the Eq. (25):

$$m_{est} = W^T \Gamma^{-1} \Theta_T [\Gamma W K^{-1} d], \tag{27}$$

where T (universal threshold criterion)$= \sigma_n \sqrt{2ln(N)}$ and σ_n is assumed standard deviation of noise and N is the number of coefficients in m. The idea here is this: solving Eq. (20) leads to noise contaminated solution. Transforming this solution to wavelet domain tends to isolate good signal into few large valued, isolated coefficients, while the noise tends to be spread around equally with smaller energy. Thus thresholding the small wavelet coefficients will tend to remove the noise and leave the coherent features untouched. Travel time inversion problem (a fundamental problem in mathematical geophysics) can be solved using the above explained method as in.[16] Couette inverse problem is solved using this wavelet vaguelette method in.[17]

4.5. *The vaguelette-wavelet decomposition method*

A natural alternative to wavelet-vaguelette decomposition is vaguelette-wavelet decomposition. In vaguelette-wavelet decomposition[18] we expand the observed data d in wavelet space, threshold the resulting coefficients and then map back by G^{-1} to obtain an estimate of m in term of vaguelette series. Hence it is $G(m)$ rather than m which is expanded in wavelet series. Suppose we have the wavelet expansion

$$G(m) = \sum_j \sum_k d_k^j \psi_k^j, \tag{28}$$

where ψ_k^j for all j and k are in the range of G. Assume the existence of β_k^j such that Eq. (12) holds for $v_k^j = G^{-1}\psi_k^j/\beta_k^j$. Then m is recovered from Eq. (28) by expanding in the vaguelette series as follows:

$$\begin{aligned} m &= \sum_j \sum_k \langle Gm, \psi_k^j \rangle \beta_k^j v_k^j \\ &= \sum_j \sum_k \langle Gm, \psi_k^j \rangle \Psi_k^j, \end{aligned} \tag{29}$$

where $\Psi_k^j = G^{-1}\psi_k^j$. As in wavelet-vaguelette decomposition the wavelet coefficients of a noisy signal y, $\hat{d}_k^j = \langle y, \psi_k^j \rangle$, are contaminated by noise

$$\hat{d}_k^j = d_k^j + w_k^j,$$

where $w_k^j = \langle \epsilon, \psi_k^j \rangle$ are the coefficients of the wavelet decomposition of white noise, and hence are themselves are white noise; note that this is not the case in the corresponding vaguelette coefficients $\hat{b}_k^j$ in Eq. (14) used in wavelet-vaguelette decomposition. Therefore $\hat{d}_k^j$ need to be denoised, for example by thresholding. The resulting vaguelette-wavelet decomposition estimator $\hat{m}_\lambda^{VWD}$ will be then

$$\hat{m}_\lambda^{VWD} = \sum_j \sum_k \delta_\lambda(\langle y, \psi_k^j \rangle)\Psi_k^j,$$

where $\delta_\lambda(\cdot)$ is soft or hard thresholding operator.

5. Recent progress and future challenges in inverse problems

- The function m to be recovered from the inverse problem $d = G(m)$ is usually assumed to be smooth apart from at the edge. Traditional methods for solving inverse problems in the presence of edges behave poorly. Qualitatively the reconstructions are blurred at the edges. Curvelets which generalize wavelets in the sense that these are localized in orientation also, can produce methods to solve inverse problems in the presence of edges.
- If the operator G in Eq. (1) is such that high powers of G has low numerical rank then we can construct multiresolution analysis which will yield scaling functions and wavelets (called diffusion wavelets[6]) on domains, manifolds, graphs and other general classes of metric spaces. Then transforming the above discussed wavelet methods to diffusion wavelet space can provide us new methods which can work well in the situations where former fails (e.g. in presence of edges or in case of inhomogeneous data.).
- Solving the inverse problem of electrocardiography (problem of describing the electrochemical activity of each cell in the heart based on body surface electrocardiograms (ECGs)), is difficult because of the non-unique relationship between the true intra-cardiac sources and the remote observations-the same set of measurements could result from more than one source configuration. So we need to design inverse problem formulations that have unique source models without much loss of generality and applicability.

References

1. V. Dicken and P. Maaß, *Inverse Ill-Posed Probl.* **4**, 203 (1996).
2. M. Mehra, *AIP Conference Proceedings* **1146**, 241 (2009).
3. I. Daubechies, *Ten Lectures on Wavelets* (SIAM, Philadelphia, 1992).

4. D. L. Donoho, *Tech. Rep. 408, Department of Mathematics, Standford University* (1992).
5. W. Sweldens, *SIAM J. Math. Anal.* **29**, 511 (1998).
6. R. R. Coifman and M. Maggioni, *Appl. Comp. Harm. Anal.* **21**, 53 (2006).
7. I. Victor, *Inverse problems for partial differential equations* (Springer, 2006).
8. H. Egger and H. W. Engl, *Inverse problems* **21**, 1027 (2005).
9. P. C. Hansen, *BIT* **27**, 534 (1987).
10. H. Egger and H. W. Engl, *SIAM Journal on Scientific and Statistical Computing* **11**, 503 (1990).
11. D. Picard and G. Kerkyacharian, *Invited article in proceedings of international congress of mathematicians: Madrid* **3**, 713 (2006).
12. F. J. H. Jonathan Kane and M. N. Toksöz, *expanded abstract in the proceedings of the 72th annual meetings of society of exploration geophysicists.* (2001).
13. Y. Meyer, *Ondelettes et operateurs II: Operateurs de Calderon-Zygmund* (Hermann, paris, 1990).
14. D. Donoho, *Appl. Comput. Harmon. Anal.* **2** (1995).
15. I. Daubechies, M. Defrise and C. D. Mol, *Comm. on pure and applied mathematics* **57**, 1413 (2004).
16. F. J. H. Jonathan Kane and M. N. Toksöz, *expanded abstract in the proceedings of the 72th annual meetings of society of exploration geophysicists.* (2001).
17. C. Ancey, *J. Rheol* **49**, 441 (2005).
18. F. Abramovich and B. W. Silverman, *Biometrika* **85**, 115 (1998).

OPTIMIZATION MODELS FOR A CLASS OF STRUCTURED STOCHASTIC GAMES

S. K. NEOGY

Indian Statistical Institute
7, S.J.S. Sansanwal Marg
New Delhi-110016, India
E. Mail: skn@isid.ac.in

SAGNIK SINHA

Jadavpur University
Kolkata-700032, India
E. Mail: sagnik62@yahoo.co.in

A. K. DAS and A. GUPTA

Indian Statistical Institute
203, B. T. Road
Kolkata-700108, India
E. Mail: akdas@isical.ac.in, agupta@isical.ac.in

Mangasarian and Stone[9] observed that the problem of computation of equilibrium points of bimatrix games is equivalent to solving a certain quadratic programming problem with linear constraints. We extend this approach of Mangasarian and Stone[9] for generalized bimatrix game problem and SC/AR-AT mixture class of stochastic game problem. We establish an alternative necessary and sufficient condition for the existence of optimal stationary strategies for a mixture class of zero-sum stochastic game in which the set of states are partitioned into sets S_1, S_2 and S_3 so that the law of motion is controlled by Player I alone when the game is played in S_1, Player II alone when the game is played in S_2 and in S_3 the reward and transition probabilities are additive. We formulate and discuss about the computation of value vector and optimal stationary strategies for SC/AR-AT mixture class of stochastic game as an optimization model for both discounted and undiscounted case.

Keywords: Optimization model; Generalized bimatrix game, Structured stochastic game; Switching control (SC) property; AR-AT property; SC/AR-AT mixture.

1. Introduction

Optimization model arises naturally for games, economics, engineering and management decision making problems. Mangasarian and Stone [9] constructed a quadratic program which has a global maximum of zero, and showed that the optimal solution of this quadratic program form Nash equilibrium points of the bimatrix game. Rothblum [20] first formulated stochastic games as an optimization model. Stochastic games concentrate on decision situations where at different time moments the players have to make a choice. Stochastic games are motivated by many practical applications and the potential future applications. A few of them are Pollution Game, Fishery Game, Inspection Game and Salary Negotiations Game. Especially these sorts of applications have motivated the study of algorithms for different classes of stochastic games with special structure.

A bimatrix game is a non-cooperative nonzero-sum two person game with payoff matries $A \in R^{m\times n}$ and $B \in R^{m\times n}$ to player I and player II respectively in which each player has a finite number pure strategies. A mixed strategy for player I is a probability vector $x \in R^m$ whose i^{th} component x_i represents the probability of choosing pure strategy i where $x_i \geq 0$ for $i = 1, \ldots, m$ and $\sum_{i=1}^{m} x_i = 1$. Similarly, a mixed strategy for player II is a probability vector $y \in R^n$. Let α and β are the expected payoffs of Player I and Player II respectively. An equilibrium point [8] for such a game is a 4-tuple $(x^*, y^*, \alpha^*, \beta^*)$ that satisfies the following relation

$$\alpha^* = {x^*}^t A y^* = \max_x \{x^t A y^*, \mid e_m^t x = 1, x \geq 0\}$$

$$\beta^* = {x^*}^t B y^* = \max_y \{{x^*}^t B y \mid e_n^t y = 1, y \geq 0\}$$

where e_m and e_n are $m \times 1$ and $n \times 1$ vector of ones respectively.

Mangasarian and Stone [9] showed that the problem of computation of equilibrium points of bimatrix games is equivalent to solving a certain quadratic programming problem with linear constraints. The equivalence theorem is stated as follows.

Theorem 1.1. *A necessary and sufficient condition that $(x^*, y^*, \alpha^*, \beta^*)$ be an equilibrium point is that it is a solution of the programming problem*

$$maximize_{x,y,\alpha,\beta} \ \{x^t(A+B)y - \alpha - \beta \mid (x,\beta) \in S, \ (y,\alpha) \in T\}$$

where S and T are the convex polyhedral sets

$$S = \{(x,\beta) \mid B^t x - \beta e_n \leq 0, \ e_m^t x = 1, \ x \geq 0\}.$$

$$T = \{(y, \alpha) \mid Ay - \alpha e_m \leq 0,\ e_n^t y = 1,\ y \geq 0\}.$$

In Section 2, we consider the generalized bimatrix game and formulate the problem of computing a Nash equilibrium point as an optimization problem in the spirit of Mangasarian and Stone [9]. In Section 3, we present some preliminaries on stochastic games which are needed for subsequent discussions. In Section 3.1, we present a necessary and sufficient condition for the existence of optimal stationary strategies for discounted SC/AR-AT *mixture class* of stochastic game in terms of finding a global minimum to a linearly constrained optimization problem with nonlinear objective function. In Section 3.2, we show that for undiscounted SC/AR-AT *mixture class* of stochastic game optimal stationary strategies exist if and only if the objective function of the optimization problem can be driven to zero and, when this occurs, a pair of optimal stationary strategies and the value vector are among the variables of the given optimization problem. Finally, in Section 4, we present concluding remarks and areas of further research.

2. Optimization Model for Generalized Bimatrix Game Problem

We require the concept of a vertical block matrix. We say that an $m \times k$ matrix N with the partitioned form $N = \left[N^1 \ldots N^k \right]^t$ is a vertical block matrix of type $(m_1, m_2, \ldots, m_k)$ if N^j is of order $m_j \times k$, $1 \leq j \leq k$ and $\sum_{j=1}^{k} m_j = m$. Let N be a vertical block matrix of type $(m_1, m_2, \ldots, m_k)$. A submatrix of size k of N is called a *representative submatrix* if its jth row is drawn from the jth block N^j of N.

Now we extend the approach of Mangasarian and Stone [9] for generalized bimatrix game. Gowda and Sznajder [7] introduced a generalization of the bimatrix game presented in previous section. This generalized version of the bimatrix game is described as follows:

Let $\mathcal{A}$ and $\mathcal{B}$ be two given finite sets of matrices, $\mathcal{A}$ containing s matrices and $\mathcal{B}$ containing r matrices, each of order $m \times n$. Player I forms his payoff matrix whose i^{th} row is chosen as the ith row of some $A \in \mathcal{A}$ and then plays his choice of a mixed strategy over $\{1, 2, \ldots, m\}$. Similarly, player II (the column player) forms his payoff matrix whose j^{th} column is chosen by him as the jth column of some $B \in \mathcal{B}$ and then plays his choice of mixed strategy over $\{1, 2, \ldots, n\}$. The rest of the description of the game is the same as that of a bimatrix game.

Now we consider the question of computing a generalized Nash equilibrium point for the generalized bimatrix game and formulate it as an optimization model. See also [15] in this connection.

Suppose, $\mathcal{A} = \{A^p \mid p = 1, 2, \ldots, s\}$ and $\mathcal{B} = \{B^p \mid p = 1, 2, \ldots, r\}$. Consider the matrices C^j, $j = 1, 2, \ldots, m$ and D^j, $j = 1, 2, \ldots, n$ defined as follows:

$$\begin{aligned} C^j_{i\cdot} &= A^i_{j\cdot}, \ \ 1 \le i \le s \\ D^j_{i\cdot} &= (B^i)^t_{j\cdot}, \ \ 1 \le i \le r. \end{aligned}$$

Without loss of generality, we may assume that each A^p, $p = 1, 2, \ldots, s$ and each B^p, $p = 1, 2, \ldots, r$ are positive matrices. Hence each C^j, $j = 1, 2, \ldots, m$ and each D^j, $j = 1, 2, \ldots, n$ are positive matrices.

$$\text{Let } C = \begin{bmatrix} C^1 \\ C^2 \\ \vdots \\ C^m \end{bmatrix} \text{ and } D = \begin{bmatrix} D^1 \\ D^2 \\ \vdots \\ D^n \end{bmatrix}$$

where each C^j is of order $s \times n$ and each D^j is of order $r \times m$ and by our assumption $C > 0$, $D > 0$. Note that C is a vertical block matrix of type $(s, \ldots, s)$ and D is a vertical block matrix of type $(r, \ldots, r)$ where the number of blocks in C is m and the number of blocks in D is n.

Given vertical block matrices C of type $(s, \ldots, s)$ and D of type $(r, \ldots, r)$ there exist representative matrices C_R and D_R so that a generalized Nash equilibrium point of a generalized bimatrix game is same as the Nash equilibrium point of a bimatrix game obatained using the representative submatrces C_R and D_R.

Now, we construct equivalent matrices $\hat{C}$ by copying $C_{\cdot k}$, r times for $k = 1, \ldots, n$ and $\hat{D}$ by copying $D_{\cdot k}$, s times for $k = 1, \ldots, m$. Note that the order of the equivalent matrix $\hat{C}$ is $ms \times nr$ and the order of the equivalent matrix $\hat{D}$ is $nr \times ms$. For convenience we denote $\bar{C} = \hat{C}$ and $\bar{D} = \hat{D}^t$.

We note that a generalized Nash equilibrium point as considered by Gowda and Sznajder [7] can be computed by obtaining a solution from the optimization problem constructed similarly as in Theorem 1.1 using equivalent matrices $\bar{C}$ and $\bar{D}$ as described above. This result is stated in the following theorem.

Theorem 2.1. *A generalized Nash equilibrium point* $(x^*, y^*, \alpha^*, \beta^*)$ *of the generalized bimatrix game can be computed from the solution of the opti-*

mization problem

$$\text{maximize}_{x,y,\alpha,\beta}\ \{x^t(\bar{C}+\bar{D})y-\alpha-\beta \mid (x,\beta)\in S,\ (y,\alpha)\in T\}$$

where S and T are the convex polyhedral sets

$$S=\{(x,\beta)\mid \bar{D}^t x-\beta e_{nr}\leq 0,\ e^t_{ms}x=1,\ x\geq 0\}.$$

$$T=\{(y,\alpha)\mid \bar{C}y-\alpha e_{ms}\leq 0,\ e^t_{nr}y=1,\ y\geq 0\}.$$

where $\bar{C}$ and $\bar{D}$ are the equivalent matrices obtained from vertical block matrices C and D as described above.

From the solution of the game problem using the optimization model in Theorem 2.1, we obtain the probability vectors x and y where

$$x_j=\sum_{i=1}^{s}x^i_j,\ j=1,\ldots,m$$

$$y_j=\sum_{i=1}^{r}y^i_j,\ j=1,\ldots,n.$$

3. Structured Stochastic Game

In 1953, Shapley [24] introduced stochastic game and established the existence of value and optimal stationary strategies for discounted stochastic games. Gillete [6] studied the undiscounted case or limiting average payoff case.

A two-player finite state/action space zero-sum stochastic game is defined by the following objects.

(1) A state space $S=\{1,2,\ldots,N\}$.
(2) For each $s\in S$, finite action sets $A(s)=\{1,2,\ldots,m_s\}$ for Player I and $B(s)=\{1,2,\ldots,n_s\}$ for Player II.
(3) A reward law $R(s)$ for $s\in S$ where $R(s)=[r(s,i,j)]$ is an $m_s\times n_s$ matrix whose $(i,j)^{th}$ entry denotes the payoff from Player II to Player I corresponding to the choices of action $i\in A(s)$, $j\in B(s)$ by Player I and Player II respectively.
(4) A transition law $q=(q_{ij}(s,s'):\ (s,s')\in S\times S,\ i\in A(s),\ j\in B(s))$, where $q_{ij}(s,s')$ denotes the probability of a transition from state s to state s' given that Player I and Player II choose actions $i\in A(s), j\in B(s)$ respectively.

The game is played in stages $t = 0, 1, 2, \ldots$ At some stage t, the players find themselves in a state $s \in S$ and independently choose actions $i \in A(s), j \in B(s)$. Player II pays Player I an amount $r(s, i, j)$ and at stage $(t+1)$, the new state is s' with probability $q_{ij}(s, s')$. Play continues at this new state.

The players guide the game via strategies and in general, strategies can depend on complete histories of the game until the current stage. However in this paper, we are concerned with the simpler class of *stationary strategies* which depend only on the current state s and not on stages. So for Player I, a stationary strategy

$$f \in F_s = \{f_i(s) \mid s \in S,\ i \in A(s),\ f_i(s) \geq 0,\ \sum_{i \in A(s)} f_i(s) = 1\}$$

indicates that the action $i \in A(s)$ should be chosen by Player I with probability $f_i(s)$ when the game is in state s.

Similarly for Player II, a stationary strategy

$$g \in G_s = \{g_j(s) \mid s \in S,\ j \in B(s),\ g_j(s) \geq 0,\ \sum_{j \in B(s)} g_j(s) = 1\}$$

indicates that the action $j \in B(s)$ should be chosen by Player II with probability $g_j(s)$ when the game is in state s.

Here F_s and G_s will denote the set of all stationary strategies for Player I and Player II, respectively. Let $f(s)$ and $g(s)$ are the m_s and n_s dimensional column vector, respectively.

Fixed stationary strategies f and g induce a Markov chain on S with transition matrix $P(f, g)$ whose $(s, s')^{th}$ entry is given by

$$P_{ss'}(f, g) = \sum_{i \in A(s)} \sum_{j \in B(s)} q_{ij}(s, s') f_i(s) g_j(s)$$

and the expected current reward vector $r(f, g)$ has entries defined by

$$r_s(f, g) = \sum_{i \in A(s)} \sum_{j \in B(s)} r(s, i, j) f_i(s) g_j(s) = f(s) R(s) g(s).$$

With fixed general strategies f, g and an initial state s, the stream of expected payoff to Player I at stage t, denoted by $v_s^t(f, g)$, $t = 0, 1, 2, \ldots$ is well defined and the resulting discounted and undiscounted payoffs are

$$\phi_s^\beta(f, g) = \sum_{t=0}^{\infty} \beta^t v_s^t(f, g) \text{ for a } \beta \in (0, 1)$$

and

$$\phi_s(f,g) = \liminf_{T\uparrow\infty} \frac{1}{T+1}\sum_{t=0}^{T} v_s^t(f,g).$$

A pair of strategies (f^*, g^*) is optimal for Player I and Player II in the undiscounted game if for all $s \in S$

$$\phi_s(f, g^*) \leq \phi_s(f^*, g^*) = v_s^* \leq \phi_s(f^*, g),$$

for any strategies f and g of Player I and Player II respectively. The number v_s^* is called the *value of the game* starting in state s and $v^* = (v_1^*, v_2^*, \ldots, v_N^*)$ is called the *value vector*. The definition for discounted case is similar. For theory and applications of stochastic games see [5].

Bewley and Kohlberg [1] obtained a very general set of sufficient conditions for a game to possess optimal stationary strategies. However, the conditions are not easily verifiable. Filar and Schultz [4] observed that the problem of finding a computationally feasible characterization of and solution methods for this class of games remains a major open problem in the theory of finite state/action stochastic games.

In the literature of stochastic game, many authors have considered stochastic games with special structures in which one can hope for finite step algorithms. We will refer to these zero-sum stochastic games with special structure collectively as the class of *structured stochastic games.* We will first describe below some known classes of games which possess ordered field property.

- **Single controller stochastic games :** In the case where player II is *single controller* this means $q(s' \mid s, i, j) = q(s' \mid s, j)\ \forall\ i, j, s, s'$.
- **Switching controlled games :** In a switching control stochastic game the law of motion is controlled by Player I alone when the game is played in a certain subset of states and Player II alone when the game is played in other states. In other words, a switching control game is a stochastic game in which the set of states are partitioned into sets S_1 and S_2 where the transition function is given by

$$q_{i,j}(s,s') = \begin{cases} q_i(s,s'), \text{ for } s' \in S, s \in S_1, i \in A(s) \text{ and } \forall j \in B(s) \\ q_j(s,s'), \text{ for } s' \in S, s \in S_2, j \in B(s) \text{ and } \forall i \in A(s) \end{cases}$$

- **AR-AT games :** A stochastic game is said to be an *Additive Reward-Additive Transition game (AR-AT game)* if

  the reward (i) $r(s,i,j) = r_i^1(s) + r_j^2(s)$ for $i \in A(s)$, $j \in B(s)$, $s \in S$

and the transition probabilities

$$\text{(ii)}\ \ q_{i,j}(s,s') = q_i^1(s,s') + q_j^2(s,s') \ \text{ for } \ i \in A(s), j \in B(s),\ (s,s') \in S \times S.$$

Filar [2] generalized the optimization model of Mangasarian and Stone [9] for two person, general sum, single controller stochastic game by showing that an optimal solution of appropriately constructed quadratic program provides a Nash equilibrium point. This generalization of Filar [2] include as special cases the known quadratic/linear programming formulations of bimatrix games, matrix games, Markovian decision processes and single controller zersum stochastic games. The model proposed by Filar [2] apply to both discounted and limiting average criteria.

The class of switching control (SC) stochastic games was introduced by Filar [3]. Even though the transition structure is a natural generalization of the single control game but from the algorithmic point of view this class of games appear to be more difficult. The game structure was used to develop a finite step algorithm in [25] but that algorithm requires solving a large number of single control stochastic games. Mohan, Neogy and Parthasarathy [10,?] formulated a single control game as solving a single linear complementarity problem and proved that Lemke's algorithm can solve such an LCP. Mohan and Raghavan [12] proposed an algorithm for discounted switching control games which is based on two linear programs. Schultz [23] formulated the discounted switching control game as a linear complementarity problem.

AR-AT games have been studied in the literature earlier by Raghavan, Tijs and Vrieze [19]. Both the discounted and the limiting average criterion of evaluation of strategies have been considered. It is known, for example, that for a β-discounted zero-sum AR-AT game, the value exists and both players have stationary optimal strategies, which may also be taken as pure strategies. As already mentioned earlier, in general, it is difficult to find a pair of equilibrium (optimal strategies) strategies. See [14], [16] and the excellent survey paper by Raghavan and Filar [18] and Mohan, Neogy and Parthasarathy [13].

Sinha [21,22] consider the mixture of the structured classes and studies the ordered field property. To be more specific, one such case is the mixture of AR-AT and the switching controller stochastic games whose data satisfy the AR-AT conditions in some state and the switching control conditions in the remaining state.

In this paper we consider the following generalization involving two classes of stochastic games in which the state space S is the union of 3 disjoint subsets S_1, S_2 and S_3 such that the law of transition is controlled by Player-I in S_1 and player -II in S_2 and all the state in S_3 of the game has AR-AT state. More specifically, a zero-sum stochastic game is in SC/AR-AT *mixture class* if

(i). $S = S_1 \cup S_2 \cup S_3,\ S_i \cap S_j = \emptyset\ \forall\ i \neq j$

(ii). $q_{i,j}(s,s') = q_i(s,s')$, for $s' \in S, s \in S_1, i \in A(s)$ and $\forall j \in B(s)$.

(iii). $q_{i,j}(s,s') = q_j(s,s')$, for $s' \in S, s \in S_2, j \in B(s)$ and $\forall i \in A(s)$

(iv). the reward $r(s,i,j) = r_i^1(s) + r_j^2(s)$ for $i \in A(s)$, $j \in B(s)$, $s \in S_3$ and the transition probabilities $q_{i,j}(s,s') = q_i^1(s,s') + q_j^2(s,s')$ for $i \in A(s), j \in B(s)$, $(s,s') \in S_3 \times S$.

Sinha [22] gives a nonconstructive proof to show that the above SC/AR-AT mixture class of game has ordered field property and raises the question that whether a finite step algorithm can be developed in SC/AR-AT mixtures. Neogy, Das, Sinha and Gupta [17] formulate the problem of computing the value vector v_s^β and optimal stationary strategies $f^\beta(s)$ for Player I and $g^\beta(s)$ for Player II for the class of discounted stochastic game with SC/AR-AT mixture as a linear complementarity problem and the class of undiscounted stochastic game with SC/AR-AT mixture is presented as a vertical linear complementarity problem. This complementarity formulation gives an alternative proof of the ordered field property.

We require the following result from Schultz [23] to prove our main result for discounted case in the next section.

Theorem 3.1. *([23][Theorem 1.1]) A β-discounted zero-sum stochastic game has values v_s^β and optimal stationary strategies f^β for Player I and g^β for Player II if and only if there exists a solution $(v^\beta, f^\beta, g^\beta)$ that solves the following nonlinear system SYS1.*

SYS1: *Find $(v^\beta, f^\beta, g^\beta)$ such that*

$$v_s^\beta - \beta \sum_{s' \in S} v_{s'}^\beta \sum_{j \in B(s)} q_{ij}(s,s') g_j^\beta(s) - [R(s)g^\beta(s)]_i \geq 0,\ i \in A(s), s \in S \tag{1}$$

$$-v_s^\beta + \beta \sum_{s' \in S} v_{s'}^\beta \sum_{i \in A(s)} q_{ij}(s,s') f_i^\beta(s) + [f^\beta(s)R(s)]_j \geq 0,\ j \in B(s), s \in S \tag{2}$$

Corollary 3.1. *If $(v^\beta, f^\beta, g^\beta)$ satisfies (1) and (2) then*

$$v_s^\beta = \beta[P(f^\beta, g^\beta)v^\beta]_s + r_s(f^\beta, g^\beta) \tag{3}$$

We require the following definition and results established by Filar and Schultz [4] to prove our subsequent results for undiscounted case.

Definition 3.1. A pair of optimal stationary strategies (f^*, g^*) for an undiscounted stochastic game is *asymptotically stable* if there exist a $\beta_0 \in (0,1)$ and stationary strategy pair (f^β, g^β) optimal in the β-discounted stochastic game for each $\beta \in (\beta_0, 1)$ such that

(i) $\lim_{\beta\uparrow 1} f^\beta = f^*$, $\lim_{\beta\uparrow 1} g^\beta = g^*$

(ii) for all $\beta \in (\beta_0, 1)$, $r(f^\beta, g^\beta) = r(f^*, g^*)$, $P(f, g^\beta) = P(f, g^*)$ for $f \in F_s$ and $P(f^\beta, g) = P(f^*, g)$ for $g \in G_s$ where $P(f,g)$ is the transition matrix and $r(f,g)$ is the current expected reward vector which are defined earlier.

Theorem 3.2. *([4][Theorem 2.1]) An undiscounted stochastic game possesses value vector v^* and optimal stationary strategies f^* for Player I and g^* for Player II if and only if there exists a solution $(v^*, t^*, u^*, f^*, g^*)$ with t^*, $u^* \in R^{|S|}$ to the following nonlinear system SYS2a.*
SYS2a: *Find (v, t, u, f, g) where $v, t, u \in R^{|S|}$, $f \in F_S$ and $g \in G_S$ such that*

$$v_s - \sum_{s' \in S} v_{s'} \sum_{j \in B(s)} q_{ij}(s, s') g_j(s) \geq 0, \; i \in A(s), s \in S \tag{4}$$

$$v_s + t_s - \sum_{s' \in S} t_{s'} \sum_{j \in B(s)} q_{ij}(s, s') g_j(s) - [R(s) g(s)]_i \geq 0, \; i \in A(s), s \in S \tag{5}$$

$$-v_s + \sum_{s' \in S} v_{s'} \sum_{i \in A(s)} q_{ij}(s, s') f_i(s) \geq 0, \; j \in B(s), s \in S \tag{6}$$

$$-v_s - u_s + \sum_{s' \in S} u_{s'} \sum_{i \in A(s)} q_{ij}(s, s') f_i(s) + [f(s) R(s)]_j \geq 0, \; j \in B(s), s \in S \tag{7}$$

Theorem 3.3. *([4][Theorem 2.2]) If a stochastic game possesses asymptotically stable stationary optimal strategies then feasibility of the nonlinear system (SYS2b) is both necessary and sufficient for existence of a stationary optimal solution.*

SYS2b: *Find* (v,t,f,g) *where* $v,t \in R^{|S|}$, $f \in F_S$ *and* $g \in G_S$ *such that* (4),(5),(6) *are satisfied and*

$$-v_s - t_s + \sum_{s' \in S} t_{s'} \sum_{i \in A(s)} q_{ij}(s,s') f_i(s) + [f(s)R(s)]_j \geq 0, \; j \in B(s), s \in S \tag{8}$$

3.1. *Discounted Zero-sum SC/AR-AT Mixture Stochastic Game*

Theorem 3.4. *A* β*-discounted zero-sum SC/AR-AT mixture stochastic game has values* v^β *where*

$$v_s^\beta = \begin{cases} v_s^\beta, & s \in S_1 \cup S_2 \\ \zeta_s^\beta + \eta_s^\beta, & s \in S_3 \end{cases}$$

and an optimal pair of stationary strategies $(f^\beta,\ g^\beta)$ *if and only if* v_s^β, $f^\beta(s)$ *and* $g^\beta(s)$ *are the solution of the following optimization problem OPT1 with global minimum objective value of zero.*
Objective function:
OPT1:

$$\min \sum_{s \in S_1} [\theta_s^\beta - \beta \sum_{s' \in S_1 \cup S_2} \sum_{i \in A(s)} v_{s'}^\beta q_i(s,s') f_i(s)$$

$$-\beta \sum_{s' \in S_3} \sum_{i \in A(s)} (\zeta_{s'}^\beta + \eta_{s'}^\beta) q_i(s,s') f_i(s)]$$

$$+ \sum_{s \in S_2} [-\theta_s^\beta + \beta \sum_{s' \in S_1 \cup S_2} \sum_{j \in B(s)} v_{s'}^\beta q_j(s,s') g_j(s)$$

$$+\beta \sum_{s' \in S_3} \sum_{j \in B(s)} (\zeta_{s'}^\beta + \eta_{s'}^\beta) q_j(s,s') g_j(s)]$$

$$+ \sum_{s \in S_3} [\eta_s^\beta - \beta \sum_{s' \in S_1 \cup S_2} \sum_{i \in A(s)} v_{s'}^\beta q_i^1(s,s') f_i(s)$$

$$-\beta \sum_{s' \in S_3} \sum_{i \in A(s)} (\zeta_{s'}^\beta + \eta_{s'}^\beta) q_i^1(s,s') f_i(s)$$

$$-\zeta_s^\beta + \beta \sum_{s'\in S_1\cup S_2} \sum_{j\in B(s)} v_{s'}^\beta q_j^2(s,s')g_j(s))$$

$$+\beta \sum_{s'\in S_3} \sum_{j\in B(s)} (\zeta_{s'}^\beta + \eta_{s'}^\beta) q_j^2(s,s')g_j(s)$$

$$-\sum_{i\in A(s)} r_i^1(s)f_i(s) + \sum_{j\in B(s)} r_j^2(s)g_j(s)]$$

Constraints:

$$v_s^\beta - \beta \sum_{s'\in S_1\cup S_2} v_{s'}^\beta q_i(s,s') - \beta \sum_{s'\in S_3} (\zeta_{s'}^\beta + \eta_{s'}^\beta) q_i(s,s')$$

$$-[R(s)g^\beta(s)]_i \geq 0,\ i \in A(s), s\in S_1 \tag{9}$$

$$v_s^\beta - \theta_s^\beta - [R(s)g^\beta(s)]_i \geq 0,\ i\in A(s)\ s\in S_2 \tag{10}$$

$$-v_s^\beta + \theta_s^\beta + [f^\beta(s)R(s)]_j \geq 0,\ j\in B(s)\ s\in S_1 \tag{11}$$

$$-v_s^\beta + \beta \sum_{s'\in S_1\cup S_2} v_{s'}^\beta q_j(s,s') + \beta \sum_{s'\in S_3} (\zeta_{s'}^\beta + \eta_{s'}^\beta) q_j(s,s')$$

$$+[f^\beta(s)R(s)]_j \geq 0,\ j\in B(s), s\in S_2 \tag{12}$$

$$-\zeta_s^\beta + \beta \sum_{s'\in S_1\cup S_2} v_{s'}^\beta q_j^2(s,s') + \beta \sum_{s'\in S_3} (\zeta_{s'}^\beta + \eta_{s'}^\beta) q_j^2(s,s')$$

$$+r_j^2(s) \geq 0,\ j\in B(s), s\in S_3 \tag{13}$$

$$\eta_s^\beta - \beta \sum_{s'\in S_1\cup S_2} v_{s'}^\beta q_i^1(s,s') - \beta \sum_{s'\in S_3} (\zeta_{s'}^\beta + \eta_{s'}^\beta) q_i^1(s,s')$$

$$-r_i^1(s) \geq 0,\ i\in A(s), s\in S_3 \tag{14}$$

$$f\in F(s),\ g\in G(s) \tag{15}$$

Proof. We prove this theorem by showing that a feasible solution to OPT1 with zero objective value is a solution of SYS1 and by Theorem 3.1, this solution solves the stochastic game with SC/AR-AT structure. Conversely, we show that any solution of SYS1 can be used to derive a solution of OPT1.

First we show that an objective function $\phi(z)$ of OPT1 with value zero must be a global minimum. Let $z = (v_s^\beta, \theta_s^\beta, f^\beta(s), g^\beta(s))$be a feasible solution of OPT1. For $s \in S_1$, constraint (9) weighted by $f(s)$ added to the constraint (11) weighted by $g(s)$ yields

$$\theta_s^\beta - \beta \sum_{s' \in S_1 \cup S_2} \sum_{i \in A(s)} v_{s'}^\beta q_i(s,s') f_i(s) - \beta \sum_{s' \in S_3} \sum_{i \in A(s)} (\zeta_{s'}^\beta + \eta_{s'}^\beta) q_i(s,s') f_i(s) \geq 0.$$

Similarly we can show that the remaining terms of $\phi(z)$ are nonnegative. Therefore, it follows that the objective function $\phi(z) \geq 0$.

Let $\hat{z} = (\hat{v}_s^\beta, \hat{\theta}_s^\beta, \hat{f}^\beta(s), \hat{g}^\beta(s))$ be a feasible solution of OPT1 with $\phi(\hat{z}) = 0$. This must be the global minimum. Since $\phi(\hat{z})$ is nonnegative term by term therefore each term of $\phi(\hat{z})$ is zero.

$$\hat{\theta}_s^\beta = \beta \sum_{s' \in S} \sum_{i \in A(s)} \hat{v}_{s'}^\beta q_i(s,s') \hat{f}_i^\beta(s), \ s \in S_1 \tag{16}$$

$$\hat{\theta}_s^\beta = \beta \sum_{s' \in S} \sum_{j \in B(s)} \hat{v}_{s'}^\beta q_j(s,s') g_j^\beta(s), \ s \in S_2 \tag{17}$$

From (17), (9), (10) and (16), (11), (12) we get (1) and (2) respectively for $s \in S_1 \cup S_2$.

For $s \in S_3$ noting that $\hat{v}_s^\beta = \hat{\zeta}_s^\beta + \hat{\eta}_s^\beta$ we obtain

$$\hat{\zeta}_s^\beta - \beta \sum_{s' \in S} \sum_{j \in B_s} \hat{v}_{s'}^\beta q_j^2(s,s') \hat{g}_j^\beta(s) - \sum_{j \in B(s)} r_j^2(s) \hat{g}_j^\beta(s) = 0 \tag{18}$$

$$\hat{\eta}_s^\beta - \beta \sum_{s' \in S} \sum_{i \in A_s} \hat{v}_{s'}^\beta q_i^1(s,s') \hat{f}_i^\beta(s) - \sum_{i \in A(s)} r_i^1(s) \hat{f}_i^\beta(s) = 0 \tag{19}$$

Adding (14) and (18) we get the inequality (1) for $s \in S_3$.

$$\hat{v}_s^\beta - \beta \sum_{s' \in S} \sum_{j \in B_s} \hat{v}_{s'}^\beta [q_i^1(s,s') + q_j^2(s,s')] g_j^\beta(s) - \sum_{j \in B(s)} [r_i^1(s) + r_j^2(s)] \hat{g}_j^\beta(s) \geq 0$$

$\Rightarrow$

$$\hat{v}_s^\beta - \beta \sum_{s' \in S} \sum_{j \in B_s} \hat{v}_{s'}^\beta q_{ij}(s,s') \hat{g}_j^\beta(s) - \sum_{j \in B(s)} r(s,i,j) \hat{g}_j^\beta(s) \geq 0, \ s \in S_3, \ i \in A(s) \tag{20}$$

Similarly, adding (13) and (19) we obtain inequality (2) for $j \in B(s), s \in S_3$ of SYS1. Therefore, by Theorem 3.1, $\hat{v}_s^\beta, \hat{f}^\beta(s), \hat{g}^\beta(s)$ is an optimal solution for OPT1.

Conversely, from any solution $(\hat{v}_s^\beta, \hat{\theta}_s^\beta, \hat{f}^\beta(s), \hat{g}^\beta(s))$ for $s \in S_1$ and $s \in S_2$ of SYS1 we define $\hat{\theta}_s^\beta$ as in (16), (17). Rewriting SYS1 using the switching control assumption, we get the inequalities (9) through (12) of OPT1. Similarly, from any solution $(\hat{v}_s^\beta, \hat{\theta}_s^\beta, \hat{f}^\beta(s), \hat{g}^\beta(s))$ of SYS1 for $s \in S_3$, we write $\hat{v}_s^\beta = \hat{\zeta}_s^\beta + \hat{\eta}_s^\beta$ and define $\hat{\zeta}_s^\beta, \hat{\eta}_s^\beta$ as in (18) and (19). Using the AR-AT structure, we rewrite SYS1 to get the inequalities (13), (14) of OPT1. Thus the vector $\hat{z}$ is a feasible solution of OPT1 and by construction $\phi(\hat{z}) = 0$. Since SYS1 is satisfied and after simplifications SYS1 is exactly the same as the constraints of OPT1 after substitution of θ_s^β as in (16), (17) and $\hat{\zeta}_s^\beta, \hat{\eta}_s^\beta$ as in (18) and (19). Using the AR-AT structure, we rewrite SYS1 to get the inequalities (13), (14). Therefore $\hat{z}$ is the global minimum of OPT1. □

3.2. *Undiscounted Zero-sum SC/AR-AT Mixture Stochastic Game*

We require the following lemma which was proved by Filar and Schultz [4].

Lemma 3.1. *([4] [Lemma 2.4])*

(i) If $(v^*, t^*, u^*, f^*, g^*)$ satisfy SYS1a, then for all $s \in S$

$$v_s^* = [P(f^*, g^*)v^*]_s$$

(ii) If $(v^*, t^*, u^*, f^*, g^*)$ solves SYS1b, then for all $s \in S$

$$v_s^* + t_s^* = [P(f^*, g^*)t^* + r(f^*, g^*)]_s$$

Theorem 3.5. *An undiscounted zero-sum SC/AR-AT mixture stochastic game has values v_s where*

$$v_s = \begin{cases} v_s, & s \in S_1 \cup S_2 \\ v_s, & s \in S_1 \cup S_2 \\ v_s = \theta_s + \phi_s, & s \in S_3 \\ t_s = \eta_s + \gamma_s - \theta_s - \phi_s, & s \in S_3 \end{cases}$$

and an optimal pair of stationary strategies $(f^\beta,\ g^\beta)$ can be derived from any global optimal solution to OPT2 with value of the objective function as zero. Conversely, for such a game, a global optimal solution to OPT2 with value of the objective function as zero can be derived from any pair of asymptotically stable stationary strategies.

OPT2:

Decision variables: $(v, t, \rho^1, \rho^2, \theta, \eta, \phi, \gamma, f, g)$ *where* $v, t, \in R^{|S|}$, $\rho^1, \rho^2 \in R^{|S_1 \cup S_2|}$, $\theta, \eta, \phi, \gamma \in R^{|S_3|}$, $f \in F_S$ *and* $g \in G_S$

Objective function:

$$\min \sum_{s \in S_1} [\rho_s^1 + \rho_s^2 - \sum_{s' \in S_1 \cup S_2} \sum_{i \in A(s)} v_{s'} q_i(s, s') f_i(s)$$

$$- \sum_{s' \in S_3} \sum_{i \in A(s)} (\theta_{s'} + \phi_{s'}) q_i(s, s') f_i(s)$$

$$- \sum_{s' \in S_1 \cup S_2} \sum_{i \in A(s)} t_{s'} q_i(s, s') f_i(s) - \sum_{s' \in S_3} \sum_{i \in A(s)} (\eta_{s'} + \gamma_{s'}$$

$$-\theta_{s'} - \phi_s') t_{s'} q_i(s, s') f_i(s)]$$

$$\sum_{s \in S_2} [-\rho_s^1 - \rho_s^2 + \sum_{s' \in S_1 \cup S_2} \sum_{j \in B(s)} v_{s'} q_j(s, s') g_j(s)$$

$$+ \sum_{s' \in S_3} \sum_{j \in B(s)} (\theta_{s'} + \phi_{s'}) q_j(s, s') g_j(s)$$

$$+ \sum_{s' \in S_1 \cup S_2} \sum_{j \in B(s)} t_{s'} q_j(s, s') g_j(s) + \sum_{s' \in S_3} \sum_{j \in B(s)} (\eta_{s'}$$

$$+\gamma_{s'} - \theta_{s'} - \phi_s') q_j(s, s') g_j(s)]$$

$$+ \sum_{s \in S_3} [\phi_s + \gamma_s - \sum_{s' \in S_1 \cup S_2} \sum_{i \in A(s)} v_{s'} q_i^1(s, s') f_i(s)$$

$$- \sum_{s' \in S_3} \sum_{i \in A(s)} (\theta_{s'} + \phi_{s'}) q_i^1(s, s') f_i(s)$$

$$- \sum_{s' \in S_1 \cup S_2} \sum_{i \in A(s)} t_{s'} q_i^1(s, s') f_i(s) - \sum_{s' \in S_3} \sum_{i \in A(s)} (\eta_{s'} + \gamma_{s'}$$

$$-\theta_{s'} - \phi_{s'}) q_i^1(s, s') f_i(s)$$

$$-\theta_s - \eta_s + \sum_{s' \in S_1 \cup S_2} \sum_{j \in B(s)} v_{s'} q_j^2(s, s') g_j(s)$$

$$+\sum_{s'\in S_3}\sum_{j\in B(s)}(\theta_{s'}+\phi_{s'})q_j^2(s,s')g_j(s)$$

$$+\sum_{s'\in S_1\cup S_2}\sum_{j\in B(s)}t_{s'}q_j^2(s,s')g_j(s)+\sum_{s'\in S_3}\sum_{j\in B(s)}(\eta_{s'}+\gamma_{s'}$$

$$-\theta_{s'}-\phi_{s'})q_j^2(s,s')g_j(s)$$

$$-\sum_{i\in A(s)}r_i^1(s)f_i(s)+\sum_{j\in B(s)}r_j^2(s)g_j(s)]$$

Constraints:

$$v_s-\sum_{s'\in S_1\cup S_2}v_{s'}q_i(s,s')-\sum_{s'\in S_3}(\theta_{s'}+\phi_{s'})q_i(s,s')\geq 0,\ i\in A(s),s\in S_1 \tag{21}$$

$$-v_s+\rho_s^1\geq 0,\ s\in S_1 \tag{22}$$

$$v_s+t_s-\sum_{s'\in S_1\cup S_2}t_{s'}q_i(s,s')-\sum_{s'\in S_3}(\eta_{s'}+\gamma_{s'}-\theta_{s'}-\phi_{s'})q_i(s,s')$$

$$-[R(s)g(s)]_i\geq 0,\ i\in A(s),s\in S_1 \tag{23}$$

$$-v_s-t_s+\rho_s^2+[f(s)R(s)]_j\geq 0,\ j\in B(s),s\in S_1 \tag{24}$$

$$-v_s+\sum_{s'\in S_1\cup S_2}v_{s'}q_j(s,s')+\sum_{s'\in S_3}(\theta_{s'}+\phi_{s'})q_j(s,s')\geq 0,\ j\in B(s),s\in S_2 \tag{25}$$

$$v_s-\rho_s^1\geq 0,\ s\in S_2 \tag{26}$$

$$v_s+t_s-\rho_s^2-[R(s)g(s)]_i\geq 0,\ i\in A(s),s\in S_2 \tag{27}$$

$$-v_s-t_s+\sum_{s'\in S_1\cup S_2}t_{s'}q_j(s,s')+\sum_{s'\in S_3}(\eta_{s'}+\gamma_{s'}-\theta_{s'}-\phi_{s'})q_j(s,s')$$

$$+[f(s)R(s)]_j\geq 0,\ j\in B(s),s\in S_2 \tag{28}$$

$$\phi_s-\sum_{s'\in S_1\cup S_2}v_{s'}q_i^1(s,s')-\sum_{s'\in S_3}(\theta_{s'}+\phi_{s'})q_i^1(s,s')\geq 0,\ i\in A(s),s\in S_3 \tag{29}$$

$$\gamma_s - \sum_{s'\in S_1\cup S_2} t_{s'}q_i^1(s,s') - \sum_{s'\in S_3}(\eta_{s'}+\gamma_{s'}-\theta_{s'}-\phi_{s'})q_i^1(s,s')$$

$$-r_i^1(s)\geq 0,\ i\in A(s), s\in S_3 \tag{30}$$

$$-\theta_s + \sum_{s'\in S_1\cup S_2} v_{s'}q_j^2(s,s') + \sum_{s'\in S_3}(\theta_{s'}+\phi_{s'})q_j^2(s,s')\geq 0,\ j\in B(s), s\in S_3 \tag{31}$$

$$-\eta_s + \sum_{s'\in S_1\cup S_2} t_{s'}q_j^2(s,s') + \sum_{s'\in S_3}(\eta_{s'}+\gamma_{s'}-\theta_{s'}-\phi_{s'})q_j^2(s,s')$$

$$+r_j^2(s)\geq 0,\ j\in B(s), s\in S_3 \tag{32}$$

$$f\in F_s,\ g\in G_s \tag{33}$$

Proof. We establish this theorem by showing that a feasible solution to OPT2 with zero objective value is a solution of SYS2b and by Theorem 3.3, it follows that this solution solves the undiscounted SC/AR-AT mixture stochastic game. Conversely, we show that any solution of SYS2b can be used to obtain a solution of OPT2. For $s\in S_1\cup S_2$, we follow a similar argument of the proof given in [4][Theorem 3.1, 4.1].

Let $z=(v,t,\rho^1,\rho^2,\theta,\eta,\phi,\gamma,f,g)$ be a feasible solution of OPT2. First we observe that an objective function $\psi(z)$ of OPT2 with value zero must be a global minimum. For $s\in S_1$, constraint (21),(22), (23)weighted by $f(s)$ added to the constraint (24) weighted by $g(s)$ yields

$$\sum_{s\in S_1}[\rho_s^1+\rho_s^2 - \sum_{s'\in S_1\cup S_2}\sum_{i\in A(s)} v_{s'}q_i(s,s')f_i(s)$$

$$-\sum_{s'\in S_3}\sum_{i\in A(s)}(\theta_{s'}+\phi_{s'})q_i(s,s')f_i(s) - \sum_{s'\in S_1\cup S_2}\sum_{i\in A(s)} t_{s'}q_i(s,s')f_i(s)$$

$-\sum_{s'\in S_3}\sum_{i\in A(s)}(\eta_{s'}+\gamma_{s'}-\theta_{s'}-\phi_s')t_{s'}q_i(s,s')f_i(s)]\geq 0$. Similarly we can show that the remaining terms of $\psi(z)$ are nonnegative. Therefore, it follows that the objective function $\psi(z)\geq 0$.

Let $z^*=(v^*,t^*,\rho^{1*},\rho^{2*},\theta^*,\eta^*,\phi^*,\gamma^*,f^*,g^*)$ be a feasible solution of OPT2 with $\psi(z^*)=0$. Since $\psi(z^*)$ is nonnegative term by term therefore each term of $\psi(\hat{z})$ is zero.

For $s \in S_1$, constraints (21),(22), (23)weighted by $f(s)$, constraint (24) weighted by $g(s)$ and for $s \in S_2$, constraints (25),(26), (28) weighted by $g(s)$, constraint (27) weighted by $f(s)$ yields

$$\rho_s^{1*} = \begin{cases} \sum_{s' \in S} \sum_{i \in A(s)} v_{s'}^* q_i(s,s') f_i^*(s), \; s \in S_1 \\ \sum_{s' \in S} \sum_{j \in B(s)} v_{s'}^* q_j(s,s') g_j^*(s), \; s \in S_2 \end{cases} \tag{34}$$

$$\rho_s^{2*} = \begin{cases} \sum_{s' \in S} \sum_{i \in A(s)} t_{s'}^* q_i(s,s') f_i^*(s), \; s \in S_1 \\ \sum_{s' \in S} \sum_{j \in B(s)} t_{s'}^* q_j(s,s') g_j^*(s), \; s \in S_2 \end{cases} \tag{35}$$

Now substituting the value of ρ_s^{1*} and ρ_s^{2*} in the system of inequalities (21) through (28) we get the system of inequalities in SYS2b. Note that the inequalities (21) and (26) yield after substitution

$$v_s^* - \sum_{s' \in S} v_{s'}^* q_i(s,s')[\sum_{j \in B(s)} g_j^*(s)] \geq 0, \; i \in A(s), s \in S_1$$

$$\text{i.e.,} v_s^* - \sum_{s' \in S} v_{s'}^* \sum_{j \in B(s)} q_i(s,s') g_j^*(s) \geq 0, \; i \in A(s), s \in S_1$$

since $\sum_{j \in B(s)} g_j^*(s) = 1$. Substituting ρ_s^1 in (26) and combining with the above using the definition of a switching control game we get

$$v_s^* - \sum_{s' \in S} v_{s'}^* \sum_{j \in B(s)} q_{i,j}(s,s') g_j^*(s) \geq 0, \; i \in A(s), s \in S_1 \cup S_2$$

which is same as (4). Similarly inequalities (5), (6) and (8) can be obtained.

We define

$$v_s^* = \theta_s^* + \phi_s^* \;\; \text{for } s \in S_3 \tag{36}$$

$$t_s^* = \eta_s^* + \gamma_s^* - \theta_s^* - \phi_s^* \;\; \text{for } s \in S_3 \tag{37}$$

From (36) and (37) we get

$$\eta_s^* + \gamma_s^* = v_s^* + t_s^* \;\; \text{for } s \in S_3$$

Substituting v_s^* for $(\theta_s^* + \phi_s^*)$ and $(v_s^* + t_s^*)$ for $(\eta_s^* + \gamma_s^*)$ in (29) through (32) and for $s \in S_3$ constraints (29),(30) weighted by $f(s)$, constraint (31), (32) weighted by $g(s)$ yields

$$\phi_s^* - \sum_{s' \in S} v_{s'}^* q_i^1(s,s') \geq 0, \ i \in A(s), s \in S_3 \tag{38}$$

$$\gamma_s^* - \sum_{s' \in S} t_{s'}^* q_i^1(s,s') - r_i^1(s) \geq 0, \ i \in A(s), s \in S_3 \tag{39}$$

$$-\theta_s^* + \sum_{s' \in S} v_{s'}^* q_j^2(s,s') \geq 0, \ j \in B(s), s \in S_3 \tag{40}$$

$$-\eta_s^* + \sum_{s' \in S} t_{s'}^* q_j^2(s,s') + r_j^2(s) \geq 0, \ j \in B(s), s \in S_3 \tag{41}$$

$$\phi_s^* = \sum_{s' \in S} \sum_{i \in A(s)} v_{s'}^* q_i^1(s,s') f_i^*(s), \ s \in S_3 \tag{42}$$

$$\gamma_s^* = \sum_{s' \in S} \sum_{j \in B(s)} t_{s'}^* q_i^1(s,s') f_i^*(s) + \sum_{i \in A(s)} r_i^1(s) f_i^*(s), \ s \in S_3 \tag{43}$$

$$\theta_s^* = \sum_{s' \in S} \sum_{j \in B(s)} v_{s'}^* q_j^2(s,s') g_j^*(s), \ s \in S_3 \tag{44}$$

$$\eta_s^* = \sum_{s' \in S} \sum_{j \in B(s)} t_{s'}^* q_j^2(s,s') g_j^*(s) + \sum_{j \in B(s)} r_j^2(s) g_j^*(s), \ s \in S_3 \tag{45}$$

Adding (38) and (44) we get

$$\theta_s^* + \phi_s^* - \sum_{s' \in S} \sum_{j \in B(s)} v_{s'}^* q_j^2(s,s') g_j^*(s) - \sum_{s' \in S} v_{s'}^* q_i^1(s,s') \geq 0, \ i \in A(s), \ s \in S_3 \tag{46}$$

Therefore

$$\theta_s^* + \phi_s^* - \sum_{s' \in S} v_{s'}^* \sum_{j \in B(s)} [q_j^2(s,s') g_j^*(s) + q_i^1(s,s') g_j^*(s)] \geq 0, \ i \in A(s), \ s \in S_3 \tag{47}$$

Substituting v_s^* for $(\theta_s^* + \phi_s^*)$ we get (4).

$$v_s^* - \sum_{s' \in S} \sum_{j \in B(s)} v_{s'}^* q_{ij}(s,s') g_j^*(s) \geq 0, \ i \in A(s), \ s \in S_3 \tag{48}$$

Adding (39) and (45) we get (5).

$$\eta_s^* + \gamma_s^* - \sum_{s' \in S} t_{s'}^* [\sum_{j \in B(s)} q_j^2(s,s') + q_i^1(s,s')] g_j^*(s) - \sum_{j \in B(s)} [r_j^2(s) + r_i^1(s)] g_j^*(s) \geq 0,$$

$$i \in A(s), s \in S \tag{49}$$

This implies

$$v_s^* + t_s^* - \sum_{s' \in S} t_{s'}^* \sum_{j \in B(s)} q_{ij}(s,s') g_j^*(s) - [R(s)g(s)]_i \geq 0, \; i \in A(s), s \in S \tag{50}$$

Subtracting (42) from (40) and subtracting (43) from (41) we get (6) and (8) respectively. Since $f \in F_s$ and $g \in G_s$ the variables satisfy SYS2b and by Theorem 3.3, this yields an optimal solution to undiscounted SC/AR-AT mixture stochastic game.

To prove the converse, we show that any solution to SYS2b which always exists for these games, since they possess asymptotically stable optimal stationary strategies can be used to derive a feasible solution for OPT2. Assume that (v^*, t^*, f^*, g^*) be a feasible solution of the SYS2b. We define ρ_s^1, ρ_s^2 as in (34), (35). Rewriting SYS2b using the switching control assumption and using (34), (35) we get (21) through (28).

From (4), (5), (6) and (8) and using the definition of AR-AT game we get

$$v_s^* - \sum_{s' \in S} \sum_{j \in B(s)} v_{s'}^* q_j^2(s,s') g_j^*(s) - \sum_{s' \in S} v_{s'}^* q_i^1(s,s') \geq 0, \; i \in A(s), \; s \in S_3 \tag{51}$$

$$v_s^* + t_s^* - \sum_{s' \in S} \sum_{j \in B(s)} t_{s'}^* q_j^2(s,s') g_j^*(s) - \sum_{s' \in S} t_{s'}^* q_i^1(s,s') - \sum_{j \in B(s)} r_j^2(s) g_j^*(s)$$

$$-r_i^1(s) \geq 0, \; i \in A(s), \; s \in S_3 \tag{52}$$

$$-v_s^* + \sum_{s' \in S} \sum_{i \in A(s)} v_{s'}^* q_i^1(s,s') f_i^*(s) + \sum_{s' \in S} v_{s'}^* q_j^2(s,s') \geq 0, \; j \in B(s), \; s \in S_3 \tag{53}$$

$$-v_s^* - t_s^* + \sum_{s' \in S} \sum_{i \in A(s)} t_{s'}^* q_i^1(s,s') f_i^*(s) + \sum_{s' \in S} t_{s'}^* q_j^2(s,s') + \sum_{i \in A(s)} r_i^1(s) f_i^*(s)$$

$$+r_j^2(s) \geq 0, \; j \in B(s), \; s \in S_3 \tag{54}$$

Take $\theta_s^*, \eta_s^*, \phi_s^*$ and γ_s^* for $s \in S_3$ as in (42) through (45). Adding (42) and (44) we get

$$\theta_s^* + \phi_s^* = \sum_{s' \in S} v_{s'}^* [\sum_{i \in A(s)} q_i^1(s,s') f_i^*(s) + \sum_{j \in B(s)} q_j^2(s,s') g_j^*(s)]$$

$$= [P(f^*, g^*)v^*]_s = v_s^* \tag{55}$$

by Lemma 3.1 (i). Similarly, using Lemma 3.1(ii) and from (43) and (45) we get

$$\eta_s^* + \gamma_s^* = [P(f^*, g^*)t^* + r(f^*, g^*)]_s = v_s^* + t_s^* \tag{56}$$

From (51), (55) and using the definition of θ_s^* in (44) we get (29).

$$\theta_s^* + \phi_s^* - \theta_s^* - \sum_{s' \in S} (\theta_{s'}^* + \phi_{s'}^*) q_i^1(s,s') \geq 0, \ i \in A(s), \ s \in S_3 \tag{57}$$

From (52),(45),(55) and (56) we get (30) of OPT2. From (53), (55) and the definition of ϕ^* in (42) yields (31) of OPT2.

$$-\theta_s^* - \phi_s^* + \sum_{s' \in S} (\theta_{s'}^* + \phi_{s'}^*) q_j^2(s,s') + \phi_s^* \geq 0, \ j \in B(s), \ s \in S_3 \tag{58}$$

Similarly from (54), (55), (56) and (43) we get (32) of OPT2. Since, $f \in F_s$ and $g \in G_s$, we obtain a feasible solution of OPT2. Note that by construction $\psi(z^*) = 0$. Since SYS2b is satisfied and after simplifications SYS2b is exactly the same as the constraints of OPT2. Therefore z^* is the global minimum of OPT2. □

4. Concluding Remarks and Areas of Further Research

Sinha [21,22] raises the question that whether a finite step algorithm can be developed for SC/AR-AT mixture class. The main results proved in this paper is the computation of optimal strategies and the value vector for both discounted and undiscounted SC/AR-AT mixture games with a optimization model in the spirit of Mangasarian and Stone [9]. The computational results using optimization approach seems to be very encouraging. Investigation regarding comparison involving different solution methods for the optimization models OPT1 and OPT2 are areas of further research.

References

1. T. Bewley and E. Kohlberg, On stochastic games with stationary optimal strategies, *Mathematics of Operations Research,* **1**, pp. 104-125 (1978).
2. J. A. Filar, Quadratic programming and the single-controller stochastic game, *Journal of Mathematical Analysis and Applications*, **113**, pp. 136-147 (1986).
3. J. A. Filar, Orderfield property for stochastic games when the player who controls transitions changes from state to state, *JOTA*, **34**, pp. 503–515 (1981).
4. J. A. Filar and T. A. Schultz, Bilinear programming and structured stochastic games, *JOTA*, **53**, pp. 85–104 (1987).
5. J. A. Filar and O. J. Vrieze, *Competitive Markov Decision Processes*, (Springer, New York, 1997).
6. D. Gillette, Stochastic game with zero step probabilities: in*Contribution to theory of games*, Edited by A. W. Tucker, M. Dresher and P. Wolfe, (Princeton University Press, Princeton, New Jersey, 1957).
7. M. S. Gowda, and R. Sznajder, A generalization of the Nash equilibrium theorem on bimatrix games, *International Journal of Game Theory* **25**, pp. 1–12 (1996).
8. O. L. Mangasarian, Equilibrium points of bimatrix games, *J. Soc. Ind. Appl. Math.*, **12**, pp. 778780 (1964).
9. O. L. Mangasarian and H. Stone, Two-person nonzero-sum games and quadratic programming, *Journal of Mathematical Analysis and Applications*, **9**, pp. 348-355 (1964).
10. S. R. Mohan, S. K. Neogy and T. Parthasarathy, Linear complementarity and discounted polystochastic game when one player controls transitions, in *Complementarity and Variational Problems*, eds: M. C. Ferris and Jong-Shi Pang, (SIAM, Philadelphia, pp. 284–294 1997).
11. S. R. Mohan, S. K. Neogy and T. Parthasarathy, Linear complementarity and the irreducible polystochastic game with the average cost criterion when one player controls transitions, in *Game theoretical applications to Economics and Operations Research*, eds. T. Parthasarathy, B. Dutta, J. A. M. Potters, T. E. S. Raghavan, D. Ray and A. Sen, (Kluwer Academic Publishers, Dordrecht, The Netherlands, pp. 153–170, 1997).
12. S. R. Mohan and T. E. S. Raghavan, An algorithm for discounted switching control games, *OR Spektrum*, **9**, pp. 41–45 (1987).
13. S. R. Mohan, S. K. Neogy and T. Parthasarathy, Pivoting algorithms for some classes of stochastic games: A survey, *International Game Theory Review*, **3**, pp. 253–281 (2001).

14. S. R. Mohan, S. K. Neogy and T. Parthasarathy, and S. Sinha, Vertical linear complementarity and discounted zero-sum stochastic games with ARAT structure, *Mathematical Programming*, Series A pp. 637–648 (1999).
15. S. R. Mohan, S. K. Neogy and R. Sridhar, The generalized linear complementarity problem revisited, *Mathematical Programming* **74**, pp. 197–218 (1996).
16. S. K. Neogy, and A. K. Das, Linear complementarity and two classes of structured stochastic games, in *Operations Research with Economic and Industrial Applications: Emerging Trends*, eds: S. R. Mohan and S. K. Neogy, (Anamaya Publishers, New Delhi, India pp. 156–180, 2005).
17. S. K. Neogy, A. K. Das, S. Sinha and A. Gupta, On a Mixture Class of Stochastic Game with Ordered Field Property, *Mathematical Programming and Game Theory for decision making*, eds. S. K. Neogy, R. B. Bapat, A. K. Das and T. Parthasarathy, ISI Platinum Jubilee Series on Statistical Science and Interdisciplinary Research, **Vol. 1**, 451–477 (2008).
18. T. E. S. Raghavan and J. A. Filar, Algorithms for stochastic games, a survey, *Zietch. Oper. Res.*, **35**, pp. 437–472 (1991).
19. T. E. S. Raghavan, S. H. Tijs, and O. J. Vrieze, On stochastic games with additive reward and transition structure, *JOTA*, **47**, pp. 375–392 (1985).
20. U. G. Rothblum, Solving stopping stochastic games by maximizing a linear function subject to quadratic constraints,in: O. Moeschlin and D. Pallashke, eds.,*Game Theory and Related Topics* (North-Holland, Amsterdam, pp. 103-105, 1979).
21. S. Sinha, *A contribution to the theory of Stochastic Games*, Ph.D thesis, Indian Statistical Institute, Delhi Centre (1989).
22. S. Sinha, *A new class of Stochastic Games having Ordered field property*, Unpublished Manuscript, (Jadavpur University, Kolkata, 2000).
23. T. A. Schultz, Linear complementarity and discounted switching controller stochastic games, *JOTA*, **73**, pp. 89–99 (1992).
24. L. S. Shapley, Stochastic games, *Proc. Nat. Acad. Sci. USA.*, **39**, pp. 1095–1100 (1953).
25. O. J. Vrieze, A finite algorithm for the switching controller stochastic game, *OR Spektrum*, **5**, pp. 15–24 (1983).

PREDATOR - PREY RELATIONS FOR MAMMALS WHERE PREY SUPPRESS BREEDING

Q.J.A. KHAN and M. AL-LAWATIA

Department of Mathematics and Statistics, Sultan Qaboos University, P.O. Box 36, Al-Khodh 123, Sultanate of Oman

In any environment, current reproduction of prey population will affect future population size, but these future changes may also affect the current reproductive decisions. We propose the dynamics of predator-prey cycles by a theoretical model based on field and laboratories experiments. These represent the suppression of breeding by prey in response to increase in predation pressure. On the other hand, non-breeding prey individuals have a better chance of avoiding predation than those in a reproductive state. The predator consumes both the breeder and suppressor individuals and this prey population is more prone to predation at higher densities. We showed that all the solutions of the first model in positive octant are bounded. The stability analysis has been carried out for the equilibrium set for both models. We found out that Hopf bifurcation will occur by varying a parameter q1 which represents the rate by which breeder population turns suppressor population. It is found that predator induced breeding suppression (PIBS) acts to destabilize the stable interaction. We discussed these finding in the light of the Fennoscandian vole cycle. The theoretical results are compared with the numerical results for different sets of parameters.

Keywords: prey, predator, Hopf bifurcation, vole cycle.

1. Introduction

Population ecologists are studying the population dynamics of small mammals like voles and snowshoe hares from many years. Several papers can be referred in the literature which reported that certain small mammals suppress breeding in response to strong predation pressure (Lima and Dill, 1990; Ylönen, 1989). It is observed in laboratories and by field experiments that small mammals like voles suppressed breeding on the exposure to predators. They believed that this happens due to change in feeding and mating behaviours due to stress in small prey mammals on the high exposure of predation (Ylönen *et al.* 1992; Ronkainen and Ylönen 1994; Ylönen and Ronkainen 1994; Heikkil *et al.*, 1993; Koskela and Ylönen 1995, Korpimki *et al.*, 1994). Oksanen and Lundberg, (1995) and Ylönen, (1994) studied the antipredator behaviour of prey and explained that the prey in non-productive state have greater chances of survival than the prey in breeding stage. The reason behind this fact is that mammal in breeding stage becomes inactive and have less antipredator capabilities to run away or to avoid themselves efficiently from predator.

Ruxton and Lima (1997) presented a mathematical model to explain the effect of predator induced breeding suppression (PIBS). They found out that the strong level of PIBS acts to stabilize predator-prey cycles and that weaker levels reduce the amplitude and increase the frequency of existing oscillations. They explained that PIBS promotes stability in the interaction between predator and prey. Ruxton et al. (2002) studied two switching models that represent the suppression of breeding by prey in response to short-term increase in predation pressure. In these models both breeding and suppression population are exposed (to varying degree) to the predator. However, the predator feeds preferentially on the abundant prey . This implies a kind of switching from breeder species to suppressor type as these prey change in numerical superiority. For both models they produced analytic conditions for the local stability of the interior steady state. They concluded the paper by a statement that no simple and general rule for the effect of the behaviours on the stability of population dynamics but these effects is system specific.

In this paper we modified the model presented in Ruxton et al. (2002) by considering that both suppressor and breeder populations contribute to the carrying capacity of the breeder class. We also consider the fact that breeder population contributes α times more than suppressor class in the growth of predator population. We extended the study of Ruxton et al. and found out that Hopf bifurcation will occur by varying the parameter q_1. We found out, contrary to Ruxton and Lima (1997) that the strong level of PIBS is a destabilizing factor in predator-prey cycles and that weaker level increases the stability. Hopf bifurcation study helped us in finding the existence of a region of Instability in the neighborhood of a nonzero equilibrium where prey-predator populations will survive undergoing regular fluctuations.

2. The mathematical model

We consider a class of Volterra predator-prey model with switching exhibition by the predator. Prey population is divided into two classes, one that always breeds and the other suppress breeding for short-term in response to increased predation pressure. The predation risk of the prey is reduced by suppressing the breeding. Both class of prey species with varying degrees are exposed to the predator. The predator will feed to that class of prey more which is numerically superior so that the chance that a predator catches a member of prey species is proportional to their abundance. This situation is described by the following system of coupled ordinary differential equations.

$$\begin{aligned} \frac{dB}{dt} &= \gamma B\left(1-\frac{S+B}{K}\right)-\frac{I_b B^2 P}{\alpha B+S}-\frac{Q_{mb}PB}{P+P_0+aB}+\frac{Q_{ms}BS}{B+B_0+bP}, \\ \frac{dS}{dt} &= \frac{Q_{mb}PB}{P+P_0+aB}-\frac{Q_{ms}BS}{B+B_0+bP}-\frac{I_s S^2 P}{\alpha B+S}, \\ \frac{dP}{dt} &= \frac{\delta_1 I_b B^2 P}{\alpha B+S}+\frac{\delta_2 I_s S^2 P}{\alpha B+S}-\mu P, \end{aligned} \tag{1}$$

where, B is the prey population that breeds regardless of predator density, S is the prey population which suppresses breeding, P is the predator population, γ is the species growth of prey individuals in the breeding sub-population, the predator that prey upon the breeder α times more often than on the suppressor class, K is the carrying capacity for prey, μ is the per capita death rate of the predator, I_b and I_s are the predator response rates towards the breeder and suppressor populations, and δ_1 and δ_2 are the efficiency with which captured breeders and suppressors respectively are converted to predators. $\frac{Q_{mb}PB}{P+P_0+aB}$ is the rate at which individuals in the breeding population move to the suppressor population. The per capita rate of movement into the suppressor population increases with predator density but is reduced with an increasing breeding population (since individuals gain protection through dilution). The positive parameters a and P_0 control the shape of the response $\frac{Q_{mb}BP}{P+P_0+aB}$ is the rate at which individuals in the suppressor population return to the breeding population. The per capita rate of movement into the breeder population increases with size of the breeder population (representing the protection afforded by dilution) and decreases with increasing predator density. The positive parameters b and B_0 control the shape of the response curve $\frac{Q_{ms}BS}{B+B_0+bP}$.

We assume all the parameters in the model are positive, and that $B(0)$, $S(0)$, $P(0) > 0$. In order to avoid mathematical complexity and to reduce number of parameters we transform the variables and parameters by

$$\frac{\gamma}{\mu} = r_1, \quad \frac{Q_{mb}}{\mu} = q_1 \ , \quad \frac{Q_{ms}}{\mu} = q_2 \ , \quad \frac{I_b}{\mu} = l_b \ , \quad \frac{I_s}{\mu} = l_s,$$

$$\frac{\delta_1}{\mu} = \alpha_1 \ , \quad \frac{\delta_2}{\mu} = \alpha_2 \quad \text{and} \quad \mu\, t = \tau,$$

so the system of equations (2.1) becomes

$$\begin{aligned}
\frac{dB}{d\tau} &= r_1 B(1 - \frac{S+B}{K}) - \frac{l_b B^2 P}{\alpha B + S} - \frac{q_1 BP}{P + P_0 + aB} + \frac{q_2 BS}{B + B_0 + bP}, \\
\frac{dS}{d\tau} &= \frac{q_1 BP}{P + P_0 + aB} - \frac{q_2 BS}{B + B_0 + bP} - \frac{l_s S^2 P}{\alpha B + S}, \\
\frac{dP}{d\tau} &= \frac{\alpha_1 l_b B^2 P}{\alpha B + S} + \frac{\alpha_2 l_s S^2 P}{\alpha B + S} - P.
\end{aligned} \tag{2}$$

3. Boundedness results

We have the following three results, on the boundedness of the system (2):

Proposition 1. The prey is always bounded above.
Proof. On adding first and second equations of system(2), we get

$$\frac{dB}{d\tau} + \frac{dS}{d\tau} \leq r_1 B \left(1 - \frac{S+B}{K}\right) \tag{3}$$

where

$$\frac{dB}{d\tau} + \frac{dS}{d\tau} > 0 \text{ because } S + B < K$$

It follows that

$$\limsup_{\tau \to \infty} \ (B+S)(\tau) \leq K. \tag{4}$$

Proposition 2. The predator are always bounded above if $a\, b\, K \leq \ 1$
Proof. Using the third equation in the system (2), we obtain

$$\frac{dP}{d\tau} + P < \frac{\alpha_1 l_b B^2 P + \alpha_2 l_s S^2 P}{B+S} \text{ where } \alpha > 1. \tag{5}$$

Let Max $\{\alpha_1, \alpha_2\} = a$ and Max $\{l_{\rm b}, l_s\} = b$ then

$$\frac{dP}{d\tau} + P < abP\left(\frac{(B+S)^2 - 2BS}{B+S}\right) \tag{6}$$

or

$$\frac{dP}{d\tau} + P < abP\left[(B+S) - \frac{2BS}{B+S}\right] < abPK \tag{7}$$

Let $abK = M$, then

$$\frac{dP}{d\tau} < P(M-1) \quad \text{and} \quad P < P(0)e^{(M-1)\tau}$$

If $M \leq 1$, then $P \leq P(0)$. It follow that $\lim_{\tau\to\infty} sup(P(\tau)) \leq P(0)$.

Proposition 3. The trajectories of system (2) are bounded if $a\,b\,K \leq 1$.

Proof. Define the function $U(\tau) = (B(\tau) + S(\tau)) + P(\tau)$. So,

$$\begin{aligned}\lim_{\tau\to\infty} [(B(\tau) + S(\tau)) + P(\tau)] &\leq \lim_{\tau\to\infty} [(B(\tau) + S(\tau))] + \lim_{\tau\to\infty} P(\tau) \\ &\leq K + P(0) = \eta\end{aligned}$$

and lim sup $U(\tau) \leq \eta$ as $\tau \to \infty$ independently of the initial conditions. Hence all the solutions of equation (2) for all nonnegative initial conditions are bounded if $a\,b\,K \leq 1$.

4. Existence and Uniqueness

The equations in the system (2) do not make sense if $B = S = 0$. If they are defined on the region

(a) $U_1 = \{(B > 0, S > 0) \text{ and } P \geq 0\}$: Equations are continuously differentiable within this region. Therefore, a solution exists and is unique for some time interval extending forward from the initial point. See Arnold (1992), p. 107, 109 and (p. 273 deals with Lipschitz condition)

(b) $U_2 = \{(B = 0, S > 0)\,\forall\tau \text{ and } P \geq 0\}$: System of equations (2) reduces to

$$\begin{aligned}\frac{dS}{d\tau} &= -I_S SP \\ \frac{dP}{d\tau} &= \alpha_2 I_S SP - P\end{aligned}$$

(c) $U_3 = \{(B > 0,\ S = 0)\, \forall \tau \text{ and } P \geq 0\}$: System of equation (2) reduces to

$$\frac{dB}{d\tau} = r_1 B\left(1 - \frac{B}{K}\right) - \frac{I_B}{\alpha} BP - \frac{q_1 BP}{P + Po + aB}$$
$$\frac{dS}{d\tau} = \frac{q_1 BP}{P + Po + aB}$$
$$\frac{dP}{d\tau} = \frac{\alpha_1}{\alpha} I_B BP - P$$

(d) $U_4 = \{(B = 0,\ S = 0)\, \forall \tau \text{ and } P \geq 0\}$: System of equations (2) reduces to $\dfrac{dP}{d\tau} = -P$

Hence in all these cases (b, c and d) equations are continuously differentiable within their respective region and so the solution exists and is unique as in case a.

5. Equilibria

We find the steady states of system (2) by equating the derivatives on the left hand sides to zero and solving the resulting algebraic equations. The ecological meaning possible equilibria of system (2) are:

(i) $\overline{E_0} = (0, 0, 0)$, where the population is extinct and this will always exist.
(ii) $\overline{E_1} = (\overline{B}, 0, 0)$, where $\overline{\mathrm{B}} = K$.
(iii) $\overline{E_2} = (0, \overline{S}, 0)$, where $\overline{S} = \dfrac{r_1 K B_0}{(r_1 B_0 - q_2 K)}$ this equilibrium point will exist if $r_1 B_0 > q_2 K$.
(iv) $\overline{E_3} = (\overline{B}, \overline{S}, \overline{P})$, where both prey and predator exist, with

$$\overline{B} = \frac{(\alpha \overline{x} + 1)\, \overline{x}}{(\alpha_1 l_b \overline{x}^2 + \alpha_2 l_s)}, \quad \overline{\mathrm{S}} = \frac{(\alpha \overline{x} + 1)}{(\alpha_1 l_b \overline{x}^2 + \alpha_2 l_s)},$$
$$\overline{P} = \frac{r_1\, \overline{x}\, R(QK - R(\overline{x} + 1))}{TRQ},$$

$$Q = \alpha_1\, l_b\, \overline{x}^2 + \alpha_2\, l_s, \quad R = \alpha\, \overline{x} + 1 \quad \text{and } T = l_b\, \overline{x}^2 + l_s.$$

For equilibrium values $(\overline{B}, \overline{S}, \overline{P})$ to be positive, we require that

$$\frac{Q\, K}{R} > \overline{x} + 1.$$

Here $\overline{x} = \frac{\overline{B}}{\overline{S}}$ is a real positive root of the 15 degree polynomial which is obtained by using second equation of (2) at equilibrium.

6. Stability

6.1. *Behaviour of the system around* $\bar{E}_0$

The stability matrix is not defined at the zero equilibrium $\bar{E}_0$. However, easily it can be proved that if $B_0 > 0$ then $\bar{E}_0$ is unstable. If $B_0 = 0$ then $\bar{E}_0$ is stable.

Lemma 6.1

(i) If $B_0 > 0$ then $\bar{E}_0$ will be unstable because all trajectories will diverge from zero equilibrium point for large times
(ii) If $B_0 = 0$ then all trajectories will converge towards the origin for large times. Hence $\bar{E}_0$ will be stable.

Proof.

(i) From the first equation of the system (2), we get

$$\frac{d}{d\tau}(\ln B) = r_1\left(1 - \frac{S+B}{K}\right) - \frac{I_b BP}{\alpha B + S} - \frac{q_1 P}{P + Po + aB} + \frac{q_2 S}{B + Bo + bP}$$

As $(B, S, P) \to (0, 0, 0)$, $\frac{d}{d\tau}(\ln B) \to r_1$, where indeterminate from $\frac{I_b BP}{\alpha B + S} = 0$. Hence there is a small sphere with center $\bar{E}_0$ and radius $r\ _1$such that within this sphere $\frac{d}{d\tau}(\ln B) \geq \frac{r_1}{2}$. If $\bar{E}_0$ is stable then $B \to 0,\ S \to 0,\ I \to 0$ as $\tau \to \infty$ then there exist τ_0 such that $B(\tau_0) = B_0$.Therefore $\tau \geq \tau_0, B(\tau) = B_0 e^{\frac{r_1}{2}(\tau - \tau_0)}$ so $B(\tau)$ will approach to infinity for large times.

(ii) If $B_0 = 0$ then $\frac{dB}{d\tau} = 0$ ie $B(\tau) = 0$ for all times.

(a) If $S_0 > 0$ and $P_0 > 0$, then from second equation of the system (2) $\frac{dS}{d\tau} < 0$. i.e. $S(\tau) \to 0$ as $\tau \to \infty$. Hence from the third equation of (2) $\frac{dP}{d\tau} < 0$. i.e. $P(\tau) \to 0$ as $\tau \to \infty$. Hence all trajectories will approach towards $\bar{E}_0$.

(b) If $S_0 > 0$ and $P_0 = 0$, then $\frac{dS}{d\tau} < 0$ so $S \to 0$ as $\tau \to \infty$ and $\frac{dP}{d\tau} = 0$ so P $(\tau) = 0$ for all times. Hence all trajectories will approach towards $\bar{E}_0$.

(c) If $S_0 = 0$ and $P_0 = 0$, then $\dfrac{dS}{d\tau} = \dfrac{dP}{d\tau} = 0$. These imply that $S(\tau)$ and $P(\tau) = 0$ for all $\tau > 0$.

(d) If $S_0 = 0$ and $P_0 > 0$, then $\dfrac{dS}{d\tau} = 0$, and so $S(\tau) = 0$ and $\dfrac{dP}{d\tau} < 0$. Hence $P(\tau) \to 0$ as $\tau \to \infty$. Hence, all trajectories will converge towards $\bar{E}_0$.

6.2. *Behaviour of the system around* $\overline{\mathbf{E}}_1$

The stability matrix of the system (2) is given by

$$\overline{J_1} = \begin{pmatrix} -r_1 - \lambda & \frac{q_2 K}{K+B_0} - r_1 & -(\frac{l_b K}{\alpha} + \frac{q_1 K}{P_0 + aK}) \\ 0 & -\frac{q_2 K}{K+B_0} - \lambda & \frac{q_1 K}{P_0 + aK} \\ 0 & 0 & \frac{\alpha_1 l_b K}{\alpha} - 1 - \lambda \end{pmatrix} \tag{8}$$

This leads to the characteristic equation

$$(r_1 + \lambda)(\frac{q_2 K}{K + B_0} + \lambda)(\frac{\alpha_1 l_b K}{\alpha} - 1 - \lambda) = 0. \tag{9}$$

It can also be shown that all these eigenvalues are negative if $\dfrac{\alpha_1 l_b K}{\alpha} < 1$.

Theorem 6.2: If $\dfrac{\alpha_1\, l_b\, K}{\alpha} < 1$, the equilibrium $\overline{E}_1$ is locally asymptotically stable and otherwise it is unstable.

6.3. *Behaviour of the system around* $\overline{\mathbf{E}}_2$

The stability matrix of the system (2) is given by

$$\overline{J_2} = \begin{pmatrix} 0 - \lambda & 0 & 0 \\ \frac{-q_2 \overline{S}}{B_0} & 0 - \lambda & -l_s \overline{S} \\ 0 & 0 & (\alpha_2 l_s \overline{S} - 1) - \lambda \end{pmatrix} \tag{10}$$

The corresponding characteristic equation is

$$\lambda^2((\alpha_2 l_s \overline{S} - 1) - \lambda) = 0. \tag{11}$$

Hence the equilibrium point $\overline{E_2} = (0, \overline{S}, 0)$ will be bounded if $\alpha_2 l_s \overline{S} < 1$.

Theorem 6.3: If $\alpha_2 l_s \overline{S} < 1$, then the equilibrium $\overline{E}_2$ will be bounded and unstable.

6.4. *Behaviour of the system around* $\overline{\mathbf{E}}_3$

The stability matrix of the system (2) is given by

$$\overline{J_3} = \begin{pmatrix} A_1 - \lambda & B_1 & C_1 \\ D_1 & E_1 - \lambda & F_1 \\ G_1 & H_1 & 0 - \lambda \end{pmatrix} \tag{12}$$

The corresponding characteristic equation is

$$\begin{aligned} &\lambda^3 - (A_1 + E_1)\lambda^2 + (A_1E_1 - H_1F_1 - B_1D_1 - C_1G_1)\lambda \\ &+H_1F_1A_1 - G_1F_1B_1 - C_1D_1H_1 + C_1G_1E_1 = 0 \end{aligned} \tag{13}$$

Equation (13) has the form

$$\lambda^3 + a_1\lambda^2 + a_2\lambda + a_3 = 0 \tag{14}$$

where

$$\begin{aligned} a_1 &= -(A_1 + E_1) \\ a_2 &= (A_1E_1 - H_1F_1 - B_1D_1 - C_1G_1) \\ a_3 &= H_1F_1A_1 - G_1F_1B_1 - C_1D_1H_1 + C_1G_1E_1 \end{aligned} \tag{15}$$

with

$$
\begin{aligned}
A_1 &= -\frac{(r_1\overline{B})}{K} - \frac{(l_b\overline{BPS})}{(\alpha\overline{B}+\overline{S})^2} + \frac{(aq_1\overline{BP})}{(\overline{P}+P_0+a\overline{B})^2} - \frac{(q_2\overline{BS})}{(\overline{B}+B_0+b\overline{P})^2} \\
B_1 &= \frac{-(r_1\overline{B})}{K} + \frac{(l_b\overline{B}^2\overline{P})}{(\alpha\overline{B}+\overline{S})^2} + \frac{(q_2\overline{B})}{(\overline{B}+B_0+b\overline{P})} \\
C_1 &= \frac{(-l_b\overline{B}^2)}{(\alpha\overline{B}+\overline{S})} - \frac{(q_1\overline{B}(P_0+a\overline{B}))}{(\overline{P}+P_0+a\overline{B})} - \frac{(q_2b\overline{BS})}{(\overline{B}+B_0+b\overline{P})^2} \\
D_1 &= \frac{l_s\alpha\overline{PS}^2}{(\alpha\overline{B}+\overline{S})^2} + \frac{q_1\overline{P}(\overline{P}+P_0)}{(\overline{P}+P_0+a\overline{B})^2} - \frac{q_2\overline{S}(B_0+b\overline{P})}{(\overline{B}+B_0+b\overline{P})^2} \\
E_1 &= -\frac{q_2\overline{B}}{(\overline{B}+B_0+b\overline{P})} - \frac{l_s\overline{PS}(2\alpha\overline{B}+\overline{S})}{(\alpha\overline{B}+\overline{S})^2} \\
F_1 &= \frac{q_1\overline{B}(P_0+a\overline{B})}{(\overline{P}+P_0+a\overline{B})^2} + \frac{q_2b\overline{BS}}{(\overline{B}+B_0+b\overline{P})^2} - \frac{l_s\overline{S}^2}{\alpha\overline{B}+\overline{S}} \\
G_1 &= \frac{(2\alpha_1 l_b\overline{B}\,\overline{P}-\alpha\overline{P})}{(\alpha\overline{B}+\overline{S})} \\
H_1 &= \frac{2\alpha_2 l_s\overline{S}\,\overline{P}-\overline{P}}{\alpha\overline{B}+\overline{S}}.
\end{aligned} \tag{16}
$$

We examine the stability of $\overline{E}_3$ using the Routh-Hurwitz criteria,

$$
\begin{aligned}
&(i)\ \ a_1 > 0,\ \ a_3 > 0, \qquad\qquad (ii)\ \ a_1\, a_2 > \ a_3. \\
&a_1\, a_2 > \ a_3 \Leftrightarrow \ \ Q(q_1) = b_0\, q_1^2 + b_1\, q_1 + b_2 > 0.
\end{aligned} \tag{17}
$$

It is difficult to list the coefficients of this equation, which has 13 different parameters. However, we have used the computer algebra system, Maple, to find the coefficient b_0, b_1, and b_2 explicitly.

Theorem 6.4: The equilibrium $\overline{E}_3$ will be locally asymptotically stable if the conditions (17) are satisfied otherwise it is unstable.

7. Hopf bifurcation around the positive interior equilibrium $\overline{E}_3$

We have taken q_1 as a bifurcation parameter to study Hopf bifurcation for the system (2).

Lemma 7.1 The System (2) undergoes a Hopf bifurcation when q_1 passes through $\overline{q}_1$.
Proof: Hopf bifurcation will occur if and only if there exists a $q_1 = \overline{q}_1$, such that

(i) $a_1(q_1)\ a_2(q_1) = a_3(q_1)$ with $a_1(q_1),\ a_2(q_1),\ a_3(q_1) > 0$, (See equation (17)), and
(ii) $\frac{d}{dq_1}(Re(\lambda_k(q_1)))|_{q_1=\overline{q}_1} \neq 0$, (Marsden and Mckracken, 1976).$k = 1, 2$.

Now, when $q_1 = \overline{q}_1$, $a_1\ a_2 = a_3$ with $a_1, a_2, a_3 > 0$.
The characteristic equation (14) is given by

$$(\lambda^2 + a_2)(\lambda + a_1) = 0 \tag{18}$$

with roots $\lambda_1 = i\sqrt{a_2}$, $\lambda_2 = -i\sqrt{a_2}$ and $\lambda_3 = -a_1$, so there is a pair of purely imaginary eigenvalues and a strictly negative real eigenvalue. For q_1 in a neighbourhood of $\overline{q}_1$ the roots have the form $\lambda_1(q_1) = u(q_1) + iv(q_1)$, $\lambda_2(q_1) = u(q_1) - iv(q_1)$, $\lambda_3(q_1) = -a_1(q_1)$.

Next, we shall verify the transversality condition

$$\frac{d}{dq_1}(Re(\lambda_k(q_1)))|_{q_1=\overline{q}_1} \neq 0 \quad k = 1, 2 \tag{19}$$

where

$$\begin{aligned} R(q_1) &= 3u^2(q_1) + 2a_1(q_1)\,u(q_1) + a_2(q_1) - 3v^2(q_1) \\ S(q_1) &= 6u(q_1)v(q_1) + 2a_1(q_1)v(q_1) \\ T(q_1) &= u^2(q_1)a_1'(q_1) + a_2'(q_1)u(q_1) + a_3'(q_1) - a_1'(q_1)v^2(q_1) \\ U(q_1) &= 2u(q_1)v(q_1)a_1'(q_1) + a_2'(q_1)v(q_1) \end{aligned} \tag{20}$$

If $SU + RT \neq 0$ at $q_1 = \overline{q}_1$, then

$$Re(\frac{d\lambda_k}{dq_1})|_{q_1=\overline{q}_1} = \frac{-(SU + RT)}{(R^2 + S^2)}|_{q_1=\overline{q}_1} \neq 0. \tag{21}$$

Now from equation (20) and $u = 0$, $v = \sqrt{a_2}$ we have

$$SU + RT = 2\,a_2(a_1a_2' + a_2a_1' - a_3') \quad \text{at} \quad q_1 = \overline{q}_1 \tag{22}$$

or

$$SU + RT = (2\,a_2)\frac{d}{dq_1}(a_1a_2 - a_3) \quad \text{at} \quad q_1 = \overline{q}_1,$$

where $a_1 a_2 - a_3 = Q(q_1)$, so $SU + RT \neq 0$ if

$$(a_1 a_2' + a_2 a_1' - a_3') = \frac{dQ}{dq_1}|_{\, q_1 = \, \bar{q}_1} \neq 0. \tag{23}$$

Now

$$\frac{d}{dq_1}(\, Re\, (\lambda_k(q_1)))|_{\, q_1 = \, \bar{q}_1} = \frac{-(SU + RT)}{(R^2 + S^2)} = \frac{-\, (2\, a_2)\, Q'(q_1)}{(R^2 + S^2)}|_{\, q_1 = \, \bar{q}_1} \neq 0. \tag{24}$$

This completes the proof.

We have performed the Hopf bifurcation analysis fo $\overline{E}_3$ as we are interested in seeing the dynamics when all three types of species are present.

8. Numerical results

The system (2) has been integrated numerically using the corresponding equilibrium values with slight perturbations as initial conditions. The integration is carried out using a fourth order Runge-Kutta method programmed in MatLab. We have observed that, the region of stability depends on the values of parameters selected. Tables 1-6, show the effect on the stability when we vary one of the parameters such as q_1, q_2, r_1, K, l_b and α and fixing the rest. Table (1) and Figure (1) show that the region of stability decreases on increasing the value of q_1. i.e. the rate at which individuals in the breeding population move to the suppressor population is a destabilizing factor. Higher birth rate, decreasing carrying capacity, more interaction of predator with breeder population, and higher rate by which suppressor population returns to breeding population, all promote the region of stability. For more details the reader can refer to Tables (2-6) and Figures (2-6). In Figure (7) we plot the polynomial (17) as a function of the parameter q_1 which has only one real root at the point $q_1 = 1.3280735$ and Figures (8) shows that there is a Hopf bifurcation for this model (2) where stable behavior changes to unstable behavior when we cross the bifurcation parameter $q_1 = 1.3280735$, (see Table (7)).

9. Discussion and Conclusion

Prey-predator model has been studied where prey population suppress breeding due to high predation and high population density. Non- breeding individuals have a better chance of avoiding predation than those in a reproductive state. Both breeding and suppressor prey populations are

Table 1. The effect on stability when value of q_1 is varying: The region of stability decreases as q_1 increases (see Figure (1))

l_b	q_2	α	$\boldsymbol{K}$	r_1	q_1
0.9	0.2	1.1	100	0.9	**0.5**
0.9	0.2	1.1	100	0.9	**1.0**
0.9	0.2	1.1	100	0.9	**1.2**

Table 2. The effect on stability when value of r_1 is varying: The region of stability increases as r_1 increases (see Figure (2))

l_b	q_2	α	$\boldsymbol{K}$	r_1	q_1
0.9	0.2	1.1	100	**0.5**	0.5
0.9	0.2	1.1	100	**1.5**	0.5
0.9	0.2	1.1	100	**3.5**	0.5

Table 3. The effect on stability when value of $\boldsymbol{K}$ is varying: The region of stability decreases as $\boldsymbol{K}$ increases (see Figure (3))

l_b	q_2	α	$\boldsymbol{K}$	r_1	q_1
0.9	0.2	1.1	**8**	0.9	0.5
0.9	0.2	1.1	**50**	0.9	0.5
0.9	0.2	1.1	**100**	0.9	0.5

exposed to predator population with varying degree. Predator feeds preferentially on the most numerous class prey species.

The results we have obtained can be summarized as follows: First of all in chapter 2, the prey population (breeder and suppressor) is found to be bounded by the environmental carrying capacity. Furthermore, we found that the predator population and the solution of the system (2) is bounded

Table 4. The effect on stability when value of α is varying: The region of stability decreases as α increases (see Figure (4))

l_b	q_2	α	$\boldsymbol{K}$	r_1	q_1
0.9	0.2	**1.1**	100	0.9	0.5
0.9	0.2	**1.9**	100	0.9	0.5
0.9	0.2	**2.8**	100	0.9	0.5

Table 5. The effect on stability when value of q_2 is varying: The region of stability decreases as q_2 increases (see Figure (5))

l_b	q_2	α	$\boldsymbol{K}$	r_1	q_1
0.9	**0.8**	1.1	100	0.9	0.5
0.9	**1.8**	1.1	100	0.9	0.5
0.9	**3.5**	1.1	100	0.9	0.5

Table 6. The effect on stability when value of l_b is varying: The region of stability decreases as l_b increases (see Figures (6))

l_b	q_2	α	$\boldsymbol{K}$	r_1	q_1
0.4	0.2	1.1	100	0.9	0.5
0.9	0.2	1.1	100	0.9	0.5
1.4	0.2	1.1	100	0.9	0.5

Table 7. Hopf bifurcation of the model with respect to q_1

Interval	**State**
$0 < q_1 < 1.3280735$	Stable Interval
$q_1 = 1.3280735$	Bifurcation point
$q_1 > 1.3280735$	Unstable Interval

if we found four equilibria. The first of these points is where all populations get extinct which is an unstable state. We also noted that as the origin is unstable, the whole ecosystem can not eventually become extinct. The suppressor and predator free equilibrium $\overline{E}_1$ is locally asymptotically

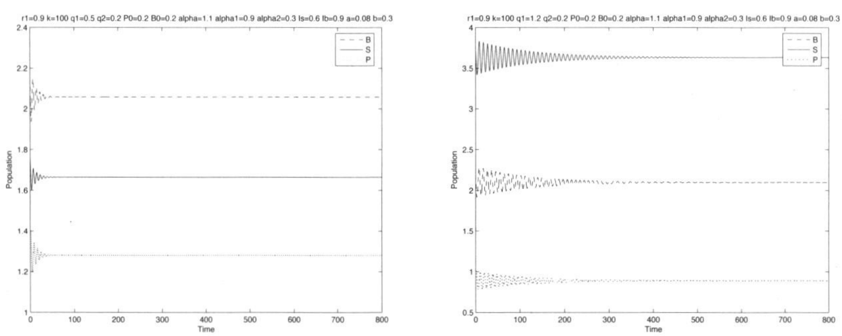

Fig. 1. The effect on stability using values of $q_1 = 0.5$ (left) and 1.2 (right).

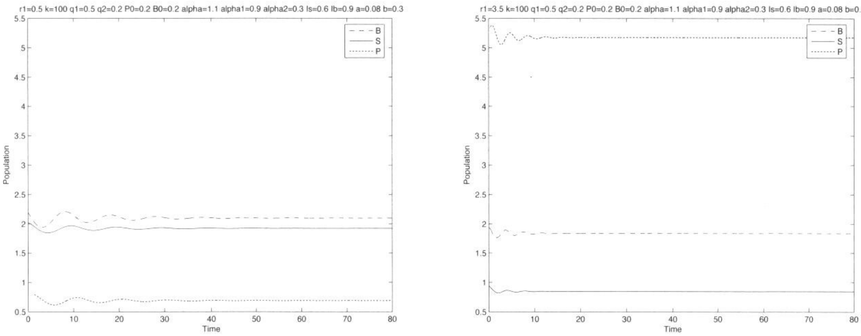

Fig. 2. The effect on stability using values of $r_1 = 0.5$ (left) and 3.5 (right).

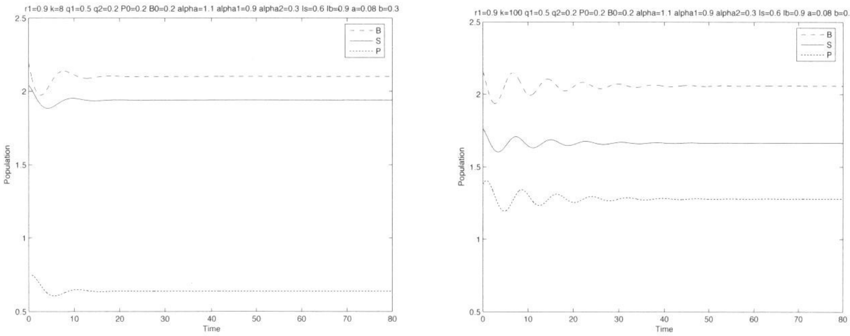

Fig. 3. The effect on stability using values of $K = 8$ (left) and 100 (right).

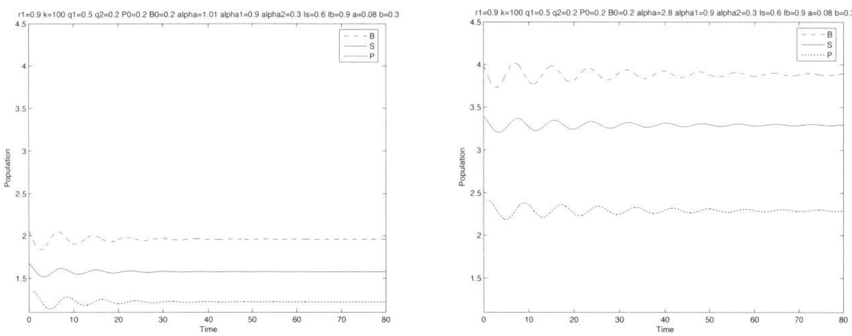

Fig. 4. The effect on stability using values of α =1.01 (left) and 2.8 (right).

stable if $\frac{\alpha_1 \, l_b \, K}{\alpha} < 1$ otherwise it is unstable. The equilibrium $\overline{E}_2$ is feasible if $r_1 \; B_0 > q_2 \, K$ and will be bounded if $\alpha_2 \, l_s \overline{S} < 1$. The non-zero equi-

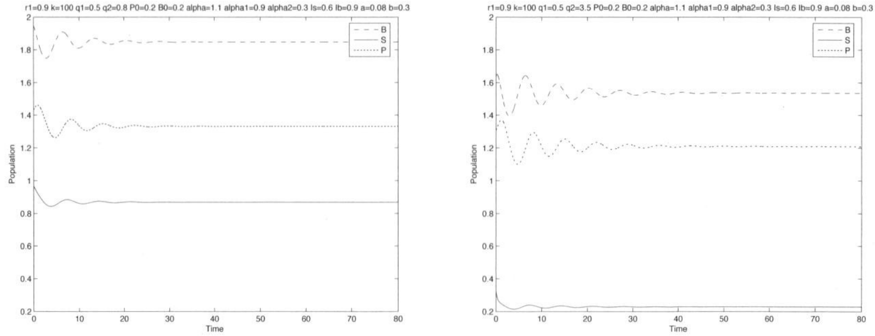

Fig. 5. The effect on stability using values of $q_2 = 0.8$ (left) and 3.5 (right).

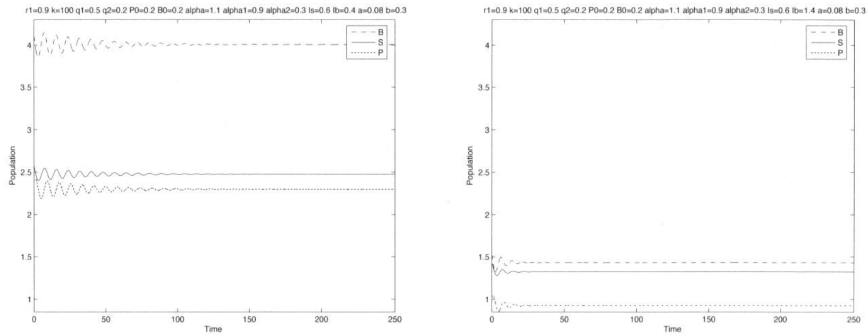

Fig. 6. The effect on stability using values of $l_b = 0.4$ (left) and 1.4 (right).

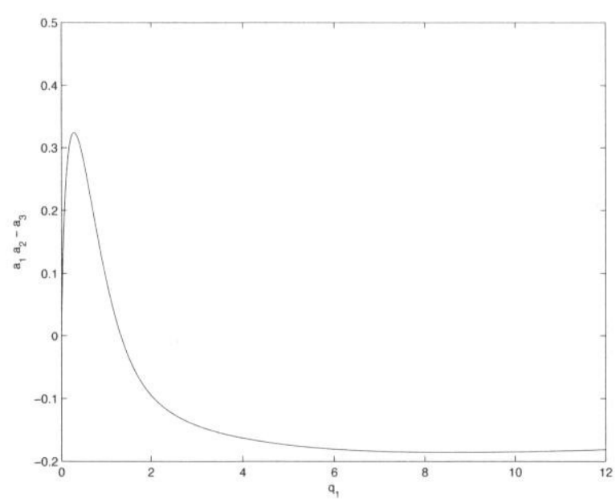

Fig. 7. Plot of the polynomial $a1a2 - a3$ as a function of q_1 on the interval [0, 12]. This function has only one real root $q_1 = 1.3280735$.

librium $\overline{E}_3$ exists if $(\alpha_1\, l_b\, \overline{X}^2 + \alpha_2\, l_s)K > (\alpha\, \overline{X} + 1)\, (\overline{X} + 1)$and is stable if it satisfies the Routh - Hurwitz criteria. Hopf bifurcation analysis

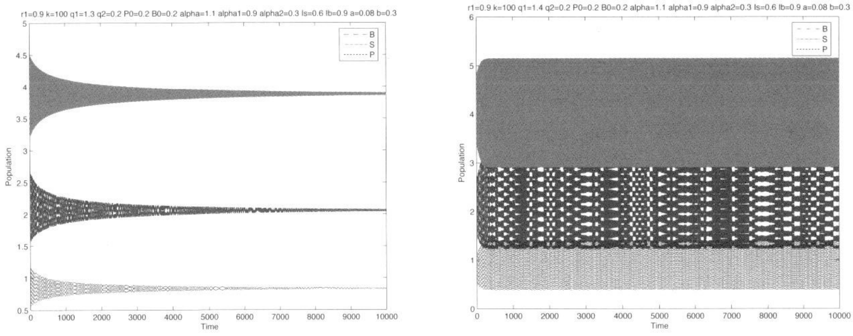

Fig. 8. The model is stable when $0 < q_1 < 1.3280735$ (left) and unstable with $q_1 > 1.3280735$ (right)

has been carried out with respect to q_1 as a parameter. In addition, there are three ecological meaningful equilibria which belong to interaction of breeder prey with predator in the absence of suppressor prey. The equilibrium $\widehat{E}_0$ is unstable, while the equilibria $\widehat{E}_1$ and $\widehat{E}_2$ are stable if $\alpha_1 \, l_b' \, K < 1$ and $r_1 > \frac{q_1 \overline{P} a \, K}{(\overline{P}+P_0+a\overline{B})^2}$ respectively, otherwise they are unstable. If there is no breeder population then ultimately the whole population of prey and predator will become extinct. The case when the predator population is absent the breeder prey will not suppress breeding so the whole prey population will be breeder.

Views on the consequences of antipredatory behavior on population dynamics have undergone rapid changes in recent years. Breeding suppression was first presented as an adaptive mechanism in individuals to avoid predation (Ylönen, 1994). Models of (Gyllenberg et al. 1996, Ruxton and Lima,1997) suggest that this suppression effect is rather likely to be stabilization of the dynamics. Our results suggest that breeding suppression in prey population is destabilizing. These results conflict with earlier models (i.e. Gyllenberg et al. 1996, Ruxton and Lima 1997) but support the result of Hik's (1995) that increasing the length of time in a suppressed state acts to reduce the strength of the stabilization caused by PIBS. PIBS in snowshoe hare exceed one or two years.

References

1. Arnold, V.I, 1971, Ordinary Differential Equations, Springer Verlag, ISBM 3-540-54813-0-387-54813-0.
2. Gyllenberg M., Hanski, I. and Lindstr, T., 1996, *A predator–prey model with*

optimal suppression of reproduction in the prey. Math Biosci, 134:119 –152.

3. Heikkil, J., Kaarsalo, K., Mustonen, O. and Pekkarinen, P., 1993, *Influence of predation risk on early development and maturation in three species of Clethrionomys voles*, Annales Zoologici Fenniei 30:153-161.
4. Hik, D. S., 1995, *Does risk of predation influence population dynamics? Evidence from the cyclic decline of snowshoe hares.* Wildl. Res. 22, 115–129.
5. Hirschfield, M. and Tinkle, D.W., 1975 , *Natural selection and the evolution of reproductive effort*, Proceedings of the National Academy of Sciences,USA 72:2227-223
6. Korpimki, E., Norrdahl, K. and Valkama, J., 1994 ,*Reproductive investment under fluctuating predation risk: Microtine rodents and small mustelids*, Evalutionary Ecology 8:357-368.
7. Koskela, E., and Ylönen, H., 1995, *Suppressed breeding in the field vole (Microtus agrestis)*, An adaptation to cyclically fluctuating predation risk Behavioral Ecology 6:311-315.
8. Lima, S.L. and Dill, L.M., 1990, *Behavioral decisions made under the risk of predation*: a review and prospectus. Can. J. Zool. 68: 619-640.
9. Marsden, J.E. and McCracken, M.,1976,*The Hopf bifurcation and its applications*, Springer-Verlag, NewYork.
10. Oksanen, L. and Lundberg , P., 1995, *optimization of reproductive effort and foraging time in mammals :the influence of resource level and predation risk.* Evolutionary Ecology 9: 45-56.
11. Ronkainen, H. and Ylönen, H.,1994. Behaviour of cyclic bank voles under risk of mustelid predation: do females avoid copulations? Oecologia, 97, 377-381.
12. Ruxton,G.D.,and Lima,S.L.,1997,*Predator-induced breeding suppression and its consequences for predator-prey population dynamics,* Proc.R.Soc.Lond B,264,409-415.
13. Ruxton,G.D., Khan,Q.J.A. and Al-Lawati ,M., 2002,*The stability of internal equilibria in predator-prey models with breeding suppression*, IMA Journal of Mathematics Applied in Medicine and Biology,19:207-219.
14. Ylönen, H., 1989, Weasels, *Mustela nivalis, suppress reproduction in the cyelic bank vales*, clethrionomys glareolus, Oikos 55:138-140.
15. Ylönen, H., Jedrzejewska,B., Jedrzejewska, W. and Heikkila, J.,1992. *Antipredatory behaviour of elethriono-mys voles.* David and Goliath arms rael Ann. Zoal.
16. Ylönen, H., and Ronkainen,H., 1994, *Breeding suppression in the bank vole as antipredatory adaptation in a predictable environment*, Evol. Ecol. 8: 658 - 666.
17. Ylönen, H., 1994. *Vole cycles and antipredatory behaviour.* Trends in Ecology and Evolution 9 :426-430.

SEI MODEL WITH VARYING TRANSMISSION AND MORTALITY RATES

GERGELY RÖST

Analysis and Stochastics Research Group, Hungarian Academy of Sciences
Bolyai Institute, University of Szeged, Hungary
Aradi vértanúk tere 1., H-6725 Szeged

An SEI model with distributed delay is proposed where the transmission and the death rates depend on the age of infection. The basic reproduction number $\mathcal{R}_0$ is identified as a threshold quantity for the stability of equilibria. If $\mathcal{R}_0 < 1$, then the disease-free equilibrium is globally asymptotically stable and the disease dies out. On the contrary, if $\mathcal{R}_0 > 1$, then a locally asymptotically stable endemic equilibrium appears, and applying a permanence theorem for infinite dimensional systems we obtain that the disease persists in the population.

AMS 2000: 92D30, 37C70, 34K10

Keywords: disease model, distributed delay, varying infectivity and death rate

1. Introduction

Many compartmental models in mathematical epidemiology assume the homogeneity of the infected class: all individuals in that compartment share the same epidemiological parameters. In reality, as time elapses and the disease develops within the host, its infectivity continuously changes. Disease induced mortality rate may also change during the course of infection. The purpose of this paper is to incorporate these features into an SEI type model. Besides multistage models[6,8] , approaches keeping track of an individual's infection-age have existed[1,2,5,7] to capture this variability. However, the model we formulate in this paper differs from the previous ones since it can be transformed into a system of differential equation with distributed delays, which is easier to deal with than integro-differential or Volterra-type models.

The paper is organized as follows. In section 2, taking into account the age of infection as a parameter, and allowing varying infectivity and death rates, we formulate an SEI model with distributed and constant delays. We identify the basic reproduction number $\mathcal{R}_0$ in terms of the model parameters as a threshold quantity in section 3. If $\mathcal{R}_0 < 1$, the disease dies out and all solutions converge to the disease free equilibrium. In section 4 we show that a stable endemic equilibrium appears if $\mathcal{R}_0 > 1$. In section 5 we prove that the disease is endemic in the sense of permanence whenever $\mathcal{R}_0 > 1$.

2. Derivation of the model

We divide a given population into the following categories: susceptibles (those who are capable of contracting the disease); exposed (those who are infected but not yet infectious); infectives (those who are infected and capable of transmitting the disease). Denote the number of individuals at time t in these classes by $S(t), E(t), I(t)$, respectively. Let $i(t,a)$ represent the density of infected individuals with respect to the age of infection a at the current time t, where $a \le \tau$, then $I(t) = \int_0^\tau i(t,a)da$. We introduce the function $0 \le \beta(a) \le \beta$ to express the infectivity according to the age of infection a. In what follows, Λ denotes the constant recruitment rate, β is the maximal transmission rate, Δ is the natural death rate, $\delta(a) \ge 0$ is the disease-induced death rate which depends on the age of infection as well, $1/\mu$ is the average incubation period. At age τ of infection, we remove all remaining individuals from the class I who has survived. Thus, $\tau > 0$ represents the maximal duration of the infectious period. All the constants above are assumed to be positive. Then, using bilinear incidence in the force of infection corrected by the infectivity factor due to the age of infection, we arrive at the SEI type model

$$\begin{aligned}
\frac{dS(t)}{dt} &= \Lambda - S(t)\int_0^\tau \beta(a)i(t,a)da - \Delta S(t),\\
\frac{dE(t)}{dt} &= S(t)\int_0^\tau \beta(a)i(t,a)da - (\mu+\Delta)E(t),\\
\frac{dI(t)}{dt} &= \mu E(t) - \Delta I(t) - \int_0^\tau \delta(a)i(t,a)da - i(t,\tau)
\end{aligned}$$

The time evolution of the density $i(t,a)$ is given by

$$\left(\frac{\partial}{\partial t} + \frac{\partial}{\partial a}\right)i(t,a) = -(\Delta + \delta(a))i(t,a), \tag{1}$$

subject to the boundary condition

$$i(t,0) = \mu E(t).$$

Solving (1) leads to

$$i(t,a) = i(t-a,0)e^{-(\Delta a+\int_0^a \delta(u)du)} = \mu E(t-a)e^{-(\Delta a+\int_0^a \delta(u)du)}, \qquad (2)$$

and we obtain the following deterministic model of delay differential equations with distributed and constant delays:

$$\frac{dS(t)}{dt} = \Lambda - S(t)\int_0^\tau \beta(a)\mu E(t-a)e^{-(\Delta a+\int_0^a \delta(u)du)}da - \Delta S(t), \qquad (3)$$

$$\frac{dE(t)}{dt} = S(t)\int_0^\tau \beta(a)\mu E(t-a)e^{-(\Delta a+\int_0^a \delta(u)du)}da - (\mu+\Delta)E(t), \quad (4)$$

$$\begin{aligned}\frac{dI(t)}{dt} &= \mu E(t) - \Delta I(t) - \int_0^\tau \delta(a)\mu E(t-a)da \\ &\quad -e^{-(\Delta\tau+\int_0^\tau \delta(u)du)}\mu E(t-\tau). \end{aligned} \qquad (5)$$

From (2) we can express $I(t)$ as a function of a solution $E(t)$:

$$I(t) = \mu\int_0^\tau E(t-a)e^{-(\Delta a+\int_0^a \delta(u)du)}da.$$

All the information (boundedness, convergence, etc.) for $I(t)$ can be obtained from the E-component of the solution, and equation (5) decouples. Therefore it is sufficient to restrict our attention to the system (3-4), and we do this in the sequel. Clearly the state of system (3-4) at time t is specified by $S(t) \in \mathbb{R}$ and $E_t \in C([-\tau,0],\mathbb{R})$, the space of continuous functions on the interval $[-\tau,0]$. It is straightforward to see that solutions of (3-4) preserve non-negativity.

Proposition 2.1. *The system (3-4) is point dissipative; that is there exists an $M > 0$ such that for any non-negative solution of (3-4), there exists a $T > 0$ such that $S(t) \leq M$ and $E(t) \leq M$ for all $t \geq T$.*

Proof. Consider an arbitrary nonnegative solution. For $W(t) = S(t) + E(t)$, we have

$$\frac{dW(t)}{dt} = \Lambda - \Delta W(t) - \mu E(t) \leq \Lambda - \Delta W(t).$$

Since any nonnegative solution of $w'(t) = \Lambda - \Delta w(t)$ satisfies

$$\lim_{t\to\infty} w(t) = \Lambda/\Delta,$$

by a standard comparison argument we obtain

$$\limsup_{t\geq 0} W(t) \leq \frac{\Lambda}{\Delta}.$$

We conclude that for any $\varepsilon > 0$, there is a $T > 0$ such that the nonnegative solution of (3-4) satisfies

$$S(t) \leq \frac{\Lambda}{\Delta} + \varepsilon, \quad E(t) \leq \frac{\Lambda}{\Delta} + \varepsilon$$

whenever $t \geq T$. Consequently, we can choose any $M > \frac{\Lambda}{\Delta}$. Additionally, we obtain the a-priori estimate

$$S(t), E(t) \leq W(t) \leq \Lambda/\Delta + \exp(-\Delta t)\big(S(0) + E(0) - \Lambda/\Delta\big). \qquad \square$$

3. Basic reproduction number and the global stability of the disease-free equilibrium

Clearly our model has a disease-free equilibrium $P_0 = (S_0, 0)$ where $S_0 = \Lambda/\Delta$. To find the basic reproduction number $\mathcal{R}_0$, we introduce a single exposed individual into a totally susceptible population in the disease-free equilibrium at $t = 0$. The probability of the presence of this individual in the E-class after time t is given by $e^{-(\mu+\Delta)t}$, so the expected number of generated secondary infections can be calculated by

$$\mathcal{R}_0 = S_0 \int_0^\infty \int_0^\mathcal{T} \beta(a)\mu e^{-(\Delta a + \int_0^a \delta(u)du)} e^{-(\mu+\Delta)t} dadt,$$

which, after interchanging the integrals, reduces to

$$\mathcal{R}_0 = \frac{S_0 \mu}{\mu + \Delta} \int_0^\mathcal{T} \beta(a) e^{-(\Delta a + \int_0^a \delta(u)du)} da. \tag{6}$$

Next we show that $\mathcal{R}_0$ determines the stability of the disease-free equilibrium and the disease dies out when $\mathcal{R}_0 < 1$.

Theorem 3.1. *The disease free equilibrium is globally asymptotically stable if $\mathcal{R}_0 < 1$, and unstable if $\mathcal{R}_0 > 1$.*

Proof. For any $\varepsilon > 0$, we define

$$\mathcal{R}_\varepsilon = \frac{\mu}{\mu + \Delta}\left(\frac{\Lambda}{\Delta} + \varepsilon\right) \int_0^\mathcal{T} \beta(a) e^{-(\Delta a + \int_0^a \delta(u)du)} da.$$

Then $\lim_{\varepsilon \to 0} \mathcal{R}_\varepsilon = \mathcal{R}_0$ and $\mathcal{R}_\varepsilon < 1$ if $\mathcal{R}_0 < 1$ and ε is sufficiently small. In Proposition 1. we have shown that for any $\varepsilon > 0$ there is a $T > 0$ such that $S(t) \leq \frac{\Lambda}{\Delta} + \varepsilon$ whenever $t > T$. Thus, without loss of generality, we can suppose that $S(t) \leq \frac{\Lambda}{\Delta} + \varepsilon$ for all $t \geq 0$. This yields that the exposed population $E(t)$ is bounded above by the solutions of the linear equation

$$\frac{dE(t)}{dt} = \left(\frac{\Lambda}{\Delta} + \varepsilon\right) \int_0^\mathcal{T} \beta(a)\mu E(t-a) e^{-(\Delta a + \int_0^a \delta(u)du)} da - (\mu + \Delta)E(t).$$

Now we show that the characteristic roots of this linear equation have negative real parts and the global stability of the disease-free equilibrium follows from the standard comparison argument. Using the exponential Ansatz $e^{\lambda t}$, we arrive at the characteristic function

$$h(\lambda) = \Big(\frac{\Lambda}{\Delta} + \varepsilon\Big)\mu \int_0^{\mathcal{T}} \beta(a) e^{-(\lambda a + \Delta a + \int_0^a \delta(u)du)} da - (\lambda + \mu + \Delta). \qquad (7)$$

We check that each characteristic root has negative real part. Suppose that $\lambda = x + iy$ is a root of $h(\lambda)$ with $x > 0$. Then $|e^{-\lambda a}| < 1$ for any $a > 0$, and

$$\begin{aligned} 1 =& \Bigg|\frac{\big(\frac{\Lambda}{\Delta} + \varepsilon\big)\mu}{\lambda + \mu + \Delta} \int_0^{\mathcal{T}} \beta(a) e^{-(-\lambda a + \Delta a + \int_0^a \delta(u)du)} da\Bigg| \\ \leq & \frac{\big(\frac{\Lambda}{\Delta} + \varepsilon\big)\mu}{|\lambda + \mu + \Delta|} \int_0^{\mathcal{T}} \beta(a) |e^{-\lambda a}| e^{-(\Delta a + \int_0^a \delta(u)du)} da < \mathcal{R}_\varepsilon, \end{aligned}$$

which is a contradiction. Therefore, if $\mathcal{R}_0 < 1$, then all roots have negative real part, thus $\lim_{t\to\infty} E(t) = 0$ and all solutions converge to the disease free equilibrium.

If $\mathcal{R}_0 > 1$, then the linearization about the disease free equilibrium gives for (4) that

$$\frac{dE(t)}{dt} = S_0\mu \int_0^{\mathcal{T}} \beta(a) E(t-a) e^{-(\Delta a + \int_0^a \delta(u)du)} da - (\mu + \Delta)E(t),$$

which leads to the characteristic function

$$\hat{h}(\lambda) = S_0\mu \int_0^{\mathcal{T}} \beta(a) e^{-(\lambda a + \Delta a + \int_0^a \delta(u)du)} da - (\lambda + \mu + \Delta). \qquad (8)$$

Clearly, $\hat{h}(\lambda)$ is a monotone decreasing continuous function for nonnegative real λ and $\hat{h}(\infty) = -\infty$. We have

$$\hat{h}(0) = S_0\mu \int_0^{\mathcal{T}} \beta(a) e^{-(\Delta a + \int_0^a \delta(u)du)} da - (\mu + \Delta) = (\mu + \Delta)(\mathcal{R}_0 - 1).$$

If $\mathcal{R}_0 > 1$, then there exists a positive real root of $\hat{h}(\lambda)$, and the disease-free equilibrium is unstable. □

4. The endemic equilibrium

Theorem 4.1. *An endemic equilibrium exists if and only if* $\mathcal{R}_0 > 1$. *Moreover, the endemic equilibrium, if exists, is unique and locally asymptotically stable.*

Proof. An endemic equilibrium $P^* = (S^*, E^*)$ must satisfy the algebraic equations

$$\Delta S^* = \Lambda - S^* \mu \int_0^{\mathcal{T}} \beta(a) E^* e^{-(\Delta a + \int_0^a \delta(u)du)} da, \tag{9}$$

$$(\mu + \Delta) E^* = S^* \mu \int_0^{\mathcal{T}} \beta(a) E^* e^{-(\Delta a + \int_0^a \delta(u)du)} da. \tag{10}$$

Since $E^* \neq 0$, (10) yields

$$S_0 / S^* = \mathcal{R}_0, \quad \text{or} \quad S^* = \frac{\Lambda}{\mathcal{R}_0 \Delta}. \tag{11}$$

Simple calculations on (9) show that

$$\frac{\Lambda}{\mathcal{R}_0} = \Lambda - (\Delta + \mu) E^*;$$

that is

$$E^* = \frac{\Lambda}{\Delta + \mu} (1 - \frac{1}{\mathcal{R}_0}).$$

So, we conclude that $E^* > 0$ if and only if $\mathcal{R}_0 > 1$.

Next we show the local asymptotic stability of the endemic equilibrium. Introducing the new variables $s(t) = S(t) - S^*$, $f(t) = E(t) - E^*$, we obtain the linearized system about the endemic equilibrium $P^* = (S^*, E^*)$

$$\frac{ds(t)}{dt} = -\int_0^{\mathcal{T}} \Big(s(t)E^* + S^* e(t-a)\Big) \mu \beta(a) e^{-(\Delta a + \int_0^a \delta(u)du)} da - \Delta s(t),$$

$$\frac{de(t)}{dt} = \int_0^{\mathcal{T}} \Big(s(t)E^* + S^* e(t-a)\Big) \mu \beta(a) e^{-(\Delta a + \int_0^a \delta(u)du)} da - (\mu + \Delta) e(t).$$

Noticing

$$\mu \int_0^{\mathcal{T}} \beta(a) E^* e^{-(\Delta a + \int_0^a \delta(u)du)} da = \mathcal{R}_0 - 1,$$

we have

$$\frac{ds(t)}{dt} = -s(t)(\mathcal{R}_0 - 1) - \frac{\Lambda \mu}{\mathcal{R}_0 \Delta} \int_0^{\mathcal{T}} \beta(a) e(t-a) e^{-(\Delta a + \int_0^a \delta(u)du)} da - \Delta s(t),$$

$$\frac{de(t)}{dt} = s(t)(\mathcal{R}_0 - 1) + \frac{\Lambda \mu}{\mathcal{R}_0 \Delta} \int_0^{\mathcal{T}} \beta(a) e(t-a) e^{-(\Delta a + \int_0^a \delta(u)du)} da - (\mu + \Delta) e(t).$$

Using the exponential Ansatz $e^{\lambda t}(s_0, e_0)$, we have

$$(\lambda + \Delta) s_0 = -(\lambda + \mu + \Delta) e_0$$

and

$$(\mathcal{R}_0 - 1 + \Delta + \lambda)s_0 = -\Big(\frac{\Lambda\mu}{\mathcal{R}_0\Delta}\int_0^\tau \beta(a)e^{-\lambda a}e^{-(\Delta a+\int_0^a \delta(u)du)}da\Big)e_0,$$

thus we obtain the characteristic equation

$$(\lambda+\mu+\Delta)(\mathcal{R}_0-1+\Delta+\lambda) = (\lambda+\Delta)\Big(\frac{\Lambda\mu}{\mathcal{R}_0\Delta}\int_0^\tau \beta(a)e^{-\lambda a}e^{-(\Delta a+\int_0^a \delta(u)du)}da\Big). \tag{12}$$

Suppose that $\lambda = x + iy$ is a root and $x \geq 0$, that implies $|e^{-\lambda a}| \leq 1$ for any $a \geq 0$. Now the inequalities

$$\Big|\frac{\Lambda\mu}{\mathcal{R}_0\Delta}\int_0^\tau \beta(a)e^{-\lambda a}e^{-(\Delta a+\int_0^a \delta(u)du)}da\Big| \leq \mu + \Delta \leq |\lambda + \mu + \Delta|$$

and

$$|\Delta + \lambda| < |\mathcal{R}_0 - 1 + \Delta + \lambda|$$

follow, contradicting to (12). Therefore, every root has negative real part and the endemic equilibrium is locally asymptotically stable if $\mathcal{R}_0 > 1$. □

5. Persistence

Denote by $T(t) : X \to X$, $t \geq 0$ the family of solution operators corresponding to (3-4), where $X = R_0^+ \times C_0^+$. Here C_0^+ denotes the set of non-negative continuous functions on the interval $[-\tau, 0]$. The ω-limit set $\omega(x)$ of x consists of $y \in X$ such that there is a sequence $t_n \to \infty$ as $n \to \infty$ with $T(t_n)x \to y$ as $n \to \infty$. We shall apply the following permanence theorem of Hale & Waltman[3] , in the spirit of Röst & Wu[4] .

Theorem. Suppose that we have the following:
(i) X^0 is open and dense in X with $X^0 \cup X_0 = X$ and $X^0 \cap X_0 = \emptyset$;
(ii) the solution operators $T(t)$ satisfy

$$T(t) : X^0 \to X^0, \quad T(t) : X_0 \to X_0;$$

(iii) $T(t)$ is point dissipative in X;
(iv) there is a $t_0 \geq 0$ such that $T(t)$ is compact for all $t \geq t_0$;
(v) $\mathcal{A} = \bigcup_{x\in A_b} \omega(x)$ is isolated and has an acyclic covering N, where A_b is the global attractor of $T(t)$ restricted to X_0 and $N = \cup_{i=1}^k N_i$;
(vi) for each $N_i \in N$,

$$W^s(N_i) \cap X^0 = \emptyset,$$

where W^s refers to the stable set.

Then $T(t)$ is a uniform repeller with respect to X^0, i.e. there is an $\eta > 0$ such that for any $x \in X^0$, $\liminf_{t\to\infty} d(T(t)x, X_0) \geq \eta$.

Theorem 5.1. *If $\mathcal{R}_0 > 1$, then the disease is endemic; more precisely, there exists an $\eta > 0$ such that*

$$\liminf_{t\to\infty} E(t) \geq \eta.$$

Proof. Let

$$X^0 = \{(S, \phi) \in X : \phi(\theta) > 0 \text{ for some } \theta < 0\}$$

$$X_0 = \{(S, \phi) \in X : \phi(\theta) = 0 \text{ for all } \theta \leq 0\}.$$

We check all the conditions of the permanence theorem. It is straightforward to see that (i) and (ii) are satisfied. The point dissipativity has been proved in Proposition 1, so we have (iii). Applying the Arzela-Ascoli theorem we obtain (iv) with $t_0 = \tau$.

Regarding (v), clearly $\mathcal{A} = \{P_0\}$ (now $P_0 = (\Lambda/\Delta, 0) \in X$) and isolated. Hence the covering is simply $N = \{P_0\}$, which is acyclic (there is no orbit which connects P_0 to itself in X_0).

It remains to show that $W^s(P_0) \cap X^0 = \emptyset$. Suppose the contrary, that is there exists a solution in X^0 such that

$$\lim_{t\to\infty} S(t) = S_0, \quad \lim_{t\to\infty} E(t) = 0.$$

Since $\mathcal{R}_0 > 1$, there exists an $\varepsilon > 0$ such that

$$(S_0 - \varepsilon)\mu \int_0^\tau \beta(a) e^{-(\Delta a + \int_0^a \delta(u)du)} da > \mu + \Delta.$$

There exists a t_0 such that for $t \geq t_0$, $S(t) > S_0 - \varepsilon$ and hence

$$E'(t) \geq (S_0 - \varepsilon)\mu \int_0^\tau \beta(a) E(t-a) e^{-(\Delta a + \int_0^a \delta(u)du)} da - (\mu + \Delta)E(t).$$

If $E(t) \to 0$, as $t \to \infty$, then by a standard comparison argument and the nonnegativity, the solution $n(t)$ of

$$n'(t) = (S_0 - \varepsilon)\mu \int_0^\tau \beta(a) n(t-a) e^{-(\Delta a + \int_0^a \delta(u)du)} da - (\mu + \Delta)n(t)$$

with initial data $n_0 = E_0$, has to converge to 0 as well. By the mean value theorem for integrals we have that for any t there is a ξ_t such that

$$\int_0^\tau \beta(a) n(t-a) e^{-(\Delta a + \int_0^a \delta(u)du)} da = n(\xi_t) \int_0^\tau \beta(a) e^{-(\Delta a + \int_0^a \delta(u)du)} da$$

and $t - \tau \le \xi_t \le t$. Define

$$V(t) := n(t) + (\mu + \Delta) \int_{\xi_t}^{t} n(s)ds.$$

Differentiating with respect to time gives

$$\frac{dV}{dt} = \Big(\beta(S_0 - \varepsilon) \int_0^{\tau} k(a)\mu e^{-(\Delta a + \int_0^a \delta(u)du)} da - (\mu + \Delta)\Big) n(\xi_t) \ge 0.$$

Therefore, $V(t)$ goes to infinity or approaches a positive limit as $t \to \infty$. On the other hand, by the definition of V, $\lim_{t\to\infty} n(t) = 0$ implies $\lim_{t\to\infty} V(t) = 0$, a contradiction. Thus $W^s(P_0) \cap X^0 = \emptyset$ and we can apply Theorem 4.2 of Hale & Waltman[3] to obtain that for some $\eta > 0$,

$$\liminf_{t\to\infty} E(t) > \eta. \qquad \square$$

Though our calculations has been done for the reduced system (3-4), and we are interested in the dynamics of the infectious class, we easily obtain that for $\mathcal{R}_0 < 1$, $E(t) \to 0$ implies $I(t) \to 0$, and for $\mathcal{R}_0 > 1$ from (2) we have the endemic equilibrium

$$I^* = \frac{\Lambda\mu}{\Delta + \mu}\big(1 - \frac{1}{\mathcal{R}_0}\big)\Big(\int_0^{\tau} e^{-(\Delta a + \int_0^a \delta(u)du)} da\Big),$$

furthermore, from (2) we obtain

$$\liminf_{t\to\infty} I(t) > \mu\eta \int_0^{\tau} e^{-(\Delta a + \int_0^a \delta(u)du)} da > 0.$$

Hence, applying the permanence theorem above, we obtain that the disease will always be present in the population when $\mathcal{R}_0 > 1$.

Acknowledgements

The author acknowledges the support of the Hungarian Scientific Research Fund, grant OTKA K75517, the Bolyai Research Scholarship of the Hungarian Academy of Sciences, and the TÁMOP-4.2.2/08/1/2008-0008 program of the Hungarian National Development Agency.

References

1. C. Castillo-Chavez, K. Cooke, W. Huang & S. A. Levin. On the role of long incubation periods in the dynamics of acquired immunodeficiency syndrome (AIDS). I. Single population models *J. Math. Biol. 27*(1989), 373–398.
2. H. R. Thieme & C. Castillo-Chavez. How may infection-age-dependent infectivity affect the dynamics of HIV/AIDS? *SIAM J. Appl. Math. 53*(1993), 1447–1479.

3. J. K. Hale & P. Waltman. Persistence in infinite-dimensional systems. *SIAM J. Math. Anal. 20*(1989), 388–395.
4. G. Röst & J. Wu SEIR epidemiological model with varying infectivity and infinite delay *Math. Biosci. Eng. 5:(2)*(2008) 389–402
5. Y. Zhou, Y. Shao, Y. Ruan, J. Xu, Z. Ma, C. Mei & J. Wu Modeling and prediction of HIV in China: transmission rates structured by infection ages. *Math. Biosci. Eng. 5:(2)*(2008) 403–418
6. A. B. Gumel, C. C. McCluskey & P. van den Driessche Mathematical study of a staged-progression HIV model with imperfect vaccine, *Bull. Math. Biol. 68:(8)*(2006) 2105–2128
7. H. Inaba Endemic threshold results in an age-duration-structured population model for HIV infection *Math. Biosci. 201:(1-2)*(2006) 15–47
8. B. D. Corbett, S. M. Moghadas & A. B. Gumel, Subthreshold domain of bistable equilibria for a model of HIV epidemiology, *Int. J. Math. Math. Sci. 58*(2003) 3679–3698

TRAJECTORIES AND STABILITY REGIONS OF THE LAGRANGIAN POINTS IN THE GENERALIZED CHERMNYKH-LIKE PROBLEM

BADAM SINGH KUSHVAH

Department of Applied Mathematics, Indian School of Mines, Dhanbad 826004, Jharkhand, India.

E-mail: bskush@gmail.com,kushvah.bs.am@ismdhanbad.ac.in

`http://ismdhanbad.ac.in`

The Lagrangian points for the Sun-Earth system are considered due to their special importance for the scientific community for the design of space missions. The location of the Lagrangian points with the trajectories and stability regions are computed numerically for the initial conditions very close to the points. The influence of belt, effect of radiation pressure due to the Sun and oblateness effect of second primary(finite body Earth) is presented for various values of parameters. The collinear points are asymptotically stable within a specific interval of time t correspond to the values of parameters and initial conditions.

Keywords: Trajectory; Stability; Equilibrium points; Radiation pressure; Oblateness; Rtbp.

1. Introduction

A modified restricted three body problem is considered which was first time studied by.[1] This problem generalizes two classical problems of Celestial mechanics: the two fixed center problem and the restricted three body problem. This gives wide perspectives for applications of the problem in celestial mechanics and astronomy. The importance of the problem in astronomy has been addressed by.[2] It is supposed that the primary bodies are moving in circular orbits about their center of mass. The well-known five equilibrium points of the planar restricted three-body problem are very important for astronautical applications, collinear points are metastable points in the sense that, like a ball sitting on top of a hill and the triangular points are conditionally stable.[3] These Lagrange points have proven to be very useful indeed since a spacecraft can be made to execute a small orbit about one of these Lagrange points with a very small expenditure of energy.[4,5] Because of the its unobstructed view of the Sun, the Sun-Earth L_1 is a good place to put instruments for doing solar science. NASA's Genesis Discovery Mission has been there, designed completely using invariant manifolds and other tools form dynamical systems theory. In 1972, the International Sun-Earth Explorer (ISEE) was established, joint project of NASA and the European Space Agency(ESA). The ISEE-3 was launched into a halo orbit around the Sun-Earth L_1 point in 1978, allowing it to collect data on solar wind conditions upstream from the Earth.[6] In the mid-1980s the Solar and Heliospheric Observatory (SOHO)[7] is places in a halo orbit around the Sun-Earth L_1 position, about a million miles the Sun ward from the Earth. They have provided useful places to "park"a spacecraft for observations.

The goal of present paper is to investigate the nature of collinear equilibrium points because of the interested point to the mission design.These results provide new information on the behavior of trajectories around the Lagrangian points for different possible set values of the parameters proposed by.[8]

2. Location of Lagrangian Points

It is supposed that the motion of an infinitesimal mass particle is influenced by the gravitational force from primaries and a belt of mass M_b. The units of the mass, the distance and the time are taken such that sum of the masses and the distance between primaries are unities, the unit of the time i.e. the time period of m_1 about m_2 consists of 2π units such that the Gaussian constant of gravitational $\mathbf{k}^2 = 1$. Then perturbed mean motion n of the primaries is given by $n^2 = 1 + \frac{3A_2}{2} + \frac{2M_b r_c}{(r_c^2+T^2)^{3/2}}$, where $T = \mathbf{a}+\mathbf{b}$, $\mathbf{a}, \mathbf{b}$ are flatness and core parameters respectively which determine the density profile of the belt, $r_c^2 = (1-\mu)q_1^{2/3} + \mu^2$, $A_2 = \frac{r_e^2 - r_p^2}{5r^2}$ is the oblateness coefficient of m_2; r_e, r_p are the equatorial and polar radii of m_2 respectively, $r = \sqrt{x^2+y^2}$ is the distance between primaries and $x = f_1(t), y = f_2(t)$ are the functions of the time t i.e. t is only independent variable. The mass parameter is $\mu = \frac{m_2}{m_1+m_2}$ (9.537×10^{-4} for the Sun-Jupiter and 3.00348×10^{-6} for the Sun-Earth mass distributions respectively), $q_1 = 1 - \frac{F_p}{F_g}$ is a mass reduction factor and F_p is the solar radiation pressure force which is exactly apposite to the gravitational attraction force F_g. The coordinates of m_1, m_2 are $(-\mu, 0)$, $(1-\mu, 0)$ respectively. In the above mentioned reference system and[9] model, the equations of motion of the infinitesimal mass particle in the xy-plane formulated as[please see[10–12]]:

$$\ddot{x} - 2n\dot{y} = \Omega_x, \tag{1}$$

$$\ddot{y} + 2n\dot{x} = \Omega_y, \tag{2}$$

where

$$\Omega_x = n^2 x - \frac{(1-\mu)q_1(x+\mu)}{r_1^3} - \frac{\mu(x+\mu-1)}{r_2^3} - \frac{3}{2}\frac{\mu A_2(x+\mu-1)}{r_2^5}$$
$$-\frac{M_b x}{(r^2+T^2)^{3/2}}$$

$$\Omega_y = n^2 y - \frac{(1-\mu)q_1 y}{r_1^3} - \frac{\mu y}{r_2^3} - \frac{3}{2}\frac{\mu A_2 y}{r_2^5}$$
$$-\frac{M_b y}{(r^2+T^2)^{3/2}}$$

$$\Omega = \frac{n^2(x^2+y^2)}{2} + \frac{(1-\mu)q_1}{r_1} + \frac{\mu}{r_2} + \frac{\mu A_2}{2r_2^3} + \frac{M_b}{(r^2+T^2)^{1/2}} \tag{3}$$
$$r_1 = \sqrt{(x+\mu)^2+y^2}, r_2 = \sqrt{(x+\mu-1)^2+y^2}.$$

From equations (1) and (2), the Jacobian integral is given by:

$$E = \frac{1}{2}\left(\dot{x}^2 + \dot{y}^2\right) - \Omega(x, y, \dot{x}, \dot{y}) = (\text{Constant}) \tag{4}$$

which is related to the Jacobian constant $C = -2E$. The location of three collinear equilibrium points and two triangular equilibrium points is computed by dividing the orbital plane into three parts $L_1, L_4(5)$: $\mu < x < (1-\mu)$, L_2: $(1-\mu) < x$ and L_3: $x < -\mu$. For the collinear points, an algebraic equation of the fifth degree is solved numerically with initial approximations to the Taylor-series as:

$$x(L_1) = 1 - (\frac{\mu}{3})^{1/3} + \frac{1}{3}(\frac{\mu}{3})^{2/3} - \frac{26\mu}{27} + \ldots \tag{5}$$

$$x(L_2) = 1 + (\frac{\mu}{3})^{1/3} + \frac{1}{3}(\frac{\mu}{3})^{2/3} - \frac{28\mu}{27} + \ldots \tag{6}$$

$$x(L_3) = -1 - \frac{5\mu}{12} + \frac{1127\mu^3}{20736} + \frac{7889\mu^4}{248832} + \ldots \tag{7}$$

$$\tag{8}$$

The solution of differential equations (1) and (2) is presented as interpolation function which is plotted for various integration intervals by substituting specific values of the time t and initial conditions i.e. $x(0) = x(L_i), y(0) = 0$ where $i = 1, 2, 3$ and $x(0) = \frac{1}{2} - \mu, y(0) = \pm\frac{\sqrt{3}}{2}$ for the

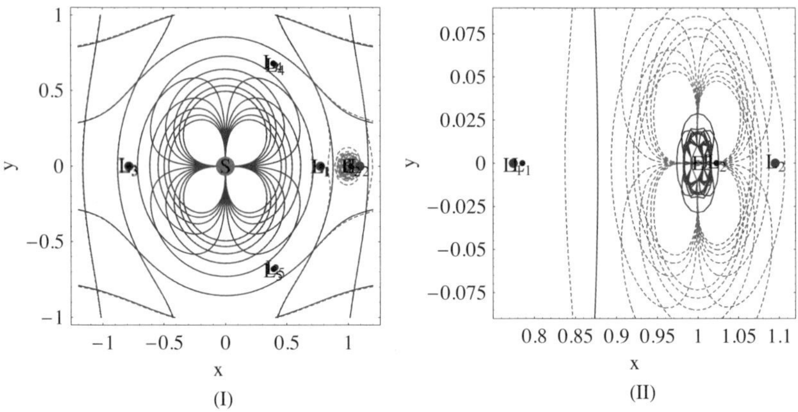

Fig. 1. The position of equilibrium points when T=0.01, $q_1 = 0.75$, $A_2 = 0.05$ and $M_b = 0.4$, panel (I):Red doted curves and blue points for Sun-Jupiter mass distribution, blue curves and black points for Sun-Earth mass distribution, (II): Position of L_1, L_2 with respect to Jupiter's and Earth's is shown in zoom

triangular equilibrium points. The equilibrium points are shown in figure 1 in which two panels i.e. (I) red solid curves and blue points correspond to the Sun-Jupiter mass distribution and blue dashed curves and black points correspond to the Sun-Earth mass distribution. Panel (II) show the zoom of the neighborhood of L_1, L_2. The numerical values of these points are presented in Table 1. One can see that the positions of L_1, L_3 appeared rightward and the positions of L_2, L_4 (L_4 is shifted downward also) are shifted leftward in the Sun-Earth system with respect to the position in the Sun-Jupiter system. The nature of the L_5 is similar to the L_4. The detail behavior of the L_1 with stability regions is discussed in sections. 3 & 4.

Sun-Jupiter			Sun-Earth	
L_i	x	y	x	y
L_1	0.774577	0	0.78569	0
L_2	1.09493	0	1.0232	0
L_3	-0.786195	0	-0.785732	0
L_4	0.410603	0.669308	0.393072	0.680342

3. Trajectory of L_1

The equations (1-2) with initial conditions $x(0) = x(L_1), y(0) = 0, x'(0) = y'(0) = 0$ are used to determine the trajectories of L_1 for different possible cases. The origin of coordinate axes is supposed to the equilibrium point at time $t = 0$ to draw the figures which show the trajectories of the point in consideration. They are shown in figure 2 with six panels i.e the panels (I-III) show the trajectory moves about the origin (L_1 at $t = 0$) with $x \in (0.990093, 1.00916)$, $y \in (-0.0061448, 0.00587171)$, the energy $E \in (-12706.5(t = 22.66), -5.08226(t = 0))$ and the distance $r(t) \in (0.990093(t = 0), 1.00916(t = 55))$. The panels (I-III:$127 < t < 129.6$) show the trajectory moves away from the origin (L_1 at $t = 0$) after a certain value of the time t, with $x \in (0.990093, 1.00916)$, $y \in (-0.0061448, 0.00587171)$. The minimum energy $E = -1447$ is found at $t = 128.52$, and $E > 0$ for $t > 128.88$.

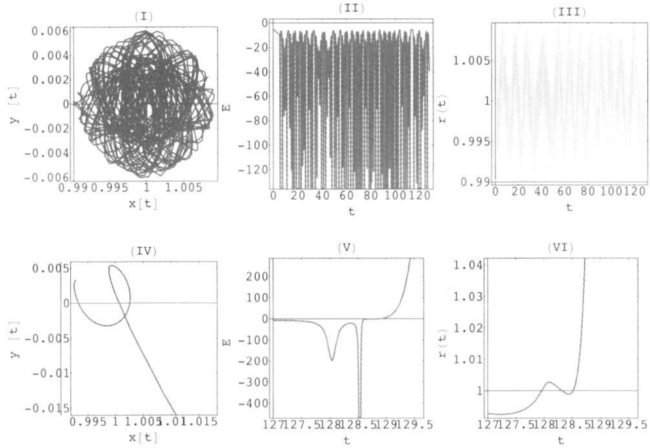

Fig. 2. The Panels (I-III):$0 < t < 128.23$ and (IV-VI):$127 < t < 129.6$ in which (I and II) show the trajectory of L_1, (II and V) show energy-versus time and (III-VI) show the local distance of trajectory at time t form the initial points i.e. $t = 0$ the other parameters are T=0.01, $q_1 = 1$, $A_2 = 0$ and $M_b = 0$.

Figure 3 is plotted for $q_1 = 1, M_b = 0$ and $A_2 = 0.05$ with six panels (I-III:$0 \leq t \leq 0.06$) and (IV-VI:$0.06 \leq t \leq 1$) which describe the effect of oblateness of Earth to the trajectory of L_1. The graphs plotted against time which describe behavior of trajectories to equilibrium points not the point

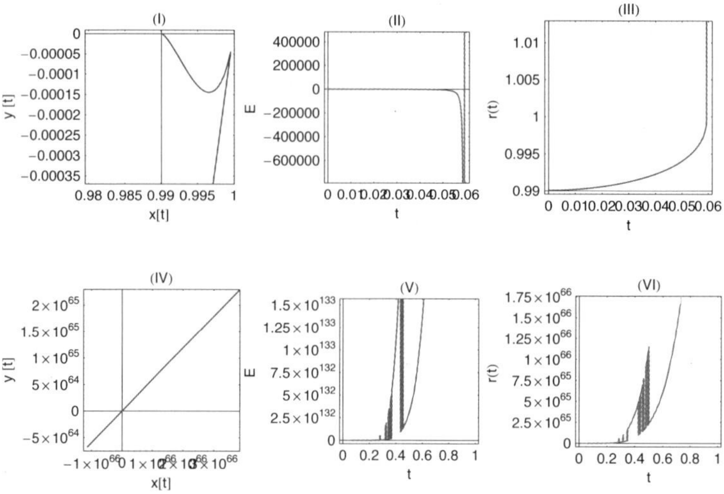

Fig. 3. The Panels (I-III):0 $< t <$ 0.06 and (IV-VI):.06 $< t <$ 1 in which (I and II) show the trajectory of L_1, (II and V) show energy-versus time and (III-VI) show the local distance of trajectory at time t form the initial points i.e. $t = 0$ the other parameters are T=0.01, $q_1 = 1$, $A_2 = 0.05$ and $M_b = 0$.

itself is moving with time. It is seen that $x = -1.91954 \times 10^{-48}(t = 0.06)$ to $x = 0.99405(t = .05)$ coordinate y is deceasing function of time that reach maxima -0.0000530614, at $t = 0.04$ and minima -0.000105662 at time $t = 0.05$ again it deceases and reach at value $-2.5677 \times 10^{47}(t = 0.06)$. Initially energy has negative values for time $0 \leq< t < 0.059$ decreases with time t which attains minimum value -2.64032 $\times 10^6 (t = 0.059)$ then strictly increases; attains positive value when $t \geq 0.0594$. When $t \in (0.2, 0.6)$, initially energy returns down and then it tend to very large (infinite) positive value. It is clear from panels (IV-VI) the trajectory moves far from the Lagrangian point L_1 when $t \geq 0.0594$. The distance $r(t)$ from this point to the trajectory increases periodically, when $0 < t < 0.6$ then it approaches to very large it $t \geq 0.6$.

The effect of radiation pressure, oblateness and mass of the belt is considered in figure 4, panels (I&III) describes the trajectory and panels (II&IV) shows the energy with respect to t. The mass reduction factor $q_1 = 0.75$ and $M_b = 0.2$ are taken to plot the graphs. In the panels, solid blue lines represent $A_2 = 0.25$, red dashed lines correspond to $A_2 = 0.50$ and dotted black lines for $A_2 = 0.75$. One can see that the trajectory move very far from the L_1 the energy is positive when time is greater than a certain value. Details of trajectory and energy is presented in Table 2 for various values of parameters. One can see that x increases but y initially decreases for certain values of the time, then strictly increases. Similarly the energy E is negative and went downward but for specific value of time it becomes positive and strictly increases lastly it attains very large positive value.

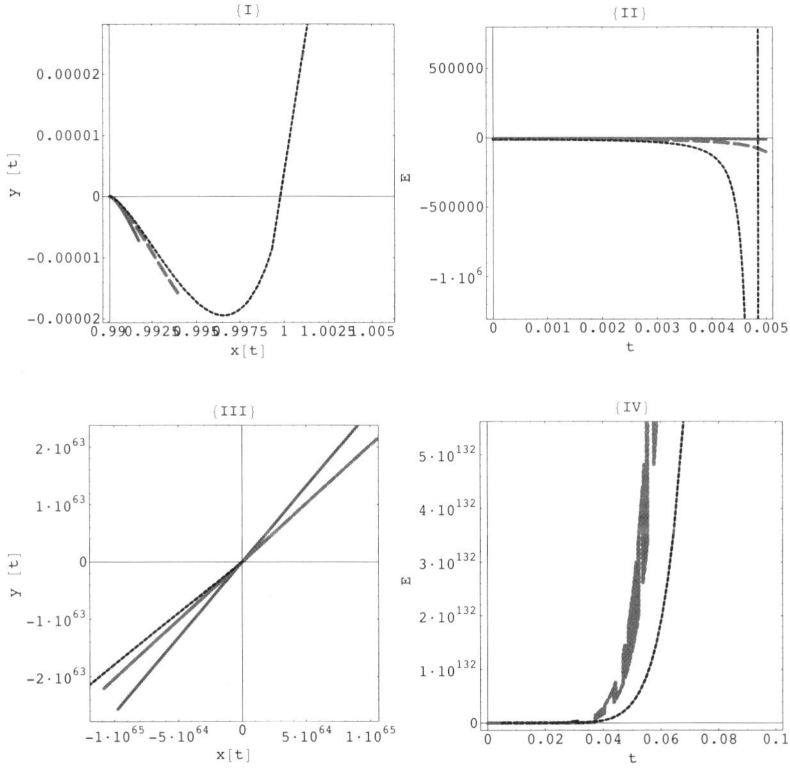

Fig. 4. The Panels (I-III):0 $< t <$ 0.005 and (II-IV):.00 $< t <$ 1 in which (I and II) show the trajectory of L_1, (II and V) show energy-versus time and (III-VI) show the local distance of trajectory at time t form the initial points i.e. $t = 0$ the other parameters are T=0.01, $q_1 = 1$, $A_2 = 0.05$ and $M_b = 0$

A_2	time t	x	y	Energy E
0.25	0.000	0.990093	-1.27593×10^{-32}	-3946.49
	0.002	0.990333	-4.4421×10^{-7}	-4460.93
	0.004	0.991106	-3.63139×10^{-6}	-6764.39
	0.006	0.992646	-1.26827×10^{-5}	-17501.7
	0.008	0.996301	-285308×10^{-5}	-544626.
	0.010	4.03799×10^{55}	-1.06467×10^{54}	7.84745×10^{117}
0.50	0.000	0.990093	7.51113×10^{-33}	-7887.36
	0.002	0.99058	-9.7398×10^{-7}	-10146.9
	0.004	0.992298	-8.12979×10^{-6}	-27763.1
	0.006	-8.35666×10^{48}	-1.70025×10^{47}	2.17394×10^{107}
	0.008	-3.05605×10^{56}	-6.21786×10^{54}	2.74317×10^{119}
	0.010	-9.05111×10^{57}	-1.84154×10^{56}	6.20464×10^{121}
0.75	0.000	0.990093	-1.80556×10^{-34}	-11828.2
	0.002	0.990836	-1.58382×10^{-6}	-17463.6
	0.004	0.993821	-1.33765×10^{-5}	-125336.
	0.006	6.67303×10^{55}	1.19389×10^{54}	4.20091×10^{118}
	0.008	1.02535×10^{58}	1.83448×10^{56}	1.32379×10^{122}
	0.010	1.19678×10^{59}	2.1412×10^{57}	6.74915×10^{123}

4. Stability of L_1

Suppose the coordinates (x_1, y_1) of L_1 are initially perturbed by changing $x(0) = x_1 + \epsilon\cos(\phi), y(0) = y_1 + \epsilon\sin(\phi)$ where $\phi = \arctan\left(\frac{y(0)-y_1}{x(0)-x_1}\right) \in (0, 2\pi), 0 \leq \epsilon = \sqrt{(x(0)-x_1)^2 + (y(0)-y_1)^2} < 1$. The ϕ indicates the direction of the initial position vector in the local frame. If the $\epsilon = 0$ means there is no perturbation. It is supposed that the $\epsilon = 0.001$ and the $\phi = \frac{\pi}{4}$ to examine the stability of L_1. Figure 5 show the path of test particle and its energy with four panels i.e. the panels (I&III):$q_1 = 0.75, 0.50, A_2 = 0.0$, in (I) trajectory of perturbed L_1 moves in chaotic-circular path around initial position without deviating far from it, then steadily move out of the region. In (III) the test particle moves in stability region and returns repeatedly on its initial position. The blue solid curves represent $M_b = 0.25$ and dashed curves represent $M_b = 0.50$. It is clear form panel (III) that bounded region for $M_b = 0.25$ is $t < 2500$ and for $M_b = 0.50, t < 2600$.

The effect of oblateness of the second primary is shown in figure 6 when

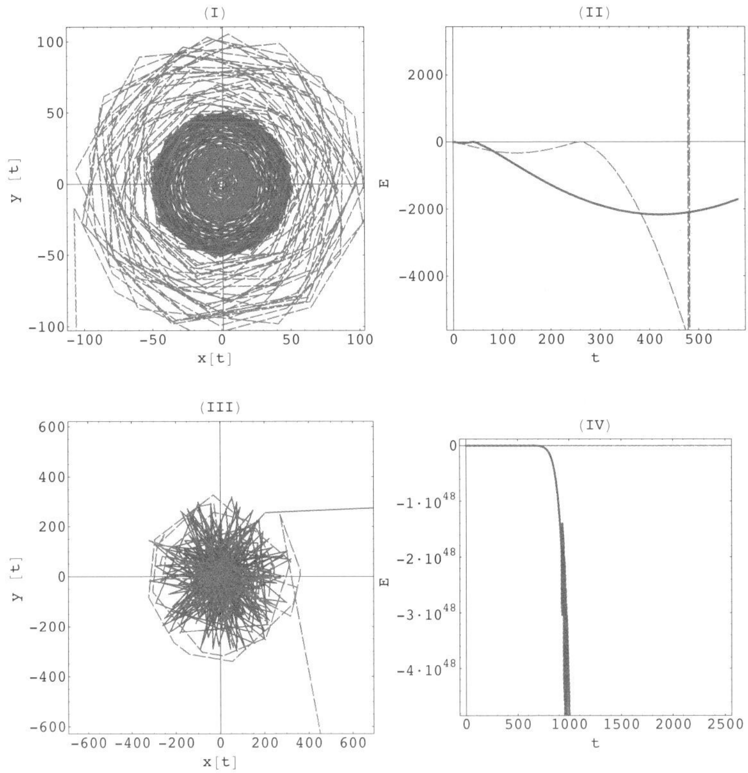

Fig. 5. Show the stability of L_1 with panels (I-II):$0 \leq t$'491, $q_1 = 0.75, A_2 = 0.0$ and (III-IV):$0 < t < 2500, q_1 = 0.50, A_2 = 0.0$ in which blue solid curves for $M_b = 0.25$, red curves for $M_b = 0.50$

$q_1 = 0.75, M_b = 0.25$. The panel (I) shows the trajectory of perturbed point L_1 and (II) shows the energy of that point. The blue doted lines correspond to $A_1 = 0.25$ and red lines for $A_2 = 0.50$. One can see that the oblate effect is very powerful on the trajectory and stability of L_1. When $A_2 = 0.0$ the L_1 is asymptotically stable for the value of t which lies within a certain interval. But if oblate effect of second primary is present($A_2 \neq 0$), the stability region of L_1 disappears as this effect increases. Further all the results presented in the manuscripts are similar to the results obtained by.[8]

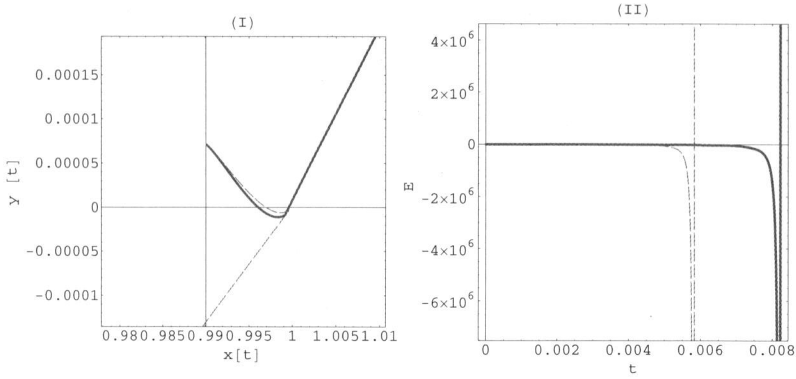

Fig. 6. Effect of oblateness coefficient A_2 on the stability of L_1 panel (I) trajectory (II) energy of perturbed point L_1 in which blue solid curves for $A_2 = 0.25$, red curves for $A_2 = 0.50$

5. Conclusion

We obtained intervals of time where trajectory continuously moves around the L_1, does not deviate far from the point but tend to approach it, the energy of perturbed point is negative for these intervals, so we conclude that the point is asymptotical stable. More over we have seen that after the specific time intervals the trajectory of perturbed point depart from the neighborhood and goes away from it, in this case the energy also becomes positive, so the Lagrangian point L_1 is unstable.

Acknowledgments

The author wishes to express his thanks to Indian School of Mines, Dhanbad (India), for providing financial support through Minor Research Project (No.2010/MRP/04/Acad. dated 30th June 2010).

References

1. S. V. Chermnykh, *Vest. Leningrad Mat. Astron.* **2**, 73 (1987).
2. I. Jiang and L. Yeh, *International Journal of Bifurcation and Chaos* **14**, 3153(September 2004).
3. V. Szebehely, *Theory of orbits. The restricted problem of three bodies* (New York: Academic Press, 1967).
4. R. W. Farquhar, *Journal of Spacecraft and Rockets* **4**, 1383(Octtober 1967).

5. R. W. Farquhar, *Astronautics Aeronautics* , 52 (1969).
6. R. Farquhar, D. Muhonen and L. C. Church, *Journal of the Astronautical Sciences* **33**, 235 (1985).
7. V. Domingo, B. Fleck and A. I. Poland, *Sol. Phys.* **162**, 1(December 1995).
8. B. S. Kushvah, *Ap&SS* , 286(Octtober 2010).
9. M. Miyamoto and R. Nagai, *PASJ* **27**, 533 (1975).
10. B. S. Kushvah, *Ap&SS* , 191(September 2008).
11. B. S. Kushvah, *Ap&SS* **323**, 57(September 2009).
12. B. S. Kushvah, *Research in Astronomy and Astrophysics* **9**, 1049(September 2009).

MHD FLOW PAST AN INFINITE PLATE UNDER THE EFFECT OF GRAVITY MODULATION

SARGAM WASU* AND S. C. RAJVANSHI

Department of Applied Sciences,
Gurukul Vidyapeeth Institute of Engineering & Technology,
Sector # 7, Banur, Distt Patiala, Punjab, India
e-mail : sargam15@rediffmail.com

Unsteady mixed convection flow under the influence of gravity modulation and magnetic field has been investigated. The conducting fluid flows past a vertical porous plate of infinite length in a porous medium subjected to oscillating suction and temperature. The solution has been obtained by using regular perturbation method. Velocity profiles, temperature profiles, skin friction and heat transfer coefficients have been derived and shown graphically. It is noted that the fluid flow and heat transfer are significantly affected by gravity modulation.

Keywords: Gravity modulation; Mixed convection; MHD; Porous medium; Suction.

1. Introduction

There has been a lot of research in the area of free convection in the presence of porous media due to its application in oil exploration, nuclear waste disposal, geothermal energy etc. Many of the engineering applications involve periodic temperature variations. These include daily or seasonal temperature changes in the earth's crust. Raptis [1] studied the free convective flow through a porous medium bounded by an infinite vertical plate with oscillating plate temperature and constant suction. Singh et al [2] investigated the effect of permeability variation on free convection flow and heat transfer in porous medium bounded by vertical porous wall. In many situations liquid metals that occur in nature and industry are electrically conducting, hence the study of fluid flow under the effect of magnetic field is important. The unsteady convection free flow past a vertical porous plate under the effect of magnetic field has been studied by Helmy [3]. The effect of magnetic field on the onset of convection in porous medium has been studied by Das et al [4] and Vassuer & Bilgen [5]. Jaiswal & Soundalgekar [6,7] has investigated the effect of free and forced convection on a MHD flow past an infinite vertical porous plate under oscillating plate temperature. Saini & Sharma [8] investigated the effect of permeability variation and oscillatory suction velocity on free convection and mass transfer of MHD flow of a viscous fluid past an infinite vertical porous plate bounded by a porous medium. The plate is subjected to oscillatory suction velocity normal to the plate in the presence of a uniform transverse magnetic field with prescribed rate of change of temperature and concentration of species on the boundary.

Over the years with the advancement of technology, oscillatory flows with gravity modulation have made their presence felt in the vast field of research. These have applications in the areas of space technology, in large scale convection in atmosphere, crystal growth and in space laboratory experiments. The effect of fluctuating gravity is a major area of interest aboard an orbiting spacecraft which experiences perturbed acceleration due to vibrations of the equipment, movement of crew members and other factors. Jules et al [9] found that the international space station (ISS) is characterized by low mean accelerations which are $O(10^{-6})g_e$ - the gravity on earth and fluctuations that are two or three order of magnitude above the mean. The studies under the microgravity conditions aboard an orbiting spacecraft have shown that gravitational field can be resolved into a mean and fluctuating component. The presence of temperature gradient and a gravitational field generate convective flows in viscous fluids and porous media. In a recent study, the effect of periodic oscillation of gravity on free convection over a vertical flat plate has been considered by Saeid [10]. He has used fully implicit finite difference scheme and concluded that heat transfer follows g-jitter forcing function. Deka and Soundalgekar [11] have studied the effect of gravity modulation on transient free convection flow past an infinite vertical isothermal plate using Laplace Transform technique. They analyzed that with increasing frequency of gravity modulation the transient velocity decreases.

In the present paper, the effect of fluctuating gravity on the flow past a vertical porous plate of infinite length in a porous medium subjected to oscillating suction and temperature field is studied using perturbation method. It is found that fluid flow and heat transfer are significantly affected by gravity modulation in presence of magnetic field. The velocity profiles and skin friction show a marked variation under the effect of gravity.

2. Formulation of the problem

We consider the flow of an electrically conducting viscous incompressible fluid through a porous medium bounded by an infinite flat plate. The vertically upward direction of flow along the plate is taken as x^* - axis and y^* - axis is normal to it. The vertical plate is of infinite length in the direction of x^* - axis; hence all the physical quantities are independent of x^* and are functions of y^* and t^* (time) only. The free stream velocity is taken uniformly as U. The fluid velocities along x^* - axis and y^* - axis are taken as u^* and v^* respectively. The induced magnetic field B_0 is negligible. The suction velocity and gravitational field are assumed to oscillate in the following form

$$v^*(t^*) = -V(1 + \varepsilon e^{\iota\omega^* t^*}), g(t^*) = g_0 + g_1 \cos \omega^* t^* \tag{1}$$

the negative sign of v^* indicates that suction is taking place at the plate. $V > 0$ is the constant mean velocity and $\varepsilon << 1$ is a constant. g_0 is the constant gravity level in the environment, g_1 is the amplitude of the oscillating component of acceleration and ω^* is the frequency of gravitational oscillation. The gravitational acceleration can be rewritten in the form

$$g(t^*) = g_0(1 + \varepsilon\alpha e^{\iota\omega^* t^*}) \tag{2}$$

where $\dfrac{g_1}{g_0} = \varepsilon\alpha$ is the gravity modulation parameter. It is assumed that the real part alone is physically relevant.

Using Boussinesq approximation, the equation of continuity, momentum and energy conservation are written as

$$\frac{\partial v^*}{\partial y^*} = 0 \tag{3}$$

$$\frac{\partial u^*}{\partial t^*} + v^* \frac{\partial u^*}{\partial y^*} = \nu \frac{\partial^2 u^*}{\partial y^{*2}} + g_0(1 + \varepsilon\alpha e^{\iota\omega^* t^*})\beta(T^* - T^*_\infty) + \frac{\sigma(U - u^*)}{\rho} B_0^2 + \frac{\nu(U - u^*)}{K^*} \tag{4}$$

$$\rho C_p \left(\frac{\partial T^*}{\partial t^*} + v^* \frac{\partial T^*}{\partial y^*} \right) = \kappa \frac{\partial^2 T^*}{\partial y^{*2}} + \mu \left(\frac{\partial u^*}{\partial y^*} \right)^2 \tag{5}$$

The boundary conditions are given by

$$\begin{aligned} y^* = 0: \quad & u^* = 0, \quad T^* = T^*_\omega + \varepsilon(T^*_\omega - T^*_\infty)e^{\iota\omega^* t^*} \\ y^* \to \infty: \quad & u^* = U, \quad T^* = T^*_\infty \end{aligned} \tag{6}$$

The physical variables used are defined in the Nomenclature.The superscript $(*)$ represents dimensional quantities and the subscript (∞) denotes free stream velocity. The following non-dimensional quantities are introduced into the equations:

$$y = \frac{y^* V}{\nu}, \ \omega = \nu \frac{\omega^*}{V^2},$$

$$t = \frac{t^* V^2}{\nu}, \ u = \frac{u^*}{U},$$

$$K = \frac{K^* V^2}{\nu^2},$$

$$\theta = \frac{(T^* - T^*_\infty)}{(T^*_\omega - T^*_\infty)}$$

The following parameters have also been introduced:

$$\text{M (Hartmann Number)} = \frac{\sigma\nu(B_0^2)}{\rho V^2}$$

$$\text{Pr (Prandtl Number)} = \frac{\mu C_p}{\kappa}$$

$$\text{Gr (Grashof Number)} = \frac{\nu\beta g_0(T^*_\omega - T^*_\infty)}{(UV^2)}$$

$$\text{Ec (Eckert Number)} = \frac{U^2}{C_p(T^*_\omega - T^*_\infty)}$$

Using these quantities the non-dimensional form of the governing equations reduce to

$$\frac{\partial u}{\partial t} - (1 + \varepsilon e^{\iota\omega t})\frac{\partial u}{\partial y} = \frac{\partial^2 u}{\partial y^2} + Gr\theta(1 + \varepsilon\alpha e^{\iota\omega t}) + M(1 - u) + \frac{1 - u}{K} \quad (7)$$

$$\frac{\partial \theta}{\partial t} - (1 + \varepsilon e^{\iota\omega t})\frac{\partial \theta}{\partial y} = \frac{1}{Pr}\frac{\partial^2 \theta}{\partial y^2} + Ec\left(\frac{\partial u}{\partial y}\right)^2 \quad (8)$$

with boundary conditions

$$\begin{aligned} y = 0\ ; &\quad u = 0, \quad \theta = 1 + \varepsilon e^{i\omega t} \\ y \to \infty\ ; &\quad u = 1, \quad \theta = 0 \end{aligned} \quad (9)$$

3. Solution of the problem

It is assumed that the amplitude of oscillation of the velocity ε is very small. Using this assumption, the steady and unsteady components of velocity and temperature are separated in the following form

$$u(y,t) = u_0(y) + \varepsilon e^{\iota\omega t} u_1(y)$$

$$\theta(y,t) = \theta_0(y) + \varepsilon e^{\iota\omega t} \theta_1(y) \tag{10}$$

Substituting (10) into (7) and (8), and equating the coefficients of harmonic and non-harmonic terms, we get

$$u_0'' + u_0' - (M + \frac{1}{K})u_0 = -Gr\theta_0 - (M + \frac{1}{K}) \tag{11}$$

$$u_1'' + u_1' - (M + \frac{1}{K} + \iota\omega)u_1 = -u_0' - Gr(\alpha\theta_0 + \theta_1) \tag{12}$$

$$\theta_0'' + Pr\theta_0' = -Pr\ Ec\ u_0'^2 \tag{13}$$

$$\theta_1'' + Pr\theta_1' - \iota\omega Pr\theta_1 = -Pr\theta_0' - 2PrEcu_0'u_1' \tag{14}$$

The modified boundary conditions are given by

$$\begin{aligned} y = 0: &\quad u_0 = 0,\ u_1 = 0,\ \theta_0 = 1,\ \theta_1 = 1 \\ y \to \infty: &\quad u_0 = 1,\ u_1 = 0,\ \theta_0 = 0,\ \theta_1 = 0 \end{aligned} \tag{15}$$

where prime denotes derivative with respect to y. The equations are still coupled non-linear differential equations. The solution is obtained using regular perturbation technique. It is assumed that $Ec \ll 1$ and $u_0,\ u_1,\ \theta_0,\ \theta_1$ are written in the following form

$$(u_0, u_1, \theta_0, \theta_1) = (u_{01}, u_{11}, \theta_{01}, \theta_{11}) + Ec(u_{02}, u_{12}, \theta_{02}, \theta_{12}) \tag{16}$$

Substituting (16) in (11), (12), (13) and (14) and equating coefficients of Ec we get the following equations

$$u_{01}'' + u_{01}' - \tilde{M}u_{01} = -Gr\theta_{01} - \tilde{M} \tag{17}$$

$$u_{02}'' + u_{02}' - \tilde{M}u_{02} = -Gr\theta_{02} \tag{18}$$

$$\theta_{01}'' + Pr\theta_{01}' = 0 \tag{19}$$

$$\theta_{02}'' + Pr\theta_{02}' = -Pru_{01}'^2 \tag{20}$$

$$u_{11}'' + u_{11}' - (\tilde{M} + \iota\omega)u_{11} = -u_{01}' - Gr(\alpha\theta_{01} + \theta_{11}) \tag{21}$$

$$u_{12}'' + u_{12}' - (\tilde{M} + \iota\omega)u_{12} = -u_{02}' - Gr(\alpha\theta_{02} + \theta_{12}) \tag{22}$$

$$\theta_{11}'' + Pr\theta_{11}' - \iota\omega Pr\theta_{11} = -Pr\theta_{01}' \tag{23}$$

$$\theta_{12}'' + Pr\theta_{12}' - \iota\omega Pr\theta_{12} = -Pr\theta_{02}' - 2Pru_{01}'u_{11}' \tag{24}$$

where $\tilde{M} = M + \frac{1}{K}$ The boundary conditions are modified to

$$\begin{aligned} y = 0 : u_{01} &= 0, u_{02} = 0, u_{11} = 0, u_{12} = 0, \\ \theta_{01} &= 1, \theta_{02} = 0, \theta_{11} = 1, \theta_{12} = 0 \\ y \to \infty : u_{01} &= 1, u_{02} = 0, u_{11} = 0, u_{12} = 0, \\ \theta_{01} &= 0, \theta_{02} = 0, \theta_{11} = 0, \theta_{12} = 0 \end{aligned} \tag{25}$$

The solutions of these ordinary differential equations under the modified boundary conditions are obtained in the following form

$$\begin{aligned} u_0(y) = {} & A_1(e^{-Ly} - e^{-Pry}) + (1 - e^{-Ly}) + Ec[A_5(e^{-Ly} - e^{-Pry}) - \\ & A_6(e^{-Ly} - e^{-2Pry}) - A_7(e^{-Ly} - e^{-2Ly}) - \\ & A_8(e^{-Ly} - e^{-(Pr+L)y})] \end{aligned} \tag{26}$$

$$\begin{aligned} \theta_0(y) = {} & e^{-Pry} + Ec[A_2(e^{-Pry} - e^{-2Pry}) + A_3(e^{-Pry} - e^{-2Ly}) + \\ & A_4(e^{-Pry} - e^{-(Pr+L)y})] \end{aligned} \tag{27}$$

$$\begin{aligned} u_1(y) = {} & A_{10}(e^{-ny} - e^{-Pry}) + A_{11}(e^{-ny} - e^{-Ly}) + \\ & A_{12}(e^{-ny} - e^{-my}) + Ec[A_{21}(e^{-Ly} - e^{-ny}) + \\ & A_{22}(e^{-ny} - e^{-Pry}) - A_{23}(e^{-ny} - e^{-2Pry}) - \\ & A_{24}(e^{-ny} - e^{-2Ly}) - A_{25}(e^{-ny} - e^{-(Pr+L)y}) + \\ & A_{26}(e^{-ny} - e^{-my}) + A_{27}(e^{-ny} - e^{-(Pr+n)y}) - \\ & A_{28}(e^{-ny} - e^{-(Pr+m)y}) + A_{29}(e^{-ny} - e^{-(L+n)y}) - \\ & A_{30}(e^{-ny} - e^{-(L+m)y})] \end{aligned} \tag{28}$$

$$\begin{aligned} \theta_1(y) = {} & e^{-my} - A_9(e^{-my} - e^{-Pry}) + Ec[A_{13}(e^{-Pry} - e^{-my}) + \\ & A_{14}(e^{-my} - e^{-2Pry}) + A_{15}(e^{-my} - e^{-2Ly}) + \\ & A_{16}(e^{-my} - e^{-(Pr+L)y}) - A_{17}(e^{-my} - e^{-(Pr+n)y}) + \\ & A_{18}(e^{-my} - e^{-(Pr+m)y}) - A_{19}(e^{-my} - e^{-(L+n)y}) - \\ & A_{20}(e^{-my} - e^{-(L+m)y})] \end{aligned} \tag{29}$$

where the constants L ,m,n,A_1, ..., A_{30} are recorded in the Appendix. Substituting (26) - (29) in (10) we get the expression for velocity and temperature profiles.

4. Results and Discussion

4.1. *Velocity profiles*

The transient velocity from (10) has been shown in fig. 1 for $Ec = 0.001$, $Pr = 0.71$, $\varepsilon\alpha = 10$, $t = \frac{\pi}{10}$. With an increase in values of permeability K and Grashof number Gr the velocity increases under the effect of gravity modulation parameter taken as $\varepsilon\alpha = 10$. The curve IV shows that the velocity attains a maximum value of 25.3743 at $y = 0.153$. As M increases there is a considerable change in the velocity profile. Also with increase in ω the velocity decreases. The effect of increase in gravity modulation parameter has also been shown. It has been observed that the velocity increases when the parameter is taken as $\varepsilon\alpha = 15$. The velocity profile of the present study agrees with the previous result [12] for a particular case with $\alpha = 0$, $K = 5$, $\omega = 10$, $Pr = 0.71$, $Gr = 10$, $Ec = 0.001$.

4.2. *Temperature profiles*

The effect of temperature has been shown in fig. 2 with $Ec = 0.001$, $K = 5$, $\varepsilon\alpha = 10$. As the distance from the plate increases there is decrease in transient temperature. With an increase in the value of gravity modulation parameter there is slight decrease in the temperature profile. The value of transient temperature is less in water ($Pr = 7$) as compared to air ($Pr = 0.71$). The temperature increases with increase in M and ω, while it falls with increase in Gr. Due to absence of gravity term explicitly in the temperature profile, the gravity modulation has less impact on it as compared to the velocity profile.

4.3. *Skin friction*

The skin friction is expressed as,

$$\tau^*_{xy} = \mu \left(\frac{\partial u^*}{\partial y^*} \right) \tag{30}$$

where μ is the viscosity.

The non-dimensional form of the skin friction on the plate $y = 0$ is given by

$$\tau = \frac{\tau^*_{xy}}{\rho UV} = -\tau_m - \varepsilon|B|(\cos\omega t + \phi) \tag{31}$$

where τ_m is the steady part of skin friction, $\varepsilon|B|$ is the amplitude of unsteady part and ϕ is the phase difference. Using (10), (26),(28), (30) and

(31), we get

$$\tau_m = \left(\frac{\partial u_0}{\partial y}\right)_{y=0} = A_1(Pr - L) + L + Ec[A_5(Pr - L) \\ -A_6(2Pr - L) - A_7(L) - A_8(Pr)] \tag{32}$$

and

$$\begin{aligned} B &= \left(\frac{\partial u_1}{\partial y}\right)_{y=0} \\ &= A_{10}(Pr - n) + A_{11}(L - n) + A_{12}(m - n) + \\ &\quad Ec[A_{21}(n - L) + A_{22}(Pr - n) - A_{23}(2Pr - n) - \\ &\quad A_{24}(2L - n) - A_{25}(Pr + L - n) + A_{26}(m - n) + \\ &\quad A_{27}(Pr) - A_{28}(Pr + m - n) + A_{29}(L) - A_{30}(L + m - n)] \end{aligned} \tag{33}$$

where ϕ is the phase angle.

The amplitude of skin fraction in fig. 3 for $Ec = 0.001,\ Pr = 0.71,\ \alpha = 100,\ \varepsilon = 0.01,\ t = \frac{\pi}{10}$ increases with increase in permeability K and Grashof Number Gr, but decreases when M is increased. There is an increase in the amplitude of skin friction with increase in ω under the effect of gravity. In fig. 4, the phase of skin friction is shown for $Gr = 10,\ Ec = 0.001,\ \alpha = 10,\ t = \frac{\pi}{10}$ versus ω. As permeability K is increased the phase change takes place and it shows a phase lead. It has been observed that even a small increase in M results in a significant decrease in the value of the phase angle of skin friction. There is a phase lead in water ($Pr = 7$) as compared to air ($Pr = 0.071$).

4.4. *Rate of heat transfer*

The rate of heat transfer is given by

$$q^* = -\kappa \frac{\partial T^*}{\partial y^*} \tag{34}$$

In non dimensional form heat transfer coefficient is given by,

$$Nu = \frac{q^* L}{(T^*_\omega - T^*_\infty)\kappa} = \left(\frac{\partial \theta}{\partial y}\right)_{y=0} = q_m + \varepsilon |Q| \cos(\omega t + \delta) \tag{35}$$

where $q_m = -Pr + Ec[A_2(Pr) + A_3(2L - Pr) + A_4(L)]$
and

$$\begin{aligned} Q = m(A_9 - 1) - A_9 Pr + Ec[A_{13}(m - Pr) + A_{14}(2Pr - m) + \\ A_{15}(2L - m) + A_{16}(Pr + L - m) - A_{17}(Pr + n - m) + \\ A_{18}(Pr) - A_{19}(L + n - m) + A_{20}(L)] \end{aligned} \tag{36}$$

and δ is the phase angle.

Fig. 5 shows that the variation of heat with ω for $Pr = 0.71$, $Ec = 0.001$, $\varepsilon\alpha = 10$. The heat transfer increases with increase in the value of Gr. When either the permeability K or M in increased, the heat transfer shows a decreasing effect. In Fig. 6, phase of heat is depicted. The phase difference increases with increase in Gr and M. It increases in air with $(Pr = 0.71)$ as compared to water $(Pr = 0.71)$.

References

1. A. Raptis, Unsteady free convection flow through a porous medium, *International Journal of Engg. Science*, **21**, (1983).
2. P. Singh, J. K. Mishra, K. A. Narayan, Free convection along a vertical wall in a porous medium with periodic permeability variation, *Int. Journal for Numerical and Analytic Methods in Geomechanics*, **13**, (2005).
3. K. A. Helmy, MHD unsteady free convection flow past a vertical porous plate, *ZAMM*, **78(4)**, (1998).
4. U. N. Das, A. Aziz, M. Rahman, Unsteady MHD free convection flow through a porous medium, *Far East Journal of Applied Mathematics*, **3(3)**, (1999).
5. S. A. Vasseur, E. Bilgen, Effect of magnetic field on the onset of convection in a porous medium, *Heat and Mass Transfer*, **30(4)**, (1995).
6. B. S. Jaiswal, V. M. Soundalgekar, Oscillating plate temperature effects on a flow past an infinite vertical porous plate with constant suction and embedded in porous medium, *Heat and Mass Transfer*, **37**, (2001).
7. B. S. Jaiswal, V. M Soundalgekar, Unsteady free and forced convection MHD flow past an infinite vertical porous plate with variable suction and oscillating plate temperature, *Bulletin of the Allahabad Mathematical Society*, **16**, (2001).
8. Saini, Sharma, Unsteady heat and mass transfer in MHD flow past a vertical plate embedded in porous media, *Ganit Sandesh*, **20(2)**, (2006).
9. K. Jules, K. Hrovat, E. Kelly, K. Mcpherson, T. Reckart, International space increment -2 microgravity environment summary report, *NASA*, **TM 211335**, (2002).
10. N.H. Saeid, G-Jitter induced free convection over a vertical flat plate, *ASEAN Journal on Science and Technology for Development*, **23**, (2006).
11. R. K. Deka, V. M. Soundalgekar, Gravity modulation effect on transient free convection flow past an infinite vertical isothermal plate, *Defence Science Journal*, **56**, (2006).
12. Tara Chand, Ph.D Thesis, *University of Rajasthan*, (2007).

5. Nomenclature

C_p=specific heat at constant pressure

Ec=Eckert number

g=acceleration due to gravity

g_0=constant gravity level
g_1=amplitude of oscillating component of acceleration due to gravity
Gr=Grashof number
K=dimensionless permeability of porous medium
K^*=permeability of porous medium
M=Hartmann number
Nu=dimensionless coefficient of heat transfer
Pr=Prandtl number
q^*=rate of heat transfer
T^*=temperature of the fluid
T^*_∞=temperature of fluid in free stream
T^*_ω=temperature of the plate
t^*=time
t=dimensionless time
U=dimensionless velocity of the moving vertical porous plate
u^*=velocity of fluid
u=dimensionless velocity of fluid
V=constant mean velocity
β=coefficient of volume expansion
ρ=density
ν=kinematic viscosity
κ=thermal conductivity
μ=coefficient of viscosity
σ=electrical conductivity
θ=dimensionless temperature
ω=dimensionless frequency of gravitational oscillation
ω^*=frequency of gravitational oscillation
τ=skin friction
τ_m=mean skin friction

6. Appendix

$L = \frac{1+\sqrt{1+4\tilde{M}}}{2}$ $\qquad \tilde{M} = M + \frac{1}{K}$

$m = \frac{Pr+\sqrt{Pr^2+4\iota\omega Pr}}{2}$ $\qquad n = \frac{1+\sqrt{1+4(\tilde{M}+\iota\omega)}}{2}$

$A_1 = \frac{G}{(Pr^2-Pr-\tilde{M})}$ $\qquad A_2 = \frac{((A_1)^2)Pr)}{2}$

$A_3 = \frac{(PrL(1-A_1)(1-A_1))}{(2(2L-Pr))}$ $A_4 = \frac{(2A_1Pr^2(1-A_1))}{(Pr+L)}$

$A_5 = \frac{(GrB_1)}{(Pr^2-Pr-\tilde{M})}$ $A_6 = \frac{GrA_2}{4Pr^2-2Pr-\tilde{M}}$

$A_7 = \frac{GrA_3}{4L^2-2L-\tilde{M}}$

$A_8 = \frac{GrA_4}{(Pr+L)^2-(Pr+L)-\tilde{M}}$ $A_9 = \frac{\iota Pr}{\omega}$

$A_{10} = \frac{A_1Pr+GrA_9+\alpha Gr}{(Pr^2-P-(\tilde{M}+\iota\omega))}$ $A_{11} = \frac{(L(1-A_1))}{(L^2-L-\iota\omega-\tilde{M})}$

$A_{12} = \frac{Gr(1-A_9)}{((m^2)-m-\tilde{M}+\iota\omega)}$ $A_{13} = \frac{(\iota PrB_1)}{\omega}$

$A_{14} = \frac{(2PrA_2+2A_1Pr^2A_{10})}{(2Pr-\iota\omega}$

$A_{15} = \frac{(2PrL(A_3+L(1-A_1)A_{11}))}{(4L^2-2PrL-\iota\omega Pr)}$

$A_{16} = \frac{(Pr(Pr+L)A_4+2A_1Pr^2LA_{11}+2P^2L(1-A_1)A_{10})}{((P+L)^2-Pr(Pr+L)-\iota\omega Pr)}$

$A_{17} = \frac{(2A_1nB_3P^2)}{(n^2-nPr-\iota\omega Pr)}$

$A_{18} = \frac{(2A_1mA_12Pr^2)}{(m^2-mPr-Pr\iota\omega)}$

$A_{19} = \frac{(2nB_3PrL(1-A_1))}{((L+n)^2-Pr(L+n)-\iota\omega Pr)}$

$A_{20} = \frac{(2PrLmA_{12}(1-A_1))}{((L+m)^2-Pr(L+m)-\iota\omega Pr)}$

$A_{21} = \frac{(LB_2)}{(L^2-L-(\tilde{M}+\iota\omega))}$

$A_{22} = \frac{(PrA_5+GrA_{13}+Gr\alpha B_1)}{(Pr^2-Pr-(\tilde{M}+\iota\omega))}$

$A_{23} = \frac{(2PrA_6+GrA_{14}+Gr\alpha A_2)}{(4Pr^2-2Pr-(\tilde{M}+\iota\omega))}$

$A_{24} = \frac{(2LA_7+GrA_{15}+\alpha GrA_3)}{(4L^2-2L-(\tilde{M}+\iota\omega))}$

$A_{25} = \frac{((Pr+L)A_8+GrA_{16}+\alpha GrA_4)}{((Pr+L)^2-(Pr+L)-(\tilde{M}+\iota\omega))}$

$A_{26} = \frac{(GrB_4)}{(m^2-m-(\tilde{M}+\iota\omega))}$

$A_{27} = \frac{(GrA_{17})}{((Pr+n)^2-(P+n)-(\tilde{M}+\iota\omega))}$

$A_{28} = \frac{(GrA_{18})}{((Pr+m)^2-(Pr+m)-(\tilde{M}+\iota\omega))}$

$A_{29} = \frac{(GrA_{19})}{((L+n)^2-(L+n)-(\tilde{M}+\iota\omega))}$

$A_{30} = \frac{(GrA_{20})}{((L+m)^2-(L+m)-(\tilde{M}+\iota\omega))}$

$B_1 = A_2 + A_3 + A_4$

$B_2 = A_5 - A_6 - A_7 - A_8$

$B_3 = A_{10} + A_{11} + A_{12}$

$B_4 = -A13 + A_{14} + A_{15} + A_{16} - A_{17} + A_{18} - A_{19} + A_{20}$

$B_5 = -A_{21} + A_{22} - A_{23} - A_{24} - A_{25} + A_{26} + A_{27} - A_{28} + A_{29} - A_{30}$

Fig.1 Transient Velocity with $Ec = 0.001, \varepsilon = 0.01, Pr=0.71, t = \pi/10$

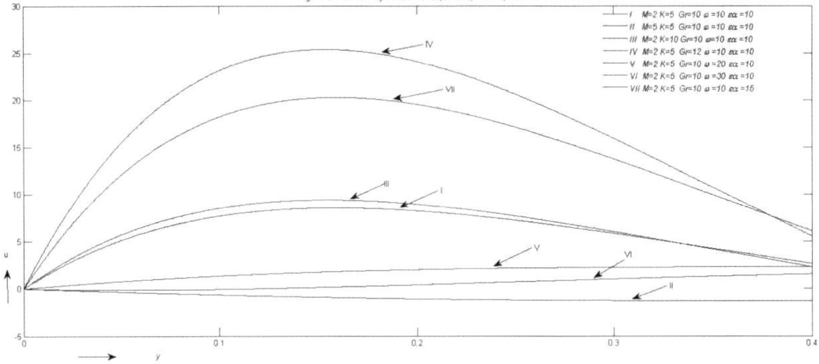

Fig.2 Transient Temperature with $Ec = 0.001, K = 5, \varepsilon = 0.01, t = \pi/10$.

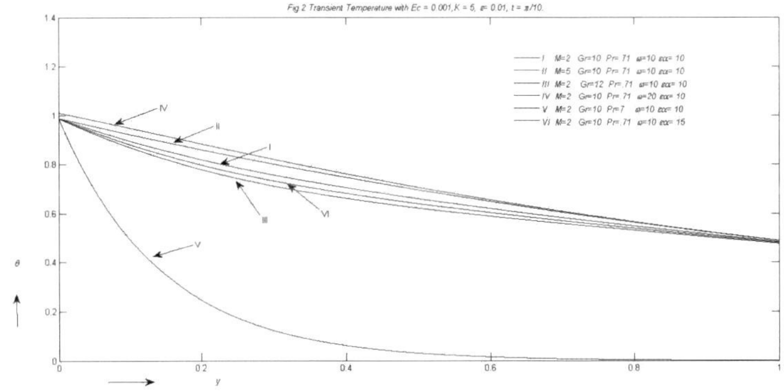

Fig 3. Amplitude of Skin Friction with Ec = 0.001, Pr = 0.71, ε = 0.01, α = 100

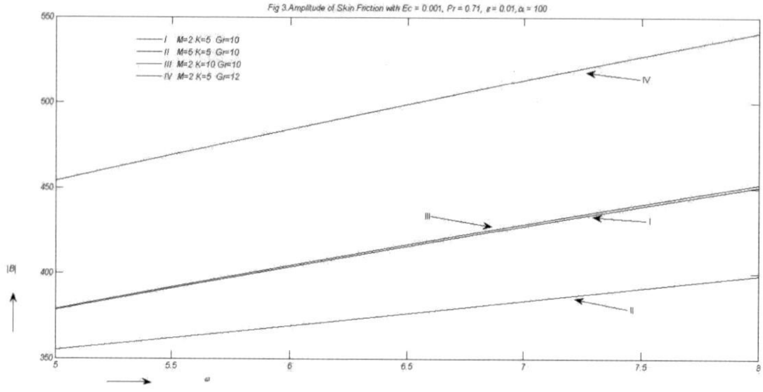

Fig 4 Phase of Skin Friction with Ec = 0.001, Gr = 10, α = 1000, ε = 0.01

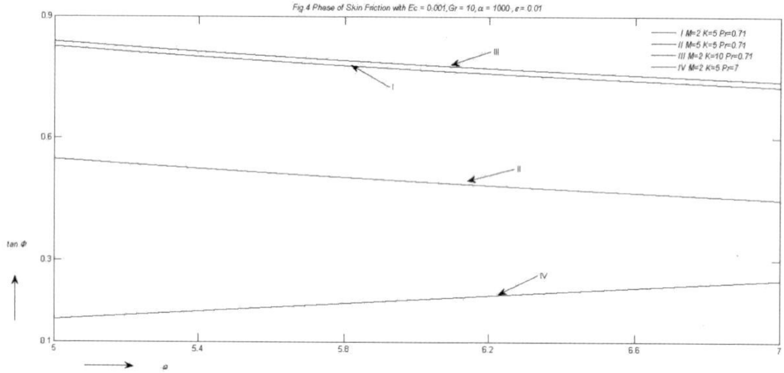

Fig. 5 Amplitude of Heat Transfer with $Ec = 0.001, Pr = 0.71, \varepsilon = 0.01, t = \pi/10$

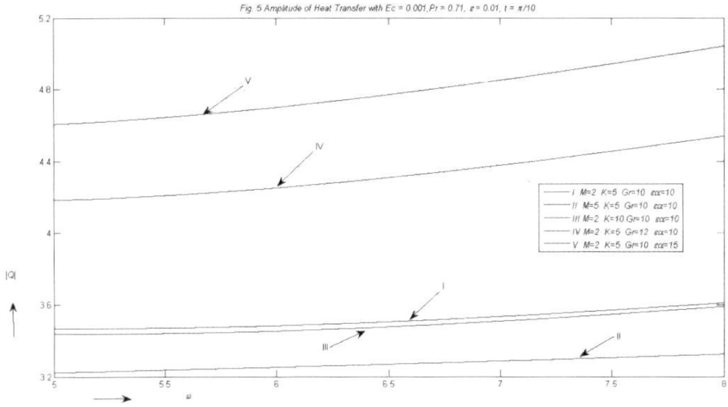

Fig. 6 Phase of Rate of Heat Transfer with $Ec = 0.001, K = 5, Gr = 10, \varepsilon = 0.01, \alpha = 100$

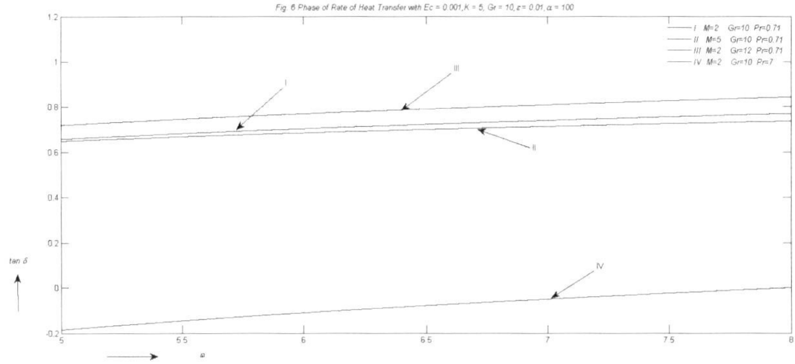

Hon'ble Mr. Salman Khurshid, Union Minister of State, Corporate Affairs Minority Affairs was the Chief Guest. Dr D.P.Agrawal,Chairman Union Public Service Commission Inaugurated a Workshop on Wavelets and Inverse Problems on this occassion

Parton	Mr. P. K. Gupta, Chancellor, Sharda University Mr. Y. K. Gupta, Pro Chancellor, Sharda University Dr. R. P. Singh, Vice Chancellor, Sharda University
Steering Committee	Prof. H.P. Dikshit (ISIAM) Prof. N.K. Gupta (IIT Delhi,ISIAM) Prof. Dinesh Singh Prof. Karmeshu (JNU) Prof. P. Manchanda (GNDU, ISIAM) Prof. O.P. Bhutani (INSA) Dr. M. Mehra (IIT, Delhi) Dr. A.K. Gupta (BMAS) Prof. R.C. Singh (Sharda University) Prof. C.T. Bhunia (Bengal Institute of Technology) Prof. A.H. Siddiqi (ISIAM & Sharda University)

LIST OF INVITED SPEAKERS & PARTICIPANTS

Sr.No.	Name	University/Address
1.	Prof. K.R. Sreenivasan	Courant Institute and Physics Department, New York University, USA katepalli.sreenivasan@nyu.edu
2.	Prof. Rolf Jeltsch,	Seminar for Applied Mathematics, ETH Zurich, CH-8092 Zurich, Switzerland Phone: +41 44 632 3452 jeltsch@sam.math.ethz.ch
3.	Prof. Douglas N. Arnold	McKnight Presidential Professor of Mathematics, University of Minnesota President, Society for Industrial and Applied Mathematics arnold@umn.edu
4.	Prof. Barbara Lee	Keyfitz, Awm, Department of mathematics, The Ohio State University, 100 Math Tower, 231 Wes+B1t 18th Avenue, Columbus OH 43210-117, USA Ph: (+1)-614-292-5583, bkeyfitz@math.ohio-state.edu
5.	Prof. Martin Golubitsky,	Director, Mathematical Biosciences Intitute, Columbus,USA Ph: +1 614 247 4758, Fax: +1 614 247 6643 mg@mbi.ohio-state.edu
6.	Prof. Alistair Fitt,	Secretary (2007-2011), School of Mathematics, University of Southampton, UK Ph: (+44) (0)23 8059 5141 a.d.fitt@maths.soton.ac.uk
7.	Prof. Pavel Exner,	Ems, Department of Theoretical Physics, Nuclear Physics Institute, Czech Republic Ph: (+420) 2 6617 3293 exner@ujf.cas.cz
8.	Prof.Shin'ichi Oishi,	Department of Information and Computer Science, School of Science anf Engineering, Waseda University, Tokyo, Japan Ph: (+81) 3 5286 3414, Fax: (+81) 3 5286 3414 represent_iciam@mail.jsiam.org, orishi@waseda.jp
9.	Prof. Hisashi Okamoto,	JSIAM, Research Institute for Mathematics Science, Kyoto University, JapanPh:(+81) 75(753) 7226, represent_iciam@mail.jsiam.org
10.	Prof. M. Zuhair Nashed	Department of Mathematics, University of Central Florida ,Orlando, FL, USA Phone: (407) 823-0445 znashed@mail.ucf.edu
11.	Prof. Dr. Martin Brokate	Zentrum Mathematik/M6, TU Muenchen, Germany Phone: ++ 49 - (0)89 - 289 - 16806, Fax – 16809 brokate@ma.tum.de
12.	Prof. R. Lozi	Laboratiire J.A Dieudonne, UMR du CNRS No 6621 University of Nice-Sophiya – Antipolis, Parc Valrose, Evan, France R.LOZI@unice.fr
13.	Prof. José Alberto	Cuminato University of Sao Paulo ,Vila Pureza , SP - Brazil - Caixa-Postal: 668, Phone: (16) 33739695 Fax: (16) 33739751 jacumina@icmc.usp.br
14.	Prof. S. Kesavan	Institute of Mathematical Sciences, C.I.T. Campus, Taramani, Chennai, India Ph: (91)-(0)44-2254 3209; (91)-(0)44-3298 3441/2kesh@imsc.res.in
15.	Prof. Dinesh Singh	Director, South Campus Delhi University, Delhi India & now Vice-Chancellor Delhi University dineshsingh1@gmail.com
16.	Prof. Osni A. Marques	Sbmac Cylotron Berkely National Laboratory, Berkeley, CA 94720-8139, USA Ph: (+510) 486 5290, Fax: (+510) 486 5812, OAMarques@lbl.gov
17.	Prof. Wesley P. Petersen	Seminar for Applied Mathematics,HG G-52.1, ETHZ CH8092, Zurich Ph: +41 (0) 44 632-5575 wpp@math.ethz.ch
18.	Prof. Dr. Volker Mehrmann	Fachbereich Mathematik, TU Berlin, Strasse des 17, Juni 136, D-10623, Berlin, Germanymehrmann@mail.math.tu-berlin.de
19.	Prof. Andreas Griewank	Humboldt-University Berlin, Institute of Mathematics, Berlin, Germany Phone: +49 +30 2093 5833 kerger@mathematik.hu-berlin.de
20.	Prof. (Dr.) Alain Damlamian	Université Paris-Est à Créteil (UPEC),Créteil Cedex, France Tel: office +33-(0)1 45171653, damla@compuserve.com
21.	Prof. Peter Maass	Director of the Center for Applied Mathematics, University of

	(Bermen)	Beremen,Mathematics and Computer Science, Bremen, Germany Ph: +49-421-218-63801 pmass@math.uni-bremen.de
22.	Dr. Ranjan K. Mallik	Department of Electrical Engineering IIT Delhi, New Delhi, India Ph: +91(11)2659-1049, 2659-1572, 2658-1170, rkmallik@ee.iitd.ernet.in,
23.	Prof. Alemdar Hasanoglu (Hasanov)	Department of Mathematics and Computer Sciences, Izmir University, Gursel Aksel Bulvari, Uckuyular, Izmir – Turkey Ph: +90 232 246 49 49, alemdar.hasanoglu@izmir.edu.com; alemdar.hasanoglu@gmail.com
24.	Dr. Akhtar A. Khan	Rochester Institute of Technology, Rochester, New York 14623, USA aaksma@rit.edu
25.	Dr. Hulya Kodal Sevindir	Department of Mathematics, Kocaeli University, Izmit/Kocaeli Turkey Ph: +90 262 303 21 15 Fax: +90 262 303 20 03 hkodal@kocaeli.edu.tr
26.	Prof. Bernard Helffer	Département de Mathématiques,Université Paris Sud, Orsay Cedex – France Ph: +33 (0)1 69 15 60 20 Bernard.Helffer@math.u-psud.fr
27.	Prof. Ian H. Sloan Ao	FAA, School of Mathematics and Statistics,University of New South Wales, Australia, Ph: +61 2 9385 7038 I.Sloan@unsw.edu.au
28.	Prof. Pingwen Zhang	Department of Scientific & Engineering Computing, School of Mathematical Sciences, Peking University, Beijing, China Phone: 86-10-6275-9851, Fax: 86-10-6275-1801, pzhang@pku.edu.cn
29.	Dr. Luis Vega	Departamento de MatematicasUniversidad del Pais Vasco, Apdo 644, 48080 Bilbao, Spain ,Phone: (34) 946015475 luis.vega@ehu.es
30.	Prof. R. P. Singh	Vice- Chancellor, Sharda University, Greater Noida, U.P. , India
31.	Prof. Abul Hasan Siddiqi	Professor Emeritus, School of Engineering & Technology, Sharda University,Greater Noida, India Ph: +91 9837069944 Siddiqi.abulhasan@gmail.com
32.	Prof. Pammy Manchanda	GNDU, 15 Doctors Avenue, Amritsar-143001, India Ph: 0091-9815010067 pmanch2k1@yahoo.co.in
33.	Prof. R.C. Singh	Department of Physics, Sharda University, Greater Noida, U.P. , India Ph: 0091-9555595533 rcsingh_physics@yahoo.com
34.	Dr. Mani Mehra	Dept. of Mathematics, IIT New Delhi, India mmehra@maths.iitd.ac.in
35.	Dr. Khwaja M. Shahid	Director, Institute of Secretarial Training & Management, Ministry of Personnel, Public Grievances & Pensions, Government of India, JNU(Old) Campus, New Delhi Ph: 011-26185308 khwajashahid@hotmail.com
36.	Prof. Phoolan Prasad	Honorary Professor and DAE Raja Ramanna Fellow Department of Mathematics, Indian Institute of Science, Bangalore, India Ph: 0091-80-22933205/2267 prasad@math.iisc.ernet.in
37.	Prof. N. Rudraiah	Honorary Professor, UGC-Centre for Advanced Studies in Fluid Mechanics, Dept. of Mathematics, Bangalore University, Bangalore rudraiahn@hotmail.com
38.	Prof. Uday B. Desai	Director IIT Hyderabad, Hyderabad, India ubdesai@iith.ac.in
39.	Prof. O. P. Bhutani	Honorary Scientist, INSA New Delhi, India bhutaniop@hotmail.com
40.	Prof. Q.H. Ansari	Department of Mathematics, AMU, India qhansari@gmail.com
41.	Dr. A.K. Agrawal	Director, Buddha Institute of Technology, Gorakhpur, India anil_agrawal_bhu@yahoo.co.in
42.	Prof. Joydeep Dutta	Department of Mathematics, IIT, Kanpur India jdutta@iitk.ac.in
43.	Dr. S.K. Mishra	Department of Mathematics, BHU, Varanasi India bhu.skmishra@gmail.com
44.	Prof. S. K. Neogy	Indian Statistical Institute, New Delhi-110016 India skn@isid.ac.in
45.	Prof. Bhola Ishwar	P.I.DST Project, BRA Bihar university, Muzaffarpur, Muzaffarpur-842001 India, Mob: 9430051967 ishwar_bhola@hotmail.com
46.	Dr. Nityanand Singh	Scientist-F & Head, Climatology & Hydrometeorology Division, Indian Institute of Tropical Meteorology, India Ph: +91-20-2589-3600 Ext. 351 nsingh@tropmet.res.in
47.	Mr. V.K. Parashar	Professor of Maths, AMU, Aligarh, India vkparasharamu@yahoo.co.in
48.	Prof. Mushahid Husain	Co-ordinator, Nanotechnology Programme, Department of Physics, Jamia Millia Islamia, Jamia Nagar, India mush_phys@rediffmail.com
49.	Prof. N. Sontakke	IITM, Pune, India neelimasontakke@hotmail.com
50.	Prof. Syed B. Qadri	Ph. D.Senior Physicist and Professor ,US Naval Research Laboratory, Washington, DC 20375 USA qadri@anvil.nrl.navy.mil
51.	Dr. Ayub Khan	Zakir Husain College (Delhi University) India
52.	Dr. Harsh	Scientist 'G', Associate Director India 011-23818581(Off) harshnd51@yahoo.com
53.	Dr. S. Sundar	Department of Mathematics, IIT Madras, Chennai, India Phone: +91-44-22574618, slnt@iitm.ac.in,
54.	Ms. Yeliz Karaca	Istanbul Commerce University, Faculty of Commercial Sciences, Department of International Relationship Science, Istanbul, Turkey

		ykaraca@iticu.edu.tr
55.	Prof. Luis J. Boya	Theoretical Physics Dept.,University of Zaragoza, Zaragoza, Spain luisjo@unizar.es
56	Dr . Mrs. Robabeh Sahandi Torogh	University of Pune Department of Mathematics, Islamic Azad University-Varamin Branch, Iran9850272766, 9623888684 sahandi_1352@yahoo.com
57	Dr. Zaheer Abbas	Assistant Professor Department of Applied Mathematics, Baba Ghulam Shah Badshah University Rajouri, Jammu and Kashmir , India Ph: 9419277404az11292000@yahoo.co.in
58	Dr. Mohammad Yahya Abbasi	Assistant Professor Jamia Millia Islamia, New Delhi 110 025 India 9911969154, 011-26984682 9911969154 yahya_jmi@rediffmail.com
59	Dr. Kuzmanz Adzievski	Department of Mathematics and Computer Science, South Carolina State University, N.E. Orangeburg, South Carolina, USA Kadzievski@scsu.edu
60.	Mr. Izhar Ahmad	Research Scholar Amu Aligarh Izhar Uddin 69/A , Rm Hall Amu Aliagrh India 9634849020 izharuddin_amu@yahoo.co.in
61	Dr . M. Kalimuddi Ahmad	Associate Professor, Dept. of Mathematics, A M U Aligarh India ahmad_kalimuddin@yahoo.co.in
6.2	Mr. Zuber Akhter	M.PHIL.Dept. Of Statistics,A.M.U., India akhterzuber022@gmail.com
63	Mr. Touseef Ahmad	JIIT Noida, India touseef.smile@gmail.com
64.	Dr. Mansaf Alam	Asst. Professor, Dept. of Computer Science, Jamia Millia Islamia, New Delhi-25 India 9810650497 mansaf_alam2002@yahoo.com
65	Mr. Javid Ali	JRF & Research Scholar, Department of Physics, Jamia Millia Islamia, New Delhi-110025 India javidphy@gmail.com
66.	Mr. Sohrab Ali	Reasearch Schloar, Jamia Millia Islamia, New Delhi m.sohrabali@gmail.com
67.	Ms. Nisha Arora	Research Schlor, J. N. V. University, Jodhpur India nishaarora4@gmail.com
68	Ms. Pooja Arora	Research Scholar,Department of OR,University of Delhi, Delhi India
69	Ms. Ruchi Arora	Ph.D Student University of Delhi Department of Mathematics, University of Delhi, Delhi India, 011-9654242409 ruchiiq@yahoo.co.in
70.	Dr. Shelly Arora	Lecturer Punjabi University, Patiala Department of Mathematics,Punjabi University, Patiala, India 0175-3046189 aroshelly@gmail.com
71	Mr. Vilas Baliram Avhale	Student, DRDO, Ministry of Defence, Govt. of India vilas_avhale2000@yahoo.com
72	Ms. Stuti Awasthi	Research Scholar, Dept. of Mathematics, Motilala Nehru Mational Institute of Technology, Allahabad, India 9.19456E+11 astuti@mnnit.ac.in
73	Mr. Kuldeep Baderia	Student, PDPM-Indian Institute of Information Technology, Design and Manufacturing Jabalpur India 9424958369 kuldeep_gec2000@yahoo.co.in
74.	Ms. Anupma	Department of Applied Sciences, Ferozepur College of Engineering and Technology, Ferozshah, Ferozepur, India anupma2512@yahoo.co.in
75	Dr. C.M. Batra	Associate Professor Krishna Institute of Engineering & Technology,Ghaziabad India 9873725677 cmbkrish@yahoo.com
76	Prof. Luis J. Boya	Theoretical Physics Department, University of Zaragoza, Zaragoza luisjo@unizar.es
77.	Ms. Ratikanta Behera	Research Scholar,Dept. of Mathematics IIT Delhi, New Delhi, India 9990252520 ratikanta555@gmail.com
78.	Mr. Anuj Bhardwaj	Research Scholar, U.P.Technical University , Lakhnow anujbhardwaj8@gmail.com.
79	Dr. Prabhas Bhardwaj	Associate Professor Mechanical Engineering Dept., Institute of Technology, BHU, Varanasi, U.P. India 0542-6702835 prabhasbhardwaj@gmail.com
80	Ms. Guneet Bhatia	Ph.D Student, University of Delh,i Department of Mathematics, University of Delhi, Delhi India 011-9811609114 guneet172@yahoo.co.in
81	Ms. Meetu Bhatia	Asistant Professor, Miranda House, University of Delhi Delhi , 9811279551 bhatia.meetu@gmail.com
82	Ms. Smriti Bhatnagar	Sr. Lecturer Jaypee Institute of Information Technology, A-10 Secor-62 Noida India 9871290169 smriti.bhatnagar@jiit.ac.in
83.	Mr. Mehar Chand	Lect. Mathematics Yadvindra College of Engg., Pbi Uni. Gurukasshi Campus Talwandi Sabo (BTI). 9780920053 mc_global@rediffmail.com
84	Mr. Prashanto Chatterjee	Asistant Professor St. Stephen's College, University of Delhi, Delhi 1145587226 9810767341 chatterjee.prashanto9@gmail.com
85	Ms. Meenu Chawla	Student Punjabi University, Patiala meenuchawla86@gmail.com
86	Ms. Garima Chopra	Research Scholar, G. B. Pant University of Agriculture and Technology, Pantnagar, India 9897792099 garima.chopra@gmail.com
87	Ms. Preeti Choudhary	School of Basic and Applied Sciences, Shobhit University, Meerut,India preeticl37@gmail.com
88	Mrs. Suma Dawn	Department of CSE/IT, Jaypee Institute of Information Technology, Noida, U.P., India 9818218720 suma.dawn@gmail.ac.in

89.	Prof. Dr. Gennadiy Demidenko	Head of Laboratory Sobolev Institute of Mathematics, Novosibirsk 630090, Russian Federation. 7-3833634684 demidenk@math.nsc.ru
90.	Ms. S. Dhawan	Research Scholar N.I.T.J., Batala dhawan311@gmail.com
91	Ms. Mansi Dhingara	Research Scholar, Delhi University,New Delhi, mansidhingra7@gmail.com
92	Ms. Dipti Dubey	Research Scholar, IIT Delhi,India diptidubey@gmail.com
93.	Dr. Rakesh Dube	Rkgit, Ghaziabad, Up. India 120-2963262 drrakeshdube@gmail.com
94	Dr. Arunava Goswami	Associate Professor, Indian Statistical Institute, Biological Science Division Indian Statistical Institute,Kolkata 700108 India agoswami@iscical.ac.in
95.	Mr. Ankit Kumar Goyal	Lecturer, IIMT Engineering College Meerut, ankitkumargoyal@rediffmail.com
96	Ms. Nisha Goyal	Student Department of Computer Science, University of Delhi, Delhi, India 9968570664 nisha.goyal1987@gmail.com
97	Ms. Bharti Gupta	Research Scholar,Department of Mathematics SLIET, Longowal - 148106 (Punjab) India 9417727643 goyal_bharti@rediffmail.com
98	Dr. Pankaj Gupta	Department of OR, Delhi University, Delhi India pankajgpta@gmail.com
99.	Ms. Puneet Gupta	LECTURER B-XIII/657,GOBIND PURA BASTI, SANGRUR India 01672-240841 Mob: 9814388501 puneetjagat@yahoo.co.in
100	Ms. Rachana Gupta	Student, Department of Mathematics, IIT- Delhi, New Delhi-16 India 9711969637 +91- 9711969637 rachanagupta07@gmail.com
101	Ms. Rajani Rani Gupta	Research Scholar, Kurukshetra University, Kurukshetra rajani_gupta_83@rediffmail.com
103	Mr. Rajan Gupta	Student, Department of Computer Science,University of Delhi 011-27313732 Mob: +91-9818236739 guptarajan2000@gmail.com
104.	Mrs. Rashmi Gupta	Assistant Professor Hans Raj College, Pitam Pura, Delhi India 9971023600 9717027115 smsrashmigupta@gmail.com
105	Miss Rekha Gupta	Research Scholar , Department of Mathematics, University of Delhi, Delhi
106	Dr. Ritu Gupta	Associate Professor, Krishna Institute of Engineering & Technology, Delhi, India Mob: 9810335686 rgkrish88@yahoo.co.in
107	Mrs. Tanu Gupta	Assistant Professor, University of Delhi, India tanugupta10@gmail.com
108	Mr. Vikram Gupta	Manager Telephia BU 206, Delhi India smsvikramgupta@gmail.com
109.	Ms. Niyati Gurudwan	Research Scholar Pt. Ravishankar Shukla University, Raipur, Chhattisgarh, India 9871898299 niyati.kuhu@gmail.com
110	Mr. Sk Sarif Hassan	Project Linked Personnel, Applied Statistics Unit, Indian Statistical Institute, Kolkata, India, 9609174055, sarimif@gmail.com
111.	Dr. Syed Shakaib Irfan	Assistant Professor, Qassim University, Kingdom of Saudi Arabia, College of Engineering. 966541882054 shakaib@qec.edu.sa
112	Dr. Roop Chand Jain	Head Of Department, ECE Jaypee Institute of Information Technology, Noida India 9953140016 rc.jain@jiit.ac.in
113	Ms. Monika Jaiswal	Research Scholar B.H.U., Varanasi, 9889577727 monikajsmj@gmail.com
114.	Ms. Jyoti	Student, Delhi University, India 9953239423 deepshahj_p@yahoo.co.in
115.	Ms. D. Kamboj	Lecturer, M.L.N.College, Yamunanagar rajani_gupta_83@rediffmail.com
116	Mr. Srinivasa Raghava Kanduru	RESEARCH SCHOLAR, K.L.UNIVERSITY FED-II, GUNTUR(DT), ANDHRA PRADESH, India 9052594763 raghava.srinivasa@gmail.com
117	Mrs. Muskan Kapoor	Research Scholar, Department of Mathematics, University of Delhi, Delhi I
118	Mr. Saurabh Kapoor	Research Scholar, IIT Roorkee, Uttarakhnad saurabh09.iitr@gmail.com
119	Ms. Shruti Kapoor	Assistant Professor Jesus and Mary College, Kalkaji, New Delhi 9811326345 shruti.kapoor85@yahoo.in
120	Ms. Deepti Kaur	Research Scholar, University Of Delhi, deepti.kaur2008@rediffmail.com
121	Ms. Harpreet Kaur	Research Schloar, Sliet, Langowal hmaan80@yahoo.com
122	Ms. Lakhveer Kaur	Research Scholar Thapar University,PUNJAB lakhveer712@gmail.com
123	Ms. Kavita	Research Scholar, IIT Delhi,New Delhi, goyalkavita9@gmail.com
124	Mr. Navin Ketu	Student, Delhi University, New Delhi. navin16ias@gmail.com
125.	Dr. Arshad Khan	Astt. Professor, Department of Mathematics, Jamia Millia Islamia, New Delhi, 9990156354 akhan1234in@rediffmail.com
126	Mrf. Kashif Khan	Reasearch Schloar, Jamia Millia Islamia,kashifqamarkhan@gmail.com
127	Dr. Qamar Jalil Ahmad Khan	Associate Prof., Domas College Of Science, Sultan Qaboos University, Muscat, Oman 96824414028 Email: qjalil@squ.edu.om
128	Dr. Zubair Khan	Asstt. Professor, Deptt of Mathematics, Integral University Kursi Road, Lucknow, 9628605862 zkhan_123@yahoo.co.in
129	Ms. P. Khandelwal	Research Scholar, Jamia Millia Islamia, New Delhi pooja2n@gmail.com
130	Mr. Abhijit Konch	Lecturer,Deptt of Mathematics Dhemaji College,,Assam, 9954666609 9954666609 abhijitkonch100@yahoo.com
131	Dr. Vijay Kumar Kukreja	Associate Professor, Department of Mathematics, SLIET, Longowal, 9463017135 vkkukreja@gmail.com
132	Dr. Ajit Kumar	Assistant Professor, Department of Mathematics, Institute of Chemical

		Technology Matunga, Mumbai, 022-3611111, ajit72@gmail.com
133.	Mr. Amit Kumar	M.Tech student, PDPM Indian Institute of Information Technology, Design & Manufacturing, Jabalpur 9575432113 amitku@iiitdmj.ac.in
134	Mr. Amod Kumar	Dept. of Mathematics, BHU, Varanasi, kumaramod1983@gmail.com
135	Dr. Anil Kumar	Assistant Professor, PDPM Indian Institute of Information Technology, Design & Manufacturing Jabalpur, Dumna,Jabalpur anilkdee@gmail.com
136	Dr. Avdhesh Kumar	Lecturer, Chandigarh Engineering College, Mohali, 9023744405 avdheshsahani@yahoo.com
137	Mr. Deepak Kumar	Research scholar, Department of paper Technology, IIT Roorkee, Saharanpur Campus, Saharnpur 9.18126E+11 dkr2009@gmail.com
138	Mr.Dhruvesh Kumar	Student, University of Delhi, Delhi, 9911314113 dhruveshkmr@gmail.com
139	Mr. Happy Kumar	Research Scholar, Punjabi University, Patiala happygarg85@gmail.com
140.	Dr.Hitesh Kumar	Lecturer IBRI College of Tecc.,Jodhpur, hiteshsharma@gmail.com
141	Mr. Jatinder Kumar	Lecturer, Department of Mathematics, G.N.D. University, Amritsar143005 0183-2273569 9779007401 bhatiajkumar@yahoo.com
142	Mr. Jitendra Kumar	Research Scholar, Department of Mathematics, Indian Institute of Technology Roorkee, Uttarakhand, 9012380252, jitendark@gmail.com
143	Mr. Koshlendra Kumar Singh	M. Tech. Student, PDPM Indian Institute of Information Technology Design & Manufacturing, Dumna, Jabalpur,M. P. kks04540@gmail.com
144	Dr. Krishna Dev Kumar	P. Eng. Associate Professor & Canada Research Chair in Space Systems, Department of Aerospace Engineering, Ryerson University, Toronto, Ontario, Canada ,+1 416 979 5000 Ext. 4908 kdkumar@ryerson.ca
145	Mr. Manish Kumar	Research Scholar, Department of Applied Mathematics Indian School of Mines, Dhanbad, 9708124852 manish.math.bhu@gmail.com
146	Dr. Manoj Kumar	Assistant Professor, Department of Mathematics, Motilal Nehru National Institute of Technology, Allahabad, India 9451369162 manoj@mnnit.ac.in
147	Mr. Narender Kumar	Assistant Professor, Deshbandhu College, Dist.-Ghaziabzd(U.P) 0120-2766485 9268444890 nkumariitd@gmail.com
148	Mr. Rakesh Kumar	Deptt of Maths, Hindu College, Moradabad, U.P. rakeshnaini@yahoo.co.in
149	Mr. Ravish Kumar	Student University of Delhi ,New Delhi. ravishkkumar@gmail.com
150	Mr. Sachin Kumar	Student Thapar University, Patiala(Pb.), sachin1jan@yahoo.com
151	Mr. Sanjay Kumar	Student Department of Mathematics, University of Delhi, New Delhi 9899001922 9899001922 skumar.patna@gmail.com
152.	Mr. Santosh Kumar	Research Scholar , Department of OR, University of Delhi, Delhi
153	Mr. Suresh Kumar	lect. Mathematics Yadv Indra College of Engineering ,Vill Talwandi Sabo 9780471563 Mob: 9780471563 mc75global@yahoomail.com
154	Mr. Umesh Kumar	Student, University of Delhi, MCA 08-11 Batch, New Academic Block University of Delhi ,Delhi , 9958120503 kumar.umesh285@gmail.com
155	Prof. V. K. Kumbi	Professor, Dept of Mathematics, Karnatak University's Karnatak Arts College Karnataka, Dharwad-580001naregal.sharada@gmail.com
156	Dr. Badam Singh Kushvah	Department of Applied Mathematics, Indian School of Mines, Dhanbad, India bskush@gmail.com, kushvah.bs.am@ismdhanbad.ac.in
157	Prof. Alexander Kuzichev	Faculty of Mechanics and Mathematics, Moscow State University, 119992, Moscow, Leninskie Gory, Russia askuzichev@rambler.ru
158	Mr. Vivek Laha	Research Scholar Banaras Hindu University, Faculty of Science, Department of Mathematics, Varanasi 221005, India laha.vivek333@gmail.com
159	Ms. Pooja Louhan	Jrf, Delhi University, Delhi, 9718166667 poojalouhan@gmail.com
160.	Mr. Maheswaran	Research Scholar, Dept of civil engg IIT Delhi maheswaran27@yahoo.co.in
161	Mr.Ashutosh Mahato	Research Scholar, ISM Dhanbad ashumahato@gmail.com
162	Mr. Anandadeep Mandal	Assistant Professor School of Management, Kiit University, Bhubaneswar, Orissa 0674-2305246 Mob: 9776187712 anandadeep@ksom.ac.in
163.	Ms. Manisha	Research Scholar IIT Delhi, New Delhi, manisha.maths@yahoo.com
164.	Prof. Dr. Inessa Matveeva	Senior Scientific Researcher , Sobolev Institute of Mathematics Acad.,Novosibirsk, Russian Federation matveeva@math.nsc.ru
165	Ms. Meenakshi	Lecturer, Department of Mathematics, Dev Samaj College for Women, Ferozepur City, Mob: 9815277516 meenakshi_wavelets@yahoo.com
166	Mr. M. Mehlawat	Research Scholar , Department of O.R., University of Delhi, Delhi
167	Dr. Vinod Mishra	Associate Professor, SLIET, Longowal vinodmishrasliet@rediffmail.com
168	Mrs. Garima Mittal	Research Scholar, Department of O.R, University of Delhi, Delhi
169	Ms. Nikita Mittal	Research Scholar Delhi University, 9811384124 niki192006@yahoo.co.in
170	Prof. Dr. R C Mittal	Professor, Department of Mathematics, IIT Roorkee, ROORKEE 01332-285193 Fax: 01332-273560 Mob:9319912030 mittalrc@gmail.com
171.	Mr. Saiful Rahman Mondal	Research scholar, Department of Mathematics, Indian Institute of Technology, Roorkee Mob: 09410165496 saiful786@gmail.com
172	Dr. Debasis	Reader, Head of the Department of Mathematics Vivekananda College,

	Mukherjee	Thakurpukur, Kolkata, 09432876108 debasis_mukherjee2000@yahoo.co.in
173	Prof. B.S. Mudagi	Nowrosjee Wadıa College, Pune(India)basaweshwarmudagi@yahoo.co.in
174	Mr. Abdullah Murad	Teacher/ M.Phil. Research Fellow , University of Chittagong , Department of Mathematics, Chittagong, Bangladesh ammathcu1@yahoo.com
175.	Ms. Pallavi Nanda	Student Department of Computer Science,University of Delhi, Delhi 9968310542 npallavi1988@gmail.com
176	Mrs. Sharada S. Naregal	Research Scholar, Dept Of Mathematics Karnatak University's Karnatak Arts College,Karnataka, Dharwad naregal.sharada@gmail.com
177	Dr. Mohamed Haniffa Nasir	Senior Lecturer, Department of Mathematics, University of Peradeniya, Peradeniya, Sri Lanka 094812394551, nasirh@pdn.ac.lk
178	Mr. Shobhit Nigam	Research Scholar, Dept. of Applied Maths, Indian School of Mines Dhanbad, Jharkhand 9199874390, shobhitngm@gmail.com
179	Dr. Sanjay Oli	Assistant Professor, Dept. of Mathematics Lingaya's University Nachuli, Faridabad Mobile: 9999717328 sanjayoli@rediffmail.com
180	Mr. Ashutosh Pandey	Asst. Prof.(maths) D.P.Vipra College Bilaspur (c.g.), Bilaspur (c.g.) 9754867695 Mob: 9754867695 antu_28jd@yahoo.co.in
181.	Dr. Rajesh Kumar Pandey	Assistant Professor, PDPM Indian Institute of Information Technology, Design & Manufacturing,Dumna, Jabalpur wavelet_r@yahoo.co.in
182.	Dr. Mrs.Kanti Pandey	Prof. in Mathematics, Lucknow University Department of Mathematics & Astronomy, Lucknow , 9838729534 pandey_kanti@yahoo.co.in
183.	Mr. Ankit Patel	Lecturer, nirma uni., ghatlodiaya, 9898234567 major_ankit@yahoo.com
184.	Dr. Govind Pathak	Assistant Professor, Head, Dept. of MathematicsGovt. P. G. College, Pauri Garhwal, Uttrakhand, 01386-276641 pathakgovind@rediffmail.com
185.	Mr. Vasudeo Rajaram Patil	Lecturer and Head, Mathematics, Arts Science and Commerce College, Chikhaldara Dist Amravati (M.S.), vrpatil2007@rediffmail.com
186	Dr. Manish Prabha	Lecturer, Dept. of Mathematics MDDM College Muzaffarpur 0621-2279540 9199943670 shamshirmail@gmail.com
187.	Dr. Chandra Shekhar Prasad	Sr. Lecturer, Lalit Narayan Tirhut Mahavidhyalay, Muzaffarpur 9431033287, csprasad.maths@gmail.com
188.	Mr. Sarosh Mumtaz Quraishi	PhD student (JRF), Dept of Mechanical Engg,Institute of Technology, BHU,Varanasi, 0542-2316527, sarosh.quraishi@gmail.com
189	Mr. Anirudh Singh Rana	P.hd. Student, Department of Mechanical Eng. University of Victoria, Canada, 12504725641 anirudh@uvic.ca
190.	Dr.Shakeel Ahmed Raina	Associate Professor and HOD Mathematics, Govt.Degree College ThannaMandi, J&K,India, 9906131464 shakeelar2002@yahoo.co.in
191	Dr. Rajeev	Lecturer DAV College, Jalandhar, 0181-2255641 mtcst_2005@yahoo.co.in
192	Dr. Gergely Rost	University of Szeged, Aradi vertanuk tere, Szeged HUNGARY Mob: =36-30-5343494 rost@math.u-szeged.hu
193	Prof. Dr. Juri Rappoport	Senior scientific staff member, Russian Academy of Sciences, Russia 4991204382 Fax: 4959381844 Mob: 903-7474243 jmrap@landau.ac.ru
194	Dr. A. P. Reddy	Deptt. Of mathematics, Karnataka University, 09902900745, paddu7_math@rediffmail.com
195	Ms. Sabina	Research Scholar, Department of Mathematics,SLIET, Longowal.(Punjab) India abinajindal8@gmail.com
196	Dr. Nirmal C. Sacheti	Associate Prof, Department of Mathematics & Statistics, College of Science, Sultan Qaboos University, Muscat, Sultanate of Oman nirmal@squ.edu.com
197	Dr. Mohammad Sajid	Assistant Professor, College of Engineering Qassim University, Buraidah, Al-Qassim Saudi Arabia , 96663246932 msajid@qec.edu.sa
198	Mr. Baljinder Singh Saini	Research Scholar S.V.Institute of Emerging Technology, Gurukul Vidyapeeth Campus, Banur, Distt Mohali Punjab bssmaths@rediffmail.com
199	Ms. Pinky Saxena	Lecturer, Krishna Institute of Engineering & Technology, Ghaziabad 0120-2870282 Mob: 9654638581 pinkysaxena@gmail.com
200	Dr. Pratiksha Saxena	Assistant Professor BLS Institute of Management, Ghaziabad 0120-4124589 Mob: 9990097720 mathematics_pratiksha@yahoo.com
201	Mr. Rakesh Kumar Saxena	Research Scholar, Rajeev Gandhi Tech. University, Bhopal (M.P.) India. saxenark06@rediffmail.com
202	Dr. Mohammad Shahzad	Department of Information Technology, Higher College of Technology, Muscat, Sultanate of Oman. dmsinfinite@gmail.com
203	Dr. Gunjeshwar Shukla	Mathematics Deptt , Post Gradate College, Padrona gunjeshwarshukla@yahoo.com
204	Mr. Ayan Sen	Research Scholar, Jadavpur University,P.S.-Thakurpukur, Kolkata West Bengal, India, 3324532144, Mob: 9433347048 ayansen2003@yahoo.co.in
205	Ms. Nandini Seth	Student, hansraj college, New Delhi nandini_cancerian@yahoo.com
206	Ms. Ruchika Sethi	Assistant Professor, Dualat ram college, New Delhi, ruchikasethi83@gmail.com

207.	Dr. Firdous Ahmad Shah	Assistant Professor, University of Kashmir, South Campus, Anantnag, Jammu and Kashmir, India, Mob: -9622496453 fashah79@gmail.com
208	Ms.Anamika Sharma	Assistant Professor, IILM, Greater Noida, anamika-sharama@iilm.ac.in
209.	Ms. Kanika Sharma	Research Scholar Delhi University, Delhi, kanika.divika@gmail.com
210	Ms. Megha Sharma	Research Scholar Delhi University, Delhi, mathmeghasharma@gmail.com
211	Dr. Meenu Sharma	Principal, S.D.P. College for Women,Ludhiana, Punjab meenusharma03@yahoo.co.in.
212.	Mr. Pankaj Sharma	Department of Mathematics, Jamia Millia Islamia, New Delhi, pankajsharma.jmi@gmail.com
213.	Mrs. Rajni Sharma	Lect. in Mathematics, Department of Applied Sciences, DAV Institute of Engg. and Technology, Jalandhar, rajni_gandher@yahoo.co.in
214.	Ms. Richa Sharma	Department of Mathematics, St. Johns College, Agra-282002 (India) aligarh.richa@gmail.com
215.	Mr. Sumeet Kumar Sharma	Assistant Professor, P.G. Department of Applied Mathematics. BGSB University, Rajouri (J&K), 09797324181, sharmasumeet766@gmail.com
216.	Ms. Sunita Sharma	Research scholar, Delhi University , Delhi, sunita_m78@rediffmail.com
217.	Mr. Vikram Sharma	Junior Research Fellow Guru Nanak Dev University, Amritsar 0183-2566728 Mob: 9988156540 vikramsharma_6898@yahoo.com
218.	Dr. Yogesh Sharma	Associate Profeesor, Jodhpur National University, Jodhpur 2912651588 2931281550 9829366710 dryogesh121@rediffmail.com
219.	Mr. V.kumar R. Sheelavant	Asst. Professor.Electrical Engg. Department, SDM College of Engg. &Tech.Dhavalagiri.Dharwad, 9480553975 sheel125@gmail.com
220.	Dr. K. S. Shekhawat	Lecturer Selection Scale S K College,Sikar Rajasthan, 0157-2270155 9414237725 dr.kishanshekhawat@yahoo.in
221.	Dr. S. C. Shiralashetti	Asst. Professor Department of Mathematics Karnatak Universitys Karnatak Arts College, Karnataka, Dharwad, shiralashettisc@yahoo.com
222.	Ms.Kalpana Shukla	Research scholar, BHU, Varanasi bhukalpana02@gmail.com
223.	Prof. Bani Singh	Department of Mathematics, Jaypee Institute of Information Technology, Sector-62, Noida, U.P. India bani.singh@jiit.ac.in
224.	Dr. Deepti Singh	Reader Mahila, Pg College Lucknow , 2/508 Vikas Nagar Lucknow Ph: 5224073374 Mob: 9450300315 rajdeepti_singh@yahoo.co.in
225.	Mr. Jeetendra Bahadur Singh	Research Scholar, Dept. of Applied Maths, Indian School of Mines Dhanbad, Dhanbad, Jharkhand Mob:9430316909 jeetendra01@gmail.com
226	Dr. Ashok K. Singh	Professor, Department of Mathematics, Banaras Hindu University,Varanasi, India Mob: (+91)9450530760 ashok@bhu.ac.in
227	Prof. Koushlendra K. Singh	PDPM Indian Institute of Information Technology Design & Manufacturing, Jabalpur, India kks04540@gmail.com
228	Mr. Lavneet Singh	16/9,balkeshwar colony,Agra(U.P), 09411083787,lavneetagra@gmail.com
229	Mr. Mehakpreet Singh	Junior Research Fellow IIT Kharagpur Deptt of Mathematics, IIT Kharagpur Mob: 9046970571 Email: sarin_mehakpreet@yahoo.com
230	Dr.Pankaj Singh	Associate Profesor Maharana Pratap Engg. College,Kanpur MPEC Ph: 0512-2770092 Mob: 9415174562 ps_mintu@yahoo.com
231.	Mr. Ram Singh	Research Scholar St. John,s College, Agra, 9018410200 singh_ram2008@hotmail.com
232.	Mr. R.K. Singh	Student, BMAS Engg.College, Agra, ranjeetsingh_don@yahoo.co.in
233	Prof. Shikha Singh	Head, Department of Mathematics, PPN College, CSJM University,Kanpur-208024, INDIA sshikha22976@yahoo.co.in
234.	Ms. Shweta Singh	Student Pt.Ravishanker Shukla University shwetasingh828@gmail.com
235.	Mr. Tarlok Singh	Deptt. Of Mathematics, G.N.D.U, Amritsar tarlok_kheewa@yahoo.com
236.	Mr. Vinay Singh	Research Scholar, Department of Mathematics Banaras Hindu University, Varanasi-221005, India 9532456687 vinaybhu1981@gmail.com
237	Mr. V.K.Sinha	Ex-Asstt. Prof., Glaitm, Mathura svinit83@gmail.com
238.	Mr. Sripathy	Research Scholar Anna University, Anakaputhur, Chennai-600070 0-9789812546 sripathi_51180@yahoo.co.in
239.	Dr. Manjari Srivastava	Associate Prof., Department of Maths, Delhi University, Delhi-7 27666836 9899228986 manjari123@yahoomail.com
240	Mrs. Vandana Srivastava	Researcher, B-169 Avas Vikas Colony Nandanpura Jhansi (U.P.) India Ph: 5102481058, Mob: 9429379617,vandy_bikes@yahoo.com
241	Dr. Weiwei Sun	Department of Mathematics, City University of Hong Kong Hong Kong, China Ph: 27887155 maweiw@math.cityu.edu.hk
242	Dr. S.K. Suneja	Deptt of Mathematics, Miranda House, University of Delhi.
243	Ms. Talat Sultana	Research Scholar Dept. of Mathematics, Jamia Millia Islamia, New Delhi, 9213056722 talat.simran.sultana@gmail.com
244.	Dr.Sushil	GBU University sushil12@gmail.com
245	Mr. Mohd Tanveer	SRF, School of Computer & Systems Sciences, Jawharlal Nehru University

		New Delhi, 9873056356 tanveergouri@gmail.com
246	Dr. Balwant Singh Thakur	Reader, Raipur India School of Studies in Mathematics, Pt. Ravishankar Shukla University, Raipur +91 9827955810 balwantst@gmail.com
247	Mrs. Krishna Thareja	Reader, Department of Mathematics ,Rajdhani College, University of Delhi
248	Miss Maria Thomas	Adhoc teacher St.Stephen's College, University of Delhi, Delhi
249	Dr. A.Mohamed Toumi	Prof , College of Engineering, Qassim Universiy, Saudi Arabia 541882054 abdel.toum@gmail.com
250	Mr. Padmesh Tripathi	Assistant Professor I.T.S. Engg. College, Greater Noida padmesh01@rediffmail.com
251.	Mr. R. K. Tripathi	Research Scholar, B.S.A. College, Mathura,U.P. rktripathi25@gmail.com
252	Mr. Balendu Bhooshan Upadhyay	Research scholar, Banaras Hindu University, Varanasi, India 9005858587 bhooshanbhu@gmail.com
253	Ms. Shabana Urooj	Department of Electronics & Instrumentation Engineering, Galgotias College of Engineering & Technology, Gr. Noida shabanabilal@gmail.com
254.	Ms. Vandana	Student Delhi University, Delhi, 9910926377 rajpal27.iitd@gmail.com
255	Ms. Suja Varghese	Research Scholar, Pt.R.S. Shukla University suja_reji@rediffmail.com
256.	Mr. Sag Ram Verma	JRF DST-CIMS, Department of Applied Mathematics Institute of Technology Banaras Hindu University Varanasi v.sagram@gmail.com
257	Mrs. Shilpi Verma	Research Scholar , University of Delhi, Delhi
258	Ms. Swati Verma	Research scholar c\o Omkar Shukla Shiksha Niketan, Near Science College,RSU Campus, Raipur(C.G.) swativerma15@gmail.com
259	Ms. Kriti Wadhwa	Assistant professor, Jesus and Mary College ,New Delhi-110060 001-42436970 9811981580 kritiwadhwa7@gmail.com
260.	Ms. Sargam Wasu	Research Scholar S.V.Institute of Emerging Technology, Banur,Patiala, Mob: Mob:9988466090 sargam15@rediffmail.com
261.	Mr. Abdul Abbas	Department of Stats & OR, AMU, Aligarh abbasmaths@gmail.com
262.	Mr. Irfan Ali	Research scholar Aigarh Muslim University, Aligarh rfii.ali@gmail.com
262.	Mr. Musavvir Ali	Research Scholar , AMU Aligarh himusavvir@gmail.com
263	Mr. Abu Zaid Ansari	Research scholar Aligarh Muslim University, ansari.abuzaid@gmail.com
264.	Mr. S. I. Ansari	Research Scholar, AMU, Aligarh, saiful.islam.ansari@gmail.com
265	Mr. Z.H. Bakhshi	Research Scholar, AMU, Aligarh, bakhshistat@gmail.com
266	Mr. Mohd Bilal	Research Scholar, AMU,Aligarh, mohd7bilal@gmail.com
267	Ms. Tarannum Bano	AMU, Aligarh. Ph: 0933622280, tarannumdlw@gmail.com
268.	Mr. M. Dilshad	Research Scholar, A.M.U. Aligarh, mdilshaad@gmail.com
269	Mr. Mohd Furkan	Research scholar AMU Aligarh, mohdfurkan786@gmail.com
270	Ms. Shazia Ghufran	Research Scholar, AMU, Aligarh, India itsshaziaghufran@gmail.com
271	Mr. Himanshu Gupta	Research Scholar, AMU, Aligarh, hmnshu08@gmail.com
272.	Dr. S.C. Gupta	Incharge Mathematics Section,Women's Coolege, AMU, Aligarh
273.	Mr. Sanjeev Gupta	Research Scholar, AMU Aligarh, guptasanmp@gmail.com
274	Ms. Hibah Islahi	Research Scholar AMU,Aligarh, hibahshaikh@gmail.com
275	Mr. Akhlad Iqbal	Research scholar, AMU Aligarh akhlad6star@gmail.com
276	Mr. Izharuddin	Research scholar, AMU, Aligarh
277.	Mr. A. A. Khan	Dept. of Statistics & O.R., AMU, Aligarh. ashfaqkhan202@gmail.com
278	Mr. Mustafa Kamal	AMU, Aligarh (U.P.) Ph: 09319868763
279	Mr. Asif Khan	Research Scholar, A.M.U, Aligarh asifJNU07@gmail.com
280	Ms. Nazneen Khan	Research Scholar, AMU, Aligarh naaz4allah@yahoo.com
281	Miss. Sana Khan	Research Scholar, AMU Aligarh ahmad_kalimuddin@yahoo.co.in
282	Mr. Shamsher Khan	Research Scholar, AMU, Aligarh, shamsherstats@gmail.com
283	Dr. Subuhi Khan	Associate Professor, Department of Mathematics, Aligarh Muslim University, Aligarh 0571-2703934 9.19413E+11 subuhi2006@gmail.com
284	Ms. Saman Khowaja	Research Scholar, AMU, Aligarh samankhowaja@gmail.com
285	Mr.Devendra Kumar	Research Scholar, AMU, Aligarh. devendrastats@gmail.com
286	Mr. Nayabuddin	Department of Stats & OR, AMU, Aligarh. nayabstats@gmail.com
287	Mr. Y. S. Raghav	Research Scholar(Ph.D.) Aligarh Muslim University ,Aligarh
288.	Ms. Nusrat Raza	Research Scholar, AMU,Aligarh nraza.maths@gmail.com
289	Mr.S. Haider Rizvi	Ph.D. Scholar, AMU, Aligarh. shujarizvi07@gmail.com
290.	Mrs.S. Shrama	Dept. of Statistics & O.R., AMU, Aligarh , sharma.sangh@gmail.com
291.	Mr. Rahul Varshney	Research Scholar, AMU, Aligarh itsrahulvarshney@gmail.com
292	Ms. Shazia Zarrin	Research Scholar ,AMU, Aligarh, shaziazarrin@yahoo.in
293	Ms. Benazir Zia	Research Scholar, AMU, Aligarh benazir.stats@gmail.com
294.	Prof. Khurshid Alam	Assistant Professor, Sharda University,Plot no. 32-34, Knowledge Park, Greater Noida, U.P., Mob:09873097866 khursheed_alam55@yahoo.com
295	Mhod. Shahid Baboo	Assistant Professor, Sharda University, Plot no. 32-34, Knowledge Park, Greater Noida - 201306 U.P., Mob:09999214783 mesub007@yahoo.co.in

296	Dr. Sonali Bhandari	Assistant Professor, Sharda University, Plot no. 32-34, Knowledge Park, Greater Noida, U.P., Mob:09650154795 sonali.bhandari@sharda.ac.in
297.	Dr. Sangeeta Gupta	Assitant Prof. , Sharda Universty, Greater Noida, sangeeta147@gmail.com
298	Mr. Rohit Khokher	Asst. Prof., Sharda University, Greater Noida, khokher_rohit@yahoo.com
299.	Mr. Anshu Kumar	Assistant Professor, Sharda University, Plot no. 32-34, Knowledge Park, Greater Noida, U.P. Ph: 09711826692, anshukumar.sgi@gmail.com
300	Mr. Anuj Kumar	Research Scholar Sharda University, anujmaths@gmail.com
301	Mr. Rajiv Kumar	Asst. Professor, SET, Sharda University, Greater Noida. Mob: 09968227298, rajiv.kumar@sharda.ac.in
302	Dr. Viveka Kumar	Associate Professor, Sharda University, Greater Noida, UP, Ph:8800474750, Mob:09953184897, vivekakumar@yahoo.com
303	Ms. Monica Kumari	Assistant Professor, Sharda University, Knowledge Park III, Greater Noida, UP, Ph: 09015033303, monika.kit@gmail.com
304	Ms. Suman Lata	Electronics and Instrumentation , Assistant Prof.& Incharge Section HIT, Greater Noida , UP, India smn_bhat@yahoo.co.in
305.	Ms. Archana Kumari Prasad	Assistant Professor., Sharda University, Plot no. 32-34, Knowledge Park, Greater Noida - 201306 U.P., Ph: 09871428387, kumari_archu@yahoo.com
306.	Mrs. Rashmi Priyadarshini	Asst. Professor Sharda University A-95,pocket-11, 0120 2323602 9911450073 priyadarshini.rashmi@rediffmail.com
307.	Ms. Manisha Rajoriya	Assistant Professor Sharda University Set, 32-34,Knowledge park 3, Gr. Noida 9718187552 manisha_rajoriya2004@yahoo.co.in
308.	Ms. Ranjeeta Roy	Electronics & Instrumentation Department, School Of Engineering & Technology,Sharda University, Greater Noida, India. ranjita.06@gmail.com
309	Ms. Shiwani Saini	Elecrical and Electronics Department, Assistant Professor, HIT, Greater Noida, UP, India shiwani_saini76@yahoo.com
310	Mr. Anju Kumar Singh	Assistant Professor, Sharda University,Plot no. 32-34, Knowledge Park, Greater Noida, U.P., Mob:09999611759 anjukumar.singh@sharda.ac.in
311	Mr. S. Kr. Singh	Asst. Professor Sharda University, 9871744973 hit.ssk@gmail.com
312	Dr. Sweta Srivastava	Associate Professor, SET, Sharda University, Greater Noida Ph: 09650996156, dr.srishweta@gmail.com
313.	Mrs. Noor E Zahra	ASST.PROFESSOR SHARDA UNIVERSITY, 09871744973 noor_zahra_india@yahoo.co.in
314	Mr. Ashish Mittal	P.hd Student, IP university, New Delhi akmittal21@gmail.com
315	Mr. Hasan Ejtaba	MCA, Student, Sikkim Manipal University, ejtaba@yahoo.com
316-345 All talented students of Sharda University	Ms. Megha Mathur, Ms. Shivani Belwal Ms. Vishakha Suyal, Ms. Swati Sabharwal, Ms. Apoorv Shree Sharma, Ms. Nida Naseem Bhat, Ms. Neha Singh, Ms. Mynakshi Majundar Md. Salman, Mr. Shubham Saxena Mr. Shubham Saxena Mr. Amir Iqbal, Mr. Savan Trehara, Mr. Raghav Dixit, Mr. Seemant Bisht Mr. Rohan Wahi, Mr. Saurabh Sharma, Mr. Puneet Kumar Mr. Mayank Arora, Mr. Sushant Mishra, Mr. Varun Verma, Mr. Vinit Viduran Mr. Vishal Singh, Ms. Richa Rathore, Ms. Ruchita Misra, Ms. Priya Mr. Abhishake Chand Saxena, Mr. Anurag Tiwari	C/O Ms Noor-E-Zahra and Ms Suman Lata, Faculty Members,Sharda University, Plot no. 32-34, Knowledge Park, Greater Noida - 201306 U.P

INDEX